COOPERATIVE
COMMUNICATIONS

COOPERATIVE COMMUNICATIONS

HARDWARE, CHANNEL & PHY

Mischa Dohler
CTTC, Spain

Yonghui Li
The University of Sydney, Australia

A John Wiley and Sons, Ltd., Publication

Library of Congress Cataloging-in-Publication Data

Dohler, Mischa.
  Cooperative communications : hardware, channel & PHY / Mischa Dohler, Yonghui Li.
    p. cm.
  Includes bibliographical references and index.
  ISBN 978-0-470-99768-0 (cloth)
  1. Cognitive radio networks. 2. internetworking (Telecommunication) 3. MIMO systems. 4.
Multiuser detection (Telecommunication) 5. Ad-hoc networks (Computer networks) I. Li,
Yonghui, 1975– II. Title.
  TK5103.4815.D64 2010
  621.384 – dc22

                                                                      2009040161

A catalogue record for this book is available from the British Library.

ISBN 978-0-470-99768-0 (H/B)

Typeset in 9/11 Times by Laserwords Private Limited, Chennai, India
Printed and Bound in Great Britain by CPI Antony Rowe, Chippenham, Wiltshire

To my girls, Gemma, Noa and Dàlia

Mischa Dohler

To my parents Kangzhi Li, Zhenyu Jin and my beloved wife Na Zou

Yonghui Li

About the Authors

Mischa Dohler is now senior researcher with CTTC in Barcelona. Prior to this, from June 2005 to February 2008, he was senior research expert in the R&D division of France Telecom, working on cooperative communication systems, cognitive radios and wireless sensor networks. From September 2003 to June 2005, he was lecturer at King's College London, Centre for Telecommunications Research. At that time, he was London Technology Network Business Fellow for King's College London, as well as student representative of the IEEE UKRI Section and member of the Student Activity Committee of IEEE Region 8 (Europe, Africa, Middle-East and Russia). He obtained his PhD in Telecommunications from King's College London, UK, in 2003, his diploma in electrical engineering from Dresden University of Technology, Germany, in 2000, and his MSc degree in Telecommunications from King's College London, UK, in 1999. Prior to telecommunications, he studied physics in Moscow. He has won various competitions in mathematics and physics, and participated in the third round of the International Physics Olympics for Germany. In the framework of the Mobile VCE, he has pioneered research on distributed cooperative space–time encoded communication systems, dating back to December 1999. He has published more than 110 technical journal and conference papers at a citation h-index of 20 and citation g-index of 41 [December 2009], holds several patents, co-edited and contributed to several books, has given numerous international short courses, and participated in standardization activities. He has been TPC member and co-chair of various conferences, such as technical chair of IEEE PIMRC 2008 held in Cannes, France. He is editor for numerous IEEE and non-IEEE journals and a senior member of the IEEE.

Yonghui Li received his PhD degree in November 2002 from Beijing University of Aeronautics and Astronautics. From 1999–2003, he was affiliated with Linkair Communication Inc., where he held a position of project manager with responsibility for the design of physical layer solutions for LAS-CDMA system. Since 2003, he has been with the Telecommunications Lab, University of Sydney, Australia. He is now a senior lecturer in the School of Electrical and Information Engineering, University of Sydney. He is also currently the Australian Queen Elizabeth II fellow. His current research interests are in the area of wireless communications, with a particular focus on MIMO, cooperative communications, coding techniques and wireless sensor networks. He holds a number of patents granted and pending in these fields.

Contents

Preface

Technological Landscape of the 21st Century

Cooperation is not a natural characteristic attributed to humans. The typical human horizon is focused on short-term gains, which might be due to our instinct-driven subconscious occupying a grander importance than we dare to admit [1]. Cooperating with other individuals or entities, however, usually means that short-term losses may translate into long-term gains – something history has proved to hold true but humans for some reason rarely ever understand. Any cooperative technology depending solely on human decisions is hence *a priori* doomed to fail. By contrast, if machines have access to some computerized decision making engines only, cooperative schemes become viable communication techniques and are likely to occupy an important place in the technological landscape of the 21st century [2].

For this reason, wireless cooperative communication systems have received significant attention in the past decade and – due to their theoretically infinite design degrees of freedom – a large body of highly useful but also often confusing and contradicting research papers has emerged. Indeed, when we commenced research in this area in 1999, online search engines yielded a handful of papers; today, Google yields almost one million hits when searching for 'cooperative wireless communications'.

Yet Another Book on Cooperation?

Clearly not! Even though the material of this book can be further complemented by the excellent edited treatises [3–5] as well as the authored book [6] from our colleagues in this field, this book not only offers complete taxonomies on this rather disperse field but also some entirely novel unpublished material.

The aim of this book is thus to shed some light on cooperative communication systems from a hardware, channel and physical layer design perspective. To this end, we introduce some thorough taxonomies in this fairly diversified field and expose early, recently emerged and entirely novel cooperative techniques. We will deal with both basic and advanced topics, ranging from, for example, complexity analysis and channel modeling to distributed PHY transceiver structures. Note that although the published results of many of our colleagues working in this field have been treated, we could not possibly cover all available material. Having said this, we have tried hard to include the latest materials published up until summer 2009. Furthermore, we will try to publish updates, corrections, changes on the accompanying website www.cttc.es/books/phycoop.

This book forms part of an entire series on cooperative communications, where more books on, for example, Shannon capacity, medium access control and routing protocols for cooperative systems will be made available at a later stage. The endeavor of this particular book is to give you, the reader, a sufficient overview of cooperative communication concepts, give some 'feeling' for certain approaches as well as detailed knowledge on some techniques and some general tools that allow the analysis to be extended to more general and more complex topologies. Ideally, this book should inspire you to apply prior knowledge to wireless cooperative relaying systems!

Book Content

This book is essentially composed of four important subject areas related to cooperative communication systems. The first is a thorough taxonomy of said systems; the second the wireless channel; the third the physical layer, and the fourth the hardware realization. To this end, this book is split into six main chapters with the following contents:

Chapter 1 contains a basic introduction to cooperative systems in general. We will then discuss typical application scenarios of cooperative techniques. Furthermore, we will discuss pros and cons of relaying and briefly quantify the capacity gains. We will also revise definitions and terminologies typically used in the context of cooperative systems. Finally, some historical background as well as key milestones are discussed.

Chapter 2 deals with the wireless relaying and space–time channel. In contrast to traditional wireless systems, the wireless channel becomes part of the cooperative system, which has a profound impact on channel statistics, including temporal and correlation behaviors. We will hence discuss in detail the channel behavior of regenerative, transparent and distributed multiple-input multiple-output (MIMO) systems. This is pivotal in understanding and quantifying the performance of cooperative protocols at the physical (PHY) layer.

Chapter 3 is dedicated to transparent PHY algorithms, and we will discuss and quantify the performance of different transparent architectures here. We will introduce some tools that facilitate the characterization of general architectures, which range from transparent relaying to transparent distributed space–time block and trellis coding, distributed multiplexing and beamforming. We will also dwell on issues pertaining to distributed system optimization, such as distributed power allocation and distributed relay selection.

Chapter 4 deals solely with regenerative PHY algorithms and we will discuss and quantify the performance of a plethora of regenerative architectures here. We will mainly deal with different estimate-and-forward, decode-and-forward, compress-and-forward and soft-information relaying protocols, among others, with the aim of characterizing their performance. We will also dwell on recently emerged advanced topics related to distributed coding, such as distributed space–time block, trellis and turbo coding, distributed network-channel coding for single as well as multiple source–destination pairs, etc.

Chapter 5 pertains to hardware and we will discuss how hardware facilitates as well as limits any implementation of cooperative relaying schemes. These limitations in hardware render, for example, the implementation of some of the recently proposed cooperative protocols infeasible. We will discuss the implementation of transparent and regenerative schemes. We will also compare their costs and implementation complexities. We then derive complexities and power consumption of various relaying architectures based on 3G and 4G standards. Finally, we discuss some available hardware platforms and testbeds implementing relaying or distributed space–time processing techniques.

Chapter 6 concludes this book by summarizing its contributions and also highlighting important open research directions and still-to-be-explored impact areas. We also describe how to include real-world impairments into the analysis, such as shadowing, interference, etc. Finally, some business issues are briefly dealt with.

Acknowledgments

The writing of this book would not have been possible without the support of many friends and colleagues.

In particular, we would like to thank Sarah Tilley, Birgit Gruber, Sarah Hinton and Mark Hammond at Wiley, for their continuous support throughout the preparation and production of this book. We are also very grateful to the reviewers for their suggestions in improving the content of this book.

Similarly, we are grateful for the comments received from the copy editor and typesetter that played a vital role in preparing a professional and coherent finish to the book.

We would also like to thank those who have proofread the contents of this book as well as provided vital material: Angel Lozano, Claude Oestges, Aggelos Bletsas, Raymond Louie, Philippe Mary, Christian Lerau, Patrick Marsch, Malte Schellmann and Stefan Berger. We also owe special thanks to the numerous friends and colleagues, with whom we had lengthy discussions relating to the topic of cooperative systems since 1999. These discussions have been fundamental in developing the ideas and techniques contained in this book. There are too many individuals to mention, but we thank them all!

Mischa Dohler would like to express his gratitude to Yonghui Li for having been such a professional and diligent coauthor. He is also infinitely grateful to his family for their understanding and support during the time he devoted to writing this book. He particularly wishes to thank Gemma, his wife, for her kind understanding and support during the final preparations of this book. A smile goes to Noa and Dàlia, his daughters, for making days night and nights day, and generally turning life upside-down. Furthermore, he would like to thank his colleagues at CTTC in Barcelona – 500 meters from the beach – for creating such a fantastic working environment.

Yonghui Li would like to thank all his colleagues and friends, Branka Vucetic, Jinhong Yuan, Zhuo Chen, Zihuai Lin, Zhendong Zhou for the creative and harmonious atmosphere and close collaborations, and students, Raymond Louie, Kun Pang, Judy Chu, Wibowo Hardjawana, Shahriar Rahman, and Rui Li. Special thanks go to Jinhong Yuan, Judy Chu and Zhuo Chen for the joint work in distributed space–time trellis codes in Chapter 4 and Raymond Louie for his work reported in the distributed spatial multiplexing in Chapter 3. He is particularly grateful to his beloved wife, Na Zou, for her love, encouragement and continuing support. She has made many sacrifices throughout the preparation of this book.

Mischa Dohler
CTTC, Barcelona, Catalonia

Yonghui Li
The University of Sydney, Sydney, Australia

Abbreviations

	Abbreviations and acronyms used throughout this book
Abbreviation	Connotation
2D	Two Dimensional
3D	Three Dimensional
3G	Third Generation
3GPP	Third Generation Partnership Project
4G	Fourth Generation
AAA	Authentication, Authorization and Accounting
ACF	Auto-Correlation Function
ACK	Acknowledgement
ACM	Adaptive Coding and Modulation
ADC	Analog-to-Digital (converter)
AF	Amplify-and-Forward Protocol
AFD	Average Fade Duration
AAOA	Azimuth AOA
AAOD	Azimuth AOD
ANCC	Adaptive Network Coded Cooperation
AOA	Angle of Arrival
AOD	Angle of Departure
AP	Access Point
APP	A Posteriori Probability
ARP	Adaptive Relaying Protocol
ARQ	Automatic Repeat Request
ASIC	Application-Specific Integrated Circuit
AWGN	Additive White Gaussian Noise
BAW	Bulk Acoustic Wave (filter)
BEP	Bit Error Probability
BER	Bit Error Rate
BLAST	Bell Labs Layered Space Time
BPF	Bandpass Filter
BPSK	Binary Phase Shift Keying
BS	Base Station
EF	Estimate-and-Forward Protocol
EM	Electro-Magnetic
CCF	Cross-Correlation Function
CDF	Cumulative Density Function
CDMA	Code Division Multiple Access
CEIGU	Channel Estimation and Interference Replica Generation Unit
CF	Compress-and-Forward Protocol
CFO	Channel Frequency Offset

Abbreviations and acronyms used throughout this book (*cont.*)

Abbreviation	Connotation
CLT	Central Limit Theorem
COST	Cooperation in Science and Technology
CP	Cyclic Prefix
CPICH	Common Pilot Channel
CRC	Cyclic Redundancy Check
CRLB	CramérRao Lower Bound
CSI	Channel State Information
CSMA	Carrier Sensing Multiple Access
CTS	Clear to Send
DAC	Digital-to-Analog (converter)
dB	Decibel
DB	Double Bounce
dec	decade
DF	Decode-and-Forward Protocol
DFD	Division Free Duplex
DFR	Division Free Relaying
DLD	Linear Dispersion Code
DMT	Diversity Multiplexing Tradeoff
DNCC	Distributed Network-Channel Coding
DPCH	Dedicated Physical Channel
DSM	Distributed Spatial Multiplexing
DSP	Digital Signal Processor
DSTBC	Distributed STBC
DSTTC	Distributed STTC
DTC	Distributed TC
EAOA	Elevation AOA
EAOD	Elevation AOD
EGC	Equal Gain Combining
EMEDS	Enhanced MEDS
ETSI	European Telecommunications Standards Institute
FDD	Frequency Division Duplex
FDMA	Frequency Division Multiple Access
FDR	Frequency Division Relaying
FER	Frame Error Rate
FFT	Fast Fourier Transform
FL	Floor Loss
FPGA	Field Programmable Gate Array
FUSC	Fully Loaded Subcarriers
GDTC	Generalized DTC
GF	Gather-and-Forward Protocol
GHz	Gigahertz
GSM	Global System for Mobile Communications
HSDPA	High Speed Downlink Packet Access
ICS	Interference Canceling System
IEEE	Institute of Electrical and Electronics Engineers
IF	Interim/Intermediate Frequency
IFFT	Inverse Fast Fourier Transform
ISI	Intersymbol Interference
ISM	Industrial, Scientific and Medical (band)
ITU	International Telecommunication Union
JCF	Joint-Correlation Function
LAN	Local Area Network
LCR	Level Crossing Rate
LDGM	Lower Density Generator Matrix

Abbreviations and acronyms used throughout this book (*cont.*)

Abbreviation	Connotation
LDPC	Lower Density Parity Check
LF	Linear Process-and-Forward Protocol
LLR	Log Likelihood Ratio
LMSNR	Linear Maximum SNR
LNA	Low Noise Amplifier
LOS	Line of Sight
LPNM	L_p-Norm Method
LR-MMSE	Low Rank MMSE
LS	Least Square
LSTC	Layered Space-Time Codes
LTE	Long Term Evolution
M-PSK	M-ary Phase Shift Keying
M-QAM	M-ary Quadrature Amplitude Modulation
M-VCE	Mobile Virtual Centre of Excellence
MAC	Medium Access Control
MAP	Maximum A Posteriori
MARC	Multi-Access Relay Channel
MARIC	Multi-Access Relay Interference Channel
MC-CDMA	Multi-Carrier CDMA
MCU	Microcontroller Unit
MEA	Method of Equal Areas
MEDS	Method of Exact Doppler Spreads
MGF	Moment Generating Function
MHz	Megahertz
MIMO	Multiple-Input Multiple-Output
MIPS	Million Instructions Per Second
MISO	Multiple-Input Single-Output
ML	Maximum Likelihood
MMEA	Modified MEA
MMEDS	Modified MEDS
MMSE	Minimum Mean Square Error
MPC	Multipath Component
MPIC	Multipath Interference Canceler
MSE	Mean Square Error
MRC	Maximum Ratio Combining
MS	Mobile Station
nLF	Nonlinear Process-and-Forward Protocol
NLOS	Non-Line of Sight
O-MIMO	Orthogonalized MIMO
OCR	On-Channel Repeater
ODMA	Opportunity Driven Multiple Access
OFDM	Orthogonal Frequency Division Multiplexing
OFDMA	Orthogonal Frequency Division Multiple Access
OFR	On-Frequency Repeater
ORC	Orthogonality Restoring Combining
OSF	Orthogonal Spreading Factor
OSI	Open Systems Interconnection
PA	Power Amplifier
PAPR	Peak-to-Average Power Ratio
PCCC	Parallel Concatenated Convolutional Code
PCG	Product Coding Gain
PDF	Probability Density Function
PDP	Power Delay Profile

Abbreviations and acronyms used throughout this book (*cont.*)

Abbreviation	Connotation
PEO	Packet Error Outage
PEP	Pairwise Error Probability
PER	Packet Error Rate
PF	Purge-and-Forward Protocol
PHY	Physical Layer
PLL	Phased Locked Loop
POTS	Plain Old Telephone Service
PSD	Power Spectral Density
PUSC	Partially Loaded Subcarriers
QAM	Quadrature Amplitude Modulation
QoS	Quality of Service
QPSK	Quadrature Phase Shift Keying
R99	Release 99
RD	Relay – Destination
RLS	Recursive Least Squares
RMS	Root Mean Square
RRC	Root-Raised Cosine
RRM	Radio Resource Management
RS	Relay Station
RS	Relay Selection
RSC	Recursive Systematic Convolutional (code)
RSCC	Recursive Systematic Convolutional Code
RSSI	Received Signal Strength Indicator
RTR	Request to Relay
RTS	Request to Send
Rx	Receiver
S-DF	Selective DF
S-R-D	Source-Relay-Destination
SAW	Surface Acoustic Wave (filter)
SB	Single Bounce
SBR	Single Bounce Receiver
SBT	Single Bounce Transmitter
SC	Selection Combining
SC	Single Carrier
SCCP	Single Carrier Cyclic Prefix
SD	Source-Destination
SEP	Symbol Error Probability
SER	Symbol Error Rate
SIMO	Single-Input Multiple-Output
SINR	Signal-to-Interference and Noise Ratio
SIR	Soft Information Relaying
SISO	Single-Input Single-Output
SNR	Signal-to-Noise Ratio
SOS	Sum-of-Sinusoids
SR	Source-Relay
SSE	Soft Symbol Estimation
STBC	Space–Time Block Code
STC	Space–Time Code
STCF	Space–Time Correlation Function
STF-CF	Space–Time–Frequency Correlation Function
STTC	Space–Time Trellis Code
SWC	Slepian–Wolf Coding
TC	Turbo Coding
TDD	Time Division Duplex

Abbreviations and acronyms used throughout this book (*cont.*)

Abbreviation	Connotation
TDMA	Time Division Multiple Access
TDR	Time Division Relaying
TENCE	Tensor-Based Channel Estimation
TTI	Time Transmission Interval
TTM	Time To Market
Tx	Transmitter
UAV	Unmanned Aerial Vehicle
UK	United Kingdom
UMTS	Universal Mobile Telecommunications System
UTD	Uniform Theory of Diffraction
UTRA	UMTS Terrestrial Radio Access
VAA	Virtual Antenna Array
VBLAST	Vertical BLAST
VCO	Voltage Controlled Oscillator
WCDMA	Wideband CDMA
WiMAX	Worldwide Interoperability for Microwave Access
WINNER	Wireless World Initiative New Radio
WLAN	Wireless LAN
WPAN	Wireless Personal Area Network
w.r.t.	with respect to
WSN	Wireless Sensor Network
WZC	Wyner–Ziv Coding
ZF	Zero Forcing

Functions

General and special functions used throughout this book

Function	Connotation
$_2F_1(\cdot,\cdot;\cdot;\cdot)$	Gauss hypergeometric function
$_nF_m$	generalized hypergeometric function
$\delta(\cdot)$	Dirac delta function
$\Gamma(\cdot)$	complete gamma function
$\gamma(\cdot,\cdot)$	lower incomplete gamma function
$\phi(\cdot)$	moment generating function
$\Psi(\cdot)$	Doppler (power) spectrum
$\Psi(\cdot,\cdot;\cdot)$	Tricomi confluent hypergeometric function
$\mathbb{E}(\cdot)$	expectation
$\mathrm{Ei}(\cdot)$	exponential integral
$F(\cdot)$	cumulative distribution function
$\mathcal{F}\{\cdot\}$	Fourier transform
$\mathcal{F}^{-1}\{\cdot\}$	inverse Fourier transform
$G(\cdot|\cdot)$	Meijer G function
$I_\nu(\cdot)$	νth order modified Bessel function of the first kind
$J_\nu(\cdot)$	νth order Bessel function of the first kind
$K_\nu(\cdot)$	νth order modified Bessel function of the second kind
$M(\cdot)$	moment generating function
$\mathcal{M}^{-1}(\cdot)$	inverse cumulative von Mises distribution
$N(\cdot)$	level crossing rate
$p(\cdot)$	probability density function
$p(\cdot,\cdot)$	joint probability density function
$P(\cdot)$	probability
$P_{\text{out}}(\cdot)$	outage probability
$Q(\cdot)$	Q Function
$Q(\cdot,\cdot)$	Marcum Q Function
$R(\cdot)$	correlation function
$T(\cdot)$	average fade duration
$\mathbb{VAR}(\cdot)$	variance
A^{H}	Hermitian transpose of A
$a*b$	convolution of a and b
$\|\mathbf{H}\|$	Frobenius norm of $\mathbf{H}$
arg max	maximum among arguments
det	determinant
diag	diagonal of matrix
inf	infimum of argument
lim	limit

General and special functions used throughout this book (*cont.*)

Function	Connotation
max	maximum of argument
min	minimum of argument
sup	supremum of argument
tr	trace of matrix

Symbols

Key symbols and variables used throughout Chapter 1

Symbol	Connotation
γ	SNR
$\overline{\gamma}$	average SNR
C	capacity
d	diversity gain
F	fading gain
g	instantaneous channel gain
$\overline{g}$	average channel gain
L	pathloss gain
M	number of nodes
n	pathloss coefficient
N	noise power
n_r	number of receive antennas
n_t	number of transmit antennas
P	power
P_e	error probability
r	rate gain, multiplexing gain
R	rate
S	shadowing gain
S	signal power
S	symbol power
x	distance

Key symbols and variables used throughout Chapter 2

Symbol	Connotation
α	LOS/NLOS weighting factor in pathloss equation
α	azimuth angle of wave departure/arrival
β	elevation angle of wave departure/arrival
γ	angle of movement
γ	SNR
$\overline{\gamma}$	average SNR
ϵ	permittivity
$\Delta\phi$	angular shift

Key symbols and variables used throughout Chapter 2 (*cont.*)

Symbol	Connotation
Δd	spatial shift/displacement
Δf	spectral shift
Δh	clutter undulation
Δt	temporal shift
η	fractional/relative power
η	intrinsic impedance
θ	azimuth antenna array orientation
κ	concentration of von Mises/Tikhonov distribution
λ	wavelength
μ	mean of von Mises/Tikhonov distribution
μ	permeability
ξ_{thr}	threshold value
ω	angular frequency
Ω	channeled power
σ_{dB}	shadow standard deviation
τ	MPC delay
τ_{RMS}	root-mean-square delay spread
ϕ	channel phase
ϕ	random phase
σ^2	AWGN power
ψ	elevation antenna array orientation
a	instantaneous channel amplitude
$\bar{a}$	average channel amplitude
A	amplification factor
A	empirical factor in pathloss equation
b	empirical factor in pathloss equation
B	empirical factor in pathloss equation
B_{c}	coherence bandwidth
B_{s}	symbol bandwidth
c	speed of light
C	empirical factor in pathloss equation
d	transmitter–receiver distance
D	transmitter–receiver distance
d_0	reference distance
d_{bp}	breakpoint distance
d_{c}	coherence distance
d_{corr}	shadowing correlation distance
D_{S}	empirical factor in shadowing equation
E	electric field
E_{s}	symbol energy
f	frequency
F	fading
$\hat{f}$	maximum Doppler shift
F_n	frequency bins
g	instantaneous channel gain
$\bar{g}$	average channel gain
G	antenna gain
G	pathloss and shadowing gains
h	terminal height
H	magnetic field
$\mathbf{H}$	MIMO channel matrix

Key symbols and variables used throughout Chapter 2 (*cont.*)

Symbol	Connotation
$H(f)$	complex channel in frequency
$h(t)$	complex channel in time
k	wave vector
K	Rice fading factor
l	antenna length
L	number of cylindric surfaces at transmitter
L	pathloss
m	Nakagami fading factor
M	number of clutter around transmitter
n	pathloss coefficient
N	number of clutter around receiver
N	number of relaying segments
N_0	noise power spectral density
n_f	number of floors
n_r	number of receiver antennas
n_t	number of transmit antennas
n_w	number of walls
p	transmit antenna index
P	number of cylindrical surfaces at receiver
P	power
P_r	received power
P_t	transmitted power
q	receive antenna index
R	clutter radius
R	reflection coefficient
s	clutter size
S	shadowing
Ss	empirical factor in shadowing equation
t	time
T	period
T	refraction coefficient
T_c	coherence time
T_s	symbol duration
$T(t, f)$	time-variant frequency transfer function
v	speed of mobile terminal
x	transmitted signal
X	empirical factor in pathloss equation
y	received signal

Key symbols and variables used throughout Chapter 3

Symbol	Connotation
β	general fading coefficients
γ	instantaneous SNR
$\overline{\gamma}$	average SNR
γ_{thr}	threshold SNR
κ	pathloss factor
λ	eigenvalue
σ	noise variance

Key symbols and variables used throughout Chapter 3 (*cont.*)

Symbol	Connotation
$\mathbf{A}$	code distance matrix
$\mathbf{B}$	code difference matrix
$f_{(i)}$	PDF of ordered variable $X_{(i)}$
$F_{(i)}$	CDF of ordered variable $X_{(i)}$
$f(x)$	probability distribution function (PDF)
$F(x) = Pr(X \leq x)$	cumulative distribution function (CDF)
G	pathloss channel gain
h	complex channel
l	codeword length
m	number of receive antennas
n	number of transmit antennas
n	number of relay
N_0	Noise power spectral density
n_{RD}	noise at destination for signal transmitted from relay
n_{SD}	noise at destination for signal transmitted from source
n_{SR}	noise at the relay
p	transmission power
P_b	bit error probability
$P_b(e)$	bit error probability
r	rank
w	combining coefficients at the destination
$X_{(1)} \leq X_{(2)} \cdots \leq X_{(n)}$	ordered variables of $X_1, X_2, \ldots, X_n$

Key symbols and variables used throughout Chapter 4

Symbol	Connotation
α'	fractional frame duration
$\alpha_k(m)$	feedforward recursive variable in BCJR (MAP) decoding
β	amplification factor at the relay
β'	fractional power
$\beta_k(m)$	feedback recursive variable in BCJR MAP decoding
γ	instantaneous SNR
$\overline{\gamma}$	average SNR
$\gamma_k(m, m')$	state transition probability in BCJR MAP decoding
Θ	end-to-end throughput
κ	pathloss factor
σ	noise variance
$\mathbf{0}_{m \times m}$	$m \times n$ zero matrix
b	number of bits
$d_h(\mathbf{X}, \mathbf{Y})$	Euclidean distance between $\mathbf{X}$ and $\mathbf{Y}$
E_b/N_0	signal-to-noise ratio per information bit
E_s	symbol energy
f	probability density distribution
F	cumulative density distribution
G	pathloss channel gain
$\mathbf{g}$	generator matrix of a linear channel code
$\mathbf{g}(D)$ or $\mathbf{G}(D)$	generator polynomial of a linear channel code
$\mathbf{G}_2^n$	n dimensional binary space

Key symbols and variables used throughout Chapter 4 (*cont.*)

Symbol	Connotation
h	complex channel
$\mathbf{H}$	channel matrix
H_2^i	harmonic mean of two random variables
$H(x\|y)$	conditional entropy
$\mathbf{I}_k$	$k \times k$ identity matrix
K	number of relaying stages
L_r	log likelihood ratio (LLR)
M	modulation index
n	number of relay
$\mathbf{N}$	noise matrix
N_0	noise power spectral density
n_{RD}	noise at destination for signal transmitted from relay
n_{SR}	noise at the relay
p	transmission power
P_b	bit error probability
$P_{f,e2e}$	end-to-end frame error rate
$p(x\|y)$	probability of x conditioned on y
Q	number of cooperating clusters
r	number of receive antennas
R	STBC rate
s	number of symbols
t	number of transmit antennas
$\mathbf{X}$	transmitted signal matrix
x_i	ith symbol
$\mathbf{Y}$	received signal matrix

Key symbols and variables used throughout Chapter 5

Symbol	Connotation
β	roll-off factor
γ	signal-to-noise ratio
C	capacitance
D_p	pilot density
DS	channel delay spread
f	frequency
h	complex channel
L	number of paths
$\mathcal{L}$	number of load/store operations
L_{RRC}	pulse-shaping filter length
$\mathcal{M}$	memory requirements
MC	number of multicodes
N	(I)FFT size
N_{burst}	number of encoded data bursts
N_{chips}	number of chips
N_{data}	number of date symbols
N_{iter}	number of turbo decoding iterations
N_{sub}	number of useful subcarriers
$\mathcal{O}$	number of operations

Key symbols and variables used throughout Chapter 5 (*cont.*)

Symbol	Connotation
OSF	oversampling factor
P	number of stages
R	number of iterations
R_c	channel code rate
$RC_0(t)$	RRC filter response
SF	spreading factor
T_c	chip duration
V	voltage
w	weight
W	maximum channel delay

1

Introduction

Wireless has become as much part of our lives as have houses, cars and computers. Mobile phones, an example application of wireless technologies, are indispensable today as they allow us to be connected anywhere at any time. We have, in fact, taken so much liking to wireless technologies that system capacity is reaching saturation levels. This is aggravated by recently emerged bandwidth hungry applications ranging from web browsing to multimedia transmissions. Network designers are struggling to meet this ever increasing demand in capacity and any means to increase capacity is hence welcome.

Interestingly, Martin Cooper of Arraycomm has observed that 'the wireless capacity has doubled every 30 months over the last 104 years.' This translates into an approximately million-fold capacity increase since the 1960s, which has been broken down [7] to yield a 25-times improvement from wider spectrum, a fivefold improvement by dividing the spectrum into smaller slices, a fivefold improvement by designing better modulation schemes, and an impressive 1600-fold gain through reduced cell sizes and transmit distance.

Among the many possible approaches to capitalize on these enticing gains, this book focuses on cooperation and we will show how cooperative techniques at the physical layer can potentially be of great benefit to capacity as well as coverage.

1.1 Book Structure

To facilitate a coherent understanding of cooperation for people working in the field as well as people acquainting themselves with the topic, the book is structured as follows:

- **Chapter 1: Introduction.** A basic introduction to cooperative systems is given first. We will then discuss typical application scenarios of cooperative techniques. Furthermore, we will discuss pros and cons of relaying and briefly quantify the capacity gains. We will also revise definitions and terminologies typically used in the context of cooperative systems. Finally, some historical background as well as key milestones are discussed.
- **Chapter 2: Wireless Channel.** In contrast to traditional wireless systems, the wireless channel becomes part of the cooperative system, which has a profound impact on channel statistics, including temporal and correlation behaviors. We will hence discuss in detail the channel behavior of regenerative, transparent and distributed multiple-input, multiple-output (MIMO) systems. This is pivotal in understanding and quantifying the performance of cooperative protocols at the physical (PHY) layer.

Cooperative Communications   Mischa Dohler and Yonghui Li
© 2010 John Wiley & Sons, Ltd

- **Chapter 3: Transparent PHY.** We will discuss and quantify the performance of different transparent architectures here. We will introduce some tools that facilitate the characterization of general architectures, which range from transparent relaying to transparent distributed space–time block coding, distributed space–time trellis coding, distributed multiplexing and distributed beamforming. We will also dwell on issues pertaining to distributed system optimization, such as distributed power allocation and distributed relay selection.
- **Chapter 4: Regenerative PHY.** We will discuss and quantify the performance of a plethora of regenerative architectures here. We will mainly deal with different estimate-and-forward, decode-and-forward, compress-and-forward and soft-information relaying protocols, among others, with the aim characterizing their performance. We will also dwell on recently emerged advanced topics related to distributed coding, such as distributed space–time block coding, distributed space–time trellis coding, distributed turbo coding, distributed network-channel coding for single as well as multiple source–destination pairs, etc.
- **Chapter 5: Hardware.** We will discuss how hardware facilitates as well as limits any implementation of cooperative relaying schemes. These limitations in hardware render, for instance, the implementation of some of the recently proposed cooperative protocols infeasible. We will discuss the implementation of transparent and regenerative schemes. We will also compare their costs and implementation complexities. We then derive complexities and power consumption of various relaying architectures based on 3G and 4G standards. Finally, we discuss some available hardware platforms and testbeds implementing relaying or distributed space–time processing techniques.
- **Chapter 6: Road Ahead.** Finally, the book is concluded by summarizing its contributions and also highlighting important open research directions and still-to-be-explored impact areas. We also describe how to include real-world impairments into the analysis, such as interference, etc. Finally, some business issues are briefly dealt with.

1.2   Quick Introduction

Wireless communication systems are traditionally conceptualized such that users individually communicate with the associated base station and vice versa. Cooperation is referred to as any architecture that deviates from this traditional approach, that is where a user's communication link is enhanced in a supportive way by relays or in a cooperative way by other users. Due to the many available degrees of freedom of such systems, that is the many ways in which supportive relays and cooperative users can be deployed, an enormous amount of different architectures exist. Some canonical architectures are briefly introduced below, after having discussed the wireless channel they experience and the gains one can expect from such deployments.

1.2.1   Channel

The wireless channel is pivotal to the understanding of the gains of cooperative systems. Whilst discussed in more detail in Chapter 2, we shall expose here some of its fundamental properties. As such, the transmitted signal is impaired by three effects:

- **Pathloss.** Averaging the received power at a particular distance over a sufficiently large area, yields the loss in power or the pathloss versus distance. The pathloss law is (almost) deterministic and traditionally behaves linearly in decibels, that is as an inverse law in linear scale. Pathloss limits interference but also rapidly diminishes the useful signal power. Any technique improving on the pathloss is hence highly appreciated by network planners.
- **Shadowing.** Averaging the received power at a particular distance over an area of radius of approximately shadowing coherence distance yields a variation in the received power around the pathloss. This variation is referred to as shadowing. The shadowing law is random and is traditionally modeled as Gaussian in decibels, that is lognormal in linear scale. Shadowing is one of the most

detrimental performance factors in modern communications systems since it cannot be absorbed by suitable channel codes, thus causing non-availabilities of links referred to as outages. Any technique improving on the shadowing outage is hence highly appreciated.

- **Fading.** Not averaging the signal at all allows one to observe fading as a signal fluctuation around pathloss and shadowing. It is caused by the constructive and destructive addition of the signal traveling via multiple propagation paths.

  If channel fading realizations change from symbol to symbol due to high mobility in the channel, then the channel is referred to as fast-fading; otherwise, it is referred to as slow-fading. A fast-fading channel offers temporal diversity, which can be picked up by a suitable channel coder; a slow-fading channel does not offer temporal diversity, which yields large outages.

  If the multiple symbol copies can be resolved because the mutual propagation delays are larger than the symbol duration, then the channel is referred to as a frequency-selective fading channel; otherwise, it is referred to as a frequency-flat or simply flat fading channel. A frequency-selective fading channel offers frequency diversity, which can be picked up by suitable signal processing coupled with a channel coder; a flat fading channel does not offer frequency diversity, which again yields large outages.

  In terms of outages, the worst case scenario is hence a slow and flat fading channel and the best case is a fast and frequency-selective fading channel. Modern communications systems can be fast or slow fading, depending on the mobility they are subjected to; however, due to the increasingly short symbol durations, they are typically frequency selective and hence inherently offer a healthy amount of diversity. Techniques improving on diversity therefore seem unnecessary. Yet, as will be shown in Section 1.5.3, diversity (reliability) can be traded against communication rate (capacity), thereby encouraging the development of techniques that can improve on either.

In decibels (dB), these effects add up, whereas in linear scale they are multiplicative. Denoting pathloss, shadowing and fading by L, S and F respectively, the received signal power is hence given in dB as $P_r = P_t + L + S + F$. Cooperative communications generally evolves around diminishing these losses associated with the wireless channel as well as using the detrimental effects to its benefit.

1.2.2 Typical Gains

As will be quantified throughout this book, cooperative communications yields several gains. Three of the most important gains are given below:

- **Pathloss Gain.** Due to the non-linear pathloss behavior, splitting the propagation path yields transmit power gains because the aggregate pathloss of the split path is less than the pathloss of the full path. The pathloss, and hence the signal-to-noise ratio (SNR), is known to be inversely proportional to the propagation distance. More precisely,

$$\text{SNR} \propto \frac{1}{x^n}, \tag{1.1}$$

  where x is the distance between transmitter and receiver, and n is the pathloss exponent. The latter typically differs for different propagation environments between 2 (line-of-sight) and 6 (highly cluttered environments). Splitting, for example, the communication path between source and destination into two equal distances and allocating to each part half of the power, the gain w.r.t. direct communication assuming a pathloss coefficient of $n = 4$ is quantified as

$$\frac{\frac{1/2}{(x/2)^4} + \frac{1/2}{(x/2)^4}}{\frac{1}{x^4}} = 16, \tag{1.2}$$

that is yielding a $10\log_{10} 16 = 12\,\text{dB}$ power saving. This is a significant gain and one of the main incentives for using cooperative techniques.

- **Diversity Gain.** Providing additional independent copies of the *same information* via independent shadowing and fading channels yields diversity gains. These gains stem from the fact that with the amount of additional copies increasing, the probability of all of them being illegible decreases. The provision of such copies can be achieved, for example, by having a relay provide a copy in addition to the information received already via the direct link; or by having several relays provide copies in parallel. Diversity gains improve the performance of the system, such as the probability of error P_e or outage P_{out}. Either probability depends on the rate R in a similar fashion and is known to be related to the diversity gain at asymptotically high SNRs as

$$P_{out}(R) = \frac{const(R)}{SNR^d}, \tag{1.3}$$

where d is the diversity gain or diversity order, and $const(R)$ is some constant depending on R. It effectively equals the number of copies of the same information; for instance, having one relay and a direct link providing two independent copies of the same information, yields $d = 2$. At the same level of performance metric, for example, requiring an outage probability of 1%, increasing the diversity order allows a saving in transmit power; for instance, with $d = 2$ about 3 dB transmission power can be saved in a flat Rayleigh fading channel.

- **Multiplexing Gain.** In an asymptotic SNR regime, the achievable data rate R is known to be proportional to the logarithm of the SNR, that is [8]

$$R = r \log_2 SNR + const. \tag{1.4}$$

Here, r is referred to as the rate or multiplexing gain and is equal to the degrees of freedom of the channel, that is the number of independent channels over which *different information* can be sent. In a system with n_t transmit and n_r receive antennas, the maximum number of independent channels under favorable propagation conditions is $r = \min(n_t, n_r)$. For instance, it is clear that, if a source node possesses one transmit and a destination node two receive antennas or vice versa, they can communicate at a multiplexing gain of one only since only one independent channel can be provided. Using a relay in addition to the direct link, allows under certain conditions the creation of a second channel, hence doubling the multiplexing gain and thus the communication rate, assuming the SNR is kept constant.

These gains are not independent and hence can be traded one against another:

- **Power versus Data Rate.** As per Equation (3.77), the power gains due to pathloss and diversity gains can be used to increase the data rate R at a constant multiplexing gain r. This is typically done by increasing the cardinality of the signal constellation. For instance, gaining about 6 dB in power, one could switch from QPSK to 16QAM.

- **Diversity versus Multiplexing Gain.** This fundamental trade-off, also known as the diversity-multiplexing-tradeoff (DMT), will be discussed in Section 1.5.3. The DMT can be quantified by inserting Equation (3.77) into Equation (3.76), solving w.r.t. d and letting the SNR grow asymptotically large to arrive at [8]

$$d = - \lim_{SNR \to \infty} \frac{\log P_{out}(r \log_2 SNR)}{\log SNR}. \tag{1.5}$$

Due to the minus sign, the logarithm being a monotonically increasing function and the probability P_{out} increasing with r, increasing the multiplexing gain r necessarily requires the diversity gain d to be diminished. Each system configuration will hence have a maximum diversity gain d_{max} at minimum multiplexing gain r_{min}, and conversely a maximum multiplexing gain r_{max} at minimum diversity gain d_{min}. The above relationship quantifies these maxima and minima and also how they trade in-between, that is how reliability (diversity gain) trades against capacity (multiplexing gain) for a given channel and system configuration.

The above gains and trade-offs justify the research into different cooperative architectures, some canonical forms of which are described below.

1.2.3 Canonical Architectures

Whilst the degree of freedom of cooperative systems is very large, we list below some choices/ parameters impacting most the realization of a particular architecture and thus used protocols:

- **Transparent versus Regenerative Relaying.** One of the foremost design drivers in cooperative systems is due to the choice between transparent and regenerative relaying approaches. Transparent relaying generally implies that the relay only amplifies the signal before retransmitting it. It is also possible, however, that the relay performs some other linear and non-linear operations in the analog domain, such as phase shifting, etc. Regenerative relaying, on the other hand, requires the relay to change the waveform and/or the information contents by performing some processing in the digital domain. An example is the relay receiving the information from the source, decoding, re-encoding and finally retransmitting it.
- **Traditional versus Distributed Space–Time Relaying.** Another important factor is the choice between traditional relaying and spatially distributed space–time processing relaying architectures. Traditional relaying has been around for some decades already and is realized by means of an arbitrary number of serial and/or parallel relays delivering the information from source towards destination. Space–time processing relaying, however, is realized by means of a distributed deployment of arbitrary number of (typically but not necessarily) synchronized nodes performing one of the many possible forms of distributed space–time processing. Known space–time techniques are thus applicable either directly or in modified form to these architectures, such as space–time coding, BLAST type algorithms or beamforming.
- **Dual-Hop versus Multi-Hop Networks.** The choice of the number of relaying stages is very important to system designers. As such, relays can be connected in series or operated in parallel. Increasing the number of serial relaying nodes increases the pathloss gain. Increasing the number of parallel relaying nodes increases the maximum diversity gain. Note that the relay channels ought to but do not necessarily have to be orthogonal so as to minimize interference; this can be achieved by using different frequencies, time slots, codes, etc.
- **Availability of Direct Link.** Depending on the propagation conditions, there may or may not be a direct link between source and destination or various relaying stages that is sufficiently strong to facilitate data transmission. Without the direct link, only pathloss gains can be achieved; with the direct link, the maximum diversity gain can also be increased. The direct link is usually available in situations where the system is capacity limited and not available where coverage limited.
- **Degree of Cooperation.** One generally distinguishes between the cases of supportive relaying and cooperative relaying. Typically, placing a relay node in-between a source and destination node is referred to as supportive relaying or simply relaying. Supportive relaying can be extended to cooperative communications, where at least two cooperative nodes are each other's respective relays at the same time to boost the other's communication links. Cooperative deployments clearly boost the maximum diversity and maximum multiplexing gains and, albeit not for every node, also the pathloss gain.

These choices and parameters have been visualized in Figure 1.1, where we have not shown the choice between transparent and regenerative architectures as any could be either of these two. The application of combinations of these canonical cooperative architectures to some practical scenarios is briefly discussed in the subsequent section.

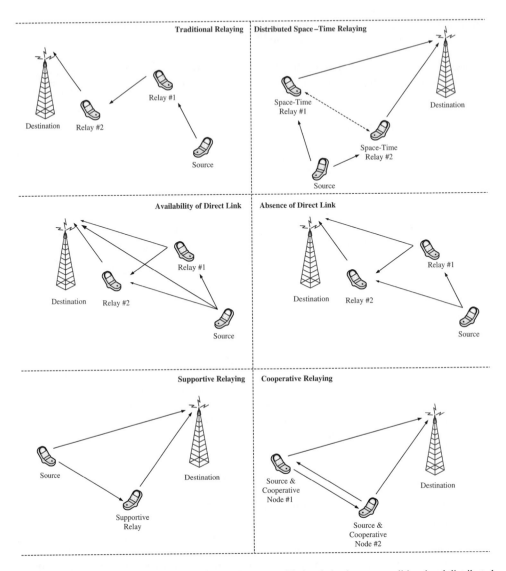

Figure 1.1 Exemplification of canonical relay architectures with the choice between traditional and distributed space–time processing relaying (top); between the availability of a direct link versus its absence (middle); and between supportive and cooperative relaying (bottom). All of these example architectures could realize transparent or regenerative relaying as well as any combination thereof

1.3 Application Scenarios

Before discussing the theoretical nature of cooperative communications, we shall give a few practical applications of cooperation by means of a set of chosen scenarios, ranging from cellular to wireless sensor networks.

1.3.1 Cellular Capacity and Coverage Extension

A network composed of adjacent cells, each served by a base station, is known as cellular network. Typical examples of today's cellular systems are GSM and UMTS; future roll-outs will include technologies from 3GPP LTE and WiMAX. Cellular networks generally suffer from three fundamental problems:

1. **Capacity:** Each cell is assigned a limited amount of resources in terms of bandwidth and allowed transmission power. These resources suffice to serve a given number of users; however, due to the ever-increasing number of users connecting to the same base station, the offered cell capacity is starting to become insufficient w.r.t. the required accumulated capacity of all associated users. The system is said to be capacity limited.
2. **Coverage:** The limit in transmission power has a profound impact on the coverage of the cell. Users at the cell edge experience insufficient power levels to support communication due to the weak signal of interest from the associated base station. The system is said to be coverage limited.
3. **Interference:** The necessity to roll out and support more than one cell leads to interference between these cells. Users at the cell edge hence not only experience insufficient power levels from their associated base station but also interference power from adjacent cells using the same frequency starts to become a dominant detrimental performance factor. The system is said to be interference limited.

These three effects are not independent, that is interference impacts coverage and capacity, etc. To alleviate these problems, however, the use of relays has been proposed where communication between a base station (BS) and a mobile station (MS) not only happens directly but also or exclusively via a relay station (RS). As explained in previous sections, such a deployment yields significant gains which, as per Figure 1.2, translates into the following [9]:

1. **Capacity Gains:** MSs not suffering from coverage problems can gain extra capacity since the BS can switch to higher modulation orders and thus boost system capacity.

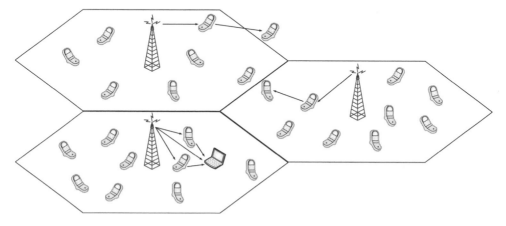

Figure 1.2 Cellular scenario: relaying boost performance of users that are capacity limited (bottom-left cell); coverage limited due to range (top-left cell) or interference (middle-right cell)

2. **Coverage Gains:** MSs at or beyond the cell edge as well as in coverage holes receive a signal of sufficient quality to communicate and hence the cell coverage is increased.
3. **Interference Gains:** Part of the capacity and/or coverage gains can be used to decrease the transmission power and hence the general interference temperature in the system. This reduction in interference, however, has to be gauged against the increased interference caused by the additional relay traffic.

Due to these obvious and potential gains, Vodafone pushed to include such a relaying approach into the 3GPP standard via ETSI's UTRA air interface technology proposals group Epsilon. The proposed access method was termed Opportunity Driven Multiple Access (ODMA) relaying protocol [10–15]. For various reasons, however, it was not selected as a candidate for 3GPP but included as an optional protocol in several early standard releases of 3GPP; it was eventually discontinued in the R99 standard release.

Many lessons have been learned ever since and current standards developing bodies are picking up on the relaying concept. In particular, IEEE 802.16j has developed a relay enabled mode to WiMAX in the hope of giving it a competitive edge in 3GPP LTE developments. The latter currently only standardizes repeaters to leverage coverage problems; however, it is envisaged that LTE Advanced will include cooperative relay features. This and related topics are discussed in [4, 5] and will also be dealt with in Chapter 5 of this book.

1.3.2 WLAN Capacity and Coverage Extension

Built on IEEE 802.11 technology, WLANs are a unique communication phenomenon of recent times. This is corroborated by the millions of WLAN stations in use today in homes, offices, cafes, train stations, airports, etc. [16–27]. In fact, usage in urban environments is already so dense that almost complete coverage could be provided and interference commences to become a serious performance-limiting factor.

Similarly to the cellular case, cooperative techniques are promising to boost capacity and increase coverage even further. This is exemplified by means of Figure 1.3, where a home WLAN access point (AP) communicates with a user in the street via another user acting as a relay. The question on the optimum number of allowed hops arises which, according to some preliminary studies using realistic system parameters, is apparently only two on average.

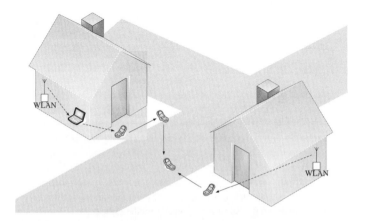

Figure 1.3   A WLAN station installed inside a house provides access to users in the street via relays

1.3.3   Vehicle-to-Vehicle Communication

Future vehicles will allow for platooning (automated steering within a group of cars), in-vehicle internet access, inter-vehicle communication, etc. [28–45]. The increasing density of vehicles allows for the deployment of cooperative vehicle systems. The advantage of cooperative techniques, as per Figure 1.4, becomes apparent here since the redundant links established in cooperation offer high link stability in such volatile and dynamic propagation conditions.

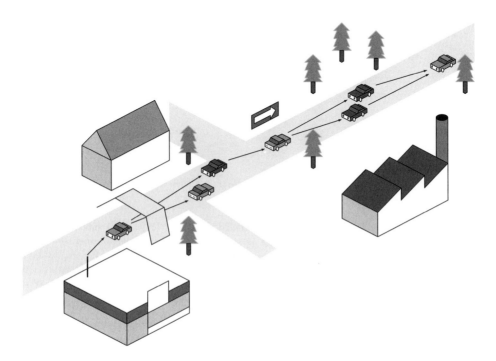

Figure 1.4   Distributed vehicle-to-vehicle communication scenario, where cars cooperate to facilitate the relay and delivery of information

1.3.4   Wireless Sensor Networks

Sensor networks have been researched and deployed for decades; their wireless extension, however, has witnessed a tremendous upsurge in recent years [25, 46–62]. This is mainly attributed to the unprecedented operating conditions of WSNs, that is an enormous number of nodes of extremely low complexity and cost reliably operating under stringent energy constraints.

As of today, a major problem in deploying WSNs is their dependence on limited battery power. A main design criterion is to optimize the lifetime of the network without jeopardizing reliable and efficient communications from sensor nodes to other nodes as well as data sinks. Limiting the transmission power is hence a must and coverage is therefore a predominant problem. As exemplified in Figure 1.5, cooperative techniques are beneficial if not crucial in closing coverage gaps and therefore maintaining network integrity.

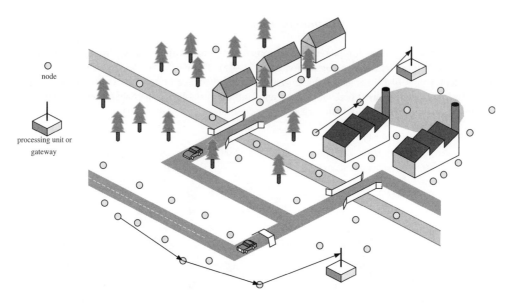

Figure 1.5   WSN application where sensors are heavily coverage limited and hence rely on relaying to deliver sensed information

Hybrid solutions are also foreseen, such as unmanned aerial vehicles (UAVs) and wireless sensor networks. As in Figure 1.6, a set of sensors in areas difficult to reach communicates with a set of UAVs, which in turn cooperate between each other prior to sending the data to a processing unit. In [63–65], it has been shown that such cooperative UAVs considerably increase the reliability of the transmission of sensor readings under real-world impairments.

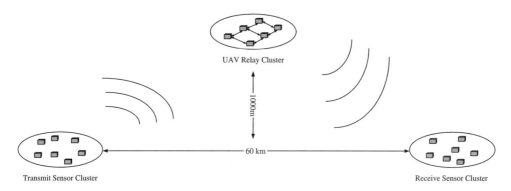

Figure 1.6   Distributed and cooperative UAVs acting as relays, which can utilize beamforming, STCs, multiplexing, etc., to relay sensor readings

1.4   Pros and Cons of Cooperation

Although partially discussed before, the advantages and disadvantages of the canonical cooperative architectures of Section 1.2.3 are summarized below. We will not only discuss the PHY layer but also dwell on some system deployment aspects.

1.4.1   Advantages of Cooperation

The key advantages of using supportive, cooperative or space–time relays in the system can be summarized as follows:

- **Performance Gains.** Large system-wide performance gains can be achieved due to pathloss gains as well as diversity and multiplexing gains. These translate into decreased transmission powers, higher capacity or better cell coverage.
- **Balanced Quality of Service.** Whilst in traditional systems users at the cell edge or in shadowed areas suffered from capacity and/or coverage problems, relaying allows to balance this discrepancy and hence give (almost) equal quality of service (QoS) to all users.
- **Infrastructure-Less Deployment.** The use of relays allows the roll-out of a system that has minimal or no infrastructure available prior to deployment. For instance, in disaster-struck areas, relaying can be used to facilitate communications even though the cellular system is nonfunctioning. For hybrid deployments, that is a cellular system coupled with relays, the deployment and maintenance costs can be lowered as has been shown in [66].
- **Reduced Costs.** Compared to a purely cellular approach to provide a given level of QoS to all users in the cell, relaying is a more cost effective solution. In [66], however, it has also been shown that whilst savings are not as dramatic as hoped for, the capital and operational expenditures are generally lower when relays are used.

1.4.2   Disadvantages of Cooperation

Some major disadvantages of using supportive, cooperative or space–time relays in the system are given below:

- **Complex Schedulers.** Whilst maintaining a single cooperative relaying link is a fairly trivial task, at system level with many users and relays this quickly becomes an arduous task. As such, relaying requires more sophisticated schedulers since not only traffic of different users and applications needs to be scheduled but also the relayed data flows. Any gains due to cooperation at the physical layer dissipate rapidly if not handled properly at medium access and network layers.
- **Increased Overhead.** A full system functioning requires handovers, synchronization, extra security, etc. This clearly induces an increased overhead w.r.t to a system that does not use relaying.
- **Partner Choice.** To determine the optimum relaying and cooperative partner(s) is a fairly intricate task. Also, the complexity of maintaining such cooperative partnership is higher w.r.t. noncooperative relaying.
- **Increased Interference.** If the offered power savings are not used to decrease the transmission power of the relay nodes but rather to boost capacity or coverage, then relaying will certainly generate extra intra- and inter-cell interference, which potentially causes the system performance to deteriorate. An optimum trade-off needs, therefore, to be found at system level.
- **Extra Relay Traffic.** The relayed traffic is, from a system throughput point of view, redundant traffic and hence decreases the effective system throughput since in most cases resources in the form of extra frequency channels or time slots need to be provided.
- **Increased End-to-End Latency.** Relaying typically involves the reception and decoding of the entire data packet before it can be re-transmitted. If delay-sensitive services are being supported, such as voice or the increasingly popular multimedia web services, then the latency induced by the decoding may become detrimental. Latency increases with the number of relays and also with the use of interleavers, such as utilized in GSM voice traffic. To circumvent this latency, either simple transparent relaying or novel decoding methods need to be used.
- **Tight Synchronization.** A tight synchronization needs generally be maintained to facilitate cooperation. This in turn requires expensive hardware and potentially large protocol overheads since nodes need to synchronize regularly by using some form of beaconing or other viable techniques.

- **More Channel Estimates.** The use of relays effectively increases the number of wireless channels. This requires the estimation of more channel coefficients and hence more pilot symbols need to be provided if coherent modulation was to be used.

The list of disadvantages is clearly longer than the list of advantages, some of which have been summarized for the most important canonical cooperative architectures in Table 1.1. A careful system design is therefore needed to realize the full gains of cooperative relaying systems and to ensure that cooperation does not cause deterioration of the system performance. This is the prime reason why many of the latest research efforts concentrate on above-listed issues.

Table 1.1   Principal pros and cons of some canonical cooperative architectures

	Main advantages	Main disadvantages
Supportive relaying	Pathloss gains	Increased interference
	Balanced user QoS	Complex schedulers
Cooperative relaying	Diversity gains	Optimum partner choice
	Balanced user QoS	Complex schedulers
Space–time relaying	Diversity gains	Increased overhead
	Multiplexing gains	Tight synchronization
	Available space–time codes	More channel estimates
	Balanced user QoS	Complex schedulers

1.4.3   System Tradeoffs

From this list of advantages and disadvantages, it is obvious that many system design parameters can be traded against one another. Some important system trade-offs are discussed below:

- **Coverage versus Capacity.** As already discussed in some detail, cooperative systems allow coverage to be traded against capacity or, equivalently, outage versus rate or diversity versus multiplexing gains. Therefore, the system designer has the choice to let a relay help boost capacity or increase the coverage range. Increasing one inherently diminishes the other.
- **Algorithmic versus Hardware Complexity.** Solving the coverage and capacity problem by means of more cellular base stations requires more complex and hence costly hardware, not to mention the expensive real estate to physically place the base stations. Relays, on the other hand, are of relatively low hardware complexity. However, the decrease in hardware complexity by using relays yields an increase in algorithmic complexity since scheduling, synchronization, handover and other algorithms become significantly more complex. An optimum solution trading algorithmic with hardware complexity hence needs to be determined, likely leading to a hybrid deployment.
- **Interference versus Performance.** As discussed in Section 1.2.2, cooperative communications yields gains which can either be used to decrease the transmission power and hence generated interference or increase capacity/coverage. Furthermore, relaying generates extra traffic, which is an additional source of interference.
- **Ease-of-Deployment versus Performance.** Relays can be deployed in a planned and unplanned manner. In the former, the network designer optimizes the placement and parameterization of the static relay node; this is a complex task but leads to superior performance. In the latter, relays are deployed in an unplanned manner and hence can be stationary or mobile; deployment is therefore significantly simplified at the cost of decreased performance w.r.t. the planned roll-out.

- **Cost versus Performance.** Being a traditional trade-off, the cost of the chosen cooperative solution has a profound impact on its performance. Clearly, deploying highly complex relaying nodes that allow, say, for cooperative space–time relaying induces high costs but also improved performance.

These trade-offs are also visualized in Figure 1.7.

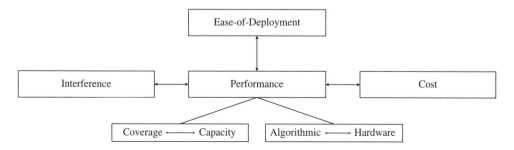

Figure 1.7 At a given performance level, coverage can be traded capacity and algorithmic with hardware complexity. Performance can also be traded against amount of interference, ease-of-deployment and cost

1.5 Cooperative Performance Bounds

So far, we have qualitatively described the potentials of cooperative relaying systems. In this section, we will only briefly quantify achievable performance bounds for cooperative relaying systems; for a more detailed treatment on this matter, the interested reader is referred to [67].

Although not the focus of this book, capacity bounds are very useful as they indicate whether a given cooperative relaying scheme performs well, that is close to the bound, or not. These bounds therefore impact the design of hardware as well as protocols of the entire OSI stack. Hence they serve as a justification and inspiration for any subsequent exposures in this book.

Staying at an introductory level and thus without going into great mathematical detail, we will discuss the achievable rate gains, outage probability gains, and the DMT in the context of cooperative systems.

1.5.1 Capacity Gains

Shannon proved that one can design codes facilitating a communication rate of R bits/symbol with arbitrarily small error [68]. His proof holds for codes that are of infinite duration (or, in practice, very long), so as to average out the effect of noise. His theory, however, is not concerned with the code construction or code complexity, nor with decoding delay.

The maximum data rate at which reliable communication is possible is referred to as capacity C of the channel. Given a signal power S and noise power N at the receiver input, the capacity offered by an additive white Gaussian noise (AWGN) channel is:

$$C = \log_2 \left(1 + \frac{S}{N} \right), \qquad (1.6)$$

which has been normalized by the bandwidth W; the units are hence [C] = bits/s/Hz. Signals, however, typically traverse a wireless channel that impacts the received signal power S and hence the capacity offered by such a channel. Whilst the deterministic effect of pathloss only scales the useful signal power

S in Equation (1.6), the randomness introduced by the wireless channel changes the capacity since the useful signal power effectively varies over the duration of the transmitted codeword. Depending on the type of channel variations, one typically distinguishes ergodic and nonergodic fading channels.

1.5.1.1   Ergodic Channel

A channel characteristic typically assumed in the context of Shannon capacity is that of an ergodic channel. Rigorously speaking, a process is ergodic if the time averages may be used to replace ensemble averages. In more practical terms, this means that the channel varies sufficiently rapidly over the duration of the transmission and hence traverses all fading states.

Ergodicity allows one to apply the concept of averages since the channel's average mutual information over all (infinitely long codewords) is the same. This means that an ergodic channel can support the following maximum error-free transmission rate with 100% reliability:

$$C = \mathbb{E}_g \left\{ \log_2 \left(1 + g \frac{S}{N} \right) \right\}, \tag{1.7}$$

where g is the instantaneous channel gain/power and $\mathbb{E}_g \{\cdot\}$ denotes the expectation operator w.r.t. λ. The random channel fluctuation g may comprise the effects of fading and shadowing but typically only includes fading due to shadowing as being a fairly slowly varying effect. The notion of an ergodic channel w.r.t. two transmitted codewords of infinite duration is illustrated in Figure 1.8.

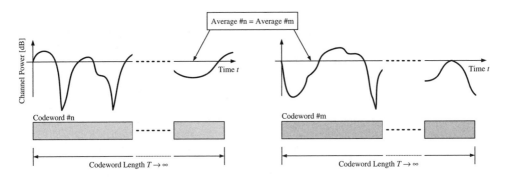

Figure 1.8   In the case of an ergodic channel all fading states are traversed over the duration of a Shannon codeword, thereby facilitating the provision of an average channel capacity

For example, if the channel obeys flat Rayleigh fading with its channel gain hence having a probability density function (PDF) $p_g(g) = 1/\overline{g} \cdot \exp(-g/\overline{g})$, the average capacity can be calculated as [69, 70]:

$$C = \mathbb{E}_g \left\{ \log_2 \left(1 + g \frac{S}{N} \right) \right\} \tag{1.8a}$$

$$= \int_0^\infty \log_2 \left(1 + g \frac{S}{N} \right) \cdot \frac{1}{\overline{g}} e^{-g/\overline{g}} \, dg \tag{1.8b}$$

$$= -e^{1/\overline{\gamma}} \text{Ei}(-1/\overline{\gamma}) / \log(2), \tag{1.8c}$$

where $\text{Ei}(\cdot)$ is the exponential integral, $\overline{g}$ is the average fading power usually normalized to unity, $\gamma = gS/N$ is the instantaneously experienced SNR at the decoder, and $\overline{\gamma} = \overline{g}S/N$ its average. The step from Equation (1.8b) to Equation (1.8c) follows from [71, Section 4.337.2].

1.5.1.2 Capacity Gains

To illustrate the capacity gains, we will follow the analysis exposed in [72–74]. To this end, the transceiver architecture as shown in Figure 1.1 (bottom-right) is assumed. The capacity gains of this simple cooperative relaying schemes assuming Rayleigh fading channels are exemplified by means of Figures 1.9 and 1.10, where the former assumes a symmetric communication scenario and the latter assumes the average fading power between the first node and destination to be significantly better than between the second node and destination.

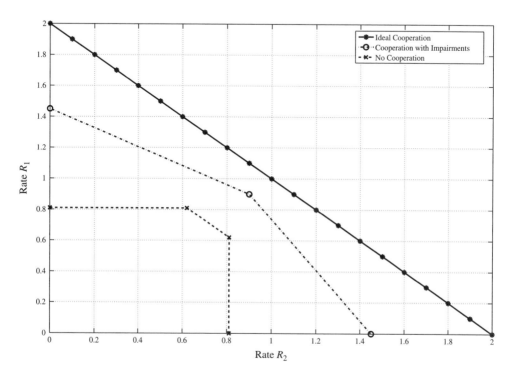

Figure 1.9 Symmetric rate region, where both users have equal channel conditions to the destination, assuming no cooperation, ideal cooperation with error-free inter-user channel, and cooperation with good inter-user channel [72]

With reference to these figures, the case of ideal cooperation is for a noiseless inter-user channel and serves as an upper bound of cooperation. No cooperation between the nodes yields the typical multiple access channel. In the cooperative case, as the inter-user channel degrades, performance approaches that of no cooperation. Typical points of interest are listed below:

- **Equal Rate Point $R_1 = R_2$.** These rates are achieved along the 45° line from the origin. It is a useful indicator if the system treats both users with equal priority. With reference to the symmetric and asymmetric communication scenarios of Figures 1.9 and 1.10, cooperation yields an equal rate gain of about 20% and 200%, respectively.
- **Maximum Rate Sum Point $\max(R_1 + R_2)$.** These rates are achieved along the points where the sum of both nodes is maximal. It is a useful indicator if the total system capacity is of importance and not how much rate each user is guaranteed. With reference to the symmetric and asymmetric communication cases, cooperation yields a maximum rate sum gain of about 20% and 10%, respectively.

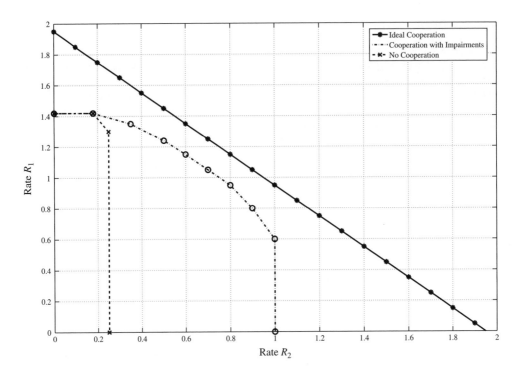

Figure 1.10 Asymmetric rate region, where the first user has much better channel conditions to the destination than the second user, assuming no cooperation, ideal cooperation with error-free interuser channel, and realistic cooperation with bad interuser channel [72]

- **Degraded Relay Rate Points $R_1 = 0$, $R_2 \neq 0$ and $R_1 \neq 0$, $R_2 = 0$.** These rates are achieved when setting one of the rates to zero. It is a useful indicator of how much maximum rate can be delivered to a user whilst the other user is not transmitting own data but only supportively relaying. With reference to the symmetric communication scenario, cooperation almost doubles the degraded relaying rate points; in the asymmetric case, however, only the user with bad channel conditions really profits.

It becomes apparent that using a cooperative relaying approach significantly improves the capacity of each user as well as the network. The gains are of particular importance in the asymmetric case where one user suffers from bad channel conditions. These results, in fact, inspired research into practical communication schemes that are capable of achieving the promised rate gains. To this end, it has already been shown [74] that in the design region of interest, an increase in sum capacity approximately equals the increase in coverage area and that a simple repetition-based coding scheme using CDMA spreading sequences performs well within the rate regions. The remainder of this book is thus dedicated to various techniques able either to perform well within or even approach the bounds of this rate region.

1.5.2 Rate Outage Gains

Shannon's information theory is not designed to cater for communication scenarios where average channel conditions change from codeword to codeword. To this end, the concept of rate outage probability has been successfully introduced in the information theory community. Its application to the context of cooperative relaying systems is discussed below.

1.5.2.1 Nonergodic Channel

In contrast to an ergodic channel, in the case of a nonergodic channel the channel does not vary sufficiently fast to traverse all fading states over the duration of the communication. In other words, a process is nonergodic if samples help meaningfully to predict values that are very far away in time from that sample, that is the stochastic process is sensitive to initial conditions. Practical situations where this occurs are when the channel is very slow fading and/or experiences severe and long-lasting shadowing.

The concept of averages is not very meaningful in the context of nonergodic channels since it will be different for each transmitted codeword. A nonergodic channel can hence not support a maximum error-free transmission rate with 100% reliability; however, it can support any given rate R with a certain probability $P_{out}(R)$, which is referred to as the rate outage probability. The notion of a nonergodic channel w.r.t. two transmitted codewords of infinite duration is illustrated in Figure 1.11.

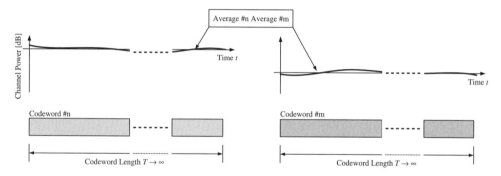

Figure 1.11 In the case of a nonergodic channel not all fading states are traversed over the duration of a Shannon codeword, thereby preventing the provision of an average channel capacity and hence requiring the concept of outage to be invoked

To quantify the outage probability, we assume an instantaneous power $\gamma = gS/N$ at the decoder input due to which an information rate of $C(\gamma) = \log_2(1 + \gamma)$ bits/s/Hz can be supported. The channel is in outage if this rate falls below a threshold information rate R; the corresponding outage event is $C(\gamma) < R$ or $\gamma < (2^R - 1)$. The outage probability is hence

$$P_{out} = \Pr\left(\gamma < (2^R - 1)\right) = \int_0^{2^R-1} p_\gamma(\gamma)d\gamma, \tag{1.9}$$

where $\Pr(\cdot)$ denotes the probability and $p_\gamma(\gamma)$ the PDF of the SNR. For instance, for a Rayleigh fading process with mean power $\overline{\gamma}$ and PDF given in the previous section, the outage probability is

$$P_{out} = 1 - \exp\left(-(2^R - 1)/\overline{\gamma}\right). \tag{1.10}$$

From Equation (1.10) it is clear that the outage probability decreases quickly with increasing SNR; since the system designer is interested in low outage probabilities, it is therefore customary to plot the outage or its compliment in logarithmic scale.

1.5.2.2 Rate Outage Gains

To illustrate the rate outage probability gains, or simply outage gains, we will utilize the same communication model as depicted in Figure 1.1 (bottom-right) and follow the analysis first exposed in [75–77]. The protocol is fairly simple: Both users send their data to the destination and hence to each other. If a user manages correctly to decode the other user's information, it relays it to the destination; if not, it continues sending its own information.

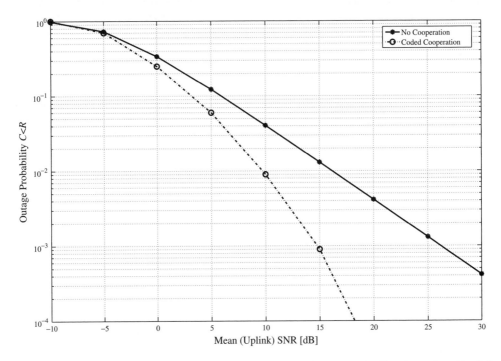

Figure 1.12 Outage versus average uplink SNR, where inter-user channel is significantly weaker; $R = 0.5$ bits/s/Hz [77]

As per Figure 1.12, the outage gains due to cooperation are significant. In the region of interest, that is typically between 1 and 10% outage probability, up to 6 dB transmission power can be saved on average. These gains are attributed to the fact that the probability of both direct and cooperative relaying link being in outage is much lower than just the direct link being in outage. More complex topologies and protocols follow the same trend and cooperation is generally able to provide significant outage gains, whether the random channel fluctuations be due to fading or shadowing.

1.5.3 Diversity-Multiplexing Tradeoff

The diversity-multiplexing trade-off (DMT) has been intuitively presumed by system designers for decades already but was first quantified in [8]. It essentially describes how quickly the probability of outage decreases and the communication rate increases with an increase in SNR. Since the concept of outage is not applicable over ergodic channels, the DMT in the Shannon sense is only applicable to nonergodic channels. However, as will be discussed below, the DMT is also applicable to any real-world system operating over slow- or fast-fading channels, which allows one to trade reliability against rate.

With reference to Equation (3.76), the diversity gain relationship can be reformulated as

$$d = -\lim_{SNR \to \infty} \frac{\log P_{out}(R, SNR)}{\log SNR}, \tag{1.11}$$

where $P_{out}(R, SNR)$ is the outage probability in the Shannon sense at a given average SNR and required rate R, and d is the diversity gain. At an asymptotically high SNR, it is equivalent to the gradient of the outage curves. If $d = 0$ then with an increasing SNR no decrease in outage is achieved, that is the gains one can potentially obtain from increasing the SNR are used somewhere else (most likely to increase the rate). Taking the example of Figure 1.12, the noncooperative case has a diversity order of $d = 1$ whereas the cooperative protocol can achieve the double diversity order of $d = 2$. Clearly, a steeper gradient yields increasing gains with an increasing SNR.

With reference to Equation (3.77), the multiplexing gain relationship can be reformulated as

$$r = \lim_{SNR \to \infty} \frac{R(SNR)}{\log_2 SNR}, \tag{1.12}$$

where r is the rate or multiplexing gain. At an asymptotically high SNR, it is equivalent to the gradient of the capacity or rate curves. If $r = 0$ then with an increasing SNR no increase in rate R is achieved, that is the gains one can potentially obtain from increasing the SNR are used somewhere else (most likely to decrease the outage).

By inserting Equation (1.12) into Equation (1.11) and solving w.r.t. d, we arrive at the general DMT expression

$$d = -\lim_{SNR \to \infty} \frac{\log P_{out}(r \log_2 SNR)}{\log SNR}, \tag{1.13}$$

which implies that increasing the rate multiplexing capabilities inherently requires the reliability of these rates to be decreased, and vice versa.

Similar arguments can be used to derive the DMT for real-world systems operating over slow- or fast-fading channels, where for slow-fading channels P_{out} in the Shannon sense needs to be replaced by the P_{out} for the given real-world scheme and for fast-fading channels by the average error rate P_e [8]. We will use the example of a system deploying quadrature amplitude modulation (QAM) with variable modulation index over fast fading channels to illustrate the DMT in such a context [8]. The average symbol error probability (SEP) of QAM over a Rayleigh fading channel at asymptotically high SNRs is given by $P_e(R, SNR) = \lim_{SNR \to \infty} 2^R / SNR$ which, having inserted (1.12), yields

$$P_e(r \log SNR) = \lim_{SNR \to \infty} \frac{2^{r \log_2 SNR}}{SNR}. \tag{1.14}$$

Finally, inserting Equation (1.14) into Equation (1.13), gives us the DMT for the real-world communication scheme using QAM:

$$d = 1 - r, \tag{1.15}$$

which is depicted in Figure 1.13. For the degraded point at $d_{max} = 1$ and $r_{min} = 0$, that is at no increase in rate, this means that with increasing SNR reliability can exhibit a maximum gradient of one; or, in other words, doubling the SNR (in dB) doubles the reliability (in log scale). For the other degraded point at $r_{max} = 1$ and $d_{min} = 0$, that is at no increase in reliability, this means that with increasing SNR the data multiplexing capability can exhibit a maximum gradient of one; or, in other words, doubling the SNR (in dB) doubles the rate capabilities. Any other point along the line and in-between these degraded points can, for example, be achieved by simple time multiplexing; for instance, if one wishes to have a diversity gain of $d = 0.5$ and a multiplexing gain of $r = 0.5$, then – for increasing SNR – one should communicate half of the time using a constant modulation, which gives the reliability benefits, and the other half of the time using an adaptive modulation that gives the rate benefits.

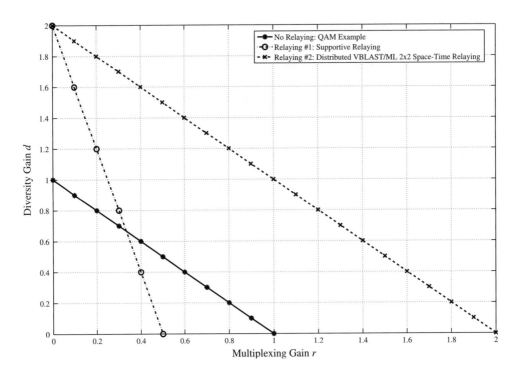

Figure 1.13 The diversity gain shows how fast the outage or error probability decreases with SNR and the multiplexing gain shows how fast the rate can be increased with SNR. A simple repetition-based relaying protocol can increase the diversity order but not the multiplexing gain. A more sophisticated 2×2 space-time relaying protocol using V-BLAST and ML can increase both diversity and multiplexing gains

Figure 1.13 also shows the impact of two example cooperative relaying protocols. The first protocol is a simple repetition-based relaying protocol that clearly yields double diversity gain but half multiplexing gain since the same data is repeated twice. The second protocol aims at increasing the multiplexing capabilities by means of relaying, which is achieved, for example, by using distributed 2×2 space–time relaying where one cooperative relay acts as a transmitting antenna array close to the source and another cooperative relay acts as a receive antenna close to the sink. If the channel between source and transmit relay as well as destination and receive relay is ideal, a V-BLAST scheme with maximum likelihood (ML) receiver achieves the depicted DMT [8].

Apart from the power and shadowing macro-diversity gains of such cooperative relaying schemes, the ability to trade reliability against rate is of great help to system designers. For instance, slow narrowband fading channels offer little diversity, which can be improved by a relaying system at the expense of a rate improvement. Wideband channels, on the other hand, already offer a healthy amount of diversity that can be used either to increase the reliability (CDMA-based transceivers) or rate (OFDM-based transceivers). Depending on this choice, cooperative relaying can then further boost the rate (CDMA-based transceivers) or reliability (OFDM-based transceivers), or both. These trade-offs have motivated research into practical cooperative relaying protocols as outlined in this book.

1.6   Definitions and Terminology

We now move on to some definitions and terminologies for cooperative communication systems. Indeed, different terms are currently being used for the same communication principle, and vice versa.

It is hence vital to harmonize an understanding of such systems, where some first excellent steps have been taken by [78]. We proceed to discuss terminologies in the context of three areas: (i) the relaying node itself; (ii) its one and two-hop neighborhood, which needs to be controlled by suitable multiple access methods; and (iii) the thus formed cooperative relaying network at system level.

1.6.1   Relaying Node

We will now characterize the different types of behavior of relaying nodes, available relaying families and how information is being handled in the relaying node. For the sake of clarity, this is also summarized in Figure 1.15.

1.6.1.1   Node Behaviors

Nodes that relay or cooperate play a central role in cooperative networks. Their behavior has a profound impact onto the system performance and, as exemplified in Figure 1.14, can generally be classified as follows:

- **Egoistic Behavior (no help).** This is the most typical node behavior found in today's wireless communication systems. Here, each node communicates with the base station separately if it has data to transmit or it remains idle if it has no own traffic to transmit, even though it could help another node that has. Other nodes are generally seen as competitors, that is increasing the resources for one node requires it to be decreased for other nodes. Nodes of networks following such design usually strongly experience the impact of the wireless channel in that nodes with good channel conditions enjoy large rates, whereas nodes with a bad channel suffer from low rates.
- **Supportive Behavior (unidirectional help).** Such a behavior is well known in the *ad hoc* community, where data is delivered from a source towards a destination via relay(s) that do not have data of their own to transmit. Supportive relaying-based designs, however, slowly find their way into cellular systems; for instance, LTE and WiMAX are likely to integrate supportive relaying

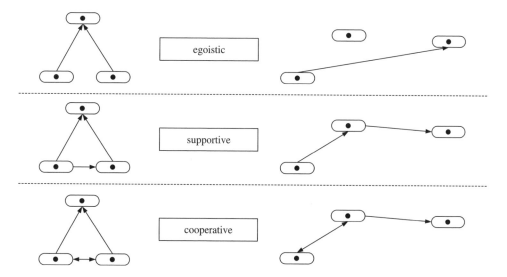

Figure 1.14   Typical forms of node behavior. In the symmetric case (left), either node could be supportive or cooperative; in the asymmetric case (right), the better option is typically chosen

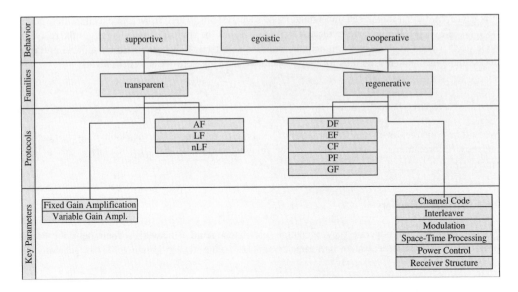

Figure 1.15   Taxonomies and definitions related to the relaying node

nodes. Clearly, the relay is not gaining in performance at that particular instant since it is only helping the source node; however, in the long run, the relay (if not part of a planned or unplanned infrastructure) may also be in the situation of requiring help and hence enjoying the supportive help of other relaying nodes.

- **Cooperative Behavior (mutual help).** The truly cooperative behavior is exhibited by nodes mutually helping each other, that is all involved nodes have data to transmit and mutually try to get it successfully delivered. System designs following a cooperative node paradigm are still a long way off. Nonetheless, cooperation generally smoothes the impact of the wireless channel in that even nodes in bad channel conditions experience an acceptable channel quality and hence sufficient communication rates.

Future networks are likely to be composed of nodes of all three behaviors. One can generally state that the higher the level of cooperation, the better the performance but also the more complicated the set-up and maintenance of the scheme.

1.6.1.2 Transparent Relaying Protocols

As is well documented throughout available literature on this subject, a whole gamut of different relaying methods exists today. They can roughly be classified into two groups, that is transparent and regenerative relaying protocols.

Using the family of transparent relaying, the relay does not modify the information represented by a chosen waveform. Very simple operations are usually performed, such as simple amplification, phase rotation, etc. Since no digital operations are performed on the signal, the analog signal is received in one frequency band, amplified and momentarily retransmitted on another frequency band. Example protocols belonging to the transparent relaying family are:

- **Amplify and Forward (AF).** Constituting one of the simplest and most popular relaying methods, the signal received by the relay is amplified, frequency translated and retransmitted. Different amplification factors can be used as will be discussed in later chapters.

- **Linear-Process and Forward (LF).** This relaying method includes some other simple linear operations, which are performed on the signal in the analog domain after amplification. An example of such a linear operation is phase shifting, which facilitates the implementation of distributed beamforming.
- **Nonlinear-Process and Forward (nLF).** Not yet fully explored, this method performs some nonlinear operations on the received analog method prior to retransmission. An example application is the nonlinear amplification of the received signal which minimizes the end-to-end error rate [79, 80].

An important design issue related to transparent relaying protocols is the choice of amplification factor in the relay, where these options are generally available:

- **Constant Output Power.** In this case, the transparent relaying node transmits at a constant output power that has been set during node manufacturing. Whilst this is the simplest way of realizing a transparent relay, it is also suboptimum compared to the fixed and variable gain amplification.
- **Fixed Gain Amplification.** In this case, the node fixes the amplification factor over a given time window to a value that depends on long-term statistics in the channel or network. For instance, the amplification factor is typically an inverse function of the average channel gain between source and relay. In poor channel conditions, this may lead to very large amplification factors and hence high output powers; in this case, the retransmitted signal is delimited to the maximum transmission power, leading to clipping effects.
- **Variable Gain Amplification.** This case differs from the case of fixed gain amplification in that the amplification gain is adapted to instantaneous changes in the channel and network. For instance, the amplification factor is typically an inverse function of the instantaneous channel gain between source and relay. Again, clipping effects may occur due to large amplification factors. As will be shown in Chapter 3 of this book, a variable gain amplification is needed for a transparent architecture to realize its full diversity gain.

1.6.1.3 Regenerative Relaying Protocols

In the case of regenerative relaying protocols, information (bits) or waveform (samples) is modified. This requires digital baseband operations and thus more powerful hardware. Hence, regenerative relays usually outperform transparent ones. The most prominent examples of regenerative relaying are:

- **Estimate and Forward (EF).** The analog signal is amplified and down-converted to baseband, after which some detection algorithms aim at recovering the original representation of the signal. This estimate is then retransmitted. For instance, the EF relay estimates the modulated symbol and retransmits its estimate using the same or a different modulation order.
- **Compress and Forward (CF).** This protocol is similar to the above EF protocol in that it relays a compressed version of the detected information stream to the destination. This involves some form of source coding on the sampled signal samples and was shown to be capacity/performance optimum for the compressing node being close to the destination.
- **Decode and Forward (DF).** Being the prominent counter protocol to the transparent AF protocol, DF detects the signal, decodes it and re-encodes it prior to retransmission. A vast amount of different DF protocols exists today and we will dedicate large parts of Chapter 4 to this protocol. Over a wide gamut of application scenarios, DF is known to be performance optimum w.r.t. typical metrics such as error rate.
- **Purge and Forward (PF).** Modern communication systems are usually designed to be interference rather than noise limited. This design principle also applies to cooperative systems where PF allows for interference between the different relaying streams and deals with it by eliminating as much of it as possible at each relay node.

Table 1.2 Example protocols belonging to the two relaying families

Transparent relaying protocols	Regenerative relaying protocols
Amplify and forward (AF)	Decode and forward (DF)
Linear-process and forward (LF)	Estimate and forward (EF)
Nonlinear-process and forward (nLF)	Compress and forward (CF)
	Purge and forward (PF)
	Gather and forward (GF)

- **Gather and Forward (GF).** Also sometimes referred to as aggregate and forward protocol, this protocol is an extension to CF in that a relay node not only performs source coding over the sampled information but also on the information itself, which is aggregated over a few communication slots.

The two relaying protocol families are summarized in Table 1.2. The list is far from exhaustive but includes the most prominent examples, the majority of which will be dealt with later in this book.

Regenerative relaying protocols also obey a large set of parameters that can be optimized, the more important of which are listed below:

- **Choice of Channel Code.** Channel coding has a profound impact on the performance of a communication system. It basically trades encoding/decoding complexity and power with coding gains in form of transmit power reduction. Using no channel code in a relay is the least complex solution but also the worst performing one. If a channel code is used, then the network designer can choose from a variety of block codes, trellis codes and concatenations thereof. Block codes can correct a fixed amount of error but not more (for example, three errors in a block of 255 bits) and trellis codes can correct with a given probability a density of errors (for example, an error every 10 bits).
- **Choice of Interleaver.** The interleaver rearranges the output bit stream w.r.t. the input bit stream. The role of the interleaver is to break long sequences of errors so that they can be corrected more easily. Since it breaks long error bursts into several short ones, the application of interleavers is useful in block fading environments where the channel remains constant over a few symbols. Interleavers trade performance gains against memory requirements, that is larger interleavers randomize errors better at the expense of needing more memory to store the bits. One therefore has the choice of using no interleaver (preferably in static channel conditions or embedded systems with low memory), block interleaver (writing the input into the columns of the interleaver matrix and reading it from the rows) or random interleaver (randomly rearranging the order of the bits).
- **Choice of Waveform and Modulation.** An important design factor in relays is the choice of modulation, which comprises three important issues. First, one has to decide between single carrier or multicarrier modulation schemes. The former includes the traditional approach to modulation and the latter for example, orthogonal frequency division modulation (OFDM). Multicarrier modulation schemes are usually susceptible to nonlinearities in the transmitter and hence require fairly expensive amplifiers. Second, there is the choice between coherent and differential modulation. The former encodes information in amplitude and phase, whereas the latter in phase only. This has an impact on the receiver design and performance. Coherent modulation outperforms differential modulation at the expense of requiring a reliable channel estimate at the receiver. Differential modulation, conversely, does not require any channel estimates and is hence suitable in environments where the channel varies rapidly. Third, the modulation order is yet another design parameter where higher modulation orders exhibit a higher spectral efficiency at the expense of being more susceptible to noise and interference.
- **Choice of Space–Time Processing.** If the relay has multiple antennas and/or relays to realize distributed space–time relaying, a vast gamut of available space–time processing techniques is available. As such, one could use distributed space–time block codes (STBCs), space–time trellis

codes (STTCs), layered space–time codes (LSTCs) where data is multiplexed over transmit antennas, and beamforming techniques. Beamforming is the only one really requiring a feedback channel; the other schemes, however, profit from feedback since waterfilling and precoding can be applied to boost performance. STBCs yield diversity gains but no coding gains; STTCs yield both diversity and multiplexing gains; and LSTC yield multiplexing gains. More general classes of codes exist that trade diversity and multiplexing gains. These codes require a rich scattered channel in order to use the full potential of the MIMO channel. Conversely, beamforming sends the information in a given spatial direction and hence benefits from sparsely scattered channels.

- **Power Control.** In addition, the regenerative relay may use adaptive amplification factors, mainly to facilitate power control and hence manage interference. Clearly, larger transmit powers facilitate larger communication ranges but also create interference at more nodes.
- **Choice of Receiver.** A large variety of receivers and detectors is available to date, which are optimized for a given choice of transmitter and system configuration. Available techniques include simple threshold detectors, zero forcing (ZF) and minimum mean square error (MMSE) receivers, sophisticated interference cancelation receivers, etc.

The regenerative relay clearly exhibits more design degrees than the transparent relay. This is an opportunity because the regenerative generally outperforms the transparent relay; however, it is also a challenge in that optimizing regenerative relaying systems is a cumbersome task.

1.6.2 Multiple Access Resolution

The wireless medium is inherently broadcast in nature, that is if a node transmits, everybody in its vicinity overhears the signal. To facilitate communication between nodes and minimize interference, suitable multiple access methods thus need to be put in place. The prime roles of these multiple access methods are to take care of the following:

- **Duplexing Method.** Duplexing traditionally refers to the way a transmitter and a receiver communicate, that is whether they are able to communicate simultaneously or not. Since a cooperative systems utilizes relays, the traditional definition of duplexing breaks down and novel concepts need to be applied.
- **Coordination of Access.** Such coordination is vital since it guarantees that potential transmitters can access the channel and reach the receiver without causing interference or being interfered with.
- **Resource Allocation.** Another important part of the multiple access is to decide how many and which resources are allocated to each transmitter.

We will now discuss these issues in the context of cooperative relaying systems in greater detail.

1.6.2.1 Duplexing Methods

Having originated in wireline communications, a duplex communication system is a system composed of two connected devices that can communicate with one another in both directions. These said devices can either communicate simultaneously or not. The former is also referred to as full-duplex and the latter as half-duplex. A prominent example of a full-duplex system is the old plain telephone system (POTS) and of a half-duplex system an apartment's intercom system.

The extension of this concept to the wireless domain has caused a lot of confusion and with it differing duplexing definitions emerged. It is important to note, however, that the concept of simultaneity, as required by a full-duplex link, is very subjective and dependent on the traffic. For instance, POTS is a truly full-duplex system from an engineering point of view but signal delays incurred by long communication distances do not give the feeling of full-duplexity from a user perception point of view. Therefore, as long as these traffic-dependent delays stay within given bounds, two connected users feel simultaneity, which allows full-duplex systems to be emulated by different techniques in practice:

- **Frequency Division Duplex (FDD).** Reception and transmission of the information stream between the connected devices happens at different frequency bands. In theory, this allows both devices to communicate simultaneously and hence the system operates in true full-duplex. In practical systems, however, the receiver often needs to decode and process the information received on one band before responding on the other band – since this delay is often negligible, practical FDD systems efficiently emulate a full-duplex system. Furthermore, in practical contexts, these frequency bands are often referred to as frequency carriers separated by a frequency offset. Generally, a system is designed in FDD if both ends have approximately the same traffic to transmit or the signal travel duration is large, leading to long guard times if TDD is used.
- **Time Division Duplex (TDD).** Here, reception and transmission of the information stream between the connected devices happens at the same frequency band. This allows one device to communicate at a time only and hence the system theoretically to operate in half-duplex. In practical digital systems, however, the delays between both half-duplex streams is often so small that a full-duplex system is efficiently emulated. For instance, the cordless telephone system DECT relies on TDD and yet successfully manages to convey the perception of a full-duplex system. Generally, if traffic flows are strongly imbalanced, TDD approaches are preferred since one device can be allocated more time to communicate.
- **Division Free Duplex (DFD).** To communicate at the same time in the same frequency band in both directions is widely applied in the wired but so far not in the wireless domain. The main problem is related to the hardware implementation, which needs to prevent the high-power transmitted signal overshadowing the low-power received signal [81]. A possible approach is to place the transmitting and receiving antennas spatially sufficiently far apart, which is clearly not a very attractive solution. Other methods, however, are possible and will be discussed in Chapter 5. DFD, if implemented, facilitates both true and emulated full-duplex systems and also exhibits a superior spectral efficiency when compared with FDD and TDD.

The introduction of cooperative relays complicates the duplexing procedure, mainly because the term originated from the notion of two and only two devices being connected. The concept of multiplexing can also not be applied since relaying does not necessarily involve the simultaneous transmission of multiple signals over one medium. It has, however, become customary to refer to a full-duplex relay if it can receive and retransmit in the same frequency band at the same time, and half-duplex if it can only do this in the same band or the same time [82]. This is conceptually different from the duplex concept applied to a link since it bears no notion of simultaneity. In analogy to the practical implementations of the duplex principles at link level, we use the following duplex principles at the relay:

- **Time Division Relaying (TDR).** The incoming and outgoing information streams at a cooperative relay are separated in time, that is they are allocated to different slots and hence realize a half-duplex relay. The incoming and outgoing streams are typically but not necessarily in the same frequency band. This leaves enough time for processing the information stream on a slot or packet basis. TDR is hence popular with regenerative relaying protocols. Note that the incoming and relayed time slots do not necessarily need to be of equal length; for instance, if the channel from the source to the relay is worse than from the relay to the sink then a DF protocol could use a higher modulation index in the second hop and thus relay a packet of shorter duration. Furthermore, there will generally be a small time gap between received and relayed frame that accounts for the packet processing delay and the time needed for radio circuitry to switch from reception to transmission.
- **Frequency Division Relaying (FDR).** Here, reception and transmission of the relay stream happen at different frequency bands, that is realizing a half-duplex relay. The incoming and outgoing streams

are typically but not necessarily scheduled at the same time. If they are scheduled at the same time, then this reduces to the well-known out-of-band or out-of-channel relaying [83]. FDR is applicable to both regenerative and transparent relaying families. Again, incoming and relayed frequency bands and packet duration do not necessarily have to be of the same size.

- **Division Free Relaying (DFR).** Here, reception and transmission of the relay stream happen at the same time using the same band, which realizes a full-duplex relay and is referred to as in-band relaying. This technique is currently being used in the context of simple repeaters in cellular systems to provide sufficient coverage, where it is referred to as on-frequency repeater (OFR) [84–86]; it is also being used in the context of more sophisticated repeaters capable of canceling interference, where it is referred to as interference canceling system (ICS) repeaters [87]; finally, broadcasting systems have also used such an approach, which is referred to as on-channel repeater (OCR) [88]. Since available technologies spatially separate receive and transmit antennas, this technique ought to be referred to as spatial division relaying. However, since such an approach is clearly inappropriate for envisaged relays of small size, we hope that viable single antenna solutions will be found soon and will hence stick to the notion of DFR. In any case, today's DFR systems usually use a simple AF protocol, which only puts tough requirements on the hardware as will be discussed in more detail in Chapter 5. Future systems may also choose to decode and re-encode the slightly delayed retransmitted signal stream, which requires special coding and decoding techniques to be used as will be discussed in Chapter 4.

DFR is thus spectrally efficient but very demanding from a technology point of view. On the other hand, TDR and FDR require the introduction of an extra time slot and extra frequency band, respectively, which clearly diminishes the spectral efficiency of a cooperative relaying system. The spectral efficiency decreases even further with an increasing number of relays. It will however be demonstrated later in this book that the loss in spectral efficiency is more than compensated for by the gains due to cooperation.

With the aid of FDD/TDD/DFD and FDR/TDR/DFR we are able to support truly full and half duplex as well as emulated full duplex systems, which is exemplified in Figure 1.17 and to be discussed in subsequent sections.

1.6.2.2 Multiple Access Protocols

The existence of more than one user or at least one relay in the system requires suitable multiple access protocols, because otherwise interference between their transmission occurs. As per Figure 1.16, this leads to the following set of canonical channel configurations:

- **Point-to-Point (unicast).** Not actually requiring multiple access methods, the point-to-point channel is formed by means of a direct channel between source and destination.

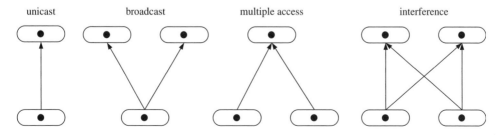

Figure 1.16 Canonical access channels: unicast, broadcast, multiple access, interference channel

- **Point-to-Multipoint (broadcast).** In this case, a single source communicates with multiple destinations in a broadcast fashion. An example is a base station communicating with several mobile stations, either transmitting the same information to all of them or different information to each of them.
- **Multipoint-to-Point (multiple access).** Here, multiple sources communicate with a single destination. A typical example is a set of users communicating with a base station.
- **Multipoint-to-Multipoint (interference channel).** This is the most general case, where multiple sources communicate with multiple destinations. This case is also sometimes referred to as interference channel because the two information streams stemming from the two sources interfere at both destinations.

The actual access for the broadcast, multiple access and interference channels can be facilitated by traditional multiple access protocols. These are generally divided into reservation-based and contention-based protocols, where the former is useful in the context of centralized systems where resources can be reserved *a priori* and where traffic is regular; the latter is applied in decentralized systems or where traffic is bursty and resources need to be contended for prior to communication. Examples of reservation-based protocols are:

- **Time Division Multiple Access (TDMA).** Different user and relay information streams are scheduled in a time slotted fashion. For instance, the direct link from source to destination is assigned the first time slot and the relay link the second time slot.
- **Frequency Division Multiple Access (FDMA).** Different user and relay information streams are scheduled at different frequency bands. For instance, the direct link from source to destination is assigned one band whereas the relay link uses another one. In practical systems, these frequency bands are often referred as frequency channels.
- **Code Division Multiple Access (CDMA).** Different user and relay information streams are assigned different (ideally orthogonal) spreading sequences. For instance, the direct link from source to destination is assigned one spreading code whereas the relay link uses another one. Whilst this method allows communication of all information streams in the same bands at the same time, power control is often needed, which is not easy to implement in a distributed cooperative network.
- **Orthogonal Frequency Division Multiple Access (OFDMA).** Different user and relay information streams are assigned different subcarriers. For instance, the relayed and own information of a cooperative node are assigned to different subcarriers. Also, the cooperative relay may decide to reshuffle the incoming subcarriers into a differently arranged set of relayed subcarriers so as to maximize the end-to-end performance.
- **Multicarrier CDMA (MC-CDMA).** This access method combines both OFDM and CDMA and could also be used in conjunction with OFDMA. The key idea is to apply a user or relay unique spreading sequence in frequency across several subcarriers or, in time, over several data symbols at a given subcarrier. Such an access method is very flexible but difficult to optimize.

In the case of contention-based protocols, it is generally assumed that users use the same frequency (and code) and contend for resources in time. This means that contention-based protocols typically use TDD and TDR. Prominent examples belonging to this access family are:

- **Aloha.** Probably the first contention-based protocol, the prime idea is that the sender transmits data when there is data in the buffer and if the message does not reach the receiver (due to unfavorable wireless channel or collision), just resend it later. Whilst very simple, the efficiency of Aloha is fairly low.
- **Carrier Sense Multiple Access (CSMA).** An improvement to Aloha is CSMA, where the transmitter first senses the channel before transmitting. If the channel is detected as occupied, it will refrain from transmission and – depending on the backoff algorithm – try again at a later moment. If the channel is detected to be free, transmission is initiated. Efficiencies of the various variants of CSMA are well above that of Aloha, but generally inferior to reservation-based protocols.

1.6.2.3 Resource Allocation Strategies

Once the duplexing methods and multiple access protocols have been determined for the system, each source and relay node can be allocated different resources in terms of time, frequency, number of codes, power, etc. The most typical approaches are discussed below:

- **Time Slots.** If one device has more traffic to transmit then it can be allocated more time slots. An example is when a device needs to transmit its own traffic in addition to relayed traffic.
- **Frequency Bands.** Devices can also be allocated more frequency bands or a larger frequency band if needed. In today's practical systems, this means that an optimum choice of frequency carriers and channels needs to be found.
- **Transmission Power.** Another way of increasing a device's ability to reach the destination or a relay is to allow it to transmit at higher transmission powers. Whilst detrimental to the overall network, since this generates excess interference, this method allows prioritizing of devices and information flows.
- **Number of Codes.** Yet another method of allocating more resources to a device in need is to assign it multiple codes in the case of CDMA. This is currently being used in the multicode transmission of HSDPA in 3GPP.
- **Choice of Subcarrier.** If some form of subcarrier scheduling has originally been performed at the source, that is allocating higher modulation indexes to subcarriers of better quality, then it is beneficial that the relay also performs such an allocation according to the subcarrier quality of the relay destination channel.
- **Probability of Persistency.** In the context of contention-based protocols, an important design parameter influencing the effective resources allocated to a device is the probability of persistency used in the backoff algorithm. This probability depends on many factors, such as the amount of contending traffic and nodes.

The discussed factors are summarized and visualized in Figure 1.17.

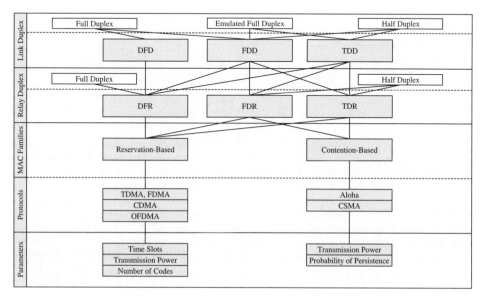

Figure 1.17 Taxonomies and definitions related to multiple access resolution

1.6.2.4 Typical Access Configurations

The degrees of freedom for applying various duplexing methods, access protocols and resource allocation algorithms to cooperative relaying networks is clearly infinite. To illustrate the above exposure, we shall demonstrate a few typical access configurations in the context of two system designs, that is a system where relays are designed to be part of the system *a priori* and a system where the relays are inserted into the system later:

- **Systems With Prior Relay Insertion.** Performing the system design with relays in mind has an impact on access mechanisms. First, to limit interference, the system ought to be power controlled in such a manner that direct links between source and destination do not occur. Whilst suboptimal at link level, this yields significant gains at system level as shown, for example, for contention-based MACs in [89] and from a capacity point of view in [90]. Second, the relay can be seen as a full system entity to which duplexing rules apply, that is the notion of full or half-duplex does not apply to source and destination but rather to any pair of communicating nodes, be it source and relay or relay and destination. This yields a typical scenario as shown in Figure 1.18 with two relaying hops, two active up and downlinks and no direct communication between BS and MSs.

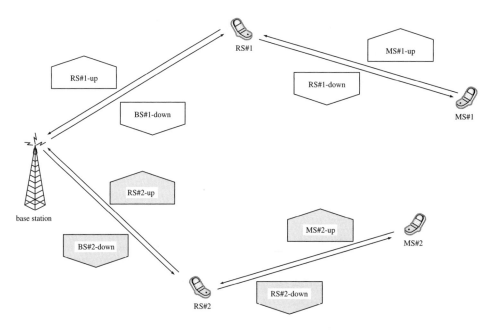

Figure 1.18 Example access scenario with one BS, two RSs and two MSs. The packet transmitted from the BS to the RS#1 has been marked 'BS#1-down'; the one from the RS#1 towards the MS#1 has been marked 'RS#1-down', etc.

The cases of TDD/TDR and particular realizations of the TDMA, FDMA and CDMA multiple access protocols are shown in Figure 1.19. Clearly, many more combinations are possible; for instance, FDMA could be implemented with more frequency channels; the uplink of user #1 and downlink of user #2 could be scheduled simultaneously in the FDMA and CDMA cases to minimize interference, etc. The cases of TDD/FDR, FDD/TDR and FDD/FDR with TDMA, FDMA and CDMA as multiple access protocols are shown in Figures 1.20–1.22, respectively. Clearly, not all combinations are optimal nor is this list exhaustive (even for this fairly simple case). The combinations with DFD and DFR have been omitted here as they are easily derived from above frame structures.

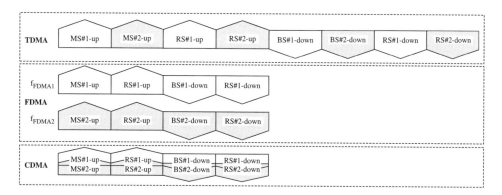

Figure 1.19 An example TDD/TDR configuration with TDMA, FDMA and CDMA as access protocols. TDMA requires only one frequency channel but takes twice the duration compared with other solutions; FDMA implementation requires two frequency channels; CDMA requires half power per user to be used to facilitate a fair comparison

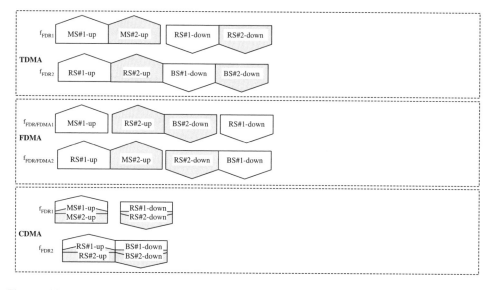

Figure 1.20 An example TDD/FDR configuration with TDMA, FDMA and CDMA as access protocols. For delay-constrained applications, CDMA is clearly the best choice

- **Systems With Subsequent Relay Insertion.** Inserting relays into an already existing and designed system with a fixed carrier and channel structure restricts the use and applicability of certain access mechanisms. This is mainly because a direct link between source and destination is needed and hence a relay has to obey the duplexing and access rules of this existing link. Assuming the same scenario as that in Figure 1.18 but with additional direct links between MSs and BS and the notion of duplexing applied to the MSs and BS, only the access configurations shown in Figures 1.19 and 1.21 are possible. Again, this is far from exhaustive since more combinations are feasible.

It shall be noted here that it is generally easier to build and maintain a mesh network with its respective schedule between cooperating nodes when only the temporal domain is used, that is TDD combined with TDR and TDMA.

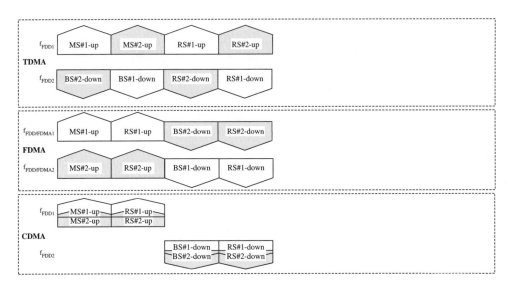

Figure 1.21 An example FDD/TDR configuration with TDMA, FDMA and CDMA as access protocols

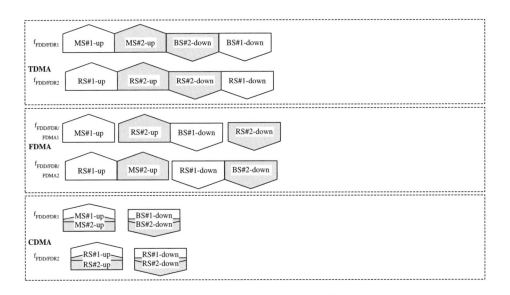

Figure 1.22 An example FDD/FDR configuration with TDMA, FDMA and CDMA as access protocols. Again, CDMA is shortening the delivery delay

1.6.3 Cooperative Networking Aspects

With the node behavior determined and the multiple access rules established, it is important to discuss how information is handled at system and network level. To this end, we will now discuss canonical information flows resulting from cooperative relaying systems. We will also discuss important design parameters that have a serious impact on the performance of such networks. Again, for the sake of completeness, we have summarized the taxonomy related to networking aspects in Figure 1.24. Note that this discussion only pertains to elements most relevant to the PHY layer and does not include a discussion on routing protocols, etc.

1.6.3.1 Canonical Information Flows

From above-discussed node behaviors, relaying protocols, duplexing and access methods, we can construct different information flows and architectures, some of which have already been discussed at the beginning of this chapter. Subsequent discussions relate to the case of a noncooperative single source, single destination systems only; the extension to the case of cooperation, multiple sources and destination follows the same recipe.

- **Direct Link.** Information from a source can of course reach the destination by means of a single direct link.
- **Serial Relaying.** As per Figure 1.23 (top left), serial relaying connects the source and the destination by means of a chain of relays that are assumed to use orthogonal channels to relay the information. Note that in this and subsequent cases a direct link may or may not be available.
- **Parallel Relaying.** As per Figure 1.23 (top right), parallel relaying connects the source and the destination by means of a parallel set of relays that are assumed to use orthogonal channels to relay the information at the same time.
- **Space–Time Relaying.** As per Figure 1.23 (bottom left), space–time relaying connects the source and the destination by means of a parallel set of relays that are assumed to use space–time encoded channels to relay the information.
- **Composites Thereof.** As per Figure 1.23 (bottom right), an information flow can be realized by building hybrids from the above-discussed relaying methods.

Such networks as just described can be infrastructure based (that is an infrastructure is available prior to deployment) or infrastructureless (that is it emerges after deployment or remains unavailable). If an infrastructure is available, it can be managed in a centralized or noncentralized fashion. Note that

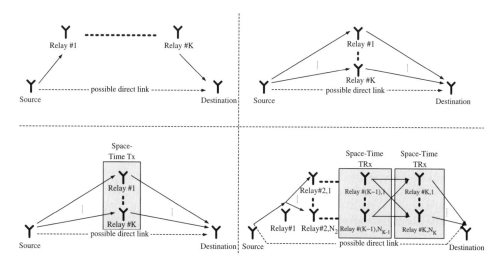

Figure 1.23 Canonical network information flows by means of serial relaying (top left), parallel relaying (top right), space–time relaying (bottom left) and hybrids thereof (bottom right)

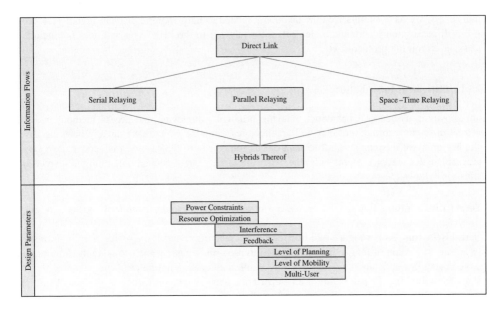

Figure 1.24   Taxonomies and definitions related to networking issues

you may have a decentralized infrastructure-based system (for example, a cooperative system using decentralized radio resource management algorithms) as well as a centralized infrastructureless system (for example, a WSN applying clustering).

1.6.3.2   Important Design Parameters

The following nonexhaustive list of issues has an impact on the optimization process as well as the performance of cooperative relaying systems:

- **Power Constraints.** It is important to know whether a power constraint applies per node or per information flow. For instance, if the power constraint applies to the flow, then introducing more relays requires the given power to be shared between these relays. Most practical applications obey a power constrained per node, that is adding a new relay node injects extra power into the system.
- **Resource Optimization.** It is also of importance to determine *a priori* whether resources for nodes in the cooperative relaying system are optimized depending on of various network parameters. For instance, one could decide to allocate more power to the relay node compared with the source if the distance between relay and destination is larger than between source and relay. Most practical applications, however, still refrain from performing such an optimization as the involved overhead is still too large.
- **Interference between Flows.** The system can be designed such that there is no interference between the information flows generated by sources and relays. From a Shannon point of view, however, designing a system to be interference limited is a better choice.
- **Feedback.** The performance will be strongly impacted by the availability of feedback channels, that is the ability of a receiver to report back to its transmitter. Useful information to be fed back is full channel state information (for example, instantaneous channel realizations) or partial channel state information (for example, averaged channel realizations), interference temperature, etc. It is generally well accepted that a feedback channel can drastically improve the performance of the system because it facilitates spatial waterfilling (that is optimal allocation of power per transmit

antenna) and precoding (that is optimal transmit code design). Practical communication systems generally have a feedback channel, even if it is only a simple acknowledgement (ACK) packet.

- **Level of Planning.** A planned roll-out of cooperative relays can significantly boost performance. However, this comes at the price of an increased planning cost and time, as well as the necessity to acquire the planned relay sites. An unplanned roll-out is clearly performance sub-optimal but saves a lot of time and money since relays can simply be placed on, say, lampposts.

- **Level of Relay Mobility.** Another important parameter is the level of mobility of the relays. For instance, if relays are fixed, they typically belong to an extended infrastructure. Alternatively, if users are used as relays, their mobility will strongly impact the performance of the cooperative system.

- **Level of Synchronization.** Synchronization generally stretches from the hardware to the network level. At hardware level, nodes may or may not be synchronized down to the carrier frequency; synchronization here is vital for distributed space–time codes to work properly. At access and network level, nodes can be synchronized by means of slots, which is known to boost system capacity but also requires suitable synchronization algorithms.

- **Multiuser Scenarios.** A recently emerging concept is the application of relaying nodes to multiuser scenarios. This has a profound impact on the design process as multiple transmitting users are not considered as competitors but as facilitators.

As they are not the focus of this book, many other design factors at system and network level have been omitted here. Nonetheless, above list serves to highlight that link and system level issues are closely interrelated in cooperative systems.

1.6.4 System Analysis and Synthesis

We also briefly summarize important quantities which are derived when analyzing cooperative systems. These quantities are often used to optimize the system performance, that is to derive optimum system configurations, protocols, etc., which we also summarize below.

1.6.4.1 Performance Analysis

The following issues and quantities play a central role when analyzing cooperative relaying systems:

- **Average Error Rates.** The 'instantaneous' bit (BER), symbol (SER) or packet error rate (PER) is calculated assuming a given channel realization and average noise power. When the channel varies sufficiently fast or, otherwise put, all channel realizations are traversed over the duration of the transmission, the average BER, SER and PER can be calculated and used for system characterization. Whilst SERs are usually easily calculated in closed form using the moment generating function (MGF) approach [91], exact BER and PER expressions are generally not easy to derive. Approximations and asymptotic expressions are hence typically involved, in particular for the PER, which is the most important quantity in real-world systems.

- **Outage Probabilities.** If the channel does not vary sufficiently fast or, in other words, not all statistical states are traversed over the duration of the transmission, then invoking the average does not make sense since it would yield a different value for each transmission. The concept of outage is thus typically involved, which quantifies the probability that a certain performance cannot be met. Typically, outage probabilities can be calculated for channel realizations, and information rates as well as bit, symbol and packet errors. Since fading is generally assumed to vary sufficiently quickly whereas shadowing not, the averages are derived for the former and its outages for the latter. For example, in a cellular system obeying fast fading and shadowing, one would first calculate the average PER over the fading statistics and then calculate the packet error outage (PEO) for the calculated average PER over the shadow statistics [92, 93]. Cell dimensioning is then typically done with the latter, which is why the quantification of outages are of prime importance in real-world systems.

- **Throughput.** Once the average rates and/or outages are calculated, the average end-to-end and system throughput can be calculated. This quantity gives an insight to the system designer of how much capacity can be offered to the user when using a particular communication scheme, be it simple direct communication or cooperative relaying.

- **Delay/Latency.** Another quantity of importance is delay or latency, which quantifies the time needed to deliver a packet to the destination from the moment it has been generated. Delay typically comprises (i) the time the packet spends in the buffer queue of the device; (ii) the time it is required to get it successfully delivered via the wireless channel to the destination; and (iii) the time needed to process it. The queuing time typically depends on the amount of traffic entering and the amount of traffic leaving the queue. The successful delivery time includes the contention or transmission time, as well as retransmissions in the case of delivery failure. If relays are present in the system, then this delay needs to be multiplied by the number of relays to obtain an average end-to-end latency.

- **Real-World Impairments.** Of augmenting importance is the quantification of the impact of real-world impairments on the performance of cooperative relaying systems. Prime concerns are the impact of channel estimation errors, synchronization errors, phase errors, erroneous feedback information, etc.

The discussed factors are summarized and visualized in Figure 1.25.

1.6.4.2 System Synthesis

The above-described performance analysis clearly gives useful insights into cooperative relaying systems; however, changing a few underlying assumptions often requires analysis and simulations to be redone. It is hence desirable to take the inverse approach, where design guidelines are synthesized

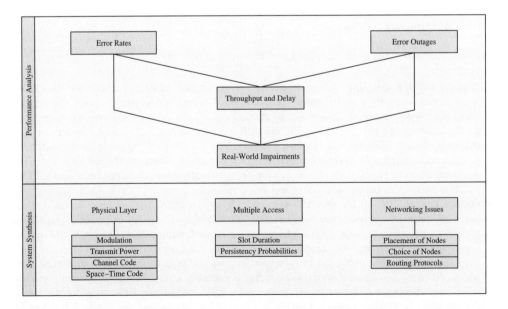

Figure 1.25 Taxonomies and definitions related to analysis and synthesis of cooperative relaying networks

for a broad range of underlying assumptions. For instance, instead of offering the spectral efficiency versus the SNR for different modulations, a synthetic approach would yield the modulation that is optimum for a given SNR and number of users. System synthesis is a very challenging area that is mainly complicated by the often very complex expressions resulting from the performance analysis. The following nonexhaustive list of issues are typically considered when performing a system synthesis:

- **Optimum Set of PHY Parameters.** Important PHY parameters, which a system synthesis should yield in explicit form, are the optimum choice of modulation order, the optimum choice of transmission power, the optimum choice of channel and space–time code and its generator matrix, etc.
- **Optimum Multiple Access Rules.** Of importance at access level are the optimum distribution of resources (for example, duration or number of time slots), the persistency probability for a given traffic load and user density, the packet duration to minimize collisions in the case of contention-based protocols, etc.
- **Optimum Networking Protocols.** Closely related to the physical and access level are the optimum choice of relaying nodes as well as their placement. Of further importance is the optimal choice of routing protocols and associated parameters.

As is clearly evident from this entire section, the number of degrees of freedom in constructing a cooperative system is virtually unbounded, somehow explaining the large amount of literature available on this subject.

1.7   Background and Milestones

This introductory chapter is concluded by summarizing key contributions to cooperative communications. Note that this does not constitute a complete state-of-the-art review but rather the exposure of early milestones that helped in shaping today's research landscape in cooperative systems. Important state-of-the-art contributions will be mentioned in the respective technical chapters.

1.7.1   First Key Milestones

Early developments concerning supportive, cooperative and space–time relaying were related but have largely emerged independently:

- **Supportive Relaying.** This simplest form of cooperation is not exactly new. Information theoretical developments stem back to the seminal contribution by van der Meulen in 1968 [94, 95] and by Cover and Gamal in 1979 [96]. Whilst some information theoretical contributions emerged here and there, the communication and protocol developments received a revival in the early 1990s with the 3GPP Concept Group Epsilon, driven by Vodafone [10]. Back then, communication engineers argued that no user would agree to relay data for another user since the short-term gains of the relaying user are nil. Harrold and Nix [97, 98] were the first to prove by means of simulations that – whilst short-term gains were indeed sometimes unfavorable – every user gained in the long run by cooperating; they also showed that by using simple relaying, coverage holes could largely be closed in a cellular deployment. Similar insights were provided a little later in [9, 99].
- **Cooperative Relaying.** Cooperative relaying, that is the case where at least two users help each other to boost each other's performance, has been pioneered by Sendonaris *et al.* in 1998 [73]. Later, around 2000, Laneman and coworkers rigorously formalized various types of supportive and cooperative relaying protocols and proved that significant performance and outage gains can be achieved [100–102]. It is largely due to Laneman's seminal work that the area of cooperative

communication systems commenced to flourish. A little later, Hunter and coworkers [75, 103, 104] and Stefanov and Erkip [105] were the first to propose a viable cooperative scheme based on channel coding and special code designs.

- **Space–Time Relaying.** Space–time relaying had been pioneered by Dohler and coworkers in 1999 and made public to the audience of the Mobile Virtual Centre of Excellence (M-VCE), a UK national research initiative, from 2000 onwards [106, 107]. Their work was based on then just emerged works on space–time codes by Foschini [119], Alamouti [120] and Tarokh [121, 122]. Subsequently and sometimes in parallel, pioneering key contributions related to distributed space–time codes and their design emerged from Laneman and Wornell [108] and Stefanov and Erkip [109, 110].

These early key contributions are summarized in a time chart in Figure 1.26. We will now describe these and other early contributions in some greater detail.

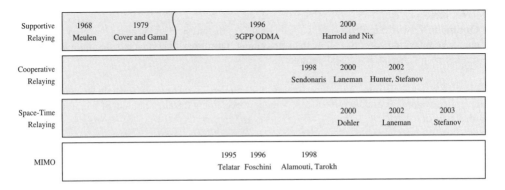

Figure 1.26   Early key contributions in the field and their timeline of publication

1.7.2   Supportive Relaying

The method of relaying has been introduced in 1968 by van der Meulen [94, 95] and has also been studied by Sato [111]. A first rigorous information theoretical analysis of the relay channel, however, has been published by Cover and Gamal [96], a more detailed description of which can be found in his excellent book [112].

In these contributions, a source MS communicates with a destination MS directly *and* via a RS. In [96] the maximum achievable communication rate was derived for various communication scenarios, including the cases with and without feedback to either source MS or RS, or both. The capacity of such a relaying configuration was shown to exceed the capacity of a simple direct link.

It should be noted that the analysis was performed for Gaussian communication channels only; therefore, neither the wireless fading channel has been considered, nor were the power gains due to shorter relaying communication distances explicitly incorporated into the analysis.

Only in the middle of the 1990s, research in and around the Concept Group Epsilon revived the idea of utilizing relaying to boost the capacity of wireless networks, thereby leading to the concept of ODMA [10]. The power gains due to the shorter relaying links were the main incentive in investigating such systems to reach MSs out of BS coverage. The emphasis of the study was its applicability to cellular systems, as well as a suitable protocol design; no theoretical investigations into capacity bounds, etc., have been performed.

1.7.3 Cooperative Relaying

Seminal work in the area of cooperative relaying has been the contribution by Sendonaris, Erkip and Aazhang, dating back to 1998 [73]. In their study, a very simple but effective user cooperation protocol was suggested in order to boost the uplink capacity and reduce the uplink outage probability for a given rate. The designed protocol stimulates a MS to broadcast its data frame to the BS and to a spatially adjacent MS, which then retransmits the frame to the BS. Such a protocol certainly yields a higher degree of diversity because the channels from both MSs to the BS can be considered as uncorrelated.

The simple cooperative protocol has been extended by the same authors to more sophisticated schemes, which can be found in the excellent contributions [72] and [74]. Note that in its original formulation [73], no distributed space–time coding has been considered.

The contributions by Laneman and Wornell in 2000 [100] are a conceptual and mathematical extension to [73], where energy-efficient multiple access protocols are suggested based on decode-and-forward and amplify-and-forward relaying technologies. It has been shown that significant diversity and outage gains are achieved by deploying the relaying protocols when compared to the direct link. Note again, that no distributed space–time coding was considered at that time.

Gupta and Kumar were the first to analyze statistically the information throughput theoretically offered for large scale relaying networks [113]. They showed that under somewhat ideal situations of no interference, hop-by-hop transmission and predefined terminal locations, capacity per MS decreases by $1/\sqrt{M}$ with an increasing number of MSs M in a fixed geographic area. They also showed that if the terminal and traffic distributions are random, then the capacity per terminal decreases even in the order of $1/\sqrt{M \log M}$. The analysis in [113] has been extended by the same authors to more general communication topologies, where the interested reader is referred to the landmark paper [114].

Furthermore, Grossglauser and Tse have shown that mobility counteracts the decrease in throughput for an increasing number of users in a fixed area [115]. The protocols suggested therein benefit from the decreased power for a hop-per-hop transmission for decreasing transmission distances. It also benefits from the location variability due to mobility, that is a packet is picked up from the source MS by any passing by RS and only re-transmitted (and hence delivered) when passing by the target MS.

1.7.4 Space–Time Relaying

To understand the contributions and timings of space–time relaying schemes, seminal works on traditional MIMO systems will be revisited first.

Contributions on MIMO systems have flourished ever since the publication of the landmark papers by Telatar [116, 117] and Foschini and Gans [118] on capacity, and by Foschini [119], Alamouti [120] and Tarokh [121, 122] on the construction of suitable space–time transceivers. In the BLAST system, introduced by Foschini in 1996 [119], a transmitter spatially multiplexes signal streams onto different transmit antennas, which are then iteratively extracted at the receiving side using the fact that the fades from any transmit to any receive antenna are uncorrelated and of different strength. The BLAST concept has ever since been extended to more sophisticated systems, a good summary of which can be found in [123]. Alamouti introduced a very appealing transmit diversity scheme by orthogonally encoding two complex signal streams from two transmit antennas, thereby achieving a rate one space–time block code [120]. His work was then mathematically enhanced by the landmark paper of Tarokh [121], who essentially exposed various important properties of space–time block codes. He also showed how to construct suitable space–time trellis codes, which were shown to yield diversity *and* coding gain [122].

A system utilizing the advantages of both MIMO and relaying was suggested by Dohler in 1999 and hence became one of the main research topics within the M-VCE from 2000 onwards. It has been suggested that spatially adjacent mobile terminals be used to form distributed antenna arrays in hot-spot areas. The thus formed distributed antenna array had been termed an artificial antenna array but – since the abbreviation AAA was already in use – was renamed as virtual antenna array or VAA. Numerous studies [106] have led to a set of patents [107], which are backed by about 20 industrial members, such as Vodafone, Nokia, Philips, Nortel Networks, Samsung, etc. The studies encompassed the following (in chrondogical order):

- downlink distributed receive diversity in cellular systems;
- downlink distributed MIMO in cellular systems;
- uplink distributed MIMO in cellular systems;
- introduction of distributed relaying to cellular systems;
- extension of the above to WLAN and hot-spot systems;
- generalization to arbitrary distributed relaying topologies.

The case of distributed space–time coding has also been analyzed by Laneman in [108] and in his PhD dissertation [102]. In his thesis, information theoretical results for distributed single-input single-output (SISO) channels with possible feedback were utilized to design simple communication protocols, taking into account systems with and without temporal diversity, as well as various forms of cooperation. He has demonstrated that cooperation yields full spatial diversity, which allows drastic transmit power savings at the same level of outage probability for a given communication rate.

Specific distributed space–time coding schemes have also been suggested, for example, by Stefanov and Erkip [109, 110]. In this publication, two spatially adjacent MSs cooperate to achieve a lower frame error rate to one or more destination(s), where a quasi-static fading channel has been assumed. Distributed space–time trellis codes have been designed that maximize the performance for the direct link from either of the MSs to the destination *and* the relaying link. Furthermore, distributed beamforming has also been introduced [124].

Finally, it should be mentioned that recently, Ozgur *et al.* [90, 125] have shown for the first time that linear transport capacity scaling is possible in a large network by means of cooperative hierarchies and space–time relaying.

1.8   Concluding Remarks

This chapter has served to introduce the topic of cooperative relaying. This area has recently received a lot of attention, which has often been coupled with a lot of hype and confusion. We therefore paid a lot of attention to introduce key notions properly and apply viable taxonomies. This, hopefully, has shed some light on an otherwise complicated topic and paved a solid basis for subsequent chapters and hardware, wireless channel and PHY algorithms.

Despite its recent revival, relaying is – strictly speaking – old hat in the information theory community where first developments emerged about four decades ago [94]. However, since the information theory community is historically not concerned with the wireless channel and usually assumes AWGN channels only, exclusively regenerative relaying families over Gaussian channels have been looked at. Other relaying protocols became popular only recently within the information theory community, mainly due to the cooperative relaying protocols' success in the communications community.

From an implementation and engineering point of view, cooperative relaying protocols are also not exactly new. In fact, the now reborn AF and DF relaying protocols have been known in parts by the satellite community for nearly five decades [126–130] and by the radio community for almost a century already [131, 132]. Here, the main incentive to use relays was to overcome coverage and range problems rather than capacity problems.

What has been truly new and responsible for the renaissance of relaying protocols in recent years is the capacity and performance benefits of cooperative relaying protocols under realistic channel and system conditions. These gains are in fact the justification and inspiration for the subsequent chapters of this book.

2

Wireless Relay Channel

2.1 Introductory Note

2.1.1 Chapter Contents

Channel models are of utmost importance in the designing process of wireless systems, because they influence power budget dimensioning, transceiver design, performance behavior, etc. Some of their impact will become evident in Chapters 3 and 4. As of 2009, however, there are only a few relaying and cooperative channel measurements and models available as compared to traditional cellular systems. Issues pertaining to channel modeling for cooperative systems will hence be dealt with in this chapter in the following order:

- general channel characterization;
- channel for regenerative relaying (decoupled channel);
- channel for transparent relaying (cascaded channel);
- distributed MIMO channel.

The properties of the regenerative relaying channel can be used to understand and quantify the performance of regenerative relaying protocols, such as the DF and CF protocols to be discussed in Chapter 4. Likewise, the properties of the transparent relaying channel can be used to understand the performance of transparent relaying protocols, such as the AF protocol to be discussed in Chapter 3. The aim of this chapter however is not to go into great channel modeling details but rather to indicate tendencies and issues that are important in modeling the cooperative relaying channel and which impact the performance of cooperative algorithms.

2.1.2 Choice of Notation

Concerning the notation, we will often but not always adopt the notation used in the open literature. We will generally treat each relaying segment separately. When there is more than one relaying stage being treated simultaneously, we will indicate occurring parameters with subscript i. Typically, the regenerative relaying case does not require such a subscript whereas the transparent one does.

If a subscript is used then, for example, the number of transmit antennas at the transmitter of the ith stage for instance becomes $n_{t,i}$, the transmit power into the ith stage is P_i, the channel of the ith stage is h_i, the number of receive antennas at the receiver of the ith stage is $n_{r,i}$, the noise variance at the receiver of the ith relaying stage is σ_i^2, etc.

Cooperative Communications Mischa Dohler and Yonghui Li
© 2010 John Wiley & Sons, Ltd

If more subscripts are required, then they simply follow the one on the relaying stage separated by a coma. For instance, if the ith relaying stage is spanned by a MIMO channel with $n_{t,i}$ transmit and $n_{r,i}$ receive antennas, then the channel spanned by the kth transmit and lth receive antenna element is denoted as $h_{i,kl}$, and so on.

With this choice of notation, it may happen that the parameters at the receiver of the ith relaying stage are the same as the parameters at the transmitter of the $(i + 1)$th stage. For instance, the number of receiver antennas is typically the same as the number of transmit antennas in a relaying node; however, they do not have to be the same, which is why the above notation is not only convenient but also more general.

The most important variables, symbols and functions used throughout this chapter are listed at the beginning of this book.

2.2 General Characteristics and Trends

We will briefly review the most important space–time–frequency characteristics of wireless fading channels. For more modeling details, the interested reader is referred to [133–135].

2.2.1 Propagation Principles

A typical scenario encountered in cooperative relaying systems is depicted in Figure 2.1, which depicts a cellular system but is easily generalized to any form of cooperative wireless system. Here, a base station (BS) communicates with associated mobile stations (MSs) through a line-of-sight (LOS) or NLOS (non-LOS) link. The MSs also communicate among themselves to facilitate a cooperative relaying system. The underlying antenna and propagation principles can be summarized as follows.

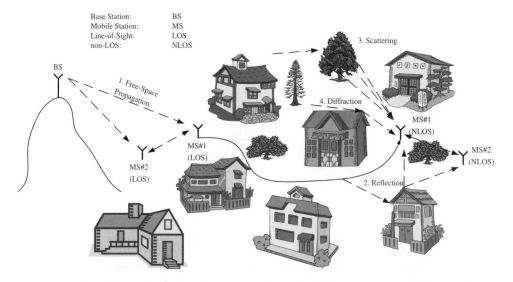

Figure 2.1 Channel scenario for LOS/NLOS traditional and cooperative links

2.2.1.1 Wave Properties

The modulated signal at the transmitter is fed into a finite-length wire antenna of length l from which it decouples in form of an electromagnetic (EM) wave of wavelength λ. The antenna is typically a $\lambda/2$

wavelength antenna or equivalent, due to favorable decoupling conditions. Generally, the efficiency of the overall radiated power is roughly proportional to $(l/\lambda)^2$.

The decoupled wave oscillates in time with angular frequency $\omega = 2\pi f = 2\pi/T$ and in space with spatial frequency $k = 2\pi/\lambda$, where f is the frequency, T the period, $\lambda = c/f$ the wavelength, and c the speed of light. It is composed of an electric component of strength E and a magnetic component H. For the distance between transmitter and receiver d being significantly larger than the wavelength, which is also known as the far-field or Fraunhofer region, the electric and magnetic field components relate via $E = \eta H$ and can respectively be represented in phasor format as $E = E_0 \cdot e^{j(\omega \cdot t - \mathbf{k} \cdot \mathbf{r})}$ and $H = H_0 \cdot e^{j(\omega \cdot t - \mathbf{k} \cdot \mathbf{r})}$. Here, $\mathbf{r} = d \cdot \mathbf{n_r}$ is the spatial translation vector, where $\mathbf{n_r}$ is a unit vector pointing from the transmitter to the observation point, and $\mathbf{k} = k \cdot \mathbf{n_k}$ is the wave vector, where $\mathbf{n_k}$ is a unit vector pointing into the direction of wave propagation. Furthermore, $\eta = \sqrt{\mu/\epsilon} = k/(\omega\epsilon)$ is the intrinsic impedance ($120\pi \approx 377\,\Omega$ for free space), μ is the permeability, and ϵ is the permittivity. Since both electric and magnetic field components oscillate in phase, the 'transport' of power $P \propto |E||H| \propto |E|^2$ is guaranteed.

This wave then propagates through free space and through/off objects cluttered throughout the environment, which leads to signal distortions to be discussed below. The clutter causes several copies of the wave to impinge upon and couple into the receive antenna, where they are again converted into an electric signal and processed by the receiver chain.

2.2.1.2 Propagation Mechanisms

Generally, the above process is governed by Maxwell's equations, which are fairly complicated to deal with when applied to a generally very complex propagation scenario. For this reason, the EM wave's behavior is typically categorized depending on the interrelationship between the wavelength λ, the distance between transmitter and receiver d, the clutter's size s and its surface undulation Δh:

- **Free Space Propagation.** Under the main condition that $d \gg \lambda$ and no clutter is encountered, the propagating wave undergoes free space propagation. It is a distance dependent effect that obeys Friis' transmission equation given by $P_r = P_t \cdot G_t \cdot G_r \cdot \lambda^2/(4\pi d)^2$, where P_r is the power of the received wave, P_t the power of the transmitted wave, G_t the transmit antenna gain in the angle of departure (AOD) of the EM wave, and G_r the receive antenna gain in the angle of arrival (AOA). The above equation only holds for perfect matching of transmit and receive antennas, no multipath propagation and aligned polarization of both antennas. In decibels (dB), this translates to

$$P_r = P_t + G_t + G_r + 148\,\text{dB} - 20\log f - 20\log d, \tag{2.1}$$

where now all variables are assumed to be given in dB. Clearly, the received power diminishes with increasing distance and/or frequency at $-20\,\text{dB/dec}$.

- **Reflection and Refraction.** Under the condition that $\lambda \gg \Delta h$ and $s \gg \lambda$, a part of the wave reflects off the clutter's surface and the remaining part refracts into the clutter. An example of the former is the reflection off building walls, where the angle of the reflected wave equates to the angle of the impinging wave. An example of the latter is the refraction into the building thereby guaranteeing indoor coverage, where the angle of the refracted wave is quantified by means of Snell's law [136]. The reflected electric field component E_{refl} is related to the impinging field component E_{imp} by means of a generally complex reflection coefficient R, that is $E_{\text{refl}} = R \cdot E_{\text{imp}}$. Equally, the refracted component E_{refr} involves a refraction coefficient T, that is $E_{\text{refr}} = T \cdot E_{\text{imp}}$. Both coefficients R and T are governed by Fresnel's law [136] and are known to depend on the clutter's material and the impinging angle. Since the latter is often not known, an average coefficient is usually assumed [106]. For instance, upon reflection, concrete invokes a loss of about $-10.5\,\text{dB}$ on a dry day and $-4.4\,\text{dB}$ on a rainy day.

- **Scattering.** Under the condition that $\lambda \approx \Delta h$ and $s \gg \lambda$, the impinging wave is known to be scattered off the clutter's surface. Depending on the clutter's surface characteristics, different scattering

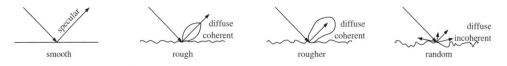

Figure 2.2 Specular reflection off smooth surface, diffuse coherent scattering off rough surface and diffuse incoherent scattering off random surface

phenomena can be observed as depicted in Figure 2.2. Generally, the rougher the surface the less the impact of the specular reflected component and the stronger the impact of the scattered component where the latter is typically assumed to be Gaussian distributed over the angle of reflection. The effect of scattering is due to the constructive and destructive addition of the reflected waves off surface elements of heights differing by Δh. For the carrier frequencies typically used in telecommunications, the well-established Rayleigh criterion can be used to determine whether a surface is smooth (that is only exhibiting a specular component) or rough (that is also exhibiting a scattered component); for $\Delta h/\lambda < 1/8$ the surface is smooth and otherwise rough. For instance, a GSM system operating at $f = 900$ MHz in an urban environment with houses having balconies of undulation $\Delta h \approx$ 1 m, reveals that the surfaces need to be considered as rough since some significant scattering occurs.

- **Diffraction.** Under the condition that the clutter has a few singular edges or curvatures that are of the size of λ or smaller, diffraction occurs. It essentially guarantees that an EM wave can still reach a zone shadowed by an object as shown in Figure 2.3. The underlying principle of the existence of a shadowed component is that, according to Maxwell's equations, the electric field must be steady, that is a hard boundary between the optical and shadow zones cannot exist. A well-established approach is to use the uniform theory of diffraction (UTD), which essentially states that the total electrical field at any point is composed of the optical components due to reflection/refraction and the diffracted

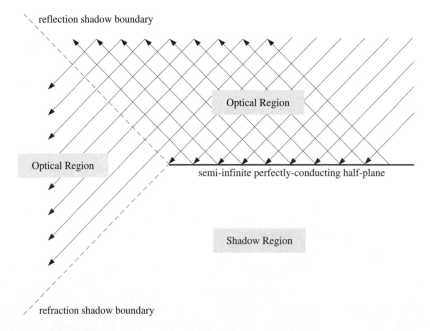

Figure 2.3 Optical regions composed of the impinging, reflected and traversed wave together with the shadow region

components. The derivation of the latter is fairly involved and heavily depends on the underlying assumptions made on the EM properties of the clutter. Generally, diffraction guarantees field strength in zones that are usually not in direct line with the transmitter, such as street canyons, etc.

A wave can undergo several of these effects prior to reaching the receiver, that is a wave propagates in free space, gets reflected, propagates in free space, gets diffracted, propagates in free space, gets scattered, propagates in free space, and couples into the receive antenna.

2.2.1.3 Signal Distortions

A signal, however, is not only impacted by the direct interaction between wave and clutter but also by other factors depending on topology and system parameters. These includes for example, the speed of transmitter and receiver as well as their direction of movement. It further includes the arrangement of the clutter, its relative optical distance, the symbol duration, etc. In short, this leads to further signal distortions in the time and frequency domain as discussed in the following:

- **Temporal Distortion – Doppler Effect.** The movement of transmitter and/or receiver (and/or clutter) causes the wave to be perceived at a different frequency than originally emitted, that is $f_{perceived} = f_{original} \cdot (1 + v/c)$, where v is the relative speed between transmitter and receiver projected onto the line connecting both. This impacts the evolution of the received signal over time. For instance, Figure 2.4 exemplifies the field strength for the case of little movement in the channel (left) and more movement in the channel (right). Waves traversing different paths will generally experience different Doppler shifts.

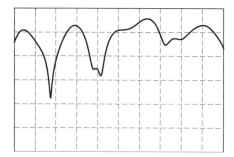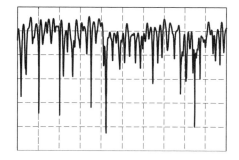

Figure 2.4 Field strength versus time, where little movement (left) causes only little variation and a lot of movement (right) causes more variation over the same time window of observation

- **Spectral Distortion – Multipath Propagation.** A multitude of clutter objects causes multiple waves to reach the receiver at different delays, where each of these waves will have undergone different free space propagation, reflection, etc. These multiple delayed copies are often referred to as multipath components (MPCs). A sum of delayed signals causes selectivity in the frequency domain, as is illustrated in Figure 2.5 by means of impulses.

  Telecommunication systems, however, use a train of pulses rather than single impulses. To this end, assume we send two pulses/symbols of duration T_s; then, objects along the ellipses with transmitter and receiver in the foci, yield the same propagation delays as illustrated in Figure 2.6. This in turn leads to intra-symbol interference, that is the overlap of symbol replicas within the symbol duration (same shading in the figure), as well as inter-symbol interference, that is the overlap of symbol replicas belonging to different symbols (different shadings in the figure). Whilst the former is generally irreversible, the latter can be reversed by means of suitable transceivers.

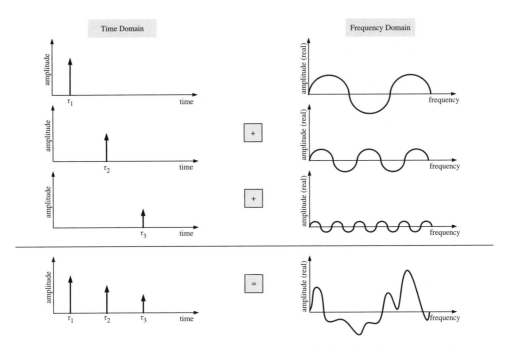

Figure 2.5 The delayed and attenuated copies of the signals in the time domain cause strong frequency selectivity in the frequency domain

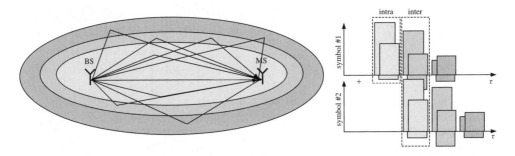

Figure 2.6 Origin of the intra-symbol and inter-symbol interference due to multipath propagation

All of the above-discussed effects can now be summarized in a single equation where the received signal is composed of the sum of the impinging MPCs with their respective distortions, that is $\sum_{\text{MPCs}}$ = (LOS) + $\sum$ once reflected + $\sum$ twice reflected + ...

The resulting signal strength and hence power are random, because trajectory/location, number of MPCs, number of reflections per MPC, reflection coefficient per reflection, speeds of transmitter, receiver and clutter, etc., are random in general. A rigorous approach to analyze and deal with this equation is difficult, if not impossible. Luckily, applying some rearrangements and approximative heuristics, we can decompose the effect the channel has on the received signal into three multiplicative components, that is pathloss, shadowing and fading. The sum in dB of the three components is depicted in Figure 2.7. Subsequent sections describe each of these phenomena in greater detail.

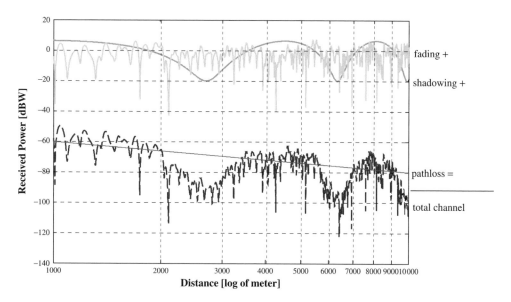

Figure 2.7 The sum in dB or product in linear scale of pathloss, shadowing and fading yield the actually perceived wireless channel

2.2.2 *Propagation Modeling*

2.2.2.1 Pathloss

As already alluded to in Section 1.2, averaging the received power at a particular distance over a sufficiently large area, yields the loss, L, in power or the pathloss versus distance. This has been exemplified in Figure 2.7. If only free space propagation prevailed, pathloss would behave strictly deterministically with a loss of 20 dB/dec. However, the average loss in power due to the random shadowing process is traditionally absorbed in the pathloss and hence yields steeper losses versus distance. To model such a behavior, different pathloss modeling approaches are available:

- **Free-Space Pathloss Model.** The simplest of all models, it assumes a loss of 20 dB/dec. It can be formalized as $L(d) = L(d_0) \cdot (d/d_0)^2$, where $L(d_0)$ is the power measured at some (far-field) reference distance d_0. It is generally not applicable to terrestrial, including cooperative, scenarios but finds some applicability in satellite channels as well as for short LOS channels.
- **Single-Slope Pathloss Model.** Taking the steeper decay due to clutter into account, this model assumes a loss of $10 \cdot n$ dB/dec, where n is the pathloss coefficient, which can range from $n = 1.5$ (waveguides), $n = 2 \ldots 4$ (LOS + clutter), $n = 4 \ldots 6$(NLOS + clutter). It can be formalized as $L(d) = L(d_0) \cdot (d/d_0)^n$. It is clearly an excellent modeling tradeoff between simplicity and accuracy. It has found application in all forms of terrestrial communication systems, including cooperative systems.
- **Dual-Slope Pathloss Model.** It can be proved that the existence of two strong wave components, for example, one direct and one reflected one, yields a dual-slope behavior [133]. This implies that for distances smaller than a given breakpoint d_{bp}, the decay follows a pathloss coefficient n_1 and thereafter a pathloss coefficient n_2. This can be formalized as $L(d) = L(d_0) \cdot (d/d_0)^{n_1}$ with typically $n_1 = 2$ for $d < d_{bp}$ and $L(d) = L(d_{bp}) \cdot (d/d_{bp})^{n_2}$ with typically $n_2 = 2 \ldots 6$ for $d \geq d_{bp}$. It is well applicable to long range communication systems, such as cellular systems. The breakpoint

distance can be calculated approximately as $d_{bp} \approx 4h_t h_r / \lambda$, where h_t and h_r are the transmitter and receiver heights respectively [135].

- **Deterministically Simulated Pathloss Behavior.** These typically include ray-tracing and ray-launching type tools, which determine the field behavior for a deterministically predefined scenario. It is generally a very complex modeling approach, and not necessarily a better model. It finds application for very specific modeling problems, such as propagation behavior close to the head, within mobile phone, etc.
- **Empirically-Fitted Pathloss Model.** This class of model is based on real measurements for a given environment. Once the measurements are taken, statistical fitting techniques are used for the pathloss coefficients so as to guarantee a good match. An example result of such a modeling approach is the Okumura–Hata pathloss model [133]. The models are generally difficult to obtain but once the statistical fitting is performed, they are very simple and fairly accurate models. They are typically used in academic simulators, planning and optimization tools.
- **Real-World Measured Pathloss.** Finally, a very accurate approach is to take real measurements for each point in the plane (or space) according to a predefined grid, let's say every 20 m in urban environments. Whilst difficult and expensive to obtain, not to mention the stringent memory requirements, this is often done by large operators so as to optimize the planning and roll-out of their real-world cellular system.

A great disadvantage of pathloss is that it rapidly diminishes the useful signal power. An advantage however is that it also limits the interference power. In the context of cooperative communication systems, this may lead to a favorable situation as the cooperative relay is often within breakpoint region and the interference often comes from beyond the breakpoint.

2.2.2.2 Shadowing

As already alluded to in Section 1.2, averaging the received power at a particular distance over an area of radius of approximately shadowing coherence distance yields a variation of the received power around the pathloss. This variation is referred to as shadowing and has been exemplified in Figure 2.7. Shadowing obeys a random distribution and under NLOS conditions is traditionally – but not always – modeled as Gaussian in dB, that is lognormal in linear scale.

The reasoning behind the distribution of shadowing is as follows. Each arriving MPC is the result of a random amount of multiple random reflections, the power of which is proportional to $\prod |R|^2$. Taking the logarithm of this relation transforms the product into a sum to which the central limit theorem (CLT) is applicable, leading to a normal distribution in dB. Performing the inverse transformation back to linear scale finally yields the lognormal distribution. A few important notes on shadowing follow:

- **Distribution.** The probability density distribution (PDF) of the received power in dB due to shadowing is given as:

$$p(S) = \frac{1}{\sqrt{2\pi}\sigma_{dB}} e^{-\left(\frac{S}{\sqrt{2}\sigma_{dB}}\right)^2},\tag{2.2}$$

where σ_{dB} is the standard deviation (not variance) of the shadowing process in dB (not linear scale). Generally, this distribution is zero mean, that is $\mu_{dB} = 0\,dB$, since the mean is absorbed into the pathloss. In linear scale, the distribution turns lognormal and is given as:

$$p(S) = \frac{1}{\sqrt{2\pi}\sigma_{dB}} \frac{10}{\ln 10} \frac{1}{S} e^{-\left(\frac{10\log_{10} S}{\sqrt{2}\sigma_{dB}}\right)^2}.\tag{2.3}$$

Note that this distribution is not zero-mean [135].

- **Auto-Correlated Shadowing.** The random shadowing components are generally correlated in space, where the degree of correlation mainly depends on the environment. The autocorrelation function (ACF) in this context quantifies how self-similar the shadowed power is over a MS spatial separation of Δd, which has been exemplified in Figure 2.8 (left). This is very important for power control algorithms in general and for macro-diversity gains in the context of cooperative relaying systems. The autocorrelation shadowing coefficient, usually applied to the power in dB, is obtained as

$$R(\Delta d) \triangleq \mathbb{E}\{S(d) \cdot S(d + \Delta d)\} \propto e^{-\frac{\Delta d}{d_{\text{corr}}}}, \tag{2.4}$$

where $\mathbb{E}\{\cdot\}$ denotes the expectation w.r.t. the random realizations of S and d_{corr} is the correlation distance which heavily depends on the distance d between transmitter and receiver. The negative-exponential dependency in Equation (2.4) has been observed experimentally and an example value in the context of cellular systems is $d_{\text{corr}} = 112\,\text{m}$ at $d = 4.8\,\text{km}$ [133].

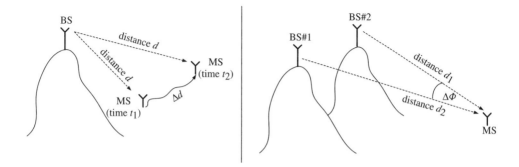

Figure 2.8   Autocorrelated shadowing (left) and site-to-site correlated shadowing (right)

- **Cross-Correlated Shadowing.** Also sometimes referred to as site-to-site correlation, the cross-correlation function (CCF) in this context quantifies how self-similar the shadowed power is over an angular BS separation of $\Delta\phi$, which has been exemplified in Figure 2.8 (right). This is highly important for quantifying interference power from adjacent cells in general and for macro-diversity gains in the context of distributed antenna systems. The autocorrelation shadowing coefficient, usually applied to the power in dB, is obtained as

$$R(\Delta\phi) \triangleq \mathbb{E}\{S(d_1) \cdot S(d_2)\} \propto \begin{cases} \sqrt{\frac{d_1}{d_2}} & \text{for} \quad 0 \leq \Delta\phi \leq \phi_{\text{thr}} \\ \frac{\phi_{\text{thr}}}{\Delta\phi}\sqrt{\frac{d_1}{d_2}} & \text{for} \quad \phi_{\text{thr}} < \Delta\phi \leq \pi, \end{cases} \tag{2.5}$$

where d_1 and d_2 are the respective distances between the MS and the BSs. Furthermore, $\phi_{\text{thr}} = 2$ arcsin$(d_{\text{corr}}/(2d_1))$.

Shadowing is probably one of the most detrimental performance factors of modern communication systems since it cannot be absorbed by suitable channel codes, thus causing nonavailabilities of links referred to as outages. A great disadvantage of shadowing is that it diminishes the useful signal power in a difficult-to-predict fashion. An advantage, however, is again that it also limits the interference power. Shadowing can also be seen as a facilitator in the context of cooperative communication systems since it allows for the capture effect in that several cooperative relaying terminals can communicate simultaneously even though theoretically in pathloss range [137–139]. Cooperative systems are certainly yielding large macro-diversity gains, as long as the cooperative terminals – be they MSs or BSs – are out of correlation distance.

2.2.2.3 Fading

Fading is typically modeled by means of channel models. The modeling process is fairly involved and depends on a multitude of assumptions and parameters. For this reason, we will dedicate the entire Section 2.2.3 to channel modeling.

2.2.3 Channel Modeling

As already alluded to in Section 1.2, not averaging the signal at all allows one to observe fading as a signal fluctuation around pathloss and shadowing. It is caused by the constructive and destructive addition of the signal traveling via multiple propagation paths and has been exemplified in Figure 2.7. It is quantified by means of the channel's impulse response which in simplified notation reads:

$$h = \sum_i a_i \cdot e^{j\phi_i} \cdot \delta(t - \tau_i), \tag{2.6}$$

where h is the generally complex channel coefficient, a_i its amplitude, ϕ_i its phase and τ_i its delay. As per Figure 2.6, the envelope a_i and phase ϕ_i for a fixed MPC i model the unresolvable intrasymbol interference. The resolvable intersymbol interference is modeled by different i's at given delays τ_i.

2.2.3.1 Important Parameters

These parameters will be discussed in some more detail:

- **Envelope a_i.** For a fixed i, the channel amplitude/envelope a_i is a random variable that is due to the random addition of many intrasymbol wave components. Different statistics of a_i have been observed that mainly depend on the operational conditions. For instance, if a strong LOS component is present, then the envelope a_i (respectively, its gain/power $g_i = |a_i|^2$) is known to be Ricean (noncentral chi-square) distributed; if all contributions are equally random, then the envelope is known to be Rayleigh (central chi-square) distributed; if some of these contributions are phase aligned, then the envelope has been observed to obey a Nakagami-m (gamma) distribution, etc. Note that whilst the first MPC for $i = 1$ may obey different distributions, the remaining MPCs for $i \geq 2$ usually obey Rayleigh fading since they are composed of equal components. The exact expressions of these envelope and power distributions are available elsewhere [135] and will be displayed when needed in this book.
- **Phase ϕ_i.** Again, the phase ϕ_i is a random variable for a fixed i, which is also due to the random addition of many intrasymbol wave components. It is typically but not necessarily uniformly distributed. The distribution of the phase impacts system behavior and is also of use when calculating an approximation of the channel's auto-correlation function as done, for example, in [140].
- **Delay Profile τ_i.** Strictly speaking, τ_i is also a random variable; however, for ease of modeling it is often assumed to be given by $\tau_i = i \cdot \Delta\tau$, where $\Delta\tau$ is usually equal to the symbol duration T_s or a multiple thereof. This yields an instantaneous delay profile as had already been depicted in Figure 2.5. Considering the power of (2.6) and taking its average yields the often utilized power delay profile (PDP) $g = \mathbb{E}\{h \cdot h^*\} = \sum_i \mathbb{E}\{|a_i|^2\} \cdot \delta(t - \tau_i) = \sum_i \mathbb{E}\{g_i\} \cdot \delta(t - \tau_i) = \sum_i \overline{g}_i \cdot \delta(t - \tau_i)$, which is depicted in Figure 2.9. Important parameters describing the PDP are the total power gain, given as $\overline{g}_{\text{total}} = \sum_i \overline{g}_i$, the excess delay, defined as $\tau_{\text{excess}} = \tau_{\text{last resolvable tap}} - \tau_1$, the mean delay $\tau_{\text{mean}} = 1/\overline{g}_{\text{total}} \cdot \sum_i \overline{g}_i \tau_i$, and most importantly the RMS delay spread, defined as

$$\tau_{\text{RMS}} = \sqrt{\frac{\sum_i \overline{g}_i \tau_i^2}{\overline{g}_{\text{total}}} - \tau_{\text{mean}}^2}. \tag{2.7}$$

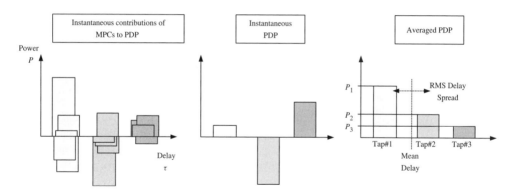

Figure 2.9 Typical power delay profile (PDP) of a wireless channel

The statistical realizations of all of these parameters, that is the envelope a_i, the phase ϕ_i and the delay profile τ_i, strongly depend on time and location. That is, for a fixed location, these values change as time evolves due to mobility in the environment. Equally, for a fixed time moment, these values also change as location is changed due to the environment's spatial arrangement.

2.2.3.2 Selectivity versus Non-Selectivity

The interplay of these parameters causes different fading behaviors of the channel h in the spectral, temporal and spatial domains:

- **Spectral Domain – Flat versus Selective Fading.** As already visualized in Figure 2.5, the channel h fades in the spectral domain due to the resolvable MPCs forming the PDP. An important notation in this context is the spectral autocorrelation function, which is defined as

$$R(\Delta f) = \mathbb{E}\left\{ H(f)H^*(f + \Delta f) \right\}, \tag{2.8}$$

where $H(f)$ is the Fourier transform of h w.r.t. τ. It determines how self-similar a signal is after an observation frequency shift of Δf. The channel is assumed to remain unchanged or insignificantly changed over the coherence frequency band B_c, that is over the shift Δf over which $R(\Delta f)$ remains (virtually) unchanged. The coherence band clearly depends on the environment through its impact onto the PDP. Whilst generally not easy to relate, some heuristics suggest that the coherence band is inversely proportional to the RMS delay spread occurring in the system, that is $B_c \approx 1/\tau_{\mathrm{RMS}}$. Related to the symbol's bandwidth $B_s \approx 1/T_s$, the channel's coherence bandwidth B_c is an important notion in wireless communications as it determines whether the channel is flat or frequency selective. As such, the system is said to undergo flat fading if $B_s < B_c$, and frequency-selective fading otherwise. With the above heuristics on the coherence bandwidth, the problem can be interpreted in the time domain where it is more intuitive to understand. That is, the channel is flat fading if $T_s > \tau_{\mathrm{RMS}}$, which implies that all MPCs arrive more or less at the same time and hence only one MPC is resolved, reducing the channel to $h = a \cdot e^{j\phi}$. On the other hand, if $T_s < \tau_{\mathrm{RMS}}$, more than one MPC is resolved which leads to ISI.

The distinction between these has a profound impact on the choice of transceiver design. Since modern communication systems have increasingly shorter symbol durations, these systems appear to be frequency-selective systems. This causes severe ISI, the combating of which is central to modern transceivers where three prominent designs are available today. First, one could use an equalizer, which effectively performs the inverse filter function to the one given in Equation (2.6); however, given the increasing number of MPCs to be equalized, this becomes more and more infeasible. Second, one could use spread spectrum techniques, which facilitates Rake receivers to be deployed that are able to gather the individual powers g_i of the MPCs. However, from an implementation point of view as is discussed in Section 5.5, there is a limit on the number of Rake fingers. Third, one uses low-complexity OFDM-like techniques that essentially transform the frequency-selective time domain into a flat frequency domain system; however, as discussed in Section 5.6, several drawbacks related to implementation prevail also here.

- **Temporal Domain – Slow versus Fast Fading.** The channel also varies in the temporal domain due to the dependency of the amplitudes a_i and phases ϕ_i on time. Important notions in this context are the temporal autocorrelation, level crossing rate and average fade duration, all of which relate to second order joint moments of the random channel h. In particular, the temporal autocorrelation function is defined as

$$R(\Delta t) = \mathbb{E}\{h(t)h^*(t + \Delta t)\} \tag{2.9}$$

and determines how self similar a signal is after an observation time shift of Δt. The channel is assumed to remain unchanged or insignificantly changed over the coherence time T_c, that is over the shift Δt over which $R(\Delta t)$ remains (virtually) unchanged. The coherence time clearly depends on the environment and its mobility is reflected in the Doppler shift. Whilst generally not easy to relate, some heuristics suggest that the coherence time is inversely proportional to the maximum Doppler shift occurring in the system, that is $T_c \approx 1/\hat{f}$. Related to the symbol duration T_s, the channel's coherence time T_c is an important notion in wireless communications as it determines whether the channel is slow or fast fading. As such, the system is said to undergo slow fading if $T_s < T_c$ and fast fading otherwise. The distinction between them has a profound impact on the choice of modulation, pilot density, etc.

Furthermore, the level crossing rate (LCR) is defined as the expected number of times per second a random even ξ crosses a threshold level ξ_{thr} in the positive direction, and is given as

$$N(\xi_{\text{thr}}) = \int_0^\infty \dot{\xi} \, p_{\xi,\dot{\xi}}(\xi_{\text{thr}}, \dot{\xi}) d\dot{\xi}, \tag{2.10}$$

where $p_{\xi,\dot{\xi}}(\xi_{\text{thr}}, \dot{\xi})$ is the joint PDF of ξ and its time derivative $\dot{\xi}$. The notation of LCR is applicable to the amplitude as well as gain of a single MPC, where in the case of the former $\xi = a$ and of the latter $\xi = a^2 = g$. If applied to the entire channel given in Equation (2.6), above expression involves the joint PDFs of different a_i and ϕ_i, which are generally difficult to obtain in analytical form. For that reason, analytical expressions are often only developed for the narrowband case.

Finally, the average fade duration (AFD) is the average time a random even ξ spends below level ξ_{thr}, and is given as

$$T(\xi_{\text{thr}}) = \frac{P_{\text{out}}(\xi_{\text{thr}})}{N(\xi_{\text{thr}})}, \tag{2.11}$$

where $P_{\text{out}}(\xi_{\text{thr}})$ is the event's outage probability given as $P_{\text{out}}(\xi_{\text{thr}}) = \int_0^{\xi_{\text{thr}}} p_\xi(\xi)d\xi$. The AFD can again be calculated for the channel's envelope and power.

- **Spatial Domain – Selective versus Non-Selective Fading.** The channel also varies in the spatial domain due to the dependency of all random variables in Equation (2.6) on the physical location. This is important for multiple-input multiple-output (MIMO) systems, which deploy several transmit and receive antennas. A viable notion in this context is the spatial autocorrelation function, which is defined as

$$R(\Delta d) = \mathbb{E}\{h(d)h^*(d + \Delta d)\} \tag{2.12}$$

and determines how self-similar a signal is after an observation distance shift of Δd. The channel is assumed to remain unchanged or insignificantly changed over the coherence distance d_c, that is over the shift Δd over which $R(\Delta d)$ remains unchanged. The coherence distance clearly depends on the environment where some heuristics suggest that it is inversely proportional to the angular spread of the incoming/outgoing signal. Related to an antenna spacing d_a, the system is said to undergo spatially nonselective fading if $d_a < d_c$ and spatially selective fading otherwise. The distinction between either has a profound impact on the performance of MIMO systems, such as depicted in Figure 2.10, where a high correlation between antenna elements incurs significant performance losses.

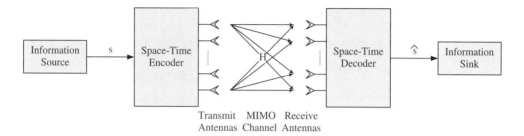

Figure 2.10   Multiple-input–multiple-output transceiver and channel

The fading behavior in frequency, time and space – most notably characterized by the respective correlation functions – probably influences transceiver design and its performance evaluation most.

2.2.3.3   Example Fading Cases

Not including a discussion on the spatial domain, we shall now briefly discuss the advantages and disadvantages of systems obeying the four possible combinations of above-discussed temporal and spectral fading cases:

- **Slow and Frequency-Flat Fading.** This case is exemplified in Figure 2.11 (first case). It is characterized by no intersymbol overlap and no amplitude change from symbol to symbol. The disadvantage of such a scenario is that no power gains are feasible and fades possibly last for a long time. Advantages are that ISI does not occur and that coherent communication is facilitated.
- **Fast and Frequency-Flat Fading.** This case is exemplified in Figure 2.11 (second case). It is characterized by no intersymbol overlap but amplitudes change from symbol to symbol. The disadvantage of such a scenario is that no power gains are possible and noncoherent transceivers need to be used. Advantages are that ISI does not occur and that the channel offers a lot of temporal diversity due to short fades which can be picked up by a suitable channel code.
- **Slow and Frequency-Selective Fading.** This case is exemplified in Figure 2.11 (third case). It is characterized by intersymbol overlap and no amplitude changes from symbol to symbol. The disadvantage of such a scenario is that ISI occurs, requiring equalizers, CDMA or OFDM to be used, and that long fades occur. Advantages are that power gains are possible and that coherent communication is facilitated. The majority of today's communication systems fall into this category.
- **Fast and Frequency-Selective Fading.** This case is exemplified in Figure 2.11 (fourth case). It is characterized by intersymbol overlap and amplitude changes from symbol to symbol. The disadvantage of such a scenario is that again ISI occurs and also that noncoherent transceivers need to be used. Advantages are that power gains are possible and that the rich time diversity can again be picked up by a suitable channel code.

This brief introduction to wireless fading channels allows us now to proceed with an overview of the peculiarities and general trends of the wireless relay channels, most notably the regenerative and transparent relay channel.

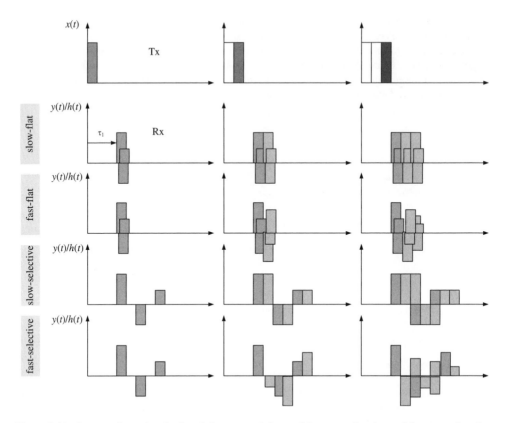

Figure 2.11   In respective order, the four fading cases of slow and frequency-flat, fast and frequency-flat, slow and frequency-selective, fast and frequency selective

2.2.4   Quick Introduction to Regenerative Relay Channels

Being an example out of an infinite amount of possible communication topologies, a typical scenario occurring in the context of regenerative relays is depicted in Figure 2.12. Here, an elevated BS communicates with a relay that is embedded into clutter at low altitude. This relay then retransmits in a regenerative fashion the signal to another terminal well embedded in the clutter environment. The two types of channel possibly occurring are the traditional link between elevated BS and embedded terminal and the cooperative relay link between embedded terminals. The former will exhibit similar characteristics as already known from cellular systems whilst the latter will behave differently as clutter is present both at the transmitter as well as receiver.

2.2.4.1   System Assumptions

A key element in this context is that the relay regenerates the signal that inherently decouples the fading channels before and after the relay. Unlike the transparent relay case to be discussed in Section 2.2.5, the channel of each segment can therefore be modeled separately which yields:

$$y_i = \sqrt{G_i} h_i x_i + n_i, \tag{2.13}$$

where x_i, y_i and n_i are respectively the transmitted signal, the received signal and the additive white Gaussian noise (AWGN) with power σ_i^2 experienced in the ith relaying segment. Furthermore,

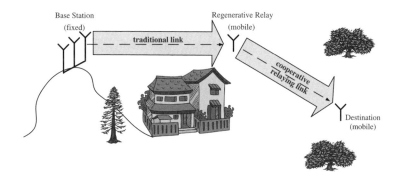

Figure 2.12 An example of a regenerative relaying channel

$G_i = L_i \cdot S_i$ is the large-scale channel gain due to pathloss and shadowing and h_i is the generally complex channel coefficient due to fading. The wireless regenerative relaying channel is hence characterized in each segment separately by L_i, S_i and h_i; the impact of each of these factors will be discussed below.

2.2.4.2 Key Channel Parameters

With this scenario in mind, each segment of the regenerative relaying channel is known to behave as follows w.r.t. the key quantities discussed in Section 2.2 where insights from [135, 141–163] have either directly been taken into account or adapted to the context of regenerative relaying:

- **Pathloss.** The pathloss coefficient of the traditional link varies between $n < 2$ in for example, wave guiding streets, $n = 2$ (LOS) and $n = 2, \ldots, 4$ (NLOS) and the segment of a cooperative link exhibits coefficients from $n = 2$ (LOS) to $n = 4, \ldots, 6$ (NLOS). Clearly, the increased amount of clutter at both ends yields a larger average pathloss.

    Already available pathloss models can be used to model the traditional and cooperative links. For instance, the traditional link can be modeled using the cellular Okumura–Hata, Walfish–Ikegami, and dual-slope models as well as the indoor COST231 and COST259-multiwall models. The cooperative link can, for instance, be modeled using the indoor COST231 and COST259-multiwall models as well as the IEEE 802.15.3a CH1-CH4 and IEEE 802.15.4a models.
- **Shadowing.** Both traditional and cooperative links suffer from shadowing, which in both cases is typically lognormally distributed, that is Gaussian in dB. Whilst the mean in both cases is usually absorbed into the pathloss, the standard deviation will differ for both link types. Typically, the traditional link exhibits a standard deviation σ_{dB} of $2, \ldots, 6\,dB$ (LOS) and $6, \ldots, 18\,dB$ (NLOS), whereas the cooperative link is in the range of $0, \ldots, 2\,dB$ (LOS) and $2, \ldots, 10\,dB$ (NLOS). The autocorrelation shadow distances d_{corr} for the traditional link is more than $100\,m$ (LOS) and around tens of meters (NLOS), whereas for the cooperative link around 40–$80\,m$ (LOS) and 20–$40\,m$ (NLOS). Clearly, the impact of shadowing is much less at the cooperative level because the distances are shorter and the amount of clutter traversed less. It should be noted however that the shadowing standard deviation was found to vary over distance, which has been quantified in [157]. Note further that many of above-mentioned pathloss models already include shadowing as an additive constant with predefined standard deviation.
- **Fading.** Whilst the class of fading distributions does not change for the cooperative link, the first and second order ordinary and joint moments may change. The reason why the class of distributions does not change is because the relay segments are decoupled and each segment obeys the same propagation principles. The reason why the moments change is because the amount of clutter changes its distribution, etc.

Fading is typically modeled by means of channel models, where already available channel models can be used to model the traditional and cooperative links. For instance, the traditional link can be modeled using the COST207, 3GPP A&B and Stanford University Interim Channels SUI1-6 models as well as the indoors ETSI-BRAN and various IEEE models. The cooperative link can, for instance, be modeled using the indoors ETSI-BRAN and various IEEE models as well as the IEEE 802.15.3a CH1-CH4 and IEEE 802.15.4a models, among many others.

Theoretical as well as empirical pathloss, shadowing and fading models which have explicitly been developed for the regenerative mobile-to-mobile case will be discussed in subsequent sections.

2.2.4.3 Impact on Fading Behavior

Since the fading behavior is impacted most by the regenerative relaying architecture, we will discuss related issues separately:

- **Envelope Distribution.** As said before, due to the decoupling between the relaying segments, the class of distributions remains unchanged. The typical envelope distributions, such as Rayleigh, Ricean, Nakagami, etc., can therefore be made use of. Generally, the statistics of the first MPC of a traditional link obeys Ricean fading with $K = 2, \ldots, 10$ (LOS) and Rayleigh fading (NLOS), whereas for the cooperative link it is typically Ricean with $K > 10$ (LOS) and Rayleigh (NLOS). The remaining MPCs are usually Rayleigh distributed. The power distributions are respectively derived from the envelope distributions.
- **Power Delay Profile.** The PDP is for both links negative-exponential distributed where the direct link often occurs in clusters, each of which is exponentially distributed. The delay spread τ_{RMS} strongly depends on the environment and, for the traditional link, can range between $\tau_{RMS} = 50\,\text{ns}, \ldots, 4\,\mu\text{s}$, whereas cooperative links are more likely to have spreads of $\tau_{RMS} = 10\,\text{ns}, \ldots, 40\,\text{ns}$. The significantly shorter delay spreads in the cooperative link are due to the distance between terminals as well as clutter being significantly reduced when compared to cellular arrangements.
- **Spectral Characteristics.** As discussed in Section 2.2, the spectral characteristics are impacted by the PDP and its delay spread. Since the delay spread is significantly reduced in the cooperative relaying link, a potentially very frequency-selective traditional communication link may turn frequency-flat when retransmitted at the same rate over the relay link. This decrease in frequency diversity has serious implications on the performance of the end-to-end system. For instance, if equalizers are used, then the relay link will perform significantly better than the main link. Contrary, if spread spectrum like techniques are used, then the relay link will not be able to provide the same (relative) power gains as the main link. Furthermore, in OFDM like techniques, the frequency diversity offered by the channel is often picked up by the channel coder, which would be less effective in the cooperative link.

  This highlights the fact that, if optimal performance is critical, link specific transceivers with appropriate parameterizations need to be developed on a case-by-case basis. For instance, a relaying node could aggregate downlink traffic and retransmit it using a higher bandwidth that ought to exhibit the same frequency diversity as the traditional link.
- **Temporal Characteristics.** One of the biggest changes w.r.t. traditional systems occurs in the temporal domain. The reason is that in traditional systems only one transceiver end is mobile whilst the other one remains static. In the context of cooperative relaying systems this is generally different as both ends are free to move around. The majority of latest scientific developments hence relate to the quantification of the temporal behavior, most notably the coherence time, which we will deal with below.

- **Spatial Characteristics.** The spatial characteristics also change significantly, which is important for the performance evaluation of distributed and/or in-built MIMO systems. The fact that both transceiver ends are well embedded into clutter guarantees that correlation distances remain low which in turn guarantees healthy gains at either end for such systems. We will hence also consider the spatial characteristics in under exposure.

2.2.4.4   Impact on End-to-End Performance

Before going into some more modeling details, we shall briefly summarize the impact of regenerative relaying onto the end-to-end performance of a system illustrated in Figure 2.12. In the regenerative relaying case, the end-to-end performance is dictated by the weakest relaying segment in the system. To this end, Figure 2.13 depicts some qualitative tendencies of the weakest channel for a given distance between BS and MS and – where applicable – a RS placed exactly midway between both.

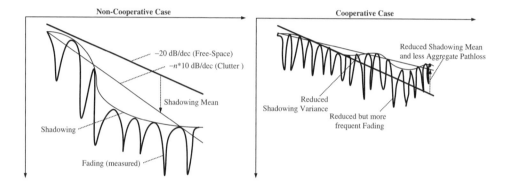

Figure 2.13   Regenerative cooperative communication impacts fading and shadowing and generally reduces pathloss

As per Figure 2.13 (left), we will take the narrowband noncooperative case with RS as a reference that exhibits a given pathloss slope, shadowing variability and fading behavior. Introducing cooperative relaying with a RS, as per Figure 2.13 (right), reduces the pathloss slope since the relaying system profits from the reduced aggregated powerloss paradigm as discussed in Section 1.2 and further quantified in Section 2.3. Furthermore, the shadowing variability is generally impacted; in this particular example it is reduced due to the reduced distance between transmitter and receiver in each segment. Importantly, fading occurs more frequently due to the higher mobility at either end of the cooperative segment. In either case, going wideband and using a suitable receiver a small power gain is achieved and the shadowing variability is also slightly reduced since the different MPCs experience different shadows.

2.2.5   Quick Introduction to Transparent Relay Channels

Being an example out of an infinite amount of possible communication topologies, a typical scenario occurring in the context of transparent relays is depicted in Figure 2.14. Here, an elevated BS communicates with a RS that is embedded into clutter at low altitude. This RS receives, amplifies and retransmits in a transparent fashion the signal to another MS well embedded in the clutter environment. The two types of channel possibly occurring, that is the traditional link between elevated BS and embedded RS and the cooperative relay link between embedded RS and MS, are now coupled.

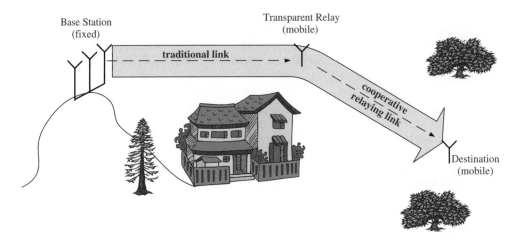

Figure 2.14   An example of a transparent relaying channel

2.2.5.1   System Assumptions

Unlike the regenerative relay case already discussed in Section 2.2.4, the fading component of each relay channel segment needs therefore be modeled jointly. In the example case of one relay leading to two relaying segments, the effective received signal is given as:

$$y_2 = \sqrt{G_2} \cdot h_2 \cdot A \cdot y_1 + n_2 \tag{2.14a}$$

$$y_1 = \sqrt{G_1} \cdot h_1 \cdot x_1 + n_1 \tag{2.14b}$$

which can be rewritten as

$$y_2 = A \cdot \sqrt{G_1 G_2} \cdot h_1 h_2 \cdot x_1 + A \cdot \sqrt{G_2} \cdot h_2 \cdot n_1 + n_2, \tag{2.15}$$

where the end-to-end wireless channel is characterized by $A \cdot \sqrt{G_1 G_2} h_1 h_2$ and the additive noise by $A \cdot \sqrt{G_2} \cdot h_2 \cdot n_1 + n_2$. Clearly, the end-to-end SNR is now a fairly complicated function of channel realizations, amplification factor and noise. In above equations, x_i, y_i and n_i are respectively the transmitted signal, the received signal and the AWGN with power σ_i^2 experienced in the ith relaying segment. Furthermore, $G_i = L_i \cdot S_i$ is the large-scale gain composed of pathloss and shadowing and h_i captures fading. A is the amplification factor, which can generally be variable, averaged or fixed:

- **Variable Amplification Factor.** Variable amplification factors refer to the case where amplification is a function of the instantaneous channel conditions. Different realizations of this case are possible. If, for example, the relaying node has perfect knowledge of the instantaneous channel conditions of both relaying segments, then the amplification could be inverse to the product of these average channel gains. A first practical problem in this context is to make these instantaneous channel realizations available in the relay. A second practical problem is that small channel realizations lead to very large amplification factors A, which in turn, as per (2.15), would lead to strong noise amplifications.

  A more common approach therefore is to assume that the relay has only knowledge of the instantaneous fading conditions of the source to relay channel and the amplification factor counterweights the impact of deep fades [164]:

$$A = \sqrt{\frac{P_2}{P_1 g_1 + \sigma_1^2}}, \tag{2.16}$$

where P_1 and P_2 are the respective average transmission powers of the sender and relay, g_1 is the instantaneous channel power in the first relaying segment, and σ_1^2 is the average power of the thermal noise at the input of the relay. Such an amplification is essentially the equivalent of an MMSE equalizer operating with instantaneous channel conditions. In subsequent analysis, this amplification factor cannot be set to unity since it significantly changes the end-to-end channel behavior. Note that his type of amplification is also sometimes referred to as CSI-assisted relaying [165].

- **Average Amplification Factor.** Average amplification factors refer to the case where amplification is a function of the average channel conditions, which change very slowly and can generally be considered constant over typical communication durations. Different realizations of this case are again possible. If, for example, the relaying node has perfect knowledge of the average channel conditions of both relaying segments, then the amplification could be inverse to the product of these average channel gains. Again, this leads to noise amplification. A more common approach is therefore to assume that the relay has only knowledge on the average fading conditions of the source to relay channel. This type of amplification is also sometimes referred to as semi-blind relaying [165].

It has been proposed [166] that an amplification factor be used that, is in notation the equivalent of (2.16), which yields:

$$A = \sqrt{\frac{P_2}{P_1 \overline{g}_1 + \sigma_1^2}},\qquad(2.17)$$

where $\overline{g}_1$ is now the average channel power in the first relaying segment.

A more common approach, however, is to assume that the amplification in the relay injects in average the same power as the amplification given in Equation (2.16), which yields the condition [167]:

$$A = \sqrt{\mathbb{E}\left\{\frac{P_2}{P_1 g_1 + \sigma_1^2}\right\}},\qquad(2.18)$$

where the expectation is taken w.r.t. the fading in the first relaying segment. Different relaying channel statistics hence require very different amplifications, a general quantification of which can be found in [168].

- **Fixed Amplification Factor.** Fixed amplification factors can, for example, be preprogrammed into the relay node. In this case, the statistics of the relaying channel does not depend on the noise of the relay terminal. Without loss of generality it can be set in subsequent analysis to unity, that is $A = 1$. Note that this type of amplification is also sometimes referred to as blind relaying [165].

The same insights on variable, average and fixed amplification factors hold for systems with multiple hops. We will now discuss the trends associated with these parameters.

2.2.5.2 Key Channel Parameters

With this scenario in mind, the end-to-end transparent relaying channel is known to behave as follows w.r.t. the key quantities discussed in Section 2.2 where again insights the literature [135, 141–163] have either directly been taken into account or adapted to the context of transparent relaying:

- **Pathloss.** The end-to-end pathloss behavior is now dependent on the pathloss of each relaying segment. As in the regenerative case, the pathloss coefficient of the traditional link varies between $n = 2$ (LOS) and $n = 2, \ldots, 4$ (NLOS) and the segment of a cooperative link exhibits coefficients from $n = 2$ (LOS) to $n = 4, \ldots, 6$ (NLOS). The end-to-end pathloss is then the product in linear scale or sum in decibels of the pathloss of each individual stage, which will be quantified in Section 2.4. Already available pathloss models can be used to model each of the segments prior to their aggregation.

- **Shadowing.** Similar to pathloss, shadowing also aggregates along the relaying segments. If each relaying segment suffers from a normal distributed shadowing in decibels, then the end-to-end aggregated shadowing will also obey a normal distribution albeit with a generally increased standard deviation. This will be quantified in Section 2.4 below. Again, each relaying segment can be modeled separately prior to aggregation where the traditional link typically exhibits a standard deviation σ_{dB} of $2, \ldots, 6\,dB$ (LOS) and $6, \ldots, 18\,dB$ (NLOS), whereas the cooperative link is in the range of $0, \ldots, 2\,dB$ (LOS) and $2, \ldots, 10\,dB$ (NLOS). As for the end-to-end shadow correlation distance d_{corr}, it is typically in the order of magnitude of the smallest correlation distance found along each relaying segment. For instance, if the traditional link exhibits a correlation distance around 80 m and the cooperative relaying link around 40 m, the end-to-end correlation distance is in the range of 40 m.
- **Fading.** The coupling between the relaying segments changes the entire statistics of the end-to-end fading channel. Therefore, the complex channel, envelope and power distributions change, as well as the second order time, space and frequency correlation functions. As per Equation (2.15), they are the result of the product of the individual fading realizations arising in each relaying segment, where each of these segments is typically modeled by already available channel models.
- **Noise.** Although not directly relating to the wireless channel it should be noted here that transparent relaying leads to thermal noise enhancement, which has a profound impact on the end-to-end performance. As per Equation (2.15), the amount of noise enhancement depends on the amplification factor, the channel realizations and/or statistics in each segment, the network topology, etc.

Pathloss, shadowing and fading models that have explicitly been developed for the transparent relaying case will be discussed in subsequent sections.

2.2.5.3 Impact on Fading Behavior

Since the fading behavior is impacted most by the transparent relaying architecture, we will discuss related issues separately:

- **Envelope Distribution.** Due to the coupling of the relaying segments, the end-to-end fading distribution changes significantly – mainly for the worse. The typical per-segment envelope distributions, such as Rayleigh, Ricean, Nakagami, etc., turn into a cascaded version thereof, which will in part be discussed in Section 2.4.2. The truly first MPC of the end-to-end PDP is typically cascaded Ricean distributed. All MPCs that result from the first MPC in the first segment and the first MPC in the second segment are cascaded Ricean–Rayleigh. All remaining MPCs are typically cascaded Rayleigh distributed.
- **Power Delay Profile.** The PDP for each segment is traditionally negative-exponential distributed where the direct link often occurs in clusters, each of which is exponentially distributed. The end-to-end PDP is the result of the cascaded channels' PDPs, their convolution to be precise, which typically leads to a channel with a delay spread τ_{RMS} being approximately the sum of the delay spreads in each segment. One can generally observe that the delay spread decreases in a nonlinear fashion with decreasing communication distance. It can therefore happen that the end-to-end delay spread of the transparent relay channel is inferior to the delay spread of the nonrelay channel due to shortened communication distances in each segment.
- **Spectral Characteristics.** As discussed in Section 2.2, the spectral characteristics are impacted by the PDP and its delay spread. Since the delay spread is generally changed in the cooperative relaying link, a potentially very frequency-flat traditional communication link may turn frequency selective when retransmitted at the same rate over the relay link, and vice versa. This change in frequency diversity has serious implications for the performance of the end-to-end system, as already discussed for the regenerative case.
- **Temporal Characteristics.** One of the biggest changes w.r.t. traditional systems occurs in the temporal domain. The reason is that in traditional systems only one transceiver end is mobile

whilst the other one remains static. In the context of cooperative relaying systems this is generally different as both ends are free to move around. In the context of transparent relaying systems, this mobility is increased even further as relays are also free to move. The majority of the latest scientific developments hence relate to the quantification of the temporal behavior, most notably the coherence time, which we will deal with below.

- **Spatial Characteristics.** The spatial characteristics also change significantly, which is important for the performance evaluation of distributed and/or in-built MIMO systems. The fact that both transceiver ends as well as the cooperative relays are well embedded into clutter guarantees that correlation distances remain low, which in turn guarantees healthy gains at both ends for such systems.

2.2.5.4 Impact on End-to-End Performance

Before going into some more modeling details, we shall briefly summarize the impact of transparent relaying onto the end-to-end performance of the example system illustrated in Figure 2.14. To this end, a qualitative trend of the end-to-end received field-strength at the target MS in dB versus distance from the BS is depicted in Figure 2.15.

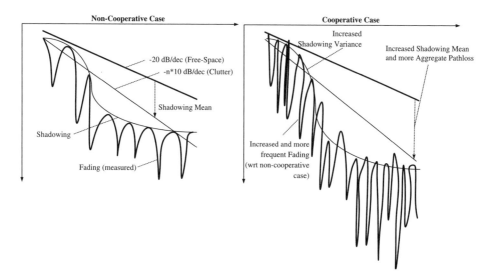

Figure 2.15 Transparent cooperative communication worsens fading behavior and also increases shadowing and pathloss

As per Figure 2.15 (left), we will take the narrowband noncooperative case as a reference that exhibits a given pathloss slope, shadowing variability and fading behavior. Introducing cooperative relaying, as per Figure 2.15 (right), impacts the pathloss slope; in this particular example with fixed amplification it is increased. Furthermore, the shadowing variability is generally impacted; in this particular example it is significantly increased due to the increased shadow variability. Importantly, fading occurs more frequently and with deeper fades, due to the higher mobility on either end of the cooperative segment and the cascaded fading channels, respectively. Generally for either case, going wideband and using a suitable receiver, a small power gain is achieved and the shadowing variability is also slightly reduced since the different MPCs experience different shadows.

This finishes the introductory section on channel modeling and subsequent sections are henceforth dedicated to more detailed modeling issues pertaining to the regenerative and transparent relaying channel.

2.3   Regenerative Relaying Channel

We will, in the following, discuss and quantify the most pertinent properties of a regenerative relay channel in greater detail. For the sake of consistency, we will follow here the same section structure as in Section 2.4 relating to transparent relays. Furthermore, the choice of notation and meaning of symbols follows that explained at the beginning of this chapter.

2.3.1   Propagation Modeling

The pathloss, shadowing and fading behaviors exhibit their peculiarities in the context of cooperative regenerative channels, which we will briefly allude to in this section. Note that we will generally henceforth refer to the mobile-to-mobile channel, which is equally applicable to the fixed RS-to-MS channel with the respective mobility set to zero.

2.3.1.1   Pathloss

There are two important issues related to pathloss in the context of regenerative relaying channels:

- **Change of Breakpoint Distance.** Most pathloss models exhibit a breakpoint behavior, as already discussed in Section 2.2.2. The breakpoint distance can in some settings be approximately calculated as $d_{bp} \approx 4 h_t h_r / \lambda$, where h_t and h_r are the transmitter and receiver heights respectively. Compared with traditional BS-MS links, the antenna height of MS-MS is reduced by an order of magnitude, for example, from 20 m to 2 m. This also means that the breakpoint distance is reduced by an order of magnitude, for example, from 100 m to 10 m, which is a striking difference to the BS-MS settings. However, given that the communication distances in cooperative settings are also reduced by about an order of magnitude, say from 1 km to 100 m, the relative behavior changes little w.r.t. the traditional BS-MS link. In summary, placing a relay between BS and some MS may advantageously move the traditional link from beyond the breakpoint to before the breakpoint but will likely cause the cooperative relaying link to be beyond breakpoint distance.
- **Aggregate Pathloss Powergains.** Another aspect is the severe impact of the system topology on the end-to-end power gains, which has already been alluded to in Section 1.2.2. This can be attributed to the nonlinear behavior of pathloss versus distance. Let us, for example, assume a system with the source and destination separated by d meters and relays placed equidistant between them so that N relay segments occur. Assuming then an example pathloss model to be discussed below in Section 2.3.7 and which has been taken from [157], the pathloss in each relaying segment can be calculated in decibel as follows:

$$L = b + 10 n \log_{10}(d/N), \tag{2.19}$$

where b is a constant and n the pathloss coefficient. The case $N = 1$ represents a direct communication scenario without relay. The aggregate power gains due to this nonlinear pathloss behavior, calculated as $10 \log_{10} N + L$, is summarized in Table 2.1 w.r.t. the case without relay in percent

Table 2.1   Absolute and relative aggregate pathloss gains with relays placed between source and destination separated by 500 m caused by the non-linear propagation model

Relay segments	1	2	3	4	5	6	7	8	9	10
Relative gain [%]	0	18	32	44	55	65	75	84	93	102
Absolute gain [dB]	0	15	23	29	34	38	41	44	46	49

and in absolute dB values, assuming $d = 500 \, \text{m}$, $b = -62.01 \, \text{dB}$ and $n = 5.86$. These gains are significant, which can be attributed to the large pathloss coefficient n. Furthermore, these gains even increase if a dual-slope pathloss model is assumed.

The major part of the performance gains in regenerative relaying systems hence originates indisputably from pathloss gains and not from fading diversity gains.

2.3.1.2 Shadowing

There are three important issues related to shadowing in the context of regenerative relaying channels:

- **Change of Shadowing Standard Deviation.** The reduced communication distances between cooperating terminals traditionally come along with a reduced shadowing standard deviation. This has, for example, been quantified in [157] and will be discussed below in Section 2.3.7.
- **Aggregate Shadowing Variations.** In the case of serial relays as shown in Figure 1.1 with independent shadowing channels in each relaying hop, the performance is impacted by the weakest relaying segment. Therefore, the decrease in shadowing variation due to shorter communication distances significantly boosts performance of regenerative relaying systems. Let us again assume a system with the source and destination separated by d meters and relays placed at equidistance between them so that N relay segments occur. Using then, for example, the model of [157], the shadowing standard deviation in each relaying segment can be calculated in decibels as follows:

$$\sigma_{\text{dB}} = S_{\text{s}} \cdot \left(1 - e^{-\frac{d/N - d_0}{D_{\text{s}}}} \right), \tag{2.20}$$

where S_{s}, d_0 and D_{s} are some model specific constants. Since shadowing is a random process, a simple power gain evaluation as done for the pathloss cannot be done here; however, a fair and typical assumption is that sufficient performance is guaranteed when the power margin due to shadowing is about three times its standard deviation σ_{dB}. The aggregate power gains due to this nonlinear shadowing behavior, calculated as $10 \log_{10} N + 3\sigma_{\text{dB}}$, is summarized in Table 2.2 w.r.t. the case without relay in percent and in absolute dB values, assuming $d = 500 \, \text{m}$, $S_{\text{s}} = 22.1 \, \text{dB}$, $d_0 = 10 \, \text{m}$ and $D_{\text{s}} = 53 \, \text{m}$. These gains are fairly high for a large number of relays but turn into losses for a small number of relays, that is in the region of practical deployment. These losses are generally not negligible and can very well diminish the gains due to the above discussed aggregate pathloss gains. They are due to a small decrease in shadowing standard deviation for large distances and a a strong decrease at very short distances.

Table 2.2 Absolute and relative aggregate shadowing gains with relays placed between source and destination separated by 500 m caused by the non-linear behavior of the shadow standard deviation

Relay segments	1	2	3	4	5	6	7	8	9	10
Relative gain [%]	0	−3	−2	2	8	15	23	31	39	47
Absolute gain [dB]	0	−2	−1	2	5	9	12	6	19	21

- **Change of Shadowing Correlation Model.** In contrast to cellular systems, a cooperative system is composed of transmitting and receiving nodes with similar characteristics in terms of antenna height, mobility pattern, etc. The shadowing fluctuations seen by the nodes at both ends of the radio link are therefore statistically identical [169]. This leads to a correlation function that is different from the ACF and CCF of the cellular case discussed in Section 2.2.2. To distinguish from the cellular case, the correlation function of the shadowing fluctuation in cooperative systems is referred to in [169] as a joint correlation function (JCF). Extensive ray-tracing simulations have revealed that the JCF is well approximated by:

$$R(\Delta d_t, \Delta d_r) \propto e^{-\frac{\Delta d_t + \Delta d_r}{d_{corr}}}, \tag{2.21}$$

where Δd_t and Δd_r are the displacement distances of the transmitter and receiver, respectively, and d_{corr} the spatial correlation distance observed in the given environment. This spatial correlation distance is often less than observed in cellular systems, where a typical mean value in urban environments is $d_{corr} = 40\,\mathrm{m}$ [162].

Depending on the system topology, regenerative cooperative relaying can hence yield shadowing gains and losses. Generally, any possible shadowing losses are inferior to the aggregate pathloss gains discussed above. However, they can be of the same order of magnitude, which requires a careful system examination before any conclusions on gains versus losses can be drawn.

2.3.1.3 Fading

To characterize the small scale behavior of the regenerative relaying channel, we shall commence with the geometrical arrangement of a general mobile-to-mobile scenario depicted in Figure 2.16. As for the system assumptions, a source and destination mobile terminal move at a given speed and at an angle w.r.t. some coordinate system. Each terminal is equipped with generally multiple antennas that are placed according to some geometrical arrangement and each of which has a generic three-dimensional (3D) radiation pattern. The clutter environment is generally 3D and nonisotropic in space. The channel can generally appear selective in time, frequency and space. A starting point is the channel impulse response Equation (2.6), that is $h = \sum_i a_i \cdot e^{j\phi_i} \cdot \delta(t - \tau_i)$, where the parameters a_i, ϕ_i and τ_i are generally jointly dependent and strongly impacted on by above system assumptions.

- **Unchanged Amplitude/Power Statistics.** Except for the cascaded keyhole fading case to be discussed below, the channel, envelope and power statistics do not change due to cooperative deployments. This is due to the regenerative nature of the system, which decouples each relaying segment.
- **Change in Correlation Functions.** In contrast, the temporal, spatial and spectral characteristics change significantly, mainly due to both transmitter and receiver being mobile and being immersed in dense clutter.

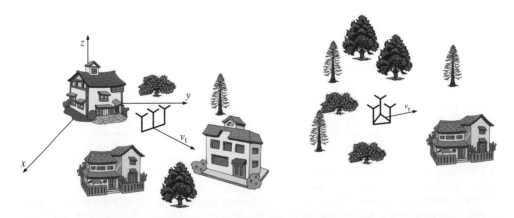

Figure 2.16 Exemplification of geometrical arrangement of regenerative relaying channel

It is the aim of subsequent sections to determine the statistics and moments associated with this random channel impulse response and we will first dwell on the statistics of the narrowband components a_i in Section 2.3.2, then quantify second order temporal, spatial and spectral moments in Sections 2.3.3–2.3.5.

2.3.2 Envelope and Power Fading Statistics

Each MPC a_i will suffer from random fluctuations due to the superposition of some unresolvable MPCs. Generally, each resolved MPC i can be modeled by a sum of n unresolved MPCs, that is $a_i \exp(j\phi_i) = \sum_n a_{i,n} \exp(j\phi_{i,n})$. If the dependency between unresolvable MPCs n is weak or nil then the traditional CLT over the sum holds and classical narrowband distributions such as Rayleigh, Nakagami or Ricean can be observed. Otherwise, if there is a strong dependency such that the sum can be decomposed into a product of sums, to each of which the CLT applies, then the resultant distribution can be characterized by keyhole distribution. We will see in Section 2.3.3 when either of the two conditions holds, and characterize the resulting statistics in Section 2.4.2 since the same insights hold in this case for the transparent relaying channel.

Note that typically the first resultant and resolvable MPC in the PDP may obey any of the known fading distributions, whereas the remaining resolvable MPCs typically obey a Rayleigh or cascaded Rayleigh fading distribution. Subsequently, we will drop the subscript i referring to a particular resolvable MPC and only quantify the statistical behavior of a given component.

2.3.2.1 PDF Transformation

Generally, it is the fading channel power distribution that is of central interest to communications engineering since it relates naturally to the SNR and hence to system performance. However, since the amplitude/envelope distribution serves as an interim step between complex channel distribution and power distribution, it has become customary to derive these too and name the distributions according to the amplitude distribution.

To obtain these, we assume that each MPC obeys an instantaneous fading amplitude or envelope of $a = |h|$ with pdf $p_a(a)$ to be quantified below. The instantaneous channel gain or power of said fading channel is thus given as $g = a^2$ with pdf $p_g(g)$; its average $\overline{g} = \mathbb{E}\{g\} = \mathbb{E}\{a^2\}$ is often normalized to unity, that is $\overline{g} = 1$. Using a simple PDF transformation, one can relate the PDF of the channel power with the one of the envelope, that is

$$p_g(g) = \frac{1}{2\sqrt{g}} \cdot p_a\left(\sqrt{g}\right) \tag{2.22}$$

and inversely

$$p_a(a) = 2a \cdot p_g\left(a^2\right). \tag{2.23}$$

We will now discuss a few typically occurring amplitude and power distributions.

2.3.2.2 Fading Distributions

A nonexhaustive list of typically encountered fading distributions includes:

- **Rayleigh Fading.** If a MPC is composed of independently impinging waves of more or less equal average amplitude, then the channel h obeys a central complex Gaussian distribution, which implies that the envelope is Rayleigh and the power central chi-square distributed with two degrees of freedom with the respective PDFs:

$$p_a(a) = \frac{2a}{\overline{g}} \exp\left(-\frac{a^2}{\overline{g}}\right) \tag{2.24a}$$

$$p_g(g) = \frac{1}{\overline{g}} \exp\left(-\frac{g}{\overline{g}}\right). \tag{2.24b}$$

The CDFs, which are useful in the calculation of outage probabilities as well as average fade durations, are given as:

$$F_a(a) = 1 - \exp\left(-\frac{a^2}{\overline{g}}\right) \tag{2.25a}$$

$$F_g(g) = 1 - \exp\left(-\frac{g}{\overline{g}}\right). \tag{2.25b}$$

Given that $g = a^2$, the CDFs are principally the same for any given fading distribution, which is due to their relationship through the PDF transformation. The Rayleigh distribution is well applicable to all MPCs which suffer from strong NLOS conditions, be it outdoors or indoors.

- **Ricean Fading.** If a MPC is composed of independently impinging waves of more or less equal average amplitude and a strong LOS component, then the channel h obeys a noncentral complex Gaussian distribution, which implies that the envelope is Ricean and the power noncentral chi-square distributed with two degrees of freedom with the respective PDFs:

$$p_a(a) = \frac{2a(1 + K)e^{-K}}{\overline{g}} \exp\left(\frac{-(1 + K)a^2}{\overline{g}}\right) I_0\left(2a\sqrt{\frac{K(1 + K)}{\overline{g}}}\right) \tag{2.26a}$$

$$p_g(g) = \frac{(1 + K)e^{-K}}{\overline{g}} \exp\left(-\frac{(1 + K)g}{\overline{g}}\right) I_0\left(2\sqrt{\frac{K(1 + K)g}{\overline{g}}}\right), \tag{2.26b}$$

where $I_0(\cdot)$ is the zeroth order modified Bessel function of the first kind and K the Ricean fading factor. The respective CDFs are given as:

$$F_a(a) = 1 - Q_1\left(\sqrt{2K}, \sqrt{2(1 + K)\frac{a^2}{\overline{g}}}\right) \tag{2.27a}$$

$$F_g(g) = 1 - Q_1\left(\sqrt{2K}, \sqrt{2(1 + K)\frac{g}{\overline{g}}}\right), \tag{2.27b}$$

where $Q_1(\cdot, \cdot)$ is the Marcum Q-Function. This reduces to the Rayleigh fading case for $K = 0$ and a nonfading channel for $K \to \infty$. The Ricean distribution is well applicable to all MPCs that enjoy strong LOS conditions, be it outdoors or indoors.

- **Nakagami-m Fading.** It is generally assumed that if a MPC is composed of bunches of phase-aligned impinging waves, then the channel's envelope is Nakagami and the power gamma distributed with the respective PDFs:

$$p_a(a) = \frac{2m^m a^{2m-1}}{(\overline{g})^m \Gamma(m)} \exp\left(-\frac{ma^2}{\overline{g}}\right) \tag{2.28a}$$

$$p_g(g) = \frac{m^m (g)^{m-1}}{(\overline{g})^m \Gamma(m)} \exp\left(-\frac{mg}{\overline{g}}\right), \tag{2.28b}$$

where $\Gamma(\cdot)$ is the complete Gamma function and m is the Nakagami fading factor. The respective CDFs are given as:

$$F_a(a) = \frac{\gamma\left(m, m\frac{a^2}{g}\right)}{\Gamma(m)} \tag{2.29a}$$

$$F_g(g) = \frac{\gamma\left(m, m\frac{g}{g}\right)}{\Gamma(m)} \tag{2.29b}$$

where $\gamma(\cdot, \cdot)$ is the lower incomplete gamma function. This reduces to the Rayleigh fading case for $m = 1/2$ and a nonfading channel for $m \to \infty$. The Gamma distribution is well applicable to all MPCs that suffer from obstructed LOS or weak NLOS conditions, that is mainly indoors.

- **Keyhole Fading.** The physical explanation behind the keyhole fading phenomenon is that of local clutter around the transmitter and receiver being connected via some non-cluttered void that leads to 'propagation pipes' or 'key holes'. Fading behavior at the transmitter is hence statistically decoupled from the fading behavior at the receiver, which requires the narrowband MPC to be modeled as the product of two sums, that is $a_i \exp(-j\phi_i) = \sum_m a_{i,m} \exp(-j\phi_{i,m}) \cdot \sum_n a_{i,n} \exp(-j\phi_{i,n})$, where each of which follows its own distribution. If more clutter islands are present on the path between transmitter and receiver, then this leads to additional multiplicative terms. This reduces this channel to the cascaded fading channel in transparent relaying system; however, the underlying physical principles are very different. We will therefore postpone the characterization of the resulting statistics to Section 2.4.2, since the same insights hold in this case for the transparent relaying channel.

Other fading distributions, such as Weibull distribution, have been reported and corroborated by measurement campaigns but are not further treated in this book.

2.3.2.3 Relationship with SNR

Although this chapter is dedicated to channel modeling issues only, we shall briefly discuss the relationship between channel power and SNR statistics. The latter is of paramount importance for the quantification of the system performance in terms of, for instance, capacity, outage, error rates, etc.

In the case of regenerative relaying, this relationship is fortunately very easy since the instantaneous SNR of a signal traversing the fading channel is given as $\gamma = a^2 E_s/N_0$ with PDF $p_\gamma(\gamma)$, where E_s is the symbol energy and N_0 the power spectral density. The average SNR is thus $\bar{\gamma} = \mathbb{E}\{a^2\}E_s/N_0$ where the average fading power is often normalized to unity, that is $\mathbb{E}\{a^2\} = 1$. Using the simple PDF transformation discussed above, one can relate the PDF of the SNR with the one of the channel power. This essentially involves replacing a^2 by γ.

For instance, the SNR for above discussed cases obeys a central chi-square distributed with two degrees of freedom in the case of Rayleigh fading, noncentral chi-square distributed with two degrees of freedom in the case of Ricean fading, and gamma distributed in the case of Nakagami fading.

Note that placing a relay in between transmitter and receiver may change the Rayleigh distribution on a traditional link to a Ricean on the cooperative relaying link and vice versa, simply because LOS conditions change.

2.3.3 Temporal Fading Characteristics

We will now quantify the temporal behavior of the regenerative relaying channel, commencing with the system assumptions and then moving to the canonical as well as more sophisticated propagation scenarios.

2.3.3.1 System Assumptions

To characterize the small scale behavior of the regenerative relaying channel, we shall commence with the geometrical two-ring scattering model of the canonical mobile-to-mobile scenario. The reason for taking a geometrical modeling approach is that it (i) reflects well the waves' realistic propagation behavior; and (ii) lends itself naturally to implementation in channel simulators, which will be discussed in Section 2.3.6. Having originated in [170], this geometrical model is based on the following assumptions:

- a wave emitted by the transmitter will impinge upon several clutter surfaces closest to the transmitter;
- each of these reflections off clutter surfaces yields a new wave that impinges upon clutter surfaces closest to the receiver;
- clutter surrounds transmitter and receiver circularly and uniformly, forming a ring of scatterers;
- waves that reflected off clutter surfaces are further away are negligible because they suffer from increased pathloss;
- waves that experience more reflections are negligible because each reflection leads to significant power losses;
- the ring of scatterers is fixed so that the mobile environment can be regarded as quasi-static for sufficiently short periods;
- the number of scatterers at either end tends towards infinity, requiring that the power of each wave is negligible compared to the total mean power.

This scenario is depicted in Figure 2.17 and formalized as follows [171, 172]. We fix a two-dimensional (2D) coordinate system with the origin in the transmitter. The transmitter is moving with speed v_t at an angle γ_t. The receiver is moving with speed v_r at an angle γ_r. The distance between transmitter and receiver is D. The clutter rings at transmitter and receiver have a radius R_t and R_r, respectively. There are M and N scatterers surrounding the transmitter and receiver respectively. The transmitting wave impinging upon the mth clutter object surrounding the transmitter has an angle of departure (AOD) of α_m. Similarly, the receiving wave having been reflected off the nth clutter object surrounding the receiver has an angle of arrival (AOA) of α_n.

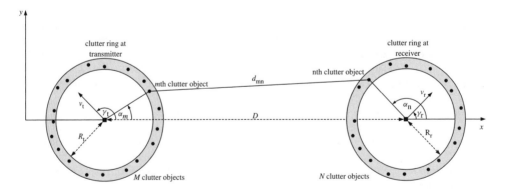

Figure 2.17 Double-bounced two-ring model for SISO regenerative relaying channel

The baseband narrowband channel realization can therefore be written as a superposition of any LOS, single bounced (SB) components that have only been bounced at the transmitter (SBT) and only bounced at the receiver (SBR), and the double bounced (DB) components. For the channel between the transmit and receive antenna this reads [172, 173]:

$$h(t) = h^{\text{LOS}}(t) + h^{\text{SBT}}(t) + h^{\text{SBR}}(t) + h^{\text{DB}}(t). \tag{2.30}$$

For ease of understanding, we will only deal with the double bounced component in this section; an extension to the other components is straightforward and illustrated in Section 2.3.4. The double bounced component hence reads [172]:

$$h^{\mathrm{DB}}(t) = \lim_{M,N\to\infty} \sum_{m=1}^{M} \sum_{n=1}^{N} a_{mn} \cdot e^{j(2\pi f_{mn}t - kd_{mn} + \phi_{mn})}, \tag{2.31}$$

where a_{mn} and ϕ_{mn} are the joint channel gain and phase shift caused by the interaction of the mth scatterers at the transmitter and nth scatterer at the receiver, respectively. Furthermore, f_{mn} is the Doppler shift induced by moving transmitter and receiver experienced by the wave reflected off the mth scatterer and nth scatterer. Finally, kd_{mn} is the phase shift induced due to the wave traversing a distance d_{mn} whilst traveling from transmitter to receiver. These quantities can be calculated as follows:

- **Amplitude a_{mn}.** The amplitude introduced by the mth clutter at the transmitter is of the same relative magnitude but independent of the amplitude due to the nth clutter at the receiver:

$$a_{mn} = a_m \cdot a_n = \frac{1}{\sqrt{MN}}, \tag{2.32}$$

where the last equation results from normalizing conditions which require that the power of Equation (2.31) remains bounded and equal to unity as $M, N \to \infty$.

- **Phase ϕ_{mn}.** The phase shift introduced by the mth clutter at the transmitter is independent of the phase shift introduced by the nth clutter at the receiver:

$$\phi_{mn} = \phi_m + \phi_n. \tag{2.33}$$

These phase shifts are random and, since the number of scatterers at either end is approaching infinity, these discrete random variables become continuous with PDFs $p_{\phi_t}(\phi_t)$ and $p_{\phi_r}(\phi_r)$, respectively. It can generally be assumed that they are uniformly distributed in $[0, 2\pi)$, that is $p_{\phi_t}(\phi_t) = 1/(2\pi)$ and $p_{\phi_r}(\phi_r) = 1/(2\pi)$.

- **Doppler Shift f_{mn}.** The Doppler shifts depend on the geometrical relationship between direction of movement of transmitter or receiver and direction of departure or arrival of the wave. The Doppler shift introduced by the mth clutter at the transmitter is generally independent of the Doppler shift introduced by the nth clutter at the receiver:

$$f_{mn} = f_m + f_n \tag{2.34a}$$

$$f_m = \hat{f}_t \cdot \cos(\alpha_m - \gamma_t), \tag{2.34b}$$

$$f_n = \hat{f}_r \cdot \cos(\alpha_n - \gamma_r), \tag{2.34c}$$

where $\hat{f}_t$ and $\hat{f}_r$ are respectively the maximum Doppler shifts experienced at the transmitter and receiver and which are given as $\hat{f}_t = f_c \cdot v_t/c = v_t/\lambda$ and $\hat{f}_r = f_c \cdot v_r/c = v_r/\lambda$, with f_c being the carrier frequency and λ the wavelength. Since the location of the scatterers is not known *a priori*, the AOD α_m and AOA α_n are discrete random variables. However, since the number of scatterers at either end is approaching infinity, these discrete random variables become continuous with PDFs $p_{\alpha_t}(\alpha_t)$ and $p_{\alpha_r}(\alpha_r)$, respectively.

- **Path Length d_{mn}.** The path length is influenced by the geometrical arrangement of clutter w.r.t. transmitter and receiver. We will now distinguish two cases, that is D is not significantly larger than $\max(R_t, R_r)$ and D is significantly larger than $\max(R_t, R_r)$. The former case is typically encountered indoors and requires the exact expression for d_{mn} to be used:

$$d_{mn} = R_t + \sqrt{(R_t \sin\alpha_m - R_r \sin\alpha_n)^2 + (D - R_t \cos\alpha_m + R_r \cos\alpha_n)^2} + R_r, \tag{2.35}$$

which essentially prevents the decoupling of the two sums in Equation (2.31) into two separate sums since the exact expression for d_{mn} can not be separated into two terms of the form $d_{mn} = d_m + d_n$.

Therefore, the CLT applies to the entire expression and the envelope $a = |h|$ is Rayleigh distributed as per Section 2.3.2.

The latter case, that is $D \gg \max(R_t, R_r)$, is typically encountered outdoors and allows the following simplification to be used [172]:

$$d_{mn} \approx D + R_t \cdot (1 - \cos \alpha_m) + R_r \cdot (1 + \cos \alpha_n), \tag{2.36}$$

which allows one to decouple the two sums in Equation (2.31) into two separate sums as per Equation (2.36). Therefore, the CLT applies to each of the sums separately and the envelope $a = |h|$ is double-Rayleigh distributed as per Section 2.3.2.

Under these assumptions, Equation (2.31) can be rewritten as:

$$h^{\mathrm{DB}}(t) = \lim_{M,N \to \infty} \sum_{m=1}^{M} \sum_{n=1}^{N} \frac{1}{\sqrt{MN}} \cdot e^{j(2\pi(f_m + f_n)t - kd_{mn} + \phi_m + \phi_n)}. \tag{2.37}$$

We are now in the position to derive temporal key quantities related to the regenerative relay fading channel.

2.3.3.2 Canonical Scenario

Following [171, 172, 174], we can obtain several temporal key characteristics related to the mobile-to-mobile narrowband fading channel.

- **Temporal Autocorrelation Function.** The temporal ACF of the complex fading channel $h(t)$ is an important quantity since it allows the quantification of the required pilot density and interleaver depths for coherent systems; the performance degradation for both coherent as well as noncoherent systems, etc. We will henceforth deal with the normalized ACF which is defined as

$$R(\Delta t) = \frac{\mathbb{E}\{h(t + \Delta t) \cdot h^*(t)\}}{\sqrt{\mathbb{VAR}\{h(t + \Delta t)\} \cdot \mathbb{VAR}\{h^*(t)\}}}, \tag{2.38}$$

where the expectation is taken w.r.t. the set of random variables, which in our case are α_t, α_r, ϕ_t and ϕ_r. Furthermore, it is henceforth assumed that the average channel power $\mathbb{VAR}\{h(t)\} = 1$, which allows us to drop it in subsequent derivations. Making use of Equation (2.37), we can calculate the ACF as follows:

$$R(\Delta t) = \mathbb{E}\left\{ \lim_{M,N \to \infty} \sum_{m=1}^{M} \sum_{n=1}^{N} \frac{1}{\sqrt{MN}} \cdot e^{j(2\pi(f_m + f_n)(t+\Delta t) + kd_{mn} + \phi_m + \phi_n)} \right.$$

$$\left. \times \lim_{M,N \to \infty} \sum_{m'=1}^{M} \sum_{n'=1}^{N} \frac{1}{\sqrt{MN}} \cdot e^{-j(2\pi(f_{m'} + f_{n'})t + kd_{m'n'} + \phi_{m'} + \phi_{n'})} \right\} \tag{2.39a}$$

$$= \mathbb{E}\left\{ \lim_{M \to \infty} \sum_{m=1}^{M} \frac{1}{M} \cdot e^{j2\pi \hat{f}_t \cdot \cos(\alpha_m - \gamma_t)\Delta t} \right\} \mathbb{E}\left\{ \lim_{N \to \infty} \sum_{n=1}^{N} \frac{1}{N} \cdot e^{j2\pi \hat{f}_r \cdot \cos(\alpha_n - \gamma_r)\Delta t} \right\} \tag{2.39b}$$

$$= \int_{-\pi}^{\pi} e^{j2\pi \hat{f}_t \cdot \cos(\alpha_t - \gamma_t)\Delta t} p_{\alpha_t}(\alpha_t)\, d\phi_t \int_{-\pi}^{\pi} e^{j2\pi \hat{f}_r \cdot \cos(\alpha_r - \gamma_r)\Delta t} p_{\alpha_r}(\phi_r)\, d\phi_r \tag{2.39c}$$

$$= J_0\left(2\pi \hat{f}_t \Delta t\right) \cdot J_0\left(2\pi \hat{f}_r \Delta t\right), \tag{2.39d}$$

where $J_0(\cdot)$ is the zeroth order Bessel function of the first kind. The random phases ϕ essentially decouple the two scattering rings and hence allow the above-invoked simplifications. The ACF is therefore the product of two classical Doppler ACFs, one originating at the transmitter and the

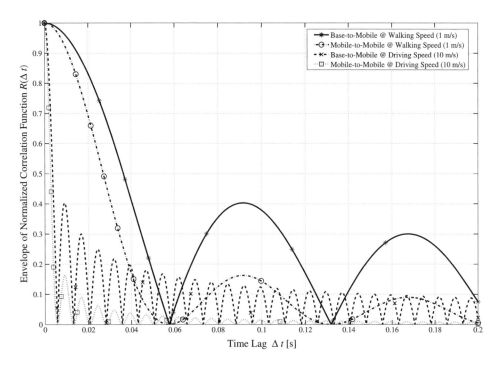

Figure 2.18 Absolute value of the temporal autocorrelation function of the complex fading channel for the base-to-mobile and mobile-to-mobile scenarios at $f_c = 2$ GHz. Observations: MS-to-MS channel decorrelates faster than BS-to-MS channel, which is good for code design but bad for channel estimation purposes

other one independently at the receiver. The derived ACF is depicted in Figure 2.18, where we compare the ACF for the base-to-mobile channel with that of the mobile-to-mobile channel at a carrier frequency of $f_c = 2$ GHz. For the former, it has been assumed that $v_t = 0$ and hence $\hat{f}_t = 0$; for both cases, it has been assumed that the mobile(s) move(s) at a speed of 1 m/s. Clearly, the MS-MS channel decorrelates significantly more quickly than the BS-MS channel. This has several implications on the system design. On the advantageous side, this guarantees that an outer channel code is offered more temporal diversity, thereby improving performance and reducing outages. On the down side, this means that a non-coherent system suffers from increased performance losses as, for example, quantified in [175]. As for coherent receivers, it may seem that pilot symbols need to be inserted more often; however, since pilot symbols are traditionally inserted every Δt_{decorr} seconds, where $R(\Delta t_{\text{decorr}}) = 1/\sqrt{2} \approx 0.7$, the difference is very small and even decreases with an increasing terminal velocity, as can be seen in Figure 2.18.

- **Doppler Power Spectrum.** The Doppler power spectrum of the complex channel $h(t)$ is determined from the Fourier transform of its temporal autocorrelation function $R(\Delta t)$, that is

$$\Psi(f) = \int_{-\infty}^{\infty} R(\Delta t) e^{-j2\pi f \Delta t} d\Delta t. \tag{2.40}$$

This transformation clearly does not yield any novel insights w.r.t. the already available autocorrelation function. However, as will be discussed below, it is of great value when simulating fading channels. Applying the Fourier transform to Equation (2.39d) yields [171]:

$$\Psi(f) = \frac{1}{\pi^2 \sqrt{\hat{f}_t \hat{f}_r}} K \left[\sqrt{\frac{(\hat{f}_t + \hat{f}_r)^2 - f^2}{4\hat{f}_t \hat{f}_r}} \right], \tag{2.41}$$

where $K(\cdot)$ is the complete elliptic integral of the first kind. This function exhibits two peaks at $\pm(\hat{f}_t - \hat{f}_r)$ and is generally symmetrical w.r.t. the motion of transmitter or receiver.

- **Level Crossing Rate.** We give here the level crossing rate of the channel's envelope, which represents the expected number of times per second the channel amplitude $a = |h|$ crosses level a_{thr} in the positive direction. It is an important notion for adaptive or opportunistic systems in that it quantifies the frequency of the channel state changes and hence the frequency for need of adaptation or opportunity. It can be calculated as [174]:

$$N(a_{thr}) = \sqrt{2\pi \left(\hat{f}_t^2 + \hat{f}_r^2 \right)} \cdot a_{thr} \cdot e^{-a_{thr}^2}, \tag{2.42}$$

assuming that $\mathbb{E}\left\{ |h|^2 \right\} = 1$. As depicted in Figure 2.19, the MS-to-MS channel varies faster than BS-to-MS channel, which is advantageous since it creates opportunities more frequently but requires also more frequent adaptations, for example, change of modulation and coding scheme.

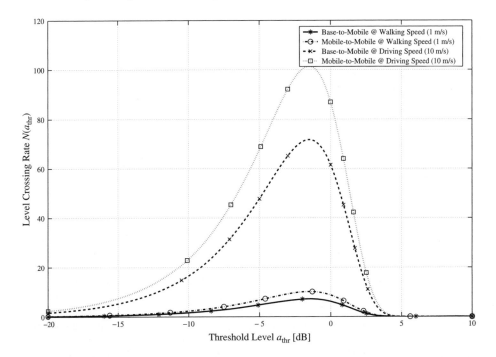

Figure 2.19 Level crossing rates of the envelope of the complex fading channel for the base-to-mobile and mobile-to-mobile scenarios at $f_c = 2\,\text{GHz}$. Observations: MS-to-MS channel varies faster than BS-to-MS channel, which is good in that there are more frequent opportunities but also more frequent need for adaptation

- **Average Fade Duration (AFD).** This is the average duration of time the envelope spends below level a_{thr}. It can be calculated as:

$$T(a_{thr}) = \frac{1}{\sqrt{2\pi \left(\hat{f}_t^2 + \hat{f}_r^2 \right)}} \left(e^{a_{thr}^2} - 1 \right), \tag{2.43}$$

where the definition given in Equation (2.11) has been made use of.

With these insights at hand, we are now able to move to more advanced narrowband SISO propagation scenarios, such as nonisotropic 2D clutter distributions as well as Ricean fading conditions.

2.3.3.3 Nonisotropic Scattering Scenario

We consider here scatter distributions that differ from the initially assumed isotropic circular distribution. Such distributions are frequently observed in real-world scenarios and include piece-wise uniform, Laplacian, Gaussian [176], cosine [177], van Mises/Tikhonov [178] distributions as well as those cases where the base station and/or mobile terminal have directional antennas [179, 180]. We will henceforth use the von Mises/Tikhonov distribution because it can be used to approximate the other distributions and has the neat property of facilitating closed form expressions for the ACF [178]. The PDF of the von Mises/Tikhonov distribution with mean μ and concentration κ is given as:

$$p_\alpha(\alpha) = \frac{e^{\kappa \cos(\alpha - \mu)}}{2\pi I_0(\kappa)}, \tag{2.44}$$

where $I_0(\cdot)$ is the zeroth order modified Bessel function of the first kind. This distribution has been illustrated in Figure 2.20 for $\mu = 0$ and various concentration factors κ.

- **Temporal Autocorrelation Function.** Applying this PDF to the AOD and AOA and inserting it in Equation (2.39c), one obtains [181]:

$$R(\Delta t) = \frac{I_0\left(\sqrt{\kappa_t^2 - 4\pi^2 \hat{f}_t^2 \Delta t^2 + j4\pi \kappa_t \hat{f}_t \Delta t \cos \mu_t}\right)}{I_0(\kappa_t)}$$

$$\times \frac{I_0\left(\sqrt{\kappa_r^2 - 4\pi^2 \hat{f}_r^2 \Delta t^2 + j4\pi \kappa_r \hat{f}_r \Delta t \cos \mu_r}\right)}{I_0(\kappa_r)}, \tag{2.45}$$

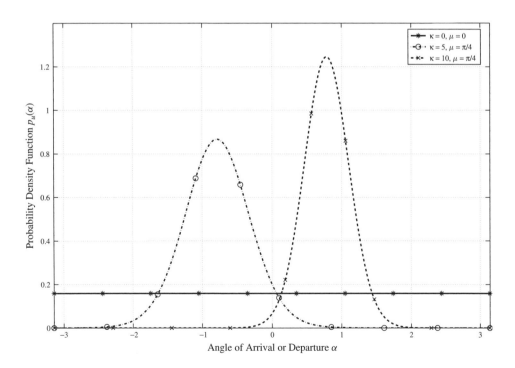

Figure 2.20 Increasing κ yields larger concentration around mean

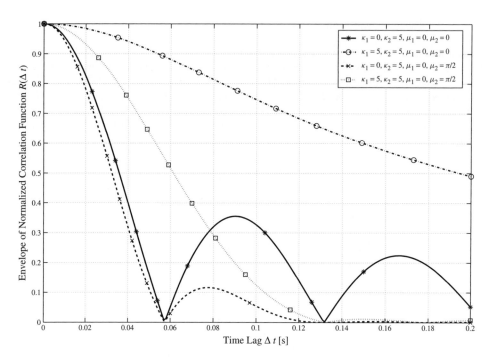

Figure 2.21 Increasing concentration κ yields higher correlation; nonaligned means yield lower correlation; $v_t = v_r = 1$ m/s

where μ_t and μ_r are the mean and κ_t and κ_r the concentration factors of the non-isotropic PDF of the AOD and AOA, respectively. Some occurring ACFs have been depicted in Figure 2.21.

Similarly, the PSD, ACF and LCR can be calculated but in most cases this leads to integral expressions that require numerical evaluation. Furthermore, expressions for the case that clutter is randomly and cylindrically distributed in 3D can be derived as shown in [182] and, for example, [183], respectively.

2.3.3.4 Ricean Fading Scenario

We now assume a SISO narrowband 2D-isotropic scatterer mobile-to-mobile Ricean fading channel that differs from the canonical scenario in that a LOS component is present. Several approaches are feasible when calculating the effective Doppler shift of the LOS component as seen by a receiver moving in a different direction and at a different speed than the transmitter. One such approach has been proposed [184, 185] by introducing the concept of relative speed. From this, the reception of this LOS component at the receiver will suffer from a fixed Doppler shift f_{LOS}, which is easily calculated by determining the relative velocity of the transmitter w.r.t. the receiver, denoted as v_{LOS}, and the angle spanned between the LOS component and v_{LOS}, denoted as α_{LOS}. Solving the trigonometric problem depicted in Figure 2.22, the following holds:

$$f_{LOS} = f_c \cdot v_{LOS}/c, \tag{2.46}$$

$$\alpha_{LOS} = \gamma_t + \cos^{-1}\left(\frac{v_t^2 + v_{LOS}^2 - v_r^2}{2v_t v_{LOS}}\right), \tag{2.47}$$

Figure 2.22   Trigonometric arrangement of the transmit, receive and relative velocity vectors

where $v_{\text{LOS}} = \sqrt{(v_t \cos(\gamma_t - \gamma_r) - v_r)^2 + (v_t \sin(\gamma_t - \gamma_r))^2}$. This allows the mobile-to-mobile LOS channel to be characterized by:

$$h_{\text{LOS}}(t) = \sqrt{\frac{1}{1+K}} \cdot h(t) + \sqrt{\frac{K}{1+K}} \cdot e^{j(2\pi f_{\text{LOS}} t \cos \alpha_{\text{LOS}} + \phi_0)}, \qquad (2.48)$$

where $h(t)$ is given in Equation (2.31) or (2.37), K is the Rice factor quantifying the ratio of the specular power to the nonspecular power, and θ_0 is some initial random phase that can be assumed to be uniformly distributed in $[0, 2\pi)$. This formulation allows us to derive important key performance metrics.

- **Temporal Autocorrelation Function.** The autocorrelation function is easily calculated following a similar approach as with the non-LOS case, leading to [184, 185]:

$$R(\Delta t) = \frac{1}{1+K} \cdot J_0\left(2\pi \hat{f}_t \Delta t\right) J_0\left(2\pi \hat{f}_r \Delta t\right) + \frac{K}{1+K} \cdot e^{j2\pi f_{\text{LOS}} \Delta t \cos \alpha_{\text{LOS}}}, \qquad (2.49)$$

which has been exemplified for different Ricean K factors in Figure 2.23 and different speeds and angles in Figure 2.24.

- **Doppler Power Spectrum.** The Doppler power spectrum of the complex channel $h_{\text{LOS}}(t)$ is determined from the Fourier transform of its ACF, which is easily derived as:

$$\Psi(f) = \frac{1}{(1+K)\pi^2 \sqrt{\hat{f}_t \hat{f}_r}} K \left[\sqrt{\frac{\left(\hat{f}_t + \hat{f}_r\right)^2 - f^2}{4 \hat{f}_t \hat{f}_r}} \right]$$

$$\times \frac{K}{1+K} \cdot \delta\left(f - f_{\text{LOS}} \cos \alpha_{\text{LOS}}\right), \qquad (2.50)$$

where $\delta(\cdot)$ is the Dirac-delta function.

- **Level Crossing Rate.** There is no general closed-form expression available for the level crossing rate of the envelope of the Ricean mobile-to-mobile channel. Some fairly complex expressions in integral form are available elsewhere [186, 187]. Under the condition that $\alpha_{\text{LOS}} = \pi/2$ or $\alpha_{\text{LOS}} = 3\pi/2$, the following simple expression however has been derived [185]:

$$N(a_{\text{thr}}) = 2(K+1)a_{\text{thr}} \sqrt{\frac{b_2}{2\pi}} e^{-K-(K+1)a_{\text{thr}}^2} \cdot I_0\left(2a_{\text{thr}}\sqrt{K(K+1)}\right), \qquad (2.51)$$

where we assume that the average channel power is normalized to unity and $b_2 = 2\pi^2(\hat{f}_t^2 + \hat{f}_r^2 + 2K \cos^2 \alpha_{\text{LOS}} f_{\text{LOS}}^2)/(K+1)$.

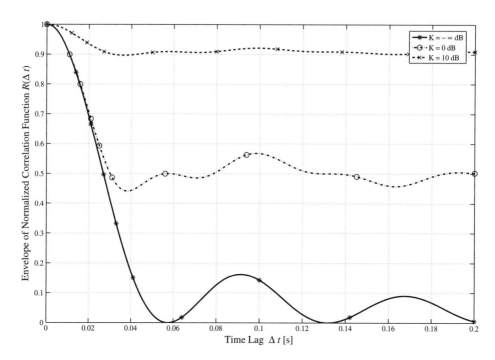

Figure 2.23   Increasing the LOS component causes increased correlation, which is good for channel estimation purposes but bad for interleaver design; $v_t = v_r = 1$ m/s; $\gamma_t = 3/4\pi$; $\gamma_r = \pi/4$

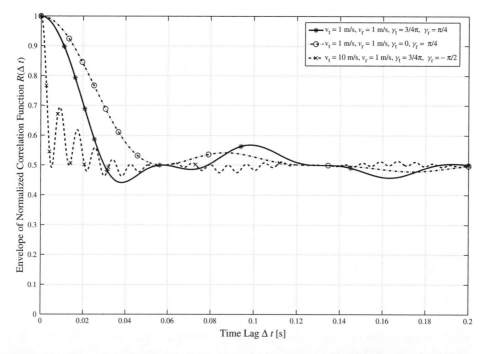

Figure 2.24   Increasing speed and/or angle between transmitter and receiver yields lower coherence, which is again good for interleaver design but bad for channel estimation purposes; $K = 0$ dB

- **Average Fade Duration.** Since the channel envelope is Ricean distributed, the average fade duration is easily derived as:

$$T(a_{\text{thr}}) = \frac{1 - Q\left(\sqrt{2K}, \sqrt{2(K+1)}a_{\text{thr}}\right)}{N(a_{\text{thr}})},\qquad (2.52)$$

where $Q(\cdot, \cdot)$ is the Marcum Q-function.

The insights gained in the above discussed canonical and advanced narrowband propagation scenarios can be used to derive the temporal characteristics of any type of propagation scenarios, which may often lead to expressions not expressible in closed form.

2.3.4 Spatial–Temporal Fading Characteristics

We now extend the temporal insights to the spatial domain where we assume that the transmitter and receiver are equipped with multiple antennas. We will first dwell on the underlying system assumptions that allow us to obtain key results for the chosen canonical configuration. We finally discuss some case studies originating from the canonical scenario.

2.3.4.1 System Assumptions

There is a healthy literature dedicated to the MIMO mobile-to-mobile case where the transmitter is equipped with L_t transmit and the receiver with L_r receive antennas, all of which yield insights into the channel's temporal and spatial behavior. A straightforward extension to the model discussed in Section 2.3.3 has been advocated [172, 188–191], which essentially extends the SISO narrowband, 2D two-ring scatter, double bounced rays model to the MIMO case. This model has further been extended [192] by taking into account the LOS component, hence leading to the Ricean fading case. This has further been extended [173] to the case of not only having the double bounced and LOS components but also single bounced components. A generalization of this latter case has been derived [193–195], where an additional elliptical scatter ring has been introduced with the transmitter and receiver as its foci; this allows the model to be easily adapted to a variety of mobile-to-mobile propagation environments, such as outdoor macro, micro, and pico-cells. The case of clutter not distributed in a two-ring fashion but rather fairly deterministically along street canyons has been discussed [196]. Finally, the extension of the narrowband, 2D two-ring scatter, double bounced rays model to the 3D case has been elaborated on [197, 198].

We will henceforth concentrate on the results derived in [173], which are a tradeoff between having more generic results than [172, 188–191] but not suffering from the complexity of [193, 197, 198]. Tuning a few parameters in this model allows us to approximate the propagation behavior of various real-world scenarios, to be discussed below. The model in [173] follows the same assumptions as already outlined in Section 2.3.3 as well as the following set of assumptions:

- a LOS component exists between MIMO transmitter and receiver;
- the existence of a LOS component increases the likelihood of also having single-bounced components;
- the resultant channel is hence modeled as a superposition of LOS, single and double bounced rays.

This scenario is depicted in Figure 2.25 and formalized [173] as follows. We will henceforth consider the transmitter to be equipped with n_t and the receiver with n_r omnidirectional antennas. Such an elementary antenna configuration can be used to construct many other types of 2D multielement antenna arrays, such as uniform linear arrays, rectangular arrays, hexagonal arrays and circular antenna

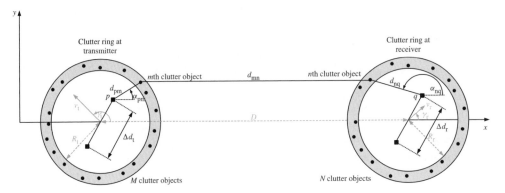

Figure 2.25 Double-bounced two-ring model for MIMO regenerative relaying channel (notation for single-bounced and LOS components are not depicted here but follow the same principles)

arrays [172]. The distance between the antenna elements at the transmitting and receiving sides is respectively denoted by Δd_t and Δd_r. The orientation of the arrays w.r.t. the coordinate systems is respectively θ_t and θ_r.

The remaining notations resemble those of the SISO case. To this end, we fix again a 2D coordinate system with the origin at the transmitter. The transmitter is moving with speed v_t at an angle γ_t. The receiver is moving with speed v_r at an angle γ_r. There are M and N scatterers surrounding the transmitter and receiver respectively. The transmitted wave from transmit antenna p impinging upon the mth clutter object surrounding the transmitter has an angle of departure (AOD) of α_{pm}. The transmitted wave from transmit antenna p impinging upon the nth clutter object surrounding the receiver has an angle of departure (AOD) of α_{pn}. The receiving wave having been reflected off the mth clutter object surrounding the transmitter and impinging onto receive antenna q has an angle of arrival (AOA) of α_{mq}. The receiving wave having been reflected off the nth clutter object surrounding the receiver and impinging onto receive antenna q has an angle of arrival (AOA) of α_{nq}. The angle of the LOS component between transmit antenna p and receive antenna q is denoted as α_{pq}.

The distance between transmitter and receiver is D. The clutter rings at transmitter and receiver have a radius R_t and R_r, respectively. The distance d_{pm} is the distance between the pth transmit antenna and the mth clutter surrounding the transmitter, d_{mq} between the mth clutter surrounding the transmitter and the qth receive antenna, d_{pn} between the pth transmit antenna and the nth clutter surrounding the receiver, d_{nq} between the nth clutter surrounding the receiver and the qth receive antenna, d_{mn} between the mth clutter surrounding the transmitter and the nth clutter surrounding the receiver, and finally d_{pq} is the distance between the pth transmit and qth receive antenna.

The baseband narrowband channel realization can therefore be written as a superposition of the LOS, single bounced (SB) components that have been bounced at the transmitter (SBT) and receiver (SBR), and the double bounced (DB) components. For the channel between the pth transmit and qth receive antenna this reads [173]:

$$h_{pq}(t) = h_{pq}^{\text{LOS}}(t) + h_{pq}^{\text{SBT}}(t) + h_{pq}^{\text{SBR}}(t) + h_{pq}^{\text{DB}}(t), \tag{2.53}$$

where the respective components read

$$h_{pq}^{\text{LOS}}(t) = a_{pq}^{\text{LOS}} e^{j\left(2\pi f_{pq}^{\text{LOS}} t - k d_{pq}^{\text{LOS}}\right)} \tag{2.54a}$$

$$h_{pq}^{\text{SBT}}(t) = \lim_{M \to \infty} \sum_{m=1}^{M} a_{pq,m}^{\text{SBT}} \cdot e^{j\left(2\pi f_{pq}^{\text{SBT}} t - k d_{pq}^{\text{SBT}} + \phi_m^{\text{SBT}}\right)} \tag{2.54b}$$

$$h_{pq}^{\text{SBR}}(t) = \lim_{N \to \infty} \sum_{n=1}^{N} a_{pq,n}^{\text{SBR}} \cdot e^{j\left(2\pi f_{pq}^{\text{SBR}} t - k d_{pq}^{\text{SBR}} + \phi_n^{\text{SBR}}\right)} \tag{2.54c}$$

$$h_{pq}^{\text{DB}}(t) = \lim_{M,N \to \infty} \sum_{m=1}^{M} \sum_{n=1}^{N} a_{pq,mn}^{\text{DB}} \cdot e^{j\left(2\pi f_{pq}^{\text{DB}} t - k d_{pq}^{\text{DB}} + \phi_{mn}^{\text{DB}}\right)}. \tag{2.54d}$$

Omitting the sub and superscripts, a and ϕ are the channel gains and phase shifts caused by the interaction of waves and scatterers. Furthermore, f are the Doppler shifts experienced by the waves due to the moving transmitter and receiver. Finally, kd is the phase shift induced due to the wave traversing a distance d whilst traveling from transmitter to receiver. These quantities can be calculated as follows:

- **Amplitudes.** The amplitudes introduced obey the same reasoning as already outlined in Section 2.3.3. In addition, the fraction of power allocated to the LOS component between transmit antenna p and receive antenna q is $K_{pq}/(1 + K_{pq})$ and to the NLOS components $1/(1 + K_{pq})$. Since $\max(\delta_{\text{t}}, \delta_{\text{r}}) \ll \min(R_{\text{t}}, R_{\text{r}})$, we can assume that the Ricean fading factor is the same between all antenna elements, that is $K_{pq} = K$ for all $1 \leq p \leq n_{\text{t}}$ and $1 \leq q \leq n_{\text{r}}$. We assume furthermore, that the relative power allocated to the SBT, SBR and DB components is respectively η_{t}, η_{r} and η_{tr} such that $\eta_{\text{t}} + \eta_{\text{r}} + \eta_{\text{tr}} = 1$. These parameters have to be either set during simulations or inferred from measurements. This allows us to write the amplitudes as follows:

$$a_{pq}^{\text{LOS}} = \sqrt{\frac{K}{K+1}} \tag{2.55a}$$

$$a_{pq,m}^{\text{SBT}} = \sqrt{\frac{1}{K+1}} \frac{\sqrt{\eta_{\text{t}}}}{\sqrt{M}} \tag{2.55b}$$

$$a_{pq,n}^{\text{SBR}} = \sqrt{\frac{1}{K+1}} \frac{\sqrt{\eta_{\text{r}}}}{\sqrt{N}} \tag{2.55c}$$

$$a_{pq,mn}^{\text{DB}} = \sqrt{\frac{1}{K+1}} \frac{\sqrt{\eta_{\text{tr}}}}{\sqrt{MN}}. \tag{2.55d}$$

Note that even if a single and double bounced wave reaching the receiver may have originated from the same mth clutter surrounding the transmitter, they are likely to be uncorrelated since we assume the surface elements on the mth clutter making the wave split into waves propagating into different direction to be uncorrelated.

- **Phases.** The phase shift introduced by the mth clutter at the transmitter is independent of the phase shift introduced by the nth clutter at the receiver:

$$\phi_{mn}^{\text{DB}} = \phi_m^{\text{DB}} + \phi_n^{\text{DB}}. \tag{2.56}$$

Furthermore, following the same argument as for the amplitudes, all involved phase shifts $\phi_m^{\text{SBT}}, \phi_n^{\text{SBR}}, \phi_m^{\text{DB}}$ and ϕ_n^{DB} can be assumed to be uniformly distributed in $[0, 2\pi)$ and independent among themselves and from any other variable.

- **Doppler Shifts.** The Doppler shifts depend on the geometrical relation between direction of movement of transmitter or receiver and direction of departure or arrival of the wave:

$$f_{pq}^{\text{LOS}} = \hat{f}_{\text{t}} \cos\left(\pi - \alpha_{pq} - \gamma_{\text{t}}\right) + \hat{f}_{\text{r}} \cos\left(\alpha_{pq} - \gamma_{\text{r}}\right) \tag{2.57a}$$

$$f_{pq}^{\text{SBT}} = \hat{f}_{\text{t}} \cos\left(\alpha_{pm} - \gamma_{\text{t}}\right) + \hat{f}_{\text{r}} \cos\left(\alpha_{mq} - \gamma_{\text{r}}\right) \tag{2.57b}$$

$$f_{pq}^{\text{SBR}} = \hat{f}_{\text{t}} \cos\left(\alpha_{pn} - \gamma_{\text{t}}\right) + \hat{f}_{\text{r}} \cos\left(\alpha_{nq} - \gamma_{\text{r}}\right) \tag{2.57c}$$

$$f_{pq}^{\text{DB}} = \hat{f}_{\text{t}} \cos\left(\alpha_{pm} - \gamma_{\text{t}}\right) + \hat{f}_{\text{r}} \cos\left(\alpha_{nq} - \gamma_{\text{r}}\right) \tag{2.57d}$$

where $\hat{f}_t$ and $\hat{f}_r$ are respectively the maximum Doppler shifts experienced at the transmitter and receiver and which are given as $\hat{f}_t = f_c \cdot v_t/c = v_t/\lambda$ and $\hat{f}_r = f_c \cdot v_r/c = v_r/\lambda$, with f_c being the carrier frequency and λ the wavelength. Since $\max(R_t, R_r) \ll D$, the LOS angles α_{pq} are approximately equal to π, that is $\alpha_{pq} \approx \pi$. Furthermore, since $\max(\delta_t, \delta_r) \ll \min(R_t, R_r)$, the angles α_{pm} will be approximately the same for all p and we will henceforth denote them by $\alpha_t^{(m)}$; similarly, $\alpha_{pn} \approx \alpha_t^{(n)}$, $\alpha_{mq} \approx \alpha_r^{(m)}$ and $\alpha_{nq} \approx \alpha_r^{(n)}$. Note that similarly to the SISO case, since the number of scatterers at either end is approaching infinity, the discrete random angles become continuous with given PDFs.

- **Path Lengths.** As per Figure 2.25, the path lengths of the respective waves are given by the following expressions:

$$d_{pq}^{\mathrm{LOS}} = d_{pq} \tag{2.58a}$$

$$d_{pq}^{\mathrm{SBT}} = d_{pm} + d_{mq} \tag{2.58b}$$

$$d_{pq}^{\mathrm{SBR}} = d_{pn} + d_{nq} \tag{2.58c}$$

$$d_{pq}^{\mathrm{DB}} = d_{pm} + d_{mn} + d_{nq}. \tag{2.58d}$$

Again, two cases can be distinguished, that is D is and is not significantly larger than $\max(R_t, R_r)$. The latter case is typically encountered indoors and requires the exact expression for all distances to be used as given in closed form in [173]. The former case, that is $D \gg \max(R_t, R_r)$, allows the following simplifications to be applied to the individual path lengths [173]:

$$d_{pq} \approx D - (0.5n_t + 0.5 - p)\Delta d_t \cos\theta_t$$
$$- (0.5n_r + 0.5 - q)\Delta d_r \cos(\pi - \theta_r) \tag{2.59a}$$

$$d_{pm} \approx R_t - (0.5n_t + 0.5 - p)\Delta d_t \cos\left(\theta_t - \alpha_t^{(m)}\right) \tag{2.59b}$$

$$d_{mq} \approx D - (0.5n_r + 0.5 - q)\Delta d_r\left(\delta_t \sin\theta_r \sin\alpha_t^{(m)} - \cos\theta_r\right) \tag{2.59c}$$

$$d_{pn} \approx D - (0.5n_t + 0.5 - p)\Delta d_t\left(\delta_r \sin\theta_t \sin\alpha_r^{(n)} + \cos\theta_t\right) \tag{2.59d}$$

$$d_{nq} \approx R_r - (0.5n_r + 0.5 - q)\Delta d_r \cos\left(\alpha_r^{(n)} - \theta_r\right) \tag{2.59e}$$

$$d_{mn} \approx D, \tag{2.59f}$$

where $p \in \{1, \ldots, n_t\}$ and $q \in \{1, \ldots, n_r\}$. Furthermore, we defined $\delta_t = R_t/D$ and $\delta_r = R_r/D$.

We are now in the position to derive temporal and spatial key quantities related to the mobile-to-mobile regenerative MIMO fading channel.

2.3.4.2 Canonical Scenario

Following [173], the space–time correlation function (STCF) and its PSD can be calculated as follows:

- **Space–Time Correlation Function.** Using above quantities and following the same approach already outlined in Equation (2.39), the STCF of the complex fading channel between any transmit–receive antenna pair pq and any other pair $\tilde{p}\tilde{q}$ is given as follows:

$$R_{pq,\tilde{p}\tilde{q}}(\Delta d_t, \Delta d_r, \Delta t) = R_{pq,\tilde{p}\tilde{q}}^{\mathrm{LOS}}(\Delta d_t, \Delta d_r, \Delta t) + R_{pq,\tilde{p}\tilde{q}}^{\mathrm{SBT}}(\Delta d_t, \Delta d_r, \Delta t)$$

$$+ R_{pq,\tilde{p}\tilde{q}}^{\mathrm{SBR}}(\Delta d_t, \Delta d_r, \Delta t) + R_{pq,\tilde{p}\tilde{q}}^{\mathrm{DB}}(\Delta d_t, \Delta d_r, \Delta t), \tag{2.60}$$

where the respective correlation functions have been derived [173] and are given as follows:

$$R_{pq,\tilde{p}\tilde{q}}^{\text{LOS}}(\Delta d_{\text{t}}, \Delta d_{\text{r}}, \Delta t) = \frac{K}{1+K} \cdot e^{j2\pi/\lambda(\Delta d_{\text{t}} \cos\theta_{\text{t}} - \Delta d_{\text{r}} \cos\theta_{\text{r}})} \times e^{j2\pi\Delta t(\hat{f}_{\text{t}} \cos\gamma_{\text{t}} - \hat{f}_{\text{r}} \cos\gamma_{\text{r}})} \quad (2.61a)$$

$$R_{pq,\tilde{p}\tilde{q}}^{\text{SBT}}(\Delta d_{\text{t}}, \Delta d_{\text{r}}, \Delta t) = \frac{\eta_{\text{t}}}{1+K} \cdot e^{-j2\pi/\lambda(\tilde{q}-q)\Delta d_{\text{r}} \cos\theta_{\text{r}} + 2\pi\Delta t \hat{f}_{\text{r}} \cos\gamma_{\text{r}}}$$

$$\times I_0\left(\sqrt{x_{\text{SBT}}^2 + y_{\text{SBT}}^2}\right) \Big/ I_0(\kappa_{\text{t}}) \quad (2.61b)$$

$$R_{pq,\tilde{p}\tilde{q}}^{\text{SBR}}(\Delta d_{\text{t}}, \Delta d_{\text{r}}, \Delta t) = \frac{\eta_{\text{r}}}{1+K} \cdot e^{j2\pi/\lambda(\tilde{p}-p)\Delta d_{\text{t}} \cos\theta_{\text{t}} - 2\pi\Delta t \hat{f}_{\text{t}} \cos\gamma_{\text{t}}}$$

$$\times I_0\left(\sqrt{x_{\text{SBR}}^2 + y_{\text{SBR}}^2}\right) \Big/ I_0(\kappa_{\text{r}}) \quad (2.61c)$$

$$R_{pq,\tilde{p}\tilde{q}}^{\text{DB}}(\Delta d_{\text{t}}, \Delta d_{\text{r}}, \Delta t) = \frac{\eta_{\text{tr}}}{1+K} \times \frac{I_0\left(\sqrt{x_{\text{DB}}^2 + y_{\text{DB}}^2}\right) I_0\left(\sqrt{z_{\text{DB}}^2 + w_{\text{DB}}^2}\right)}{I_0(\kappa_{\text{t}}) I_0(\kappa_{\text{r}})} \quad (2.61d)$$

where $I_0(\cdot)$ is the zeroth-order modified Bessel function of the first kind and the auxiliary variables are given as:

$$x_{\text{SBT}} = j2\pi/\lambda(\tilde{p}-p)\Delta d_{\text{t}} \cos\theta_{\text{t}} - j2\pi\Delta t \hat{f}_{\text{t}} \cos\gamma_{\text{t}} + \kappa_{\text{t}} \cos\mu_{\text{t}} \quad (2.62a)$$

$$y_{\text{SBT}} = j2\pi/\lambda((\tilde{p}-p)\Delta d_{\text{t}} \sin\theta_{\text{t}} + (\tilde{q}-q)\Delta d_{\text{r}}\delta_{\text{t}} \sin\theta_{\text{r}})$$

$$- j2\pi\Delta t\left(\hat{f}_{\text{t}} \sin\gamma_{\text{t}} + \hat{f}_{\text{r}}\delta_{\text{t}} \sin\gamma_{\text{r}}\right) + \kappa_{\text{t}} \sin\mu_{\text{t}} \quad (2.62b)$$

$$x_{\text{SBR}} = j2\pi/\lambda(\tilde{q}-q)\Delta d_{\text{r}} \cos\theta_{\text{r}} - j2\pi\Delta t \hat{f}_{\text{r}} \cos\gamma_{\text{r}} + \kappa_{\text{r}} \cos\mu_{\text{r}} \quad (2.62c)$$

$$y_{\text{SBR}} = j2\pi/\lambda((\tilde{p}-p)\Delta d_{\text{t}}\delta_{\text{r}} \sin\theta_{\text{t}} + (\tilde{q}-q)\Delta d_{\text{r}} \sin\theta_{\text{r}})$$

$$- j2\pi\Delta t\left(\hat{f}_{\text{t}}\delta_{\text{r}} \sin\gamma_{\text{t}} + \hat{f}_{\text{r}} \sin\gamma_{\text{r}}\right) + \kappa_{\text{r}} \sin\mu_{\text{r}} \quad (2.62d)$$

$$x_{\text{DB}} = j2\pi/\lambda(\tilde{p}-p)\Delta d_{\text{t}} \cos\theta_{\text{t}} - j2\pi\Delta t \hat{f}_{\text{t}} \cos\gamma_{\text{t}} + \kappa_{\text{t}} \cos\mu_{\text{t}} \quad (2.62e)$$

$$y_{\text{DB}} = j2\pi/\lambda(\tilde{p}-p)\Delta d_{\text{t}} \sin\theta_{\text{t}} - j2\pi\Delta t \hat{f}_{\text{t}} \sin\gamma_{\text{t}} + \kappa_{\text{t}} \sin\mu_{\text{t}} \quad (2.62f)$$

$$z_{\text{DB}} = j2\pi/\lambda(\tilde{q}-q)\Delta d_{\text{r}} \cos\theta_{\text{r}} - j2\pi\Delta t \hat{f}_{\text{r}} \cos\gamma_{\text{r}} + \kappa_{\text{r}} \cos\mu_{\text{r}} \quad (2.62g)$$

$$w_{\text{DB}} = j2\pi/\lambda(\tilde{q}-q)\Delta d_{\text{r}} \sin\theta_{\text{r}} - j2\pi\Delta t \hat{f}_{\text{r}} \sin\gamma_{\text{r}} + \kappa_{\text{r}} \sin\mu_{\text{r}} \quad (2.62h)$$

where again the von Mises distribution given in Equation (2.44) has been assumed at the transmitting and receiving side with mean μ_{t} and μ_{r} and parameter κ_{t} and κ_{r}, respectively. As pointed out [173], the above STCF can be simplified to many special cases, such as uniform scatter distribution, fixed mobile, etc.

- **Doppler Power Spectrum.** The Doppler power spectrum is obtained by taking the Fourier transform of the above STCF with respect to the time lag Δt, which essentially leads to a space-Doppler power spectral density. Applying the transform given in Equations (2.40) to (2.60) yields the following results [173]:

$$\Psi_{pq,\tilde{p}\tilde{q}}(\Delta d_{\text{t}}, \Delta d_{\text{r}}, f) = \Psi_{pq,\tilde{p}\tilde{q}}^{\text{LOS}}(\Delta d_{\text{t}}, \Delta d_{\text{r}}, f) + \Psi_{pq,\tilde{p}\tilde{q}}^{\text{SBT}}(\Delta d_{\text{t}}, \Delta d_{\text{r}}, f)$$

$$+ \Psi_{pq,\tilde{p}\tilde{q}}^{\text{SBR}}(\Delta d_{\text{t}}, \Delta d_{\text{r}}, f) + \Psi_{pq,\tilde{p}\tilde{q}}^{\text{DB}}(\Delta d_{\text{t}}, \Delta d_{\text{r}}, f), \quad (2.63)$$

where the respective spectral densities have been derived [173] as:

$$\Psi_{pq,\tilde{p}\tilde{q}}^{\text{LOS}}(\Delta d_{\text{t}}, \Delta d_{\text{r}}, f) = \frac{K}{1+K} \cdot \exp\left(j2\pi/\lambda((\tilde{p}-p)\Delta d_{\text{t}} \cos\theta_{\text{t}} - (\tilde{q}-q)\Delta d_{\text{r}} \cos\theta_{\text{r}})\right)$$

$$\times \delta\left(f - \hat{f}_{\text{t}} \cos\gamma_{\text{t}} + \hat{f}_{\text{r}} \cos\gamma_{\text{r}}\right) \quad (2.64a)$$

$$\Psi_{pq,\tilde{p}\tilde{q}}^{\text{SBT}}(\Delta d_{\text{t}}, \Delta d_{\text{r}}, f) = \frac{\eta_{\text{t}}/(1+K)}{I_0(\kappa_{\text{t}})} \cdot \frac{\exp\left(-j2\pi p_{x,\text{SBR}} - j2\pi(f - \hat{f}_{\text{r}}\cos\gamma_{\text{r}})A_{\text{SBT}}\right)}{\pi \hat{f}_{\text{t}}\sqrt{1 - \left[\left(f - \hat{f}_{\text{r}}\cos\gamma_{\text{r}}\right)\big/\hat{f}_{\text{t}}\right]^2}}$$

$$\times \cosh\left[(\kappa_{\text{t}}\sin(\mu_{\text{t}} - \gamma_{\text{t}}) + j2\pi p_{x,\text{SBT}}q_{y,\text{SBT}} - j2\pi p_{y,\text{SBT}}q_{x,\text{SBT}})\right.$$

$$\left.\times \sqrt{1 - \left[\left(f - \hat{f}_{\text{r}}\cos\gamma_{\text{r}}\right)\big/\hat{f}_{\text{t}}\right]^2}\right] \tag{2.64b}$$

$$\Psi_{pq,\tilde{p}\tilde{q}}^{\text{SBR}}(\Delta d_{\text{t}}, \Delta d_{\text{r}}, f) = \frac{\eta_{\text{r}}/(1+K)}{I_0(\kappa_{\text{r}})} \cdot \frac{\exp\left(j2\pi p_{x,\text{SBT}} - j2\pi\left(f + \hat{f}_{\text{t}}\cos\gamma_{\text{t}}\right)A_{\text{SBR}}\right)}{\pi \hat{f}_{\text{r}}\sqrt{1 - \left[\left(f + \hat{f}_{\text{t}}\cos\gamma_{\text{t}}\right)\big/\hat{f}_{\text{r}}\right]^2}}$$

$$\times \cosh\left[(\kappa_{\text{r}}\sin(\mu_{\text{r}} - \gamma_{\text{r}}) + j2\pi p_{x,\text{SBR}}q_{y,\text{SBR}} - j2\pi p_{y,\text{SBR}}q_{x,\text{SBR}})\right.$$

$$\left.\times \sqrt{1 - \left[\left(f + \hat{f}_{\text{t}}\cos\gamma_{\text{t}}\right)\big/\hat{f}_{\text{r}}\right]^2}\right] \tag{2.64c}$$

$$\Psi_{pq,\tilde{p}\tilde{q}}^{\text{DB}}(\Delta d_{\text{t}}, \Delta d_{\text{r}}, f) =$$

$$\eta_{\text{tr}}/(1+K) \cdot \frac{\exp\left(-j(2\pi p_{x,\text{DB}}q_{x,\text{DB}} + 2\pi p_{y,\text{DB}}q_{y,\text{DB}} + j\kappa_{\text{t}}\cos(\gamma_{\text{t}} - \mu_{\text{t}}))f/\hat{f}_{\text{t}}\right)}{I_0(\kappa_{\text{t}})\pi \hat{f}_{\text{t}}\sqrt{1 - \left(f/\hat{f}_{\text{t}}\right)^2}}$$

$$\times \cosh\left[(\kappa_{\text{t}}\sin(\mu_{\text{t}} - \gamma_{\text{t}}) + j2\pi p_{x,\text{DB}}q_{y,\text{DB}} - j2\pi p_{y,\text{DB}}q_{x,\text{DB}}) \times \sqrt{1 - \left(f/\hat{f}_{\text{t}}\right)^2}\right]$$

$$\star$$

$$\times \frac{\exp\left(-j(2\pi p_{z,\text{DB}}q_{z,\text{DB}} + 2\pi p_{w,\text{DB}}q_{w,\text{DB}} + j\kappa_{\text{r}}\cos(\gamma_{\text{r}} - \mu_{\text{r}}))f/\hat{f}_{\text{r}}\right)}{I_0(\kappa_{\text{r}})\pi \hat{f}_{\text{r}}\sqrt{1 - \left(f/\hat{f}_{\text{r}}\right)^2}}$$

$$\times \cosh\left[(\kappa_{\text{r}}\sin(\mu_{\text{r}} - \gamma_{\text{r}}) + j2\pi p_{z,\text{DB}}q_{w,\text{DB}} - j2\pi p_{w,\text{DB}}q_{z,\text{DB}})\right.$$

$$\left.\times \sqrt{1 - \left(f/\hat{f}_{\text{r}}\right)^2}\right] \tag{2.64d}$$

where $\star$ denotes the convolution w.r.t. the frequency f. The auxiliary variables are given below:

$$p_{x,\text{SBT}} = (\tilde{p} - p)(\Delta d_{\text{t}}/\lambda)\cos\theta_{\text{t}} \tag{2.65a}$$

$$q_{x,\text{SBT}} = \cos\gamma_{\text{t}} \tag{2.65b}$$

$$p_{y,\text{SBT}} = ((\tilde{p} - p)\Delta d_{\text{t}}\sin\theta_{\text{t}} + (\tilde{q} - q)\Delta d_{\text{r}}\delta_{\text{t}}\sin\theta_{\text{r}})/\lambda \tag{2.65c}$$

$$q_{y,\text{SBT}} \approx \sin\gamma_{\text{t}} \tag{2.65d}$$

$$A_{\text{SBT}} = \frac{2\pi p_{x,\text{SBT}}q_{x,\text{SBT}} + 2\pi p_{y,\text{SBT}}q_{y,\text{SBT}} - j\kappa_{\text{t}}\cos(\gamma_{\text{t}} - \mu_{\text{t}})}{2\pi \hat{f}_{\text{t}}} \tag{2.65e}$$

$$p_{x,\text{SBR}} = (\tilde{q} - q)(\Delta d_{\text{r}}/\lambda)\cos\theta_{\text{r}} \tag{2.65f}$$

$$q_{x,\text{SBR}} = \cos\gamma_{\text{r}} \tag{2.65g}$$

$$p_{y,\text{SBR}} = ((\tilde{p} - p)\Delta d_{\text{t}}\delta_{\text{r}}\sin\theta_{\text{t}} + (\tilde{q} - q)\Delta d_{\text{r}}\sin\theta_{\text{r}})/\lambda \tag{2.65h}$$

$$q_{y,\text{SBR}} \approx \sin \gamma_r \tag{2.65i}$$

$$A_{\text{SBR}} = \frac{2\pi p_{x,\text{SBR}} q_{x,\text{SBR}} + 2\pi p_{y,\text{SBR}} q_{y,\text{SBR}} - j\kappa_r \cos(\gamma_r - \mu_r)}{2\pi \hat{f}_r} \tag{2.65j}$$

$$p_{x,\text{DB}} = (\tilde{p} - p)(\Delta d_t / \lambda) \cos \theta_t \tag{2.65k}$$

$$q_{x,\text{DB}} = \cos \gamma_t \tag{2.65l}$$

$$p_{y,\text{DB}} = (\tilde{p} - p)(\Delta d_t / \lambda) \sin \theta_t \tag{2.65m}$$

$$q_{y,\text{DB}} = \sin \gamma_t \tag{2.65n}$$

$$p_{z,\text{DB}} = (\tilde{q} - q)(\Delta d_r / \lambda) \cos \theta_r \tag{2.65o}$$

$$q_{z,\text{DB}} = \cos \gamma_r \tag{2.65p}$$

$$p_{w,\text{DB}} = (\tilde{q} - q)(\Delta d_r / \lambda) \sin \theta_r \tag{2.65q}$$

$$q_{w,\text{DB}} = \sin \gamma_r \tag{2.65r}$$

A closed-form expression for the convolution of the double bounced term is unfortunately not available at the time of writing of this book; however, any numerical approach suffices as involved terms are fairly simple.

- **Level Crossing Rate.** We will not explicitly derive the LCR here but rather refer to the later Section 2.3.5 on the wideband case, where these expressions can be simplified to the narrowband case by simply setting all delays to zero.
- **Average Fade Duration.** Again, we will not explicitly give the expression here; however, it is easily derived by using the same approach as already outlined in Section 2.3.3.

With the above generic expressions, we will now investigate the behavior of the correlation function and its spectral density for a few parameterizations.

2.3.4.3 Case Studies

We now study a few parameterizations of above equations, where we will mainly concentrate on the space–time correlation function.

- **2 × 2 MIMO, 2D-Isotropic Scattering, DB Component.** When only isotropic scattering is assumed and there is only the double bounced component of significance, then the STCF can be reduced to [189, 190]:

$$R(\Delta d_t, \Delta d_r, \Delta t)$$

$$= J_0 \left(2\pi \sqrt{(\Delta d_t / \lambda)^2 + \left(\hat{f}_t \Delta t \right)^2 - 2\frac{\Delta d_t}{\lambda} \hat{f}_t \Delta t \cos (\gamma_t - \theta_t)} \right)$$

$$\times J_0 \left(2\pi \sqrt{(\Delta d_r / \lambda)^2 + \left(\hat{f}_r \Delta r \right)^2 - 2\frac{\Delta d_r}{\lambda} \hat{f}_r \Delta r \cos(\gamma_r - \theta_r)} \right), \tag{2.66}$$

where $J_0(\cdot)$ is the zeroth-order Bessel function of the first kind. To arrive at this result, one only needs to parameterize above equations as follows: $K = 0$, $\eta_t = 0$, $\eta_r = 0$, $\eta_{tr} = 1$, $\kappa_t = 0$, $\kappa_r = 0$, $\mu_t = 0$, $\mu_r = 0$, $p = 1$, $\tilde{p} = 2$, $q = 1$, $\tilde{q} = 2$. An example realization of two STCFs is shown in Figures 2.26 and 2.27, where further parameters are given in the respective figure captions.
- **2 × 2 MIMO, 2D-Non-Isotropic Scattering, LOS + SBT + SBR + DB Components.** The correlation function for the general case exhibits some interesting tendencies, some of which have been illustrated in Figure 2.28. To obtain these correlation functions, we have assumed the following

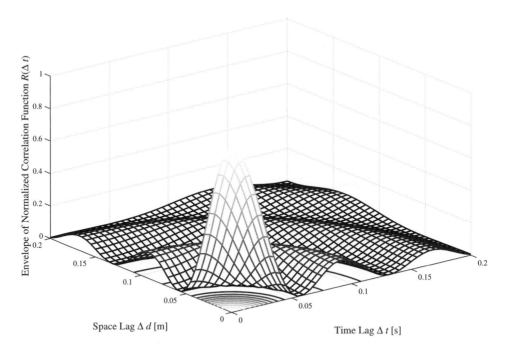

Figure 2.26 Spatial and temporal domains decorrelate similarly; $v_t = v_r = 1$ m/s; $\lambda = 15$ cm, $\gamma_t = \gamma_r = 0$, $\theta_t = \theta_r = \pi/2$, $\Delta d = \Delta d_t = \Delta d_r$

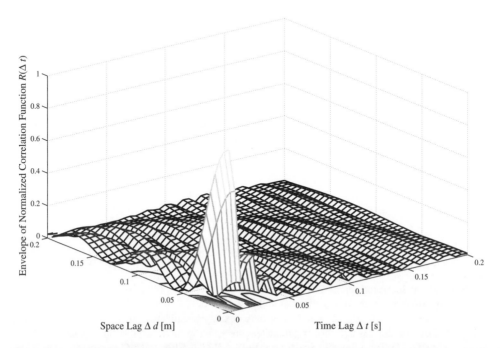

Figure 2.27 Fast terminal movements make the temporal domain decorrelate faster; $v_t = 5$ m/s, $v_r = 1$ m/s; $\lambda = 15$ cm, $\gamma_t = \gamma_r = 0$, $\theta_t = \theta_r = \pi/2$, $\Delta d = \Delta d_t = \Delta d_r$

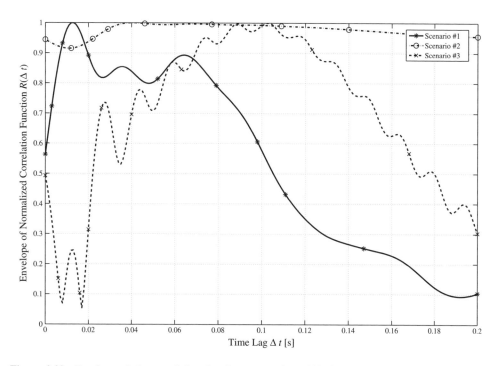

Figure 2.28 Envelope of the correlation function versus time shift for some given spatial positions at $f_c = 2\,\text{GHz}$. Scenario #1: $K = 0$, $v_t = 1\,\text{m/s}$, $v_r = 5\,\text{m/s}$, $\Delta d_t = \Delta d_r = \lambda/2$. Scenario #2: $K = 10$, $v_t = 1\,\text{m/s}$, $v_r = 5\,\text{m/s}$, $\Delta d_t = \Delta d_r = \lambda/2$. Scenario #3: $K = 0$, $v_t = 10\,\text{m/s}$, $v_r = 1\,\text{m/s}$, $\Delta d_t = \Delta d_r = \lambda$

set of parameters: $\lambda = 0.3$ m, $\theta_t = \pi/3$, $\theta_r = \pi/4$, $\gamma_t = \gamma_r = \pi/2$, $\eta_t = \eta_r = 0.4$, $\eta_{tr} = 0.2$, $\kappa_t = 2$, $\kappa_r = 5$, $\mu_t = \mu_r = 0$, $\delta_t = \delta_r = 0.01$, $p = 1$, $\tilde{p} = 2$, $q = 1$ and $\tilde{q} = 2$. The remaining parameters are scenario dependent and are given in Figure 2.28, where Scenario #1 is about high mobility with NLOS, Scenario #2 about high mobility with LOS, and Scenario #3 about low mobility with NLOS and larger antenna spacings. It can be observed that the largest space–time correlation value is not necessarily at zero delay. Furthermore, the Ricean channel causes the channel to be highly correlated.

In a similar fashion other scenarios can be studied and the impact of various parameters quantified.

2.3.5   Spectral–Spatial–Temporal Fading Characteristics

We finally extend the spatio-temporal insights to the spectral domain where we assume that the transmitter and receiver are equipped with multiple antennas and the signal components are resolvable in time. We will again first dwell on the underlying system assumptions that allow us to obtain key results for the chosen canonical configuration. We finally discuss some case studies originating from this canonical scenario.

2.3.5.1   System Assumptions

Contributions have recently emerged to the wideband MIMO mobile-to-mobile case where the transmitter is equipped with L_t transmit antennas, the receiver with L_r receive antennas and the signal undergoes a frequency-selective wideband channel, all of which yield insights into the channel's temporal, spatial and spectral behavior.

One of the earlier contributions to the mobile-to-mobile regenerative SISO wideband channel is discussed [199] in which ray-tracing tools are used to obtain the PDP in the context of intervehicle communications systems. This modeling approach was refined and its impact on system performance investigated [200]. A significant step forward w.r.t. these prior publications to the MIMO case and extending the model discussed in Section 2.3.4 has been advocated [183, 187, 201, 202], which essentially expands the prior introduced MIMO narrowband, 2D two-ring scatter model with LOS component, single and double bounced rays to the 3D wideband case. Similarly to the SISO case, the case of clutter not distributed in a two-ring fashion but rather fairly deterministically along street canyons has been discussed [203].

We will henceforth concentrate on the results derived in [187]. Tuning a few parameters in this model allows approximation of the propagation behavior of various real-world scenarios. The model in [187] follows the same assumptions as already outlined in Sections 2.3.3 and 2.3.4 as well as the following assumption:

- scatterers are distributed in 3D within concentric cylinders, which allows the PDP to be modeled more precisely in urban environments.

This scenario is depicted in Figure 2.29 and has been formalized [187] as follows. Again, we will consider a $n_t \times n_r$ antenna configuration where both the transmitter and the receiver are equipped with omnidirectional antennas. The distance between the antenna elements at transmitting and receiving side is respectively denoted by Δd_t and Δd_r. The transmit antenna array is elevated by a distance h_t and the receive antenna array by h_r.

The remaining notations resemble those of the narrowband SISO and MIMO cases. To this end, we fix a 3D coordinate system with the origin at the transmitter. The transmitter is moving with speed v_t at

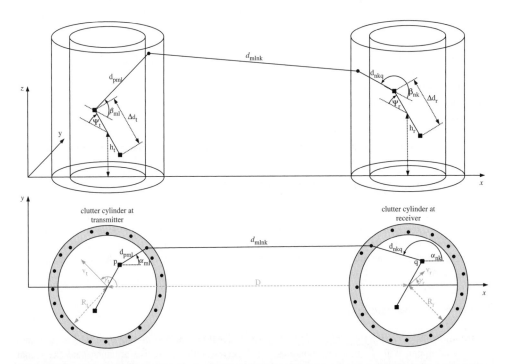

Figure 2.29 Double-bounced two-cylinder model for MIMO wideband regenerative relaying channel (notation for single-bounced and LOS components are not depicted here but follow the same principles)

an azimuth angle γ_t. The receiver is moving with speed v_r at an angle γ_r in the $x - y$ plane. Angles θ_t and θ_r are the azimuth orientation of the transmit and receive antenna arrays in the $x - y$ plane. Angles ψ_t and ψ_r are the elevation orientation of the transmit and receive antenna arrays in the $y - z$ plane.

There are M omnidirectional scatterers surrounding the transmitter occupying a volume between cylinders of radii R_{t1} and R_{t2}. These M scatterers hence lie on L cylindrical surfaces with radius $R_t^{(l)}$ such that each cylinder l accommodates $M^{(l)}$ scatterers denoted by $S_t^{(m,l)}$, where $R_{t1} \leq R_t^{(l)} \leq R_{t2}$, $1 \leq l \leq L$ and $1 \leq m \leq M^{(l)}$. Similarly, there are N omnidirectional scatterers surrounding the receiver occupying a volume between cylinders of radii R_{r1} and R_{r2}. These N scatterers hence lie on P cylindric surfaces with radius $R_r^{(k)}$ such that each cylinder k accommodates $N^{(k)}$ scatterers denoted by $S_r^{(n,k)}$, where $R_{r1} \leq R_r^{(k)} \leq R_{r2}$, $1 \leq k \leq P$ and $1 \leq n \leq N^{(k)}$.

The transmitted wave departing from transmit antenna p impinging upon the mth clutter object on the lth cylindric surface surrounding the transmitter has an azimuth angle of departure (AAOD) of $\alpha_t^{(m,l)}$ and an elevation angle of departure (EAOD) of $\beta_t^{(m,l)}$. Similarly, the transmitted wave departing from transmit antenna p impinging upon the nth clutter object on the kth cylindrical surface surrounding the receiver has an azimuth angle of arrival (AAOA) of $\alpha_t^{(n,k)}$ and an elevation angle of arrival (EAOA) of $\beta_t^{(n,k)}$. Finally, α_{Rq}^{LOS} describes the AAOAs of the LOS components.

The distance between transmitter and receiver is again D. The distance $d_{p,m,l}$ denotes the distance between the pth transmit antenna and the mth clutter laying on the lth cylindrical surface surrounding the transmitter, $d_{m,l,q}$ between the mth clutter on the lth cylindrical surface surrounding the transmitter and the qth receive antenna, $d_{p,n,k}$ between the pth transmit antenna and the nth clutter on the kth cylindrical surface surrounding the receiver, $d_{n,k,q}$ between the nth clutter on the kth cylindrical surface surrounding the receiver and the qth receive antenna, $d_{m,l,n,k}$ between the mth clutter surrounding the lth cylindrical surface of the transmitter and the nth clutter on the kth cylindrical surface surrounding the receiver, and finally $d_{p,q}$ is the distance between the pth transmit and qth receive antenna.

The baseband wideband channel realization can again be written as a superposition of the LOS, single bounced (SB) components, which have been bounced at the transmitter (SBT) and receiver (SBR), and the double bounced (DB) components. For the channel between the pth transmit and qth receive antenna this reads [187]:

$$h_{pq}(t, \tau) = h_{pq}^{LOS}(t, \tau) + h_{pq}^{SBT}(t, \tau) + h_{pq}^{SBR}(t, \tau) + h_{pq}^{DB}(t, \tau). \tag{2.67}$$

To simplify proceedings, it has been proposed [187] that the time-variant frequency transfer function be used rather than the delay-spread function Equation (2.67), where the former is the Fourier transform w.r.t. the delay τ of the latter. We can therefore write:

$$T_{pq}(t, f) = T_{pq}^{LOS}(t, f) + T_{pq}^{SBT}(t, f) + T_{pq}^{SBR}(t, f) + T_{pq}^{DB}(t, f), \tag{2.68}$$

where the respective components read

$$T_{pq}^{LOS}(t, f) = a_{pq}^{LOS} e^{j\left(2\pi f_{pq}^{LOS}t - 2\pi f \tau_{pq}^{LOS} - kd_{pq}^{LOS}\right)} \tag{2.69a}$$

$$T_{pq}^{SBT}(t, f) = \lim_{M \to \infty} \sum_{l=1}^{L} \sum_{m=1}^{M^{(l)}} a_{pq,ml}^{SBT} \cdot e^{j\left(2\pi f_{pq,ml}^{SBT}t - 2\pi f \tau_{pq,ml}^{SBT} - kd_{pq,ml}^{SBT} + \phi_{ml}^{SBT}\right)} \tag{2.69b}$$

$$T_{pq}^{SBR}(t, f) = \lim_{N \to \infty} \sum_{k=1}^{P} \sum_{n=1}^{N^{(k)}} a_{pq,nk}^{SBR} \cdot e^{j\left(2\pi f_{pq,nk}^{SBR}t - 2\pi f \tau_{pq,nk}^{SBR} - kd_{pq,nk}^{SBR} + \phi_{nk}^{SBR}\right)} \tag{2.69c}$$

$$T_{pq}^{DB}(t, f) = \lim_{M,N \to \infty} \sum_{l,m=1}^{L,M^{(l)}} \sum_{k,n=1}^{P,N^{(k)}} a_{pq,mlnk}^{DB} \cdot e^{j\left(2\pi f_{pq,mlnk}^{DB}t - 2\pi f \tau_{pq,mlnk}^{DB}\right)}$$

$$\times e^{j\left(-kd_{pq,mlnk}^{DB} + \phi_{mlnk}^{DB}\right)}. \tag{2.69d}$$

Omitting the subscripts and superscripts, a and ϕ are the channel gains and phase shifts caused by the interaction of waves and scatterers. Furthermore, f are the Doppler shifts experienced by the waves due to the moving transmitter and receiver. Then, τ are the delays of the MPCs induced by the 3D clutter. Finally, kd is the phase shift induced due to the wave traversing a distance d whilst traveling from transmitter to receiver. These quantities can be calculated as follows:

- **Amplitudes.** The amplitudes introduced follow the same reasoning as already outlined in Sections 2.3.3 and 2.3.4. In addition, the fraction of power allocated to the LOS component between transmit antenna p and receive antenna q is $K_{pq}/(1 + K_{pq})$ and to the NLOS components $1/(1 + K_{pq})$. Since $\max(\Delta d_t, \Delta d_r) \ll \min(R_t, R_r)$, we can assume that the Ricean fading factor is the same between all antenna elements, that is $K_{pq} = K$ for all $1 \leq p \leq L_t$ and $1 \leq q \leq L_r$. We assume furthermore, that the relative power allocated to the SBT, SBR and DB components is respectively η_t, η_r and η_{tr} such that $\eta_t + \eta_r + \eta_{tr} = 1$. We also assume that a power Ω_{pq} is channeled from transmit antenna p to receive antenna q. These parameters have to be either set during simulations or inferred from measurements. This allows us to write the amplitudes as follows:

$$a_{pq}^{\text{LOS}} = \sqrt{\Omega_{pq}} \sqrt{\frac{K}{K+1}} \tag{2.70a}$$

$$a_{pq,m}^{\text{SBT}} = \sqrt{\Omega_{pq}} \sqrt{\frac{1}{K+1}} \frac{\sqrt{\eta_t}}{\sqrt{M}} \tag{2.70b}$$

$$a_{pq,n}^{\text{SBR}} = \sqrt{\Omega_{pq}} \sqrt{\frac{1}{K+1}} \frac{\sqrt{\eta_r}}{\sqrt{N}} \tag{2.70c}$$

$$a_{pq,mn}^{\text{DB}} = \sqrt{\Omega_{pq}} \sqrt{\frac{1}{K+1}} \frac{\sqrt{\eta_{tr}}}{\sqrt{MN}}. \tag{2.70d}$$

The power Ω_{pq} is given as $\Omega_{pq} = P_{pq} D^{-n} (\lambda/4\pi)^2$, P_{pq} is the transmitted power from transmit antenna p towards receiver antenna q and n is the pathloss coefficient.

- **Phases.** Following the same arguments outlined in Sections 2.3.3 and 2.3.4, we assume that all phases ϕ are random variables uniformly distributed on $[0, 2\pi)$.
- **Doppler Shifts.** The Doppler shifts depend on the geometrical relationship between direction of movement of transmitter or receiver and direction of departure or arrival of the wave:

$$f_{pq}^{\text{LOS}} = \hat{f}_t \cos \gamma_t + \hat{f}_r \cos(\pi - \gamma_r) \tag{2.71a}$$

$$\begin{aligned} f_{pq,ml}^{\text{SBT}} = {} & \hat{f}_t \cos\left(\alpha_t^{(m,l)} - \gamma_t\right) \cos \beta_t^{(m,l)} \\ & + \hat{f}_r \left(\cos \gamma_r - \delta_t^{(l)} \sin \gamma_r \sin \alpha_t^{(m,l)}\right) \cos\left(\delta_t^{(l)} \beta_t^{(m,l)} + \delta_H/D\right) \end{aligned} \tag{2.71b}$$

$$\begin{aligned} f_{pq,nk}^{\text{SBR}} = {} & \hat{f}_r \cos\left(\alpha_r^{(n,k)} - \gamma_r\right) \cos \beta_r^{(n,k)} \\ & + \hat{f}_t \left(\cos \gamma_t + \delta_r^{(k)} \sin \gamma_t \sin \alpha_r^{(n,k)}\right) \cos\left(\delta_r^{(k)} \beta_r^{(n,k)} - \delta_H/D\right) \end{aligned} \tag{2.71c}$$

$$\begin{aligned} f_{pq,mlnk}^{\text{DB}} = {} & \hat{f}_t \cos\left(\alpha_t^{(m,l)} - \gamma_t\right) \cos \beta_t^{(m,l)} \\ & + \hat{f}_r \cos\left(\alpha_r^{(n,k)} - \gamma_r\right) \cos \beta_r^{(n,k)}, \end{aligned} \tag{2.71d}$$

where $\hat{f}_t$ and $\hat{f}_r$ are respectively the maximum Doppler shifts experienced at the transmitter and receiver and which are given as $\hat{f}_t = f_c \cdot v_t/c = v_t/\lambda$ and $\hat{f}_r = f_c \cdot v_r/c = v_r/\lambda$, with f_c being the carrier frequency and λ the wavelength. Furthemore, $\delta_t^{(l)} = R_t^{(l)}/D$, $\delta_r^{(k)} = R_r^{(k)}/D$, and $\delta_H = h_t - h_r$. Note that similarly to the SISO case, since the number of scatterers at either end is approaching infinity, the discrete random angles become continuous with given PDFs.

- **Path Lengths.** As per Figure 2.29, the path lengths of the respective waves are given by the following expressions:

$$d_{pq}^{LOS} = d_{pq} \tag{2.72a}$$

$$d_{pq,ml}^{SBT} = d_{p,ml} + d_{ml,q} \tag{2.72b}$$

$$d_{pq,nk}^{SBR} = d_{p,nk} + d_{nk,q} \tag{2.72c}$$

$$d_{pq,mlnk}^{DB} = d_{p,ml} + d_{mlnk} + d_{nk,q}. \tag{2.72d}$$

Again, two cases can be distinguished, that is D is on is not significantly larger than $\max(R_t, R_r)$. The latter case is typically encountered indoors and requires the exact expression for all distances to be used, which are given in closed form in [173]. The former case, that is $D \gg \max(R_t, R_r)$, allows the following simplifications to be applied to the individual path lengths [187]:

$$d_{pq} \approx D - (0.5L_t + 0.5 - p)\Delta d_t \cos\theta_t \cos\psi_t$$
$$- (0.5L_r + 0.5 - q)\Delta d_r \cos(\pi - \theta_r) \cos\psi_r \tag{2.73a}$$

$$d_{p,ml} \approx R_t^{(l)} - (0.5L_t + 0.5 - p)$$
$$\times \left(\delta_{t,x} \cos\alpha_t^{(m,l)} \cos\beta_t^{(m,l)} + \delta_{t,y} \sin\alpha_t^{(m,l)} \cos\beta_t^{(m,l)} + \delta_{t,z} \sin\beta_t^{(m,l)} \right) \tag{2.73b}$$

$$d_{ml,q} \approx D - (0.5L_r + 0.5 - q)\Delta d_r \cos\psi_r \left(\Delta_t^{(l)} \sin\theta_r \sin\alpha_t^{(m,l)} - \cos\theta_r \right) \tag{2.73c}$$

$$d_{p,nk} \approx D - (0.5L_t + 0.5 - p)\Delta d_t \cos\psi_t \left(\Delta_r^{(k)} \sin\theta_t \sin\alpha_r^{(n,k)} + \cos\theta_t \right) \tag{2.73d}$$

$$d_{nk,q} \approx R_r^{(k)} - (0.5L_r + 0.5 - q) \times (\delta_{r,x} \cos\alpha_r^{(n,k)} \cos\beta_r^{(n,k)}$$
$$+ \delta_{r,y} \sin\alpha_r^{(n,k)} \cos\beta_r^{(n,k)} + \delta_{r,z} \sin\beta_r^{(n,k)}) \tag{2.73e}$$

$$d_{mlnk} \approx D, \tag{2.73f}$$

where we have $\delta_{t,x} = \Delta d_t \cos\psi_t \cos\theta_t$, $\delta_{t,y} = \Delta d_t \cos\psi_t \sin\theta_t$, $\delta_{t,z} = \Delta d_t \sin\psi_t$, $\delta_{r,x} = \Delta d_r \cos\psi_r \cos\theta_r$, $\delta_{r,y} = \Delta d_r \cos\psi_r \sin\theta_r$, and $\delta_{r,z} = \Delta d_r \sin\psi_r$.

- **MPC Delays.** The delays of the MPCs can be inferred from the traversed distances and are given by the following expressions:

$$\tau_{pq}^{LOS} = \sqrt{D^2 + \delta_H^2}/c \tag{2.74a}$$

$$\tau_{pq,ml}^{SBT} = \left(D + R_t^{(l)} \left(1 - \cos\alpha_t^{(m,l)} \right) \right) / \cos\beta_t^{(m,l)}/c \tag{2.74b}$$

$$\tau_{pq,nk}^{SBR} = \left(D + R_r^{(k)} \left(1 + \cos\alpha_r^{(n,k)} \right) \right) / \cos\beta_r^{(n,k)}/c \tag{2.74c}$$

$$\tau_{pq,mlnk}^{DB} = \left(D + R_t^{(l)} \left(1 - \cos\alpha_t^{(m,l)} \right) \right) / \cos\beta_t^{(m,l)}$$
$$+ R_r^{(k)}(1 + \cos\alpha_r^{(n,k)})/\cos\beta_r^{(n,k)} \right) /c, \tag{2.74d}$$

where c is the speed of light.

We are now in the position to derive temporal, spatial and spectral key quantities related to the mobile-to-mobile MIMO fading channel.

2.3.5.2 Canonical Scenario

Following [187], the space-time-frequency correlation function (STF-CF) and its PSD can be calculated as follows:

- **Space–Time–Frequency Correlation Function.** Using the above quantities and following the same approach already outlined in Equation (2.39), the STF-CF of the complex fading channel between any transmit–receive antenna pair pq and any other pair $\tilde{p}\tilde{q}$ is given as follows:

$$
\begin{aligned}
R_{pq,\tilde{p}\tilde{q}}(\Delta d_{\mathrm{t}}, \Delta d_{\mathrm{r}}, \Delta t, \Delta f) = \quad & R^{\mathrm{LOS}}_{pq,\tilde{p}\tilde{q}}(\Delta d_{\mathrm{t}}, \Delta d_{\mathrm{r}}, \Delta t, \Delta f) \\
& + R^{\mathrm{SBT}}_{pq,\tilde{p}\tilde{q}}(\Delta d_{\mathrm{t}}, \Delta d_{\mathrm{r}}, \Delta t, \Delta f) \\
& + R^{\mathrm{SBR}}_{pq,\tilde{p}\tilde{q}}(\Delta d_{\mathrm{t}}, \Delta d_{\mathrm{r}}, \Delta t, \Delta f) \\
& + R^{\mathrm{DB}}_{pq,\tilde{p}\tilde{q}}(\Delta d_{\mathrm{t}}, \Delta d_{\mathrm{r}}, \Delta t, \Delta f),
\end{aligned}
\tag{2.75}
$$

where the respective correlation functions have been derived [187] and are given as follows:

$$
\begin{aligned}
R^{\mathrm{LOS}}_{pq,\tilde{p}\tilde{q}}(\Delta d_{\mathrm{t}}, \Delta d_{\mathrm{r}}, \Delta t, \Delta f) \approx \; & K/(1+K) \cdot e^{j2\pi/\lambda((p-\tilde{p})\delta_{t,x}-(q-\tilde{q})\delta_{r,x})} \\
& \times e^{j2\pi\Delta t\left(\hat{f}_{\mathrm{t}}\cos\gamma_{\mathrm{t}} - \hat{f}_{\mathrm{r}}\cos\gamma_{\mathrm{r}}\right)} \\
& \times e^{j2\pi/c\Delta f\sqrt{D^2+\delta_H^2}}
\end{aligned}
\tag{2.76a}
$$

$$
\begin{aligned}
R^{\mathrm{SBT}}_{pq,\tilde{p}\tilde{q}}(\Delta d_{\mathrm{t}}, \Delta d_{\mathrm{r}}, \Delta t, \Delta f) \approx \; & \frac{\cos\left(2\pi/\lambda\hat{\beta}_{\mathrm{t}}(p-\tilde{p})\delta_{t,z}\right)}{1-\left(\frac{4\hat{\beta}_{\mathrm{t}}(p-\tilde{p})\delta_{t,z}}{\lambda}\right)^2} \\
& \times e^{-j2\pi/\lambda(q-\tilde{q})\delta_{r,x}-j2\pi\Delta t\hat{f}_{\mathrm{r}}\cos\gamma_{\mathrm{r}}} \\
& \times \frac{\eta_{\mathrm{t}}/(1+K)}{I_0(\kappa_{\mathrm{t}})} \int_{R_{\mathrm{t}1}}^{R_{\mathrm{t}2}} (1-\gamma R_{\mathrm{t}}/D)\, e^{-j2\pi/c\Delta f(D+R_{\mathrm{t}})} \frac{2R_{\mathrm{t}}}{R_{\mathrm{t}2}^2 - R_{\mathrm{t}1}^2} \\
& \times I_0\left(\sqrt{x_{\mathrm{SBT}}^2 + y_{\mathrm{SBT}}^2}\right) d\,R_{\mathrm{t}}
\end{aligned}
\tag{2.76b}
$$

$$
\begin{aligned}
R^{\mathrm{SBR}}_{pq,\tilde{p}\tilde{q}}(\Delta d_{\mathrm{t}}, \Delta d_{\mathrm{r}}, \Delta t, \Delta f) \approx \; & \frac{\cos(2\pi/\lambda\hat{\beta}_{\mathrm{r}}(q-\tilde{q}\delta_{r,z}))}{1-\left(\frac{4\hat{\beta}_{\mathrm{r}}(q-\tilde{q})\delta_{r,z}}{\lambda}\right)^2} \\
& \times e^{-j2\pi/\lambda(p-\tilde{p})\delta_{t,x}-j2\pi\Delta t\hat{f}_{\mathrm{t}}\cos\gamma_{\mathrm{t}}} \\
& \times \frac{\eta_{\mathrm{r}}/(1+K)}{I_0(\kappa_{\mathrm{r}})} \int_{R_{\mathrm{r}1}}^{R_{\mathrm{r}2}} (1-\gamma R_{\mathrm{r}}/D)\, e^{-j2\pi/c\Delta f(D+R_{\mathrm{r}})} \frac{2R_{\mathrm{r}}}{R_{\mathrm{r}2}^2 - R_{\mathrm{r}1}^2} \\
& \times I_0\left(\sqrt{x_{\mathrm{SBR}}^2 + y_{\mathrm{SBR}}^2}\right) d\,R_{\mathrm{t}}
\end{aligned}
\tag{2.76c}
$$

$$
\begin{aligned}
R^{\mathrm{DB}}_{pq,\tilde{p}\tilde{q}}(\Delta d_{\mathrm{t}}, \Delta d_{\mathrm{r}}, \Delta t, \Delta f) \approx \; & A_{\mathrm{DB}} \\
& \times \int_{R_{\mathrm{t}1}}^{R_{\mathrm{t}2}} 2e^{-j2\pi/c\Delta f R_{\mathrm{t}}} R_{\mathrm{t}} I_0\left(\sqrt{x_{\mathrm{DB}}^2 + y_{\mathrm{DB}}^2}\right) d\,R_{\mathrm{t}} \\
& \times \int_{R_{\mathrm{r}1}}^{R_{\mathrm{r}2}} e^{-j2\pi/c\Delta f R_{\mathrm{r}}} R_{\mathrm{r}} I_0\left(\sqrt{w_{\mathrm{DB}}^2 + z_{\mathrm{DB}}^2}\right)\left(1-\gamma\frac{R_{\mathrm{r}}}{D}\right) d\,R_{\mathrm{r}} \\
& + A_{\mathrm{DB}} \int_{R_{\mathrm{r}1}}^{R_{\mathrm{r}2}} 2e^{-j2\pi/c\Delta f R_{\mathrm{r}}} R_{\mathrm{r}} I_0\left(\sqrt{w_{\mathrm{DB}}^2 + z_{\mathrm{DB}}^2}\right) d\,R_{\mathrm{r}} \\
& \times \int_{R_{\mathrm{t}1}}^{R_{\mathrm{t}2}} e^{-j2\pi/c\Delta f R_{\mathrm{t}}} R_{\mathrm{t}} I_0\left(\sqrt{x_{\mathrm{DB}}^2 + y_{\mathrm{DB}}^2}\right)\left(1-\gamma\frac{R_{\mathrm{t}}}{D}\right) d\,R_{\mathrm{t}}
\end{aligned}
\tag{2.76d}
$$

with the auxiliary variables being:

$$x_{SBT} \approx j2\pi/\lambda(\tilde{p}-p)\delta_{t,x} + j2\pi\Delta t\,\hat{f}_t\cos\gamma_t$$
$$+\,\kappa_t\cos\mu_t + j2\pi\Delta f\,R_t/c \tag{2.77a}$$

$$y_{SBT} \approx j2\pi((\tilde{p}-p)\delta_{t,y}/\lambda + (\tilde{q}-q)\delta_{r,y}\delta_t)$$
$$+\,j2\pi\Delta t\left(\hat{f}_t\sin\gamma_t + \hat{f}_r\delta_t\sin\gamma_r\right) + \kappa_t\sin\mu_t \tag{2.77b}$$

$$x_{SBR} \approx j2\pi/\lambda(\tilde{q}-q)\delta_{r,x} + j2\pi\Delta t\,\hat{f}_r\cos\gamma_r$$
$$+\,\kappa_r\cos\mu_r - j2\pi\Delta f\,R_r/c \tag{2.77c}$$

$$y_{SBR} \approx j2\pi((\tilde{q}-q)\delta_{r,y}/\lambda + (\tilde{p}-p)\delta_{t,y}\delta_r)$$
$$+\,j2\pi\Delta t\left(\hat{f}_r\sin\gamma_r + \hat{f}_t\delta_r\sin\gamma_t\right) + \kappa_r\sin\mu_r \tag{2.77d}$$

$$x_{DB} \approx j2\pi/\lambda(\tilde{p}-p)\delta_{t,x} + j2\pi\Delta t\,\hat{f}_t\cos\gamma_t$$
$$+\,\kappa_t\cos\mu_t + j2\pi\Delta f\,R_t/c \tag{2.77e}$$

$$y_{DB} \approx j2\pi/\lambda(\tilde{p}-p)\delta_{t,y} + j2\pi\Delta t\,\hat{f}_t\sin\gamma_t + \kappa_t\sin\mu_t \tag{2.77f}$$

$$z_{DB} \approx j2\pi/\lambda(\tilde{q}-q)\delta_{r,x} + j2\pi\Delta t\,\hat{f}_r\cos\gamma_r$$
$$+\,\kappa_r\cos\mu_r - j2\pi\Delta f\,R_r/c \tag{2.77g}$$

$$w_{DB} \approx j2\pi/\lambda(\tilde{q}-q)\delta_{r,y} + j2\pi\Delta t\,\hat{f}_r\sin\gamma_r + \kappa_r\sin\mu_r \tag{2.77h}$$

$$A_{DB} = \frac{\eta_{tr}/(1+K)}{I_0(\kappa_t)I_0(\kappa_r)}\frac{\cos(2\pi/\lambda\hat{\beta}_t(p-\tilde{p})\delta_{t,z})}{1-\left(\frac{4\hat{\beta}_t(p-\tilde{p})\delta_{t,z}}{\lambda}\right)^2}$$
$$\times\,\frac{\cos(2\pi/\lambda\hat{\beta}_r(q-\tilde{q})\delta_{r,z})}{1-\left(\frac{4\hat{\beta}_r(q-\tilde{q})\delta_{r,z}}{\lambda}\right)^2}\frac{e^{-j2\pi\Delta f\,D/c}}{(R_{t2}^2-R_{t1}^2)(R_{r2}^2-R_{r1}^2)}. \tag{2.77i}$$

Concerning the distance distribution of the cylindric clutter radii $R_t^{(l)}$ and $R_r^{(k)}$ at the transmitting and receiving sides, we have assumed a uniform distribution. For example, the transmitting side hence reads:

$$f(R_t) = \frac{2R_t}{R_{t2}^2-R_{t1}^2}. \tag{2.78}$$

For the azimuth distribution of the AAODs and AAOAs, again the von Mises distribution given in Equation (2.44) has been assumed at the transmitting and receiving side with mean μ_t and μ_r and parameter κ_t and κ_r, respectively. For the elevation distribution of the EAODs and EAOAs, uniform, cosine and Gaussian distributions have been used; we use the following distribution, which can typically be found in urban environments dominated by street canyons [187, 204, 205]:

$$f(\beta) = \begin{cases} \dfrac{\pi}{4|\hat{\beta}|}\cos\left(\dfrac{\pi}{2}\dfrac{\beta}{\hat{\beta}}\right) & \text{for } |\beta| \leq |\hat{\beta}| \leq \pi/2 \\ 0 & \text{otherwise} \end{cases} \tag{2.79}$$

where $\hat{\beta}$ is the maximum elevation angle and typically lies in the range of $0° \leq |\hat{\beta}| \leq 20°$; the maximum value for the transmitting and receiving antenna array is respectively denoted $\hat{\beta}_t$ and $\hat{\beta}_r$ in the above equations.

- **Doppler Power Spectrum.** The Doppler power spectrum is obtained by taking the Fourier transform of above STF-CF with respect to the time lag Δt and at $\Delta f = 0$, which essentially leads to the space-Doppler power spectral density. Applying the transform given in Equations (2.40) to (2.75) yields the following results [187]:

$$\Psi_{pq,\bar{p}\bar{q}}(\Delta d_t, \Delta d_r, f) = \Psi_{pq,\bar{p}\bar{q}}^{LOS}(\Delta d_t, \Delta d_r, f) + \Psi_{pq,\bar{p}\bar{q}}^{SBT}(\Delta d_t, \Delta d_r, f)$$

$$+ \Psi_{pq,\bar{p}\bar{q}}^{SBR}(\Delta d_t, \Delta d_r, f) + \Psi_{pq,\bar{p}\bar{q}}^{DB}(\Delta d_t, \Delta d_r, f), \qquad (2.80)$$

where the respective spectral densities have been derived [187] and are given as follows:

$$\Psi_{pq,\bar{p}\bar{q}}^{LOS}(\Delta d_t, \Delta d_r, f) = K/(1+K) \cdot e^{j2\pi/\lambda((\bar{p}-p)\delta_{t,x} - (\bar{q}-q)\delta_{r,x})}$$

$$\times \delta\left(f + \hat{f}_t \cos\gamma_t - \hat{f}_r \cos\gamma_r\right) \qquad (2.81a)$$

$$\Psi_{pq,\bar{p}\bar{q}}^{SBT}(\Delta d_t, \Delta d_r, f) = \frac{\eta_t/(1+K)}{I_0(\kappa_t)}$$

$$\times \frac{\exp\left\{-j2\pi p_{x,SBR} + j2\pi(f + \hat{f}_r \cos\gamma_r)A_{SBT}\right\}}{\pi \hat{f}_t \sqrt{1 - ((f + \hat{f}_r \cos\gamma_r)/\hat{f}_t)^2}}$$

$$\times \cosh\left((\kappa_t \sin(\mu_t - \gamma_t) + j2\pi p_{x,SBT}q_{y,SBT} - j2\pi p_{y,SBT}q_{x,SBT})\right.$$

$$\left. \times \sqrt{1 - ((f - \hat{f}_r \cos\gamma_r)/\hat{f}_t)^2}\right)$$

$$\times \frac{\cos(2\pi/\lambda\hat{\beta}_t(p - \bar{p})\delta_{t,z})}{1 - \left(\frac{4\hat{\beta}_t(p-\bar{p})\delta_{t,z}}{\lambda}\right)^2}$$

$$\times \frac{(3 - 2\gamma R_{t2}/D)R_{t2}^2 - (3 - 2\gamma R_{t1}/D)R_{t1}^2}{3(R_{t2}^2 - R_{t1}^2)} \qquad (2.81b)$$

$$\Psi_{pq,\bar{p}\bar{q}}^{SBR}(\Delta d_t, \Delta d_r, f) = \frac{\eta_r/(1+K)}{I_0(\kappa_r)}$$

$$\times \frac{\exp\left\{j2\pi p_{x,SBT} + j2\pi(f + \hat{f}_t \cos\gamma_t)A_{SBR}\right\}}{\pi \hat{f}_r \sqrt{1 - ((f + \hat{f}_t \cos\gamma_t)/\hat{f}_r)^2}}$$

$$\times \cosh\left((\kappa_r \sin(\mu_r - \gamma_r) + j2\pi p_{x,SBR}q_{y,SBR} - j2\pi p_{y,SBR}q_{x,SBR})\right.$$

$$\left. \times \sqrt{1 - ((f + \hat{f}_t \cos\gamma_t)/\hat{f}_r)^2}\right)$$

$$\times \frac{\cos(2\pi/\lambda\hat{\beta}_r(q - \tilde{q})\delta_{r,z})}{1 - \left(\frac{4\hat{\beta}_r(q-\tilde{q})\delta_{r,z}}{\lambda}\right)^2}$$

$$\times \frac{(3 - 2\gamma R_{r2}/D)R_{r2}^2 - (3 - 2\gamma R_{r1}/D)R_{r1}^2}{3(R_{r2}^2 - R_{r1}^2)} \qquad (2.81c)$$

$$\Psi_{pq,\bar{p}\bar{q}}^{DB}(\Delta d_t, \Delta d_r, f) = \frac{\eta_{tr}}{1+K}I_{DB}$$

$$\times \frac{\cos(2\pi/\lambda\hat{\beta}_t(p - \bar{p})\delta_{t,z})}{1 - \left(\frac{4\hat{\beta}_t(p-\bar{p})\delta_{t,z}}{\lambda}\right)^2} \frac{\cos(2\pi/\lambda\hat{\beta}_r(q - \tilde{q})\delta_{r,z})}{1 - \left(\frac{4\hat{\beta}_r(q-\tilde{q})\delta_{r,z}}{\lambda}\right)^2} \qquad (2.81d)$$

$$\times \frac{\exp\left\{j\left(2\pi p_{x,\mathrm{DB}}q_{x,\mathrm{DB}} + 2\pi p_{y,\mathrm{DB}}q_{y,\mathrm{DB}} - \kappa_{\mathrm{t}}\cos(\gamma_{\mathrm{t}} - \mu_{\mathrm{t}}))f/\hat{f}_{\mathrm{t}}\right\}}{I_0(\kappa_{\mathrm{t}})\pi \hat{f}_{\mathrm{t}}\sqrt{1 - (f/\hat{f}_{\mathrm{t}})^2}}$$

$$\times \cosh\left((\kappa_{\mathrm{t}}\sin(\mu_{\mathrm{t}} - \gamma_{\mathrm{t}}) + j2\pi p_{x,\mathrm{DB}}q_{y,\mathrm{DB}} - j2\pi p_{y,\mathrm{DB}}q_{x,\mathrm{DB}})\right.$$

$$\left.\times \sqrt{1 - (f/\hat{f}_{\mathrm{t}})^2}\right)$$

$\star$

$$\frac{\exp\left\{j\left(2\pi p_{z,\mathrm{DB}}q_{z,\mathrm{DB}} + 2\pi p_{w,\mathrm{DB}}q_{w,\mathrm{DB}} - \kappa_{\mathrm{r}}\cos(\mu_{\mathrm{r}} - \gamma_{\mathrm{r}}))f/\hat{f}_{\mathrm{r}}\right\}}{I_0(\kappa_{\mathrm{r}})\pi \hat{f}_{\mathrm{r}}\sqrt{1 - (f/\hat{f}_{\mathrm{r}})^2}}$$

$$\times \cosh((\kappa_{\mathrm{r}}\sin(\mu_{\mathrm{r}} - \gamma_{\mathrm{r}}) + j2\pi p_{z,\mathrm{DB}}q_{w,\mathrm{DB}} - j2\pi p_{w,\mathrm{DB}}q_{z,\mathrm{DB}})$$

$$\times \sqrt{1 - (f/\hat{f}_{\mathrm{r}})^2}),$$

where $\star$ denotes the convolution w.r.t. the frequency f. The auxiliary variables are given below:

$$p_{x,\mathrm{SBT}} = (\tilde{p} - p)(\delta_{t,x}/\lambda) \tag{2.82a}$$

$$q_{x,\mathrm{SBT}} = \cos\gamma_{\mathrm{t}} \tag{2.82b}$$

$$p_{y,\mathrm{SBT}} = ((\tilde{p} - p)\delta_{t,y} + (\tilde{q} - q)\delta_{r,y}(R_{t1} + 0.5(R_{t2} - R_{t1})))/D)/\lambda \tag{2.82c}$$

$$q_{y,\mathrm{SBT}} \approx \sin\gamma_{\mathrm{t}} \tag{2.82d}$$

$$A_{\mathrm{SBT}} = \frac{2\pi p_{x,\mathrm{SBT}}q_{x,\mathrm{SBT}} + 2\pi p_{y,\mathrm{SBT}}q_{y,\mathrm{SBT}} - j\kappa_{\mathrm{t}}\cos(\gamma_{\mathrm{t}} - \mu_{\mathrm{t}})}{2\pi \hat{f}_{\mathrm{t}}} \tag{2.82e}$$

$$p_{x,\mathrm{SBR}} = (\tilde{q} - q)(\delta_{r,x}/\lambda) \tag{2.82f}$$

$$q_{x,\mathrm{SBR}} = \cos\gamma_{\mathrm{r}} \tag{2.82g}$$

$$p_{y,\mathrm{SBR}} = ((q - \tilde{q})\delta_{r,y} + (p - \tilde{p})\delta_{t,y}(R_{r1} + 0.5(R_{r2} - R_{r1})))/D)/\lambda \tag{2.82h}$$

$$q_{y,\mathrm{SBR}} \approx \sin\gamma_{\mathrm{r}} \tag{2.82i}$$

$$A_{\mathrm{SBR}} = \frac{2\pi p_{x,\mathrm{SBR}}q_{x,\mathrm{SBR}} + 2\pi p_{y,\mathrm{SBR}}q_{y,\mathrm{SBR}} - j\kappa_{\mathrm{r}}\cos(\gamma_{\mathrm{r}} - \mu_{\mathrm{r}})}{2\pi \hat{f}_{\mathrm{r}}} \tag{2.82j}$$

$$p_{x,\mathrm{DB}} = (p - \tilde{p})(\delta_{t,x}/\lambda) \tag{2.82k}$$

$$q_{x,\mathrm{DB}} = \cos\gamma_{\mathrm{t}} \tag{2.82l}$$

$$p_{y,\mathrm{DB}} = (p - \tilde{p})(\delta_{t,y}/\lambda) \tag{2.82m}$$

$$q_{y,\mathrm{DB}} = \sin\gamma_{\mathrm{t}} \tag{2.82n}$$

$$p_{z,\mathrm{DB}} = (q - \tilde{q})(\delta_{r,x}/\lambda) \tag{2.82o}$$

$$q_{z,\mathrm{DB}} = \cos\gamma_{\mathrm{r}} \tag{2.82p}$$

$$p_{w,\mathrm{DB}} = (q - \tilde{q})(\delta_{r,y}/\lambda) \tag{2.82q}$$

$$q_{w,\mathrm{DB}} = \sin\gamma_{\mathrm{r}} \tag{2.82r}$$

$$I_{\mathrm{DB}} = (R_{t2}^2 - R_{t1}^2)(0.5R_{r2}^2 - \gamma R_{r2}^2/(3D) - 0.5R_{r1}^2 + \gamma R_{r1}^3/(3D))$$

$$+ (R_{r2}^2 - R_{r1}^2)(0.5R_{t2}^2 - \gamma R_{t2}^2/(3D) - 0.5R_{t1}^2 + \gamma R_{t1}^3/(3D)). \tag{2.82s}$$

A closed-form expression for the convolution of the double bounced term is unfortunately not available at the time of writing of this book; however, any numerical approach suffices, as involved terms are fairly simple.

- **Level Crossing Rate.** The level crossing rate of the channel envelope can be calculated using the insights from [186], which are valid in the presence of a LOS component. The final expression reads [202]:

$$N(a_{\text{thr}}) = \frac{2a_{\text{thr}}\sqrt{K+1}}{\pi^{3/2}} \sqrt{\frac{b_2}{b_0} - \frac{b_1^2}{b_0^2}} e^{-K-(K+1)a_{\text{thr}}^2}$$

$$\times \int_0^{\pi/2} \cosh\left(2\sqrt{K(K+1)}a_{\text{thr}}\cos\theta\right)$$

$$\times \left(e^{-(\chi\sin\theta)^2} + \sqrt{\pi}\chi\sin\theta\,\text{erf}(\chi\sin\theta)\right)d\theta, \tag{2.83}$$

where

$$\chi = \sqrt{Kb_1^2/(b_0 b_2 - b_1^2)} \tag{2.84a}$$

$$b_n = b_n^{\text{SBT}} + b_n^{\text{SBR}} + b_n^{\text{DB}}, \quad n \in \{0, 1, 2\} \tag{2.84b}$$

$$b_0^{\text{SBT}} = \eta_t/(K+1)/2 \tag{2.84c}$$

$$b_0^{\text{SBR}} = \eta_r/(K+1)/2 \tag{2.84d}$$

$$b_0^{\text{DB}} = \eta_{tr}/(K+1)/2 \tag{2.84e}$$

$$b_1^{\text{SBT}} = 2\pi b_0^{\text{SBT}} \left\{ \hat{f}_t \cos(\mu_t - \gamma_t) \frac{I_1(\kappa_t)}{I_0(\kappa_t)} \frac{\pi^2 \cos\hat{\beta}_t}{\pi^2 - 4\hat{\beta}_t^2} \right.$$

$$+ \int_{R_{t1}}^{R_{t2}} \left[\frac{R_t}{D}\sin\gamma_r\sin\mu_t \frac{I_1(\kappa_t)}{I_0(\kappa_t)} + \cos\gamma_r \right] \hat{f}_r \frac{2R_t}{R_{t2}^2 - R_{t1}^2}$$

$$\left. \times \frac{\pi^2[\cos(R_t/D\hat{\beta}_t + \delta_h/D) + \cos(R_t/D\hat{\beta}_t - \delta_h/D)]}{2[\pi^2 - 4(R_t/D)^2\hat{\beta}_t^2]} dR_t \right\} \tag{2.84f}$$

$$b_1^{\text{SBR}} = 2\pi b_0^{\text{SBR}} \left\{ \hat{f}_r \cos(\mu_r - \gamma_r) \frac{I_1(\kappa_r)}{I_0(\kappa_r)} \frac{\pi^2 \cos\hat{\beta}_r}{\pi^2 - 4\hat{\beta}_r^2} \right.$$

$$+ \int_{R_{r1}}^{R_{r2}} \left[\frac{R_r}{D}\sin\gamma_t\sin\mu_r \frac{I_1(\kappa_r)}{I_0(\kappa_r)} + \cos\gamma_t \right] \hat{f}_t \frac{2R_r}{R_{r2}^2 - R_{r1}^2}$$

$$\left. \times \frac{\pi^2[\cos(R_r/D\hat{\beta}_r + \delta_h/D) + \cos(R_r/D\hat{\beta}_r - \delta_h/D)]}{2[\pi^2 - 4(R_r/D)^2\hat{\beta}_r^2]} dR_r \right\} \tag{2.84g}$$

$$b_1^{\text{DB}} = 2\pi b_0^{\text{DB}} \left\{ \hat{f}_t \cos(\mu_t - \gamma_t) \frac{I_1(\kappa_t)}{I_0(\kappa_t)} \frac{\pi^2 \cos\hat{\beta}_t}{\pi^2 - 4\hat{\beta}_t^2} \right.$$

$$\left. + \hat{f}_r \cos(\mu_r - \gamma_r) \frac{I_1(\kappa_r)}{I_0(\kappa_r)} \frac{\pi^2 \cos\hat{\beta}_r}{\pi^2 - 4\hat{\beta}_r^2} \right\} \tag{2.84h}$$

$$b_2^{\text{SBT}} = (2\pi)^2 b_0^{\text{SBT}} \left\{ \hat{f}_t^2 \frac{\pi^2 \cos(2\hat{\beta}_t) + \pi^2 - 16\hat{\beta}_t^2}{2(\pi^2 - 16\hat{\beta}_t^2)} \right.$$

$$\times \frac{1 + \cos(2(\mu_t - \gamma_t))I_2(\kappa_t)}{2I_0(\kappa_t)}$$

$$+ \int_{R_{t1}}^{R_{t2}} \frac{4\hat{\beta}_t^2 + 8R_t/D\hat{\beta}_t^2 - \pi^2 + 4(R_t/D)^2\hat{\beta}_t^2}{4F_{SBT}}$$

$$\times \left[\cos((R_t/D - 1)\hat{\beta}_t + \delta_h/D) + \cos((R_t/D + 1)\hat{\beta}_t + \delta_h/D) \right.$$

$$\left. + \cos((R_t/D - 1)\hat{\beta}_t - \delta_h/D) + \cos((R_t/D + 1)\hat{\beta}_t - \delta_h/D) \right]$$

$$\times \left[\hat{f}_t \hat{f}_r R_t/D \sin \gamma_r \left(\frac{I_2(\kappa_t)}{I_0(\kappa_t)} \cos(2\mu_t - \gamma_t) \right. \right.$$

$$\left. \left. - \cos \gamma_t \right) + 2\hat{f}_t f_{r,max} \cos \gamma_r \right]$$

$$\left[\frac{\cos[2(R_t/D\hat{\beta}_t) + \delta_h/D]\pi^2 + 2\pi^2 - 32(R_t/D)^2\hat{\beta}_t^2}{4[\pi^2 - 16(R_t/D)^2\hat{\beta}_t]} \right.$$

$$\left. + \frac{\cos[2(R_t/D\hat{\beta}_t) - \delta_h/D]\pi^2}{4[\pi^2 - 16(R_t/D)^2\hat{\beta}_t]} \right] \left[\hat{f}_r^2 \cos \gamma_r^2 \right.$$

$$+ \hat{f}_r^2 \left(\frac{R_t}{D} \right)^2 \sin \gamma_r^2 \frac{1 - \cos(2\mu_t) I_2(\kappa_t)}{2I_0(\kappa_t)}$$

$$\left. + \hat{f}_r^2 \frac{R_t}{D} \sin(2\gamma_r) \sin \mu_t \frac{I_1(\kappa_t)}{I_0(\kappa_t)} \right] \frac{2R_t}{R_{t2}^2 - R_{t1}^2} d R_t \right\} \quad (2.84i)$$

$$b_2^{SBR} = (2\pi)^2 b_0^{SBR} \left\{ \hat{f}_r^2 \frac{\pi^2 \cos(2\hat{\beta}_r) + \pi^2 - 16\hat{\beta}_r^2}{2(\pi^2 - 16\hat{\beta}_r^2)} \right.$$

$$\times \frac{1 + \cos(2(\mu_r - \gamma_t)) I_2(\kappa_r)}{2I_0(\kappa_r)}$$

$$+ \int_{R_{r1}}^{R_{r2}} \frac{4\hat{\beta}_r^2 + 8R_r/D\hat{\beta}_r^2 - \pi^2 + 4(R_r/D)^2\hat{\beta}_r^2}{4F_{SBR}}$$

$$\times \left[\cos((R_r/D - 1)\hat{\beta}_r + \delta_h/D) + \cos((R_r/D + 1)\hat{\beta}_r + \delta_h/D) \right.$$

$$\left. + \cos((R_r/D - 1)\hat{\beta}_r - \delta_h/D) + \cos((R_r/D + 1)\hat{\beta}_r - \delta_h/D) \right]$$

$$\times \left[\hat{f}_r \hat{f}_t R_r/D \sin \gamma_t \left(\frac{I_2(\kappa_r)}{I_0(\kappa_r)} \cos(2\mu_r - \gamma_r) \right. \right.$$

$$\left. \left. - \cos \gamma_r \right) + 2\hat{f}_r \hat{f}_t \cos \gamma_t \right]$$

$$\left[\frac{\cos[2(R_r/D\hat{\beta}_r) + \delta_h/D]\pi^2 + 2\pi^2 - 32(R_r/D)^2\hat{\beta}_r^2}{4[\pi^2 - 16(R_r/D)^2\hat{\beta}_r]} \right.$$

$$\left. + \frac{\cos[2(R_r/D\hat{\beta}_r) - \delta_h/D]\pi^2}{4[\pi^2 - 16(R_r/D)^2\hat{\beta}_r]} \right] \left[\hat{f}_t^2 \cos \gamma_t^2 \right.$$

$$+ \hat{f}_t^2 \left(\frac{R_r}{D} \right)^2 \sin \gamma_t^2 \frac{1 - \cos(2\mu_r) I_2(\kappa_r)}{2I_0(\kappa_r)}$$

$$\left. + \hat{f}_t^2 \frac{R_r}{D} \sin(2\gamma_t) \sin \mu_r \frac{I_1(\kappa_r)}{I_0(\kappa_r)} \right] \frac{2R_r}{R_{r2}^2 - R_{r1}^2} d R_r \right\}$$

$$(2.84j)$$

$$b_2^{\mathrm{DB}} = (2\pi)^2 b_0^{\mathrm{DB}} \left\{ \hat{f}_t^2 \left[\frac{\pi^2 \cos(2\hat{\beta}_t)}{2(\pi^2 - 16\hat{\beta}_t^2)} + \frac{1}{2} \right] \right.$$

$$\times \frac{1 + \cos(2(\mu_t - \gamma_t)) I_2(\kappa_t)}{2 I_0(\kappa_t)} + 2\hat{f}_t f_{r,\max} \frac{\pi^2 \cos \hat{\beta}_t}{\pi^2 - 4\hat{\beta}_t^2}$$

$$\times \frac{\pi^2 \cos \hat{\beta}_r}{\pi^2 - 4\hat{\beta}_r^2} \cos(\mu_t - \gamma_t) \frac{I_1(\kappa_t)}{I_0(\kappa_t)} \cos(\mu_r - \gamma_r) \frac{I_1(\kappa_r)}{I_0(\kappa_r)}$$

$$\left. + \hat{f}_r^2 \left[\frac{\pi^2 \cos(2\hat{\beta}_r)}{2(\pi^2 - 16\hat{\beta}_r^2)} + \frac{1}{2} \right] \frac{1 + \cos(2(\mu_r - \gamma_r)) I_2(\kappa_r)}{2 I_0(\kappa_r)} \right\} \tag{2.84k}$$

$$F_{\mathrm{SBT}} = 8\hat{\beta}_t^2 \pi^2 - 16\hat{\beta}_t^4 - \pi^4 - 16(R_t/D)^4 \hat{\beta}_t^4$$
$$+ 8\pi^2 (R_t/D)^2 \hat{\beta}_t^2 + 32(R_t/D)^2 \hat{\beta}_t^4 \tag{2.84l}$$

$$F_{\mathrm{SBR}} = 8\hat{\beta}_r^2 \pi^2 - 16\hat{\beta}_r^4 - \pi^4 - 16(R_r/D)^4 \hat{\beta}_r^4$$
$$+ 8\pi^2 (R_r/D)^2 \hat{\beta}_r^2 + 32(R_r/D)^2 \hat{\beta}_r^4 \tag{2.84m}$$

Again, the integrals need to be evaluated numerically.

- **Average Fade Duration.** Since the channel envelope is Ricean distributed, the average fade duration is again easily derived as:

$$T(a_{\mathrm{thr}}) = \frac{1 - Q\left(\sqrt{2K}, \sqrt{2(K+1)} a_{\mathrm{thr}}\right)}{N(a_{\mathrm{thr}})}, \tag{2.85}$$

where $Q(\cdot, \cdot)$ is the Marcum Q-function.

Using the above generic expressions, we will now investigate the behavior of the correlation function and its spectral density for a few parameterizations.

2.3.5.3   Case Studies

A detailed case study is omitted here since the authors of [202] have discussed and analyzed a large set of parameterizations. They have effectively determined that the above equations reduce to simpler canonical configurations, such as the narrowband case, Jakes scenario, etc. They have also shown that the model fits well the available wideband measurements taken in Atlanta, Georgia.

2.3.6   Simulating Regenerative Fading Channels

Simulating the regenerative mobile-to-mobile relay channel is crucial in characterizing the performance of such a system by means of link layer simulations. Ideally, any simulation should mimic the statistical behavior of the channel as precisely as possible. Since the above calculations assumed an infinite number of scatterers, an identical reproduction of the channel behavior is prohibitive in terms of computational complexity. Any published channel simulation approaches for example, [157, 169, 172, 173, 187, 189–191, 198, 206–215], hence trade the precision of reproducing the channel's characteristics with the associated numerical complexity. A handy way of quantifying such tradeoff is the modeling error by means of the variance of the model output w.r.t. the channel with infinite clutter. Applied to the modeled correlation function $\hat{R}(N)$ with N scatterers and the theoretical correlation function R with infinite scatterers, this quantifier would read $\mathrm{VAR}[R] = \mathrm{E}[|\hat{R}(N) - R|^2]$, where a model for the same variance with minimum N or for the same N and minimum variance is considered to be superior to other models [208]. Note that the expression for the variance often eludes a closed form formulation.

We will now review a few modeling approaches and then show how to simulate the narrow and wideband MIMO relaying channel.

2.3.6.1 Typical Modeling Approaches

Far from being exhaustive, the following approaches are available when simulating the mobile-to-mobile fading channel:

- **Akki and Haber's Simulation Model.** The original model proposed in [171] assumes a sum of N propagation paths, each of which is distorted by an appropriately calculated Doppler shift at the transmitting and receiving end. A natural way to model this channel is hence to generate random AODs, AOAs and phases. A drawback, as quantified in [208], is that the modeling error is large compared with other models introduced below.

- **Discrete Line Spectrum Method.** This method generates the channel by discretizing the power spectral density prior to transforming it back into the time domain. An approximation to the spectrum sampling method [143] has been introduced [206] and reads:

$$\hat{h}(t) = \frac{1}{A} \sum_{n=1}^{N} \sqrt{\Psi(\hat{f}_n)} e^{-j(2\pi \hat{f}_n t + \phi_n)}, \tag{2.86}$$

where

$$\Psi(\hat{f}_n) = \sum_{n=1}^{N} \left(\int_{F_n}^{F_{n+1}} \Psi(f) d f \right) \delta(f - \hat{f}_n) \tag{2.87a}$$

$$\hat{f}_n = \frac{\int_{F_n}^{F_{n+1}} f \Psi(f) d f}{\int_{F_n}^{F_{n+1}} \Psi(f) d f} \tag{2.87b}$$

The constant A is a normalization constant that can be used to normalize the power of $\hat{h}(t)$ to unity. The frequency bins F_n are equispaced in frequency with $F_1 = -(\hat{f}_t + \hat{f}_r)$ and $F_{N+1} = (\hat{f}_t + \hat{f}_r)$. The center of mass frequencies $\hat{f}_n$ and the corresponding masses $\Psi(\hat{f}_n)$ are obtained by numerically integrating over the sampled Doppler spectra given in previous sections. These prior derived PSDs are hence of paramount use if used for this modeling approach. Drawbacks of this model, as quantified in [208], are that the correlation functions vary with simulation trials, the I and Q components are statistically correlated, the autocorrelation function exhibits periodicities, the need for numerical integration, etc.

- **Deterministic Sum-of-Sinusoids Method.** The deterministic sum-of-sinusoids (SOS) method for mobile-to-mobile channels is essentially an extension to the method of exact doppler spreads (MEDS) [216] and has been introduced [172] and enhanced [for example, by [173]. It is sometimes referred to as modified MEDS (MMEDS) or enhanced MEDS (EMEDS). The idea is to choose a suitable set of discrete and deterministic AOAs and AODs in the above equations, such as Equation (2.37), which, together with a random realization of the phases ϕ, yields desired statistical properties. It has been shown that the choice proposed in [173] is superior to that in [172] in that correlation properties match the ideal channel for longer time lags.

- **Random Sum-of-Sinusoids Method.** This method has been proposed [in for example, 173] and extends above MMEDS in that amplitudes, phases and Doppler frequencies are also random (but obeying some rules). We will discuss this approach below as it yields superior statistical properties at an acceptable level of complexity.

- **Modified Method of Equal Areas.** The method of equal areas (MEA) was first proposed in [216] and later extended to the modified MEA (MMEA) [217] to cater for any distribution of AODs and AOAs, such as the von Mises distribution given in Equation (2.44). This method also advocates the use of discrete and deterministic AODs and AOAs but which are derived from boundary conditions requiring numerical integration and root-finding techniques [172].

- **L_p-Norm Method.** The L_p-norm method (LPNM) has been treated in [134]. It has also advantageously been used in the case of general distributions for AODs and AOAs. This method also advocates the use of discrete and deterministic AODs and AOAs but which are now derived directly from some error-minimizing criteria between the simulated and theoretical correlation functions. Whilst numerical minimization is involved, the produced waveform naturally yields superior statistical properties.

All these methods have their pros and cons. Some of them also facilitate the generation of the lognormally distributed shadowing process with required correlation functions, as outlined in [169]. Subsequently, however, we will focus on the fading channel only and we will discuss both deterministic and statistical SOS methods in the context of MIMO narrow and wideband regenerative relay channels.

2.3.6.2 MIMO Narrowband Relay Channels

The narrowband MIMO channel introduced in Section 2.3.4 results from a superposition of a LOS component with single bounced and double bounced waves in a 2D scattering environment. We will further assume that the scatterers are isotropically distributed around transmitter and receiver. With this in mind, we can write the complex channel realization of the simulated channel as follows [173]:

$$h_{pq}(t) = h_{pq}^{(I)}(t) + h_{pq}^{(Q)}(t) \tag{2.88}$$

$$h_{pq}^{(I)}(t) = \sum_{m=1}^{M} \frac{2P_{\mathrm{t}}}{\sqrt{M}} \cos\left[\beta_m - K_q \cos\theta_{\mathrm{r}} - 2\pi t \hat{f}_{\mathrm{r}} \cos\gamma_{\mathrm{r}}\right]$$

$$\times \cos\left[K_p \cos(\theta_{\mathrm{t}} - \alpha_{\mathrm{t}}^{(m)}) + 2\pi t \hat{f}_{\mathrm{r}}\right]$$

$$\times \cos\left(\alpha_{\mathrm{t}}^{(m)} - \gamma_{\mathrm{t}}\right) + K_q \Delta_{\mathrm{t}} \sin\theta_{\mathrm{r}} \sin\alpha_{\mathrm{t}}^{(m)} + 2\pi t \hat{f}_{\mathrm{r}}\Delta_{\mathrm{t}} \sin\gamma_{\mathrm{r}} \sin\alpha_{\mathrm{t}}^{(m)} + \phi_m\right]$$

$$+ \sum_{n=1}^{N} \frac{2P_{\mathrm{r}}}{\sqrt{N}} \cos\left[\beta_n + K_p \cos\theta_{\mathrm{t}} + 2\pi t \hat{f}_{\mathrm{t}} \cos\gamma_{\mathrm{t}}\right]$$

$$\times \cos\left[K_q \cos(\alpha_{\mathrm{r}}^{(n)} - \theta_{\mathrm{r}}) + 2\pi t \hat{f}_{\mathrm{r}}\right]$$

$$\times \cos(\alpha_{\mathrm{r}}^{(n)} - \gamma_{\mathrm{r}}) + K_p \Delta_{\mathrm{r}} \sin\theta_{\mathrm{t}} \sin\alpha_{\mathrm{r}}^{(n)}$$

$$+ 2\pi t \hat{f}_{\mathrm{t}}\Delta_{\mathrm{r}} \sin\gamma_{\mathrm{t}} \sin\alpha_{\mathrm{r}}^{(n)} + \phi_n\right]$$

$$+ \sum_{m,n=1}^{M,N} \frac{2P_{\mathrm{tr}}}{\sqrt{MN}} \cos\left[K_p \cos(\theta_{\mathrm{t}} - \alpha_{\mathrm{t}}^{(m)}) + 2\pi t \hat{f}_{\mathrm{r}}\right] \cos(\alpha_{\mathrm{t}}^{(m)} - \gamma_{\mathrm{t}})$$

$$\times \cos\left[K_q \cos(\alpha_{\mathrm{r}}^{(n)} - \theta_{\mathrm{r}}) + 2\pi t \hat{f}_{\mathrm{r}} \cos(\alpha_{\mathrm{r}}^{(n)} - \gamma_{\mathrm{r}}) + \phi_{mn}\right]$$

$$P_{\mathrm{LOS}} \cos\left[K_p \cos\theta_{\mathrm{t}} + K_q \cos(\alpha_{Rq}^{\mathrm{LOS}}\theta_{\mathrm{r}}) + 2\pi t (f_{\mathrm{t}}^{\mathrm{LOS}} + f_{\mathrm{r}}^{\mathrm{LOS}})\right] \tag{2.89}$$

$$h_{pq}^{(Q)}(t) = \sum_{m=1}^{M} \frac{2P_{\mathrm{t}}}{\sqrt{M}} \sin\left[\beta_m - K_q \cos\theta_{\mathrm{r}} - 2\pi t \hat{f}_{\mathrm{r}} \cos\gamma_{\mathrm{r}}\right]$$

$$\times \cos\left[K_p \sin(\theta_{\mathrm{t}} - \alpha_{\mathrm{t}}^{(m)}) + 2\pi t \hat{f}_{\mathrm{r}}\right]$$

$$\times \sin(\alpha_t^{(m)} - \gamma_t) + K_q \Delta_t \sin \theta_r \cos \alpha_t^{(m)}$$

$$+ 2\pi t \hat{f}_r \Delta_t \sin \gamma_r \cos \alpha_t^{(m)} + \phi_m \Big]$$

$$+ \sum_{n=1}^{N} \frac{2P_r}{\sqrt{N}} \sin \Big[\beta_n + K_p \cos \theta_t + 2\pi t \hat{f}_t \cos \gamma_t \Big]$$

$$\times \cos \Big[K_q \sin(\alpha_r^{(n)} - \theta_r) + 2\pi t \hat{f}_r$$

$$\times \sin(\alpha_r^{(n)} - \gamma_r) + K_p \Delta_r \sin \theta_t \cos \alpha_r^{(n)}$$

$$+ 2\pi t \hat{f}_t \Delta_r \sin \gamma_t \cos \alpha_r^{(n)} + \phi_n \Big]$$

$$+ \sum_{m,n=1}^{M,N} \frac{2P_{\text{tr}}}{\sqrt{MN}} \sin \Big[K_p \cos(\theta_t - \alpha_t^{(m)}) + 2\pi t \hat{f}_t \Big] \cos(\alpha_t^{(m)} - \gamma_t)$$

$$\times \sin \Big[K_q \sin(\alpha_r^{(n)} - \theta_r) + 2\pi t \hat{f}_r \sin(\alpha_r^{(n)} - \gamma_r) + \phi_{mn} \Big]$$

$$P_{\text{LOS}} \sin \Big[K_p \cos \theta_t + K_q \cos(\alpha_{Rq}^{\text{LOS}} \theta_r) + 2\pi t (f_t^{\text{LOS}} + f_r^{\text{LOS}}) \Big]. \tag{2.90}$$

Here, $h_{pq}^{(I)}(t)$ and $h_{pq}^{(Q)}(t)$ are the in-phase and quadrature components, respectively. Then, β_m and β_n are some random path gains to be determined below. Furthermore,

$$P_t = \sqrt{\frac{\eta_t \Omega_{pq}}{(K_{pq} + 1)}} \tag{2.91}$$

$$P_r = \sqrt{\frac{\eta_r \Omega_{pq}}{(K_{pq} + 1)}} \tag{2.92}$$

$$P_{\text{tr}} = \sqrt{\frac{\eta_{\text{tr}} \Omega_{pq}}{(K_{pq} + 1)}} \tag{2.93}$$

$$P_{\text{LOS}} = \sqrt{\frac{K_{pq} \Omega_{pq}}{(K_{pq} + 1)}} \tag{2.94}$$

$$K_p = 2\pi (0.5L_t + 0.5 - p)\Delta d_t / \lambda \tag{2.95}$$

$$K_q = 2\pi (0.5L_r + 0.5 - q)\Delta d_r / \lambda \tag{2.96}$$

$$f_t^{\text{LOS}} = \hat{f}_t \cos(\pi - \alpha_{Rq}^{\text{LOS}} - \gamma_t) \tag{2.97}$$

$$f_r^{\text{LOS}} = \hat{f}_r \cos(\alpha_{Rq}^{\text{LOS}} - \gamma_r) \tag{2.98}$$

Two modeling methods are now considered that trade precision with simulation speed and complexity:

- **Deterministic Sum-of-Sinusoids Method.** The deterministic SOS needs only a single simulation trial to obtain a channel with desired statistical properties. Here, only the phases ϕ_m, ϕ_n and ϕ_{mn} are generated as independent uniformly distributed random variables in $[-\pi, \pi)$. The theoretically random AODs, AOAs and path gains are deterministically fixed to [173]:

$$\alpha_t^{(m)} = \frac{\pi}{M}(m - 0.5) + \gamma_t \tag{2.99}$$

$$\alpha_r^{(n)} = \frac{2\pi}{N}(n - 0.5) + \gamma_r \tag{2.100}$$

$$\beta_m = \frac{\pi(m - 0.5)}{M} \tag{2.101}$$

$$\beta_n = \frac{2\pi(n - 0.5)}{N}, \tag{2.102}$$

for $m = 1, \ldots, M$ and $n = 1, \ldots, N$. It can be shown [173] that even such deterministic realization of above parameters yields the desired statistical properties for $M, N \to \infty$.

- **Random Sum-of-Sinusoids Method.** The random SOS yields channel realizations with statistical properties that vary from trial to trial; however, they converge to the desired statistical properties when averaged over a sufficient channel ensemble. The advantage over the deterministic SOS is that it reflects better the random arrangements of clutter in real-world scenarios. The AODs and AOAs are here obtained as follows [173]:

$$\alpha_t^{(m)} = 0.5 \left(\frac{2\pi m}{M} + \frac{\psi - \pi}{M} \right) \tag{2.103}$$

$$\alpha_r^{(n)} = \frac{2\pi n}{N} + \frac{\phi - \pi}{N}, \tag{2.104}$$

for $m = 1, \ldots, M$ and $n = 1, \ldots, N$. The phases ϕ_m, ϕ_n and ϕ_{mn}, the path gains β_m and β_n, and the parameters ψ and θ are all independent random variables uniformly distributed in $[0, 2\pi)$.

The complexity and performance of these models and their superiority have been demonstrated [173].

2.3.6.3 MIMO Wideband Relay Channels

To finalize the modeling section, we follow [187] to show how the wideband MIMO channel introduced in Section 2.3.5 can be modeled. The model assumes that the received wave results from a superposition of a LOS component with single bounced and double bounced waves in a 3D scattering environment. In contrast to the previous section, we will here assume that the scatterers are distributed around transmitter and receiver following the von Mises distribution introduced in Equation (2.44). With this in mind, we can write the complex channel realization of the simulated channel as follows [187]:

$$h_{pq}(t, \tau) = \mathcal{F}_f^{-1}\left\{T_{pq}(t, f)\right\} \tag{2.105a}$$

$$T_{pq}(t, f) = T_{pq}^{(I)}(t, f) + T_{pq}^{(Q)}(t, f) \tag{2.105b}$$

$$
\begin{aligned}
T_{pq}^{(I)}(t, f) = {} & \frac{\rho_{SBT}}{\sqrt{M}} \sum_{l,m,i=1}^{L,M_A^{(l)},M_E^{(l)}} \left(1 - \frac{\gamma}{2}\frac{R_t^{(l)}}{D}\right) \cos\left[K_p D_t^{SBT}\right. \\
& + K_q D_r^{SBT} + 2\pi t \hat{f}_t \cos(\alpha_t^{(m,l)} - \gamma_t) \cos\beta_t^{(i,l)} + \phi_{m,i,l} \\
& + 2\pi t \hat{f}_r \left(\cos\gamma_r - \Delta_t^{(l)} \sin\gamma_r \sin\alpha_t^{(m,l)}\right) \cos\left(\Delta_t^{(l)} \beta_t^{(i,l)}\right) \\
& \left. - \frac{2\pi}{c} f \left(D + R_t^{(l)}\left(1 - \cos\alpha_t^{(m,l)}\right)\right)\right] \\
& + \frac{\rho_{SBR}}{\sqrt{N}} \sum_{k,n,g=1}^{F,N_A^{(k)},N_E^{(k)}} \left(1 - \frac{\gamma}{2}\frac{R_r^{(k)}}{D}\right) \cos\left[K_p D_t^{SBR}\right. \\
& + K_q D_r^{SBR} + 2\pi t \hat{f}_r \cos(\alpha_r^{(n,k)} - \gamma_r) \cos\beta_r^{(g,k)} + \phi_{n,g,k}
\end{aligned}
$$

$$+ 2\pi t \hat{f}_t \left(\cos \gamma_t + \Delta_r^{(k)} \sin \gamma_t \sin \alpha_r^{(n,k)} \right) \cos \left(\Delta_r^{(k)} \beta_r^{(g,k)} \right)$$

$$- \frac{2\pi}{c} f \left(D + R_r^{(k)} \left(1 + \cos \alpha_r^{(n,k)} \right) \right) \Bigg]$$

$$+ \frac{\rho_{DB}}{\sqrt{MN}} \sum_{l,m,i=1}^{L,M_A^{(l)},M_E^{(l)}} \sum_{k,n,g=1}^{F,N_A^{(k)},N_E^{(k)}} \left(1 - \frac{\gamma}{2} \frac{R_t^{(l)} + R_r^{(k)}}{2D} \right) \cos \Bigg[K_p D_t^{DB}$$

$$+ K_q D_r^{DB} + 2\pi t \hat{f}_t \cos(\alpha_t^{(m,l)} - \gamma_t) \cos \beta_t^{(i,l)}$$

$$+ 2\pi t \hat{f}_r \cos(\alpha_r^{(n,k)} - \gamma_r) \cos \beta_r^{(g,k)}$$

$$- \frac{2\pi}{c} f \left(D + R_t^{(l)} \left(1 - \cos \alpha_t^{(m,l)} \right) + R_r^{(k)} \left(1 + \cos \alpha_r^{(n,k)} \right) \right)$$

$$+ \phi_{m,i,l,n,g,k} \Bigg] + \rho_{LOS} \cos \Bigg[2\pi t (f_t^{LOS} + f_r^{LOS})$$

$$- \frac{2\pi}{c} f \sqrt{D^2 + \Delta_H^2} + K_p d_{t,x} + K_q d_r \cos \psi_r \cos(\alpha_{Rq}^{LOS} - \theta_r) \Bigg] \qquad (2.105c)$$

$$T_{pq}^{(Q)}(t, f) = \frac{\rho_{SBT}}{\sqrt{M}} \sum_{l,m,i=1}^{L,M_A^{(l)},M_E^{(l)}} \left(1 - \frac{\gamma}{2} \frac{R_t^{(l)}}{D} \right) \sin \Bigg[K_p D_t^{SBT}$$

$$+ K_q D_r^{SBT} + 2\pi t \hat{f}_t \cos(\alpha_t^{(m,l)} - \gamma_t) \cos \beta_t^{(i,l)} + \phi_{m,i,l}$$

$$+ 2\pi t \hat{f}_r \left(\cos \gamma_r - \Delta_t^{(l)} \sin \gamma_r \sin \alpha_t^{(m,l)} \right) \cos \left(\Delta_t^{(l)} \beta_t^{(i,l)} \right)$$

$$- \frac{2\pi}{c} f \left(D + R_t^{(l)} \left(1 - \cos \alpha_t^{(m,l)} \right) \right) \Bigg]$$

$$+ \frac{\rho_{SBR}}{\sqrt{N}} \sum_{k,n,g=1}^{F,N_A^{(k)},N_E^{(k)}} \left(1 - \frac{\gamma}{2} \frac{R_r^{(k)}}{D} \right) \sin \Bigg[K_p D_t^{SBR}$$

$$+ K_q D_r^{SBR} + 2\pi t \hat{f}_r \cos(\alpha_r^{(n,k)} - \gamma_r) \cos \beta_r^{(g,k)} + \phi_{n,g,k}$$

$$+ 2\pi t \hat{f}_t \left(\cos \gamma_t + \Delta_r^{(k)} \sin \gamma_t \sin \alpha_r^{(n,k)} \right) \cos \left(\Delta_r^{(k)} \beta_r^{(g,k)} \right)$$

$$- \frac{2\pi}{c} f \left(D + R_r^{(k)} \left(1 + \cos \alpha_r^{(n,k)} \right) \right) \Bigg]$$

$$+ \frac{\rho_{DB}}{\sqrt{MN}} \sum_{l,m,i=1}^{L,M_A^{(l)},M_E^{(l)}} \sum_{k,n,g=1}^{F,N_A^{(k)},N_E^{(k)}} \left(1 - \frac{\gamma}{2} \frac{R_t^{(l)} + R_r^{(k)}}{2D} \right) \sin \Bigg[K_p D_t^{DB}$$

$$+ K_q D_r^{DB} + 2\pi t \hat{f}_t \cos(\alpha_t^{(m,l)} - \gamma_t) \cos \beta_t^{(i,l)}$$

$$+ 2\pi t \hat{f}_r \cos(\alpha_r^{(n,k)} - \gamma_r) \cos \beta_r^{(g,k)}$$

$$- \frac{2\pi}{c} f \left(D + R_t^{(l)} \left(1 - \cos \alpha_t^{(m,l)} \right) + R_r^{(k)} \left(1 + \cos \alpha_r^{(n,k)} \right) \right) \Bigg]$$

$$+ \phi_{m,i,l,n,g,k}] + \rho_{LOS} \sin \Bigg[2\pi t (f_t^{LOS} + f_r^{LOS})$$

$$- \frac{2\pi}{c} f \sqrt{D^2 + \Delta_H^2} + K_p d_{t,x} + K_q d_r \cos \psi_r \cos(\alpha_{Rq}^{LOS} - \theta_r) \Bigg] \qquad (2.105d)$$

Again, $h_{pq}^{(I)}(t)$ and $h_{pq}^{(Q)}(t)$ are the in-phase and quadrature components, respectively. Furthermore, $\mathcal{F}_f^{-1}(\cdot)$ denotes the inverse Fourier transform w.r.t. f. In addition, the following parameterization has been used:

$$M = \sum_{l=1}^{L} M^{(l)} = \sum_{l=1}^{L} M_A^{(l)} M_E^{(l)} \tag{2.106a}$$

$$N = \sum_{k=1}^{F} N^{(k)} = \sum_{k=1}^{F} N_A^{(k)} N_E^{(k)} \tag{2.106b}$$

$$\rho_{SBT} = \sqrt{\frac{\eta_t}{(K+1)}} \tag{2.106c}$$

$$\rho_{SBR} = \sqrt{\frac{\eta_r}{(K+1)}} \tag{2.106d}$$

$$\rho_{DB} = \sqrt{\frac{\eta_{tr}}{(K+1)}} \tag{2.106e}$$

$$\rho_{LOS} = \sqrt{\frac{K}{(K+1)}} \tag{2.106f}$$

$$K_p = \pi(L_t + 1 - 2p)/\lambda \tag{2.106g}$$

$$K_q = \pi(L_r + 1 - 2q)/\lambda \tag{2.106h}$$

$$D_t^{SBT} = d_{t,x} \cos\alpha_t^{(m,l)} + d_{t,y} \sin\alpha_t^{(m,l)} + d_{t,z} \sin\beta_t^{(i,l)} \tag{2.106i}$$

$$D_r^{SBT} = -d_{r,x} + d_{r,y}\Delta_t^{(l)} \sin\alpha_t^{(m,l)} \tag{2.106j}$$

$$D_t^{SBR} = d_{t,x} + d_{t,y}\Delta_r^{(k)} \sin\alpha_r^{(n,k)} \tag{2.106k}$$

$$D_r^{SBR} = d_{r,x} \cos\alpha_r^{(n,k)} + d_{r,y} \sin\alpha_r^{(n,k)} + d_{r,z} \sin\beta_r^{(g,k)} \tag{2.106l}$$

$$D_t^{DB} = d_{t,x} \cos\alpha_t^{(m,l)} + d_{t,y} \sin\alpha_t^{(m,l)} + d_{t,z} \sin\beta_t^{(i,l)} \tag{2.106m}$$

$$D_r^{DB} = d_{t,x} \cos\alpha_r^{(n,k)} + d_{r,y} \sin\alpha_r^{(n,k)} + d_{r,z} \sin\beta_r^{(g,k)} \tag{2.106n}$$

$$f_t^{LOS} = \hat{f}_t \cos(\pi - \alpha_{Rq}^{LOS} - \gamma_t) \tag{2.106o}$$

$$f_r^{LOS} = \hat{f}_r \cos(\alpha_{Rq}^{LOS} - \gamma_r) \tag{2.106p}$$

Two modeling methods are again considered that trade precision with simulation speed and complexity:

- **Deterministic Sum-of-Sinusoids Method.** Again, only the phases $\phi_{m,i,l}$, $\phi_{n,g,k}$ and $\phi_{m,i,l,n,g,k}$ are generated as independent uniformly distributed random variables in $[0, 2\pi)$. The theoretically random AAODs and AAOAs are deterministically fixed to [187, 198]:

$$\alpha_t^{(m,l)} = \mathcal{M}^{-1}\left(\frac{m - 0.5}{M_A^{(l)}}\right) \tag{2.107a}$$

$$\alpha_r^{(n,k)} = \mathcal{M}^{-1}\left(\frac{n - 0.5}{N_A^{(k)}}\right) \tag{2.107b}$$

for $m = 1, \ldots, M_A^{(l)}$ and $n = 1, \ldots, N_A^{(k)}$. Here, $\mathcal{M}^{-1}(\cdot)$ is the inverse cumulative von Mises distribution function, which can be evaluated using the method described in [218]. The theoretically random EAODs and EAOAs are deterministically fixed to [187, 198]:

$$\beta_t^{(i,l)} = \frac{2\beta_{tm}}{\pi} \arcsin\left(\frac{2i-1}{M_E^{(l)}} - 1\right) \tag{2.108a}$$

$$\beta_r^{(g,k)} = \frac{2\beta_{rm}}{\pi} \arcsin\left(\frac{2g-1}{N_E^{(k)}} - 1\right) \tag{2.108b}$$

for $i = 1, \ldots, M_E^{(l)}$ and $g = 1, \ldots, N_E^{(k)}$. The theoretically random clutter radii $R_t^{(l)}$ and $R_r^{(k)}$ are deterministically fixed to [187, 198]:

$$R_t^{(l)} = \sqrt{\frac{(l-0.5)(R_{t2}^2 - R_{t1}^2)}{L} + R_{t1}^2} \tag{2.109a}$$

$$R_r^{(k)} = \sqrt{\frac{(k-0.5)(R_{r2}^2 - R_{r1}^2)}{F} + R_{r1}^2} \tag{2.109b}$$

for $l = 1, \ldots, L$ and $k = 1, \ldots, F$. Again, according to [173], it can be shown that such deterministic realization of above parameters yields the desired statistical channel properties for $M, N \to \infty$.

• **Random Sum-of-Sinusoids Method.** Again, the random SOS yields channel realizations with statistical properties that vary from trial to trial. This can be mitigated by averaging over many realizations yielding a channel realization with better statistical properties than the deterministic SOS. To this end, the AAODs, AAOAs, EAODs, EAOA and clutter radii are obtained as follows [187, 198]:

$$\alpha_t^{(m,l)} = \mathcal{M}^{-1}\left(\frac{m + \theta_A - 1}{M_A^{(l)}}\right) \tag{2.110a}$$

$$\alpha_r^{(n,k)} = \mathcal{M}^{-1}\left(\frac{n + \psi_A - 1}{N_A^{(k)}}\right) \tag{2.110b}$$

$$\beta_t^{(i,l)} = \frac{2\beta_{tm}}{\pi} \arcsin\left(\frac{2(i + \theta_E - 1)}{M_E^{(l)}} - 1\right) \tag{2.110c}$$

$$\beta_r^{(g,k)} = \frac{2\beta_{rm}}{\pi} \arcsin\left(\frac{2(g + \psi_E - 1)}{N_E^{(k)}} - 1\right) \tag{2.110d}$$

$$R_t^{(l)} = \sqrt{\frac{(l + \theta_R - 1)(R_{t2}^2 - R_{t1}^2)}{L} + R_{t1}^2} \tag{2.110e}$$

$$R_r^{(k)} = \sqrt{\frac{(k + \psi_R - 1)(R_{r2}^2 - R_{r1}^2)}{F} + R_{r1}^2} \tag{2.110f}$$

where θ_A, ψ_A, θ_E, ψ_E, θ_R and ψ_R are all independent random variables uniformly distributed in $[0, 1)$. Finally, the phases $\phi_{m,i,l}$, $\phi_{n,g,k}$ and $\phi_{m,i,l,n,g,k}$ are again generated as independent uniformly distributed random variables in $[0, 2\pi)$.

The performance and precision of above modeling approaches have been evaluated [187, 198] and are omitted here for brevity.

2.3.7 *Measurements and Empirical Models*

As of today, a fairly exhaustive amount of mobile-to-mobile channel measurements has been conducted that includes a wide frequency and bandwidth range. This has allowed propagation inherent parameters to be extracted and thus empirical propagation models to be built. In the open literature, these models are equally referred to as body-to-body, car-to-car, vehicle-to-vehicle, inter-vehicle, peer-to-peer, *ad hoc* propagation models, among others. The key difference between these models and more traditional cellular propagation models is that both transmit and receive antennas are likely to be closer to the ground than those in conventional cellular networks; in addition, they may also be located in less favorable propagation positions, such as in users' pockets, bags or desk drawers [148]. We will henceforth discuss a nonexhaustive list of conducted measurement campaigns and also some selected empirical models. To obtain a complete picture of measurement and modeling approaches, the reader is invited to consult the references in citations discussed below.

2.3.7.1 Mobile-to-Mobile Measurement Campaigns

Harley [219] was one of the first to highlight that traditional large-cell propagation models do not apply to short range communication systems with low antenna heights at both transmitting and receiving ends. He also conducted measurements and extracted a suitable propagation model.

One of the first measurements conducted explicitly for the mobile-to-mobile scenario have been published [220]. The authors conducted an extensive measurement campaign at 900 MHz with the aim of determining the pathloss and delay spread probability distribution parameters, such as Rician K.

Patwari *et al.* studied the outdoor propagation characteristics of systems using low antenna heights of 1.7 m at both the transmitter and receiver and operating over a bandwidth of 100 MHz at 1.8 GHz. The wideband measurements allowed the characterization of power delay profiles for 22 different transmitter–receiver placements in rural and urban areas, together with the fading rate variance, angular spread, etc. These measurements and models have been extended [221, 222].

The UK's M-VCE [223] has been conducting channel measurements since its founding days in 1997. Some pertinent results have been published [148], which are based on two measurement campaigns. The first campaign deals with pathloss measurements in an indoor environment and the second set of data was taken to determine variations in the terminal radiation pattern given the close proximity of the human body. These early results indicated that for the environments under consideration, an additional attenuation factor should be added to the pathloss model to compensate for the unfavorable location of the terminals in a mobile-to-mobile propagation environment. This attenuation is dependent on the heights of the antennas involved, and is in the range 0–8 dB. Further refinements and related models from these M-VCE campaigns have appeared [157, 169, 224–230].

Kivinen *et al.* [231] presented characteristics of the wideband indoor radio channel at 5.3 GHz using a channel sounder with 19 ns delay resolution. The observed pathloss exponents were in the range of 1.3–1.5 in LOS and 2.9–4.8 in NLOS conditions. The delay dispersion was characterized by the CDF of the RMS delay spread with typical values between 20 to 180 ns at 90% CDF in different setups in an office building and large hall environments. The correlation functions of the radio channel in spatial and frequency domains were also extracted. Finally, using the results of the campaign, some small scale models for five typical indoor scenarios were developed using tapped delay lines.

Kovacs *et al.* [151] investigated the mobile-to-mobile radio channel in different suburban outdoor-to-indoor environments. As already discussed in a previous section, the measured signal envelope statistics were found to be a combination of single and double Rayleigh distributions, which generally lead to a system performance inferior to the reference ETSI limits. A thorough description of conducted measurement campaigns and thus derived propagation models is available [232]. The authors have also lately extended the modeling approach [233] to body-to-body channels.

Maurer *et al.* characterized the narrowband inter-vehicle transmission channel at 5.2 GHz. Based on realistic road traffic scenarios, measurements were obtained for four different classes of environments: urban, suburban (village), motorway and highway. The resultant channel was then characterized in terms of the CDF of the received signal, the average Doppler power density spectrum and the LCR. Modeling approaches were also taken in that measured results were compared to analytical functions, and the least squares together with the conjugate gradient method were used to fit the parameters of the analytical functions to the measurement data. The measurements and model have been extended and analyzed [199, 234, 235].

A simple but nonetheless accurate pathloss equation for calculating the free-space path loss for IEEE 802.11 WLAN LOS links with antennas between 1 and 2.5 meters in height has been proposed [153]. The model is validated against empirical measurements taken by means of several WLAN systems. The pathloss model was compared to other models and was found to be superior for predicting the propagation behavior.

Acosta *et al.* [156], measured the per-tap Doppler spectra for a frequency selective mobile-to-mobile wireless communications link at 2.45 GHz. The campaign was conducted in various multipath environments in Atlanta, Georgia. Chosen for their exceptionally long delay spreads, environments such as an expressway, an urban T-intersection and an exit ramp were measured, all of which induced fairly different spectra. Interestingly, for a given channel, the spectra corresponding to different delays were different, implying a nonseparable channel model. An extension to these modeling efforts has been presented [200, 236–238].

An experimental characterization of cooperative transmission schemes discussed in this book in later chapters, where the prime difficulty was provide simultaneous characterization of at least three links has been carried out [239, 240]. In this paper, measurements conducted in an indoors office environment involving two access points and two user terminals, each of which is equipped with four antennas, were presented. The data indeed allows for the simultaneous characterization of the MIMO links between all involved parties. These measurements were then used to practically assess several cooperative transmission schemes.

An empirical radio propagation model has been proposed [162] for the relaying system with low height terminals. The model differentiates LOS and NLOS propagation paths. It takes into account the effect of transmitter height, receiver height, receiver location and environmental parameters and it is also complemented by shadow fading statistics, namely the distribution and autocorrelation function estimation.

Sen and Matolak [241] analyzed the results of a channel measurement and modeling campaign for the vehicle-to-vehicle channel at 5 GHz. They analyzed the delay spread, and amplitude statistics as well as correlations. From these measurements, several statistical channel models were developed, and using simulation results, they gauged tradeoffs between model implementation complexity and modeling precision. Related results were also presented [34, 242, 243].

A comprehensive experimental MIMO channel-sounding campaign has been conducted for mobile-to-mobile vehicular communication with vehicles that travel along surface streets and expressways in a metropolitan area [202]. To compare the first- and second-order channel statistics that were obtained from the reference model discussed in Section 2.3.5 with those obtained from the empirical measurements, a new maximum-likelihood-based stochastic estimator was derived to extract the relevant model parameters from the measured data. The close agreement between the analytically and empirically obtained channel statistics confirms the utility of the proposed reference model and the method for estimating the model parameters.

A very comprehensive measurement and modeling campaign has also been conducted by the European IST WINNER and WINNER II projects [244]. Most notably, the public deliverable WINNER II [245] covers propagation scenarios that are pertinent to mobile-to-mobile communications. An example is the indoor office scenario A1. The generic WINNER II channel models follow a geometry-based stochastic channel modeling approach where channel parameters are determined stochastically, based on statistical distributions extracted from prior channel measurement. The distributions are defined for a large set of parameters, such as delay spread, delay values, angular spread, shadow fading, and cross-polarization ratios. Different scenarios are generally modeled by using the same underlying model but with different parameters. The parameter tables for each scenario are included [245] and are summarized for scenario A1 below. Several measurement campaigns provide the background for the parameterization of the propagation scenarios for both

LOS and NLOS conditions. The models are scalable from SISO to MIMO scenarios and can cater for polarization, multiuser, multicell, and multihop networks, among others. The models can be applied to any wireless system operating in the 26 GHz frequency range with up to 100 MHz bandwidth.

Finally, note that COST 231, COST 259 and COST 273 [246] and the recommendation ITU-R P.1411-3, 'Propagation data and prediction methods for the planning of short-range outdoor radio-communication systems and radio local area networks in the frequency range 300 MHz and 100 GHz.' Propose some channel models that are also applicable to mobile-to-mobile scenarios.

2.3.7.2 Empirical Outdoors Relay Propagation Models

Almost all of above discussed measurement campaigns have resulted in propagation and channel models, where the models in the literature [157, 162, 227, 231, 245, 247] are particularly worth noting for their comprehensive approach.

We will now discuss two recent outdoor models that are applicable to the mobile-to-mobile channel:

- **Dense Urban and Suburban 2.1 GHz Propagation Model.** Measurements at the UMTS frequency of 2.1 GHz have been performed [162] for low transmit and receive antenna heights in dense urban and suburban London, UK, and have lead to the following empirically fitted models. The pathloss is a function of LOS and NLOS and is given as

$$L_{LOS}(d) = 26.6 - 20\log_{10}(\lambda) - 2.24h_t - 4.9h_r + 29.6\log_{10}d \qquad (2.111a)$$

$$L_{NLOS}(d) = 22.0 - 20\log_{10}(\lambda) - 2h_r + 40\log_{10}d + C \qquad (2.111b)$$

where λ is the wavelength, h_t and h_r the transmit and receive antenna heights, d the distance between transmitter and receiver, and C a constant which is $C = 0$ for dense urban environments where buildings are larger than 18 m and $C = -4$ for suburban and also urban environments where buildings are lower than 12 m. Shadowing was found to be log normally distributed with a standard deviation varying between 6 and 11 dB. The autocorrelation function exhibited decorrelation distances between 20 and 80 m, with a mean distance of 40 m. The viability of the development model w.r.t. obtained measurements has been demonstrated [162].

  It has been shown that [162] the pathloss slope changes significantly over the range of measured distances; the slope alters to a much steeper angle as the distance from the transmitter increases. The pathloss increases as the transmitter height decreases and this is even more evident at short distances from the transmitter. The pathloss is also larger for lower receiver heights, which is again more apparent at short distances from the transmitter. The pathloss does not depend on the transmitted height for NLOS scenarios. Best pathloss fitting is achieved if the LOS and NLOS losses are weighted and added, that is $L = \alpha L_{LOS} + (1 - \alpha)L_{NLOS}$; the weighting coefficient α decreases almost linearly from 1 to 0 in log scale from 10 to 200 m. The shadowing standard deviation here does depend very little on antenna heights.

- **Urban 2.1 and 5.2 GHz Propagation Model.** Based on ray-tracing tools with a realistic geographical database, statistical propagation models have been obtained for the UMTS frequency of 2.1 GHz and the IEEE 802.11x frequency of 5.2 GHz in urban environments under LOS and NLOS propagation conditions [157]. The model includes BS-MS, BS-RS, RS-RS, RS-MS and MS-MS channels, where we will only concentrate on the latter. The behavior of the pathloss coefficients, shadowing parameters and Ricean K-factor have been quantified.

  For the mobile-to-mobile channel, the pathloss coefficient in dB is modeled as:

$$L(d) = b + 10n\log_{10}d, \qquad (2.112)$$

where it has been found that under LOS conditions at 2.1 and 5.2 GHz $n = 2$ and $b = -27.6$, under NLOS conditions at 2.1 GHz $n = 5.86$ and $b = -62.01$, and under NLOS conditions at 5.2 GHz $n = 5.82$ and $b = -51.22$.

Under LOS conditions, shadowing was found to be more uniformly distributed in dB than Gaussian; the shadow realizations were generally very small. Under NLOS conditions, shadowing was indeed found to obey a Gaussian distribution in dB. In both cases, the standard deviation was found to increase with distance according to the following law:

$$\sigma_{dB} = S_s \cdot \left(1 - e^{-\frac{d-d_0}{D_s}}\right), \tag{2.113}$$

where it was found that under LOS conditions at 2.1 and 5.2 GHz $S_s = 2$ dB, $d_0 = 0$ m and $D_s = 31$ m, under NLOS conditions at 2.1 GHz $S_s = 22.1$ dB, $d_0 = 10$ m and $D_s = 53$ m, and under NLOS conditions at 5.2 GHz $S_s = 23.4$ dB, $d_0 = 10$ m and $D_s = 36$ m.

Under LOS conditions, the Ricean K – factor was found to be distance dependent in a fairly deterministic fashion according to:

$$K_{LOS} = b_K + 10 n_K \log_{10} d, \tag{2.114}$$

where at 2.1 GHz $b_K = 15$ and $n_K = 0.019$, and at 5.2 GHz $b_K = 23$ and $n_K = 0.029$. Under NLOS conditions, the Ricean K−factor was found to depend weakly on distance but to exhibit large log normal randomness, which can be modeled as:

$$K_{NLOS} = \log \mathcal{N}(\mu_K, \sigma_K) - 10, \tag{2.115}$$

where at 2.1 GHz $\mu_K = 2.36$ and $\sigma_K = 0.48$, and at 5.2 GHz $\mu_K = 2.43$ and $\sigma_K = 0.45$.

Both outdoor models yield similar insights, in that propagation losses per given distance are usually larger in cooperative relaying settings when compared to cellular deployments, which is mainly due to the increased clutter density and hence increased pathloss coefficient.

2.3.7.3 Empirical Indoors Relay Propagation Model

Finally, we will discuss an indoors propagation model to complete the modeling section:

- **Office 2–6 GHz Propagation Model.** Measurements conducted and analyzed [244] over a frequency range from 2–6 GHz have facilitated the parameterization of a large number of scenarios. One of these is the indoor office scenario, referred to as scenario A1 [245], where low-height transmitters are assumed to be placed in the corridor, thus yielding LOS for corridor-to-corridor and NLOS for corridor-to-room communication links. In the NLOS case, the pathloss must account for losses induced by walls and floors, if applicable. Concerning the floor loss (FL), it is assumed to be constant for the same distance between floors, but increases with the floor separation and has to be added to the pathloss calculated for the same floor. The mobile-to-mobile propagation model is constructed by following the steps below, whereas the complete channel model can be obtained by following all steps outlined in Section 4.2 of Ref. [245].

  The pathloss is obtained for each link according to the following equation:

$$L(d) = A \log_{10} d + B + C \log_{10}(f_c/5) + X, \tag{2.116}$$

where d is the distance between the transmitter and the receiver in meters, f_c is the system frequency in GHz, the fitting parameter A includes the pathloss exponent, parameter B is the intercept, parameter C describes the pathloss frequency dependence, and X is an optional, environment-specific term including, for example, the wall attenuation in the A1 NLOS scenario. The model can be applied with different antenna heights. The parameterization of above pathloss model is summarized in Table 2.3.

Table 2.3 Pathloss and shadowing parameterization of the empirical indoors relay propagation model in Ref. [245], applicable in the range of $3 < d < 100$ m, $h_t = h_r = 1 \dots 2.5$ m; n_w is the number of walls and n_f the number of floors

Scenario	Pathloss	Shadowing
LOS	$A = 18.7, B = 46.8, C = 20$	$\sigma_{dB} = 3$
NLOS (light walls)	$A = 20.0, B = 46.4, C = 20, X = 5n_w$	$\sigma_{dB} = 6$
NLOS (heavy walls)	$A = 20.0, B = 46.4, C = 20, X = 12n_w$	$\sigma_{dB} = 8$
NLOS (through floors)	add to above $17 + 4(n_f - 1)$	

Furthermore, the probability that the link between transmitter and receiver enjoys LOS conditions first is given as:

$$p_{LOS} = \begin{cases} 1 & \text{for } d \leq 2.5 \text{ m} \\ 1 - 0.9(1 - (1.24 - 0.61 \log_{10} d)^3)^{1/3} & \text{for } d > 2.5 \text{ m} \end{cases} \qquad (2.117)$$

and $p_{NLOS} = 1 - p_{LOS}$.

We now move away from modeling approaches and finally discuss issues pertaining to the estimation of the regenerative relaying channel.

2.3.8 Estimating Regenerative Fading Channels

Whether analyzed or simulated by means of theoretical or empirical models, the estimation of the relaying channel has a profound impact onto the performance of regenerative relaying protocols [248–252]. The aim of this section is not to go into great detail but rather to highlight the few available contributions in this area. It should be noted that although many estimation problems pertaining to the regenerative relaying channel can be reduced to problems related to the traditional point-to-point communication channel, the papers discussed below capitalize on its peculiarities.

Yi *et al.* [253] investigated how finite-rate feedback and imperfect channel estimation impacted a regenerative MIMO relay network. They formulated the SNR and capacity loss due to channel estimation and quantization errors in a multiantenna relay network with multiple regenerative MIMO relay hops and discuss how the end-to-end throughput is scaled by MIMO pilot size, codebook size, relay network size, among others.

Zhang *et al.* [254, 255] dealt with channel estimation in a regenerative but also transparent cooperative orthogonal frequency-division multiplexing (OFDM) network in the presence of frequency offsets. In order to eliminate the multiple access interference occurring in such systems, the maximum number of active regenerative relays is $\lceil N/L \rceil$, where N is the total number of subcarriers and L is the channel order. Various tradeoffs were then formulated in the paper.

A data-aided decision-directed Maximum A Posteriori (MAP) iterative channel estimation algorithm for a regenerative relaying scheme supporting distributed Alamouti STBCs has been proposed [256]. The algorithm profits from an optimum initialization by means of the MAP data-aided channel estimation.

Gao *et al.* [257–259] propose some optimal training designs for regenerative relay networks that have knowledge of the interference covariance. Two channel estimation methods are invoked: maximum likelihood (ML) and minimum mean square error (MMSE). For ML channel estimation, the channels are assumed to be deterministic and the optimal training results from an efficient multilevel waterfilling type solution that is derived from majorization theory. For MMSE channel estimation, the second-order statistics of the channels are assumed to be known and the general optimization problem unfortunately turns out to be nonconvex. The authors thereupon consider three special scenarios, where the problem in the first scenario is convex and can efficiently be solved by state-of-the-art optimization tools. Furthermore, closed-form waterfilling type solutions are found in the remaining two scenarios.

Lalos *et al.* dealt with a regenerative wideband relay network [260]. It was shown that all channels in the network from any node to the destination node can be blindly estimated up to a phase ambiguities vector, related to the source-to-destination channel frequency response. Therefore, by utilizing a small number of pilot symbols, phase ambiguities can be effectively resolved. The proposed methods exhibit high estimation accuracy even for a short training sequence, and outperforms direct training-based channel estimation.

Zhao *et al.* [261, 262] deal with channel estimation issues in the two-way relaying system. Instead of using pilot sequences, they exploit the self-interference that contains the data known at the receivers to get a first estimate of the channel. With this rough estimate, a decision-directed iterative estimation process is started in order to improve the accuracy of the channel estimates. The proposed scheme has an essentially similar performance to pilot-aided channel estimation schemes, albeit at a higher spectral efficiency since pilot sequences are not needed anymore.

From the above survey it becomes apparent that developments related to channel estimation in regenerative relaying channels are very recent. Many open issues pertain, mainly in the context of the two-way relaying system as well as phase estimation and channel prediction.

2.4   Transparent Relaying Channel

We will now discuss and quantify the most pertinent properties of a transparent relay channel in greater detail. For the sake of consistency, we will follow here the same section structure as in Section 2.3 related to regenerative relays. The published material devoted to transparent relaying though is fairly scarce compared to the regenerative case. Note that the choice of notation and meaning of symbols follows that explained at the beginning of this chapter.

2.4.1   Propagation Modeling

The pathloss, shadowing and fading behaviors exhibit their peculiarities in the context of transparent cooperative channels, which we will briefly allude to in this section.

2.4.1.1   Pathloss

There are two important issues related to pathloss in the context of regenerative relaying channels:

- **Change of Breakpoint Behavior.** As already discussed in the context of regenerative relays, lowering both transmit and receive antennas decreases the breakpoint distance. The breakpoint distances for the traditional and cooperative segments are therefore different. This, together with the coupling of the relaying segments, yields fairly complex pathloss behaviors that can be quantified for a given pathloss model at hand by using Equation (2.15). To exemplify such a behavior, we have depicted the pathloss versus distance in Figure 2.30 where we compare the cases of no relay with relay placed exactly between transmitter and receiver. For both cases, we have assumed that the breakpoint distance is 100 m for the traditional link; for the relay case, we have assumed that the breakpoint distance is 10 m. For all cases, we have assumed a pathloss coefficient $n_1 = 2$ before breakpoint and $n_2 = 4$ after breakpoint. It can be observed that from Equation (2.15) and the different breakpoint distances, the overall breakpoint behavior is complicated, leading in this case to two breakpoints.
- **Increased Pathloss.** From Figure 2.30 an increased pathloss w.r.t. non-cooperative case can also be observed, which can again be explained using Equation (2.15). Let us assume for simplicity, for

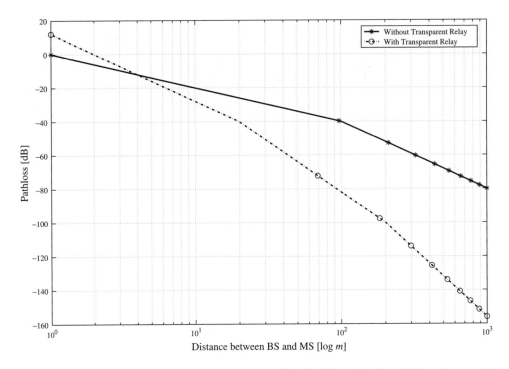

Figure 2.30 Pathloss and breakpoint behavior for system with and without relay placed midway between BS and MS, assuming fixed amplification

instance, a fixed amplification factor and a single-slope pathloss model for both relaying stages, such as introduced in [157]. It is of the form $L_i(d) = b_i + n_i \log_{10}(d)$, where b_i is some normalization constant and n_i the pathloss coefficient. The end-to-end pathloss in decibels can hence be written as the sum of each segment's contribution, that is $b_1 + b_2 + n_1 \log_{10}(d_1) + n_2 \log_{10}(d_2)$, where $d = d_1 + d_2$. If the transparent relay is placed exactly midway between transmitter and receiver and both relaying segments exhibit the same pathloss behavior, this expression simplifies to $2b + 2n \log_{10}(d/2)$. The end-to-end pathloss hence experiences a change in parameters, that is

$$L(d) = b' + n' \log_{10}(d) \tag{2.118}$$

where $b' = 2b - 2n \log_{10} 2$ and $n' = 2n$. This leads to an increased pathloss slope and shift as already observed in Figure 2.30. To quantify this loss for an arbitrary number of relaying segments, let us, for example, assume a system with the source and destination separated by $d = 500$ m and relays placed equidistant between them so that N relay segments occur. Then, with the model from [157], we have $b = -62.01$ dB and $n = 5.86$ leading to $b' = -62.01 \cdot N - 5.86 \cdot \log_{10} N$ dB and $n = 5.86 \cdot N$. The power losses due to this non-linear pathloss behavior is summarized in Table 2.4 w.r.t. the case without relay in percent. These losses diminish with variable amplification factors but are generally very large.

Table 2.4 Absolute and relative pathloss losses with relays placed between source and destination separated by 500 m caused by the nonlinear propagation model

Relay segments	1	2	3	4	5	6	7	8	9	10
Relative gain [%]	0	−50	−66	−75	−80	−83	−86	−87	−89	−90

These losses are significant and generally outweigh potential aggregated SNR gains as already quantified [263]. This fact, therefore, *a priori* discourages the use of transparent relaying techniques, despite their slightly reduced complexity.

2.4.1.2 Shadowing

There are three important issues related to shadowing in the context of regenerative relaying channels:

- **Change of Shadowing Standard Deviation.** The reduced communication distances between cooperating terminals traditionally come along with a reduced shadowing standard deviation. This has been quantified [157].
- **Increased Shadowing Variations.** In the case of transparent relays with independent shadowing channels in each relaying segment, the aggregated shadowing still obeys a Gaussian distribution in decibels, but with a changed standard deviation w.r.t. a direct link. Using Equation (2.15), the effective end-to-end shadowing standard deviation with N transparent relaying segments and a fixed amplification factor is given as $\sigma_{e2e,dB} = \sqrt{N}\sigma_{dB}$. Using a more sophisticated model [157], the aggregated end-to-end shadowing standard deviation for N equally long relaying segments can be calculated in decibels as follows:

$$\sigma_{e2e,dB} = \sqrt{N} \cdot S_s \cdot \left(1 - e^{-\frac{d/N - d_0}{D_s}}\right), \tag{2.119}$$

where S_s, d_0 and D_s are some model specific constants and d is the distance between source and destination. The increase in shadowing standard deviation w.r.t. the case without relay in percent and in absolute dB values is summarized in Table 2.5, assuming $d = 500\,\text{m}$, $Ss = 22.1\,\text{dB}$, $d_0 = 10\,\text{m}$ and $D_s = 53\,\text{m}$. These losses are generally not negligible and further aggravate above discussed pathlosses. It is interesting to observe, however, that with the above model, the shadowing losses start to diminish after around five relays and even turn into gains beyond 20 relays (not shown). This is due to a strong decrease of the shadowing standard deviation at very short distances.

Table 2.5 Absolute and relative aggregate shadowing losses with relays placed between source and destination separated by 500 m caused by the nonlinear behavior of the shadow standard deviation

Relay segments	1	2	3	4	5	6	7	8	9	10
Relative gain [%]	0	−29	−39	−44	−45	−46	−45	−44	−42	−40
Absolute gain [dB]	0	−9	−14	−17	−18	−18	−18	−17	−16	−15

- **Change of Shadowing Correlation Model.** To distinguish from the cellular case, the correlation function of the shadowing fluctuation in cooperative systems is referred to [169] as a joint correlation function (JCF). There are no models yet available for the JCF in the context of transparent relaying systems. However, based on Ref. [169] and some parallels drawn from the temporal correlation function of transparent relay channels:

$$R(\Delta d_{t1}, \Delta d_{r1}, \Delta d_{t2}, \Delta d_{r2}) \propto e^{-\frac{\Delta d_{t1} + \Delta d_{r1}}{d_{corr1}}} \cdot e^{-\frac{\Delta d_{t2} + \Delta d_{r2}}{d_{corr2}}}, \tag{2.120}$$

where Δd_{t1}, Δd_r and $\Delta d_{r1} = \Delta d_{t2}$ are the displacement distances of the transmitter, receiver and relay, respectively. Furthermore, d_{corr1} and d_{corr2} are the spatial correlation distances observed in the first and second relaying segment. This expression is dominated by the term with the smaller correlation distance.

In summary, transparent cooperative relaying yields significant shadowing losses and decreased shadow correlation distances.

2.4.1.3 Fading

To characterize the small scale behavior of the transparent relaying channel, we shall again commence with the geometrical arrangement of a general scenario as depicted in Figure 2.31. As for the system assumptions, a source, relay and destination mobile terminal move at a given speed and at an angle w.r.t. some coordinate system. Each terminal is equipped with multiple antennas, which are placed according to some geometrical arrangement and each of which has a generic 3D radiation pattern. The clutter environment is generally 3D and nonisotropic in space. The channel can generally appear selective in time, frequency and space. A starting point is again the channel impulse response Equation (2.6) of the end-to-end fading channel, that is $h = \sum_i a_i \cdot e^{j\phi_i} \cdot \delta(t - \tau_i)$, where the parameters a_i, ϕ_i and τ_i are generally jointly dependent and strongly influenced by underlying system assumptions.

- **Changed Amplitude/Power Statistics.** The coupling between the relaying segments causes the complex channel and resulting envelope and power statistics to change significantly.
- **Change in Correlation Functions.** The temporal, spatial and spectral characteristics also change significantly, mainly due to transmitter, relay(s) and receiver being mobile and immersed in dense clutter.

Figure 2.31 Exemplification of geometrical arrangement of transparent relaying channel

It is the aim of the following sections to determine the statistics and moments associated with this random channel impulse response, and we will first dwell on the statistics of the end-to-end narrowband components in Section 2.4.2, and then quantify second order temporal, spatial and spectral moments in Sections 2.4.3–2.4.5.

2.4.2 Envelope and Power Fading Statistics

As per Equation (2.15), the end-to-end fading channel can be modeled by cascading the fading channels of each individual relaying stage. If, in addition, a given relaying stage suffers from keyhole behavior then its individual channel is characterized by a cascaded fading channel, such as double Rayleigh. From a keyhole propagation point of view, an analytical treatment of this phenomenon was originally developed for the cascaded Rayleigh channel [264, 265]; a generalized form with, for example, an arbitrary number of product terms for the cascaded Rayleigh channel has been introduced [266]. From a transparent relaying point of view, an analytical treatment has also been been developed [140, 267].

We will next characterize the PDFs of the channel's envelope and power fading statistics as well as its connection to the end-to-end SNR. Note that generally all the relationships derived here are significantly more complex than for the regenerative channel since factors such as topology, choice of amplification factor, etc., heavily influence the end-to-end behavior.

2.4.2.1 Cascaded Fading Distributions with Constant Amplification

We assume here a constant amplification factor A appearing in Equation (2.15). It can be the result of a fixed amplification or an average amplification, such as given in Equations (2.17) or (2.18). We will briefly summarize the findings for some cascaded distributions:

- **Cascaded Rayleigh Fading.** The double-Rayleigh channel results from the product of two complex Gaussian processes, that is $h = h_1 \cdot h_2$, the amplitude/envelope of which is $a = |h| = |h_1| \cdot |h_2| = a_1 \cdot a_2$ and the gain/power of which is $g = |h|^2 = |h_1|^2 \cdot |h_2|^2 = g_1 \cdot g_2$. The distribution of the power of the double-Rayleigh channel therefore obeys the the product of two central chi-square random variables with two degrees of freedom, the PDF of which is given in Equation (2.24b). Using the rule for the PDF transformations of products of random variables, one easily establishes that the resultant PDF is given as:

$$p_g(g) = \int_0^\infty \frac{1}{A^2\xi} \cdot p_{g_1}(\xi) \cdot p_{g_2}\left(\frac{g}{A^2\xi}\right) d\xi \tag{2.121a}$$

$$= \int_0^\infty \frac{1}{A^2\xi} \cdot \frac{1}{\bar{g}_1} \exp\left(-\frac{\xi}{\bar{g}_1}\right) \cdot \frac{1}{\bar{g}_2} \exp\left(-\frac{g}{\bar{g}_2 A^2\xi}\right) d\xi \tag{2.121b}$$

$$= \frac{1}{A^2} \frac{2}{\bar{g}_1 \cdot \bar{g}_2} \cdot K_0\left(2\sqrt{\frac{1}{A^2}\frac{g}{\bar{g}_1 \cdot \bar{g}_2}}\right), \tag{2.121c}$$

where the last step follows from Ryzhik and Gradstein [268, Section 3.471.9) and $K_0(\cdot)$ is the zeroth order modified Bessel function of the second kind. Using the PDF transformation rule in Equation (2.23), one readily obtains the PDF of the envelope of the cascaded Rayleigh process as:

$$p_a(a) = \frac{1}{A^2} \frac{4a}{\bar{g}_1 \cdot \bar{g}_2} \cdot K_0\left(2\sqrt{\frac{1}{A^2}\frac{a^2}{\bar{g}_1 \cdot \bar{g}_2}}\right). \tag{2.122}$$

As reported [151], double-cascaded Rayleigh channels yield even more severe performance degradations than simple Rayleigh fading channels.

Using the Mellin transform or H-function technique, Salo *et al.* [266] obtained an expression for the PDF of the envelope and power of N cascaded Rayleigh fading channels, also referred to as

$N * $ Rayleigh fading channel:

$$p_a(a) = \frac{1}{A} \frac{2}{\sqrt{2^N \overline{g}}} G_{0,N}^{N,0} \left(\frac{(a/A)^2}{2^N \overline{g}} \Bigg|_{\frac{1}{2},\ldots,\frac{1}{2}}^{-} \right) \tag{2.123a}$$

$$p_g(g) = \frac{1}{A} \frac{1}{\sqrt{2^N g \overline{g}}} G_{0,N}^{N,0} \left(\frac{g/A^2}{2^N \overline{g}} \Bigg|_{\frac{1}{2},\ldots,\frac{1}{2}}^{-} \right), \tag{2.123b}$$

where $\overline{g} = \prod_{i=1}^{N} \overline{g}_i$ is the product of the average power of the individual Rayleigh fading channel segments, $A = \prod_{i=1}^{N-1} A_i$ is the product of the amplification factors in the ith relay. Furthermore, the CDFs for envelope and power of the $N *$ Rayleigh fading channel read:

$$F_a(a) = \pi^{-N/2} G_{1,N+1}^{N,1} \left((a/A)^2 (2\overline{g})^{-N} \Bigg|_{\frac{1}{2},\ldots,\frac{1}{2},0}^{1} \right) \tag{2.124a}$$

$$F_g(g) = \pi^{-N/2} G_{1,N+1}^{N,1} \left(g/A^2 (2\overline{g})^{-N} \Bigg|_{\frac{1}{2},\ldots,\frac{1}{2},0}^{1} \right). \tag{2.124b}$$

For fixed amplification as well as averaged amplification of the form in Equation (2.17) the use is straightforward; for averaged amplification of the form in Equation (2.18), the amplification factor has been calculated as [168]:

$$A_i = \sqrt{\frac{P_i}{\sigma_i^2} \cdot e^{\lambda_i} \lambda_i \Gamma(0, \lambda_i)}, \tag{2.125}$$

where σ_i^2 is the noise power at the input of the ith relay stage, P_i is the transmission power into the ith relay stage and $\lambda_i = 1/\overline{g}_i$. Finally, $G(\cdot)$ is the Meijer G-function generally defined as:

$$G_{p,q}^{m,n} \left(z \Bigg|_{b_1,\ldots,b_q}^{a_1,\ldots,a_p} \right) = \frac{1}{j 2\pi} \times \int_{\mathcal{L}} \frac{\prod_{i=1}^{m} \Gamma(b_i + s) \prod_{i=1}^{n} \Gamma(1 - a_i - s)}{\prod_{i=n+1}^{p} \Gamma(a_i + s) \prod_{i=m+1}^{q} \Gamma(1 - b_i - s)} z^{-s} \, ds, \tag{2.126}$$

where z, $\{a_i\}_{i=1}^{p}$ and $\{b_i\}_{i=1}^{q}$ are, in general, complex-valued. The contour $\mathcal{L}$ is chosen so that it separates the poles of the gamma products in the numerator [266]. The Meijer G-function has been implemented in some but not all commercial mathematics software packages, which is the reason why some very tight series approximations have been proposed [266] to facilitate numerical evaluation. In addition, the Meijer G-function can always be written in terms of the more familiar generalized hypergeometric function [268, 269].

- **Cascaded Ricean Fading.** Following the procedure outlined in Equation (2.121) and using the PDFs given in Equation (2.26), an integral expression for the resultant envelope and power distributions can be obtained [267]. Unfortunately, a closed form solution is not available today even for the simple double Ricean fading case. Karagiannidis [168] derived some bounds on the fading behavior as well as the required amplification factor. Again, for fixed amplification as well as averaged amplification of the form in Equation (2.17) the use is straightforward; for averaged amplification of the form of Equation (2.18), the amplification factor has been calculated as [168]:

$$A_i = \sqrt{\frac{P_i(1 + K_i)e^{\frac{K_i + 1}{\overline{g}_i} - K_i}}{\overline{g}_i \sigma_i^2} \times \sum_{n=0}^{\infty} \left[\frac{K_i^n (1 + K_i)^n}{n! (\overline{g}_i)^n} \Gamma\left(-n, \frac{K_i + 1}{\overline{g}_i}\right) \right]}. \tag{2.127}$$

Finally, note that for sufficiently large Ricean fading factors K_i in each relaying segment, the resulting distribution can be approximated by another Rice fading distribution with

$$K \approx \frac{\prod_{i=1}^{N} K_i}{\sum_{i=1}^{N} \prod_{j=1, j \neq i}^{N} K_j}. \tag{2.128}$$

For the double cascaded case this reduces to $K = K_1 K_2/(K_1 + K_2)$. To include the amplification factor, the PDFs given in Equation (2.26) need to be scaled by $1/A \cdot p_a(a/A)$ and $1/A^2 \cdot p_g(g/A^2)$, respectively. The same applies to the CDFs.

- **Cascaded Nakagami-m Fading.** Even though not corroborated by any measurement campaign (yet), the PDF of the envelope and power of the double-cascaded Nakagami fading case can be expressed in closed form following above double Rayleigh approach:

$$p_a(a) = \frac{1}{A} \frac{4m_1^{m_1} m_2^{m_2}}{\bar{g}_1 \Gamma(m_1) \bar{g}_2 \Gamma(m_2)} \left(\frac{m_2 \bar{g}_1}{m_1 \bar{g}_2} \right)^{\frac{m_1 - m_2}{2}} (a/A)^{m_1 + m_2 - 1}$$

$$\times \mathrm{K}_{m_1 - m_2} \left(2\sqrt{\frac{m_1}{\bar{g}_1} \frac{m_1}{\bar{g}_1}} (a/A)^2 \right) \tag{2.129a}$$

$$p_g(g) = \frac{1}{A^2} \frac{2m_1^{m_1} m_2^{m_2}}{\bar{g}_1 \Gamma(m_1) \bar{g}_2 \Gamma(m_2)} \left(\frac{m_2 \bar{g}_1}{m_1 \bar{g}_2} \right)^{\frac{m_1 - m_2}{2}} (g/A^2)^{\frac{m_1 + m_2}{2} - 1}$$

$$\times \mathrm{K}_{m_1 - m_2} \left(2\sqrt{\frac{m_1}{\bar{g}_1} \frac{m_2}{\bar{g}_2}} (g/A^2) \right), \tag{2.129b}$$

where m_1 is the Nakagami fading factor associated with the first relaying segment and m_2 with the second relaying segment. Furthermore, $\mathrm{K}_n(\cdot)$ is the nth order modified Bessel function of the second kind.

The results have been generalized [270] using a similar approach as earlier, [266], which led to the following PDFs for envelope and power of the $N *$ Nakagami fading channel:

$$p_a(a) = \frac{2}{a \prod_{i=1}^N \Gamma(m_i)} G_{0,N}^{N,0} \left((a/A)^2 \prod_{i=1}^N \left(\frac{m_i}{\bar{g}} \right) \bigg|_{m_1, \ldots, m_N}^{-} \right) \tag{2.130a}$$

$$p_g(g) = \frac{1}{g \prod_{i=1}^N \Gamma(m_i)} G_{0,N}^{N,0} \left(g/A^2 \prod_{i=1}^N \left(\frac{m_i}{\bar{g}} \right) \bigg|_{m_1, \ldots, m_N}^{-} \right), \tag{2.130b}$$

which again may require a series approximation if not built in as a stand-alone function in the used mathematical software package. Furthermore, the CDFs for envelope and power of the $N *$ Nakagami fading channel read:

$$F_a(a) = \frac{1}{\prod_{i=1}^N \Gamma(m_i)} G_{1,N+1}^{N,1} \left((a/A)^2 \prod_{i=1}^N \left(\frac{m_i}{\bar{g}} \right) \bigg|_{m_1, \ldots, m_N, 0}^{1} \right) \tag{2.131a}$$

$$F_g(g) = \frac{1}{\prod_{i=1}^N \Gamma(m_i)} G_{1,N+1}^{N,1} \left(g/A^2 \prod_{i=1}^N \left(\frac{m_i}{\bar{g}} \right) \bigg|_{m_1, \ldots, m_N, 0}^{1} \right). \tag{2.131b}$$

As for the amplification factors, for fixed amplification as well as averaged amplification of the form in Equation (2.17) the use is again straightforward; for averaged amplification of the form in Equation (2.18), the amplification factor has been calculated as [168]:

$$A = \sqrt{\frac{P_i}{\sigma_i^2} \cdot e_i^{\lambda} \lambda_i^{m_i} \Gamma(1 - m_i, \lambda_i)}, \tag{2.132}$$

where m_i is the Nakagami fading factor of the ith link.

In summary, the double cascaded fading case is usually tractable in closed form or can be easily evaluated numerically. However, the generalization to higher order cascaded channels leads to expressions which are often only tractable via infinite series expansions [266].

2.4.2.2 Cascaded Fading Distributions with Variable Amplification

We assume now a variable amplification factor A appearing in Equation (2.15). In the asymptotic regime of high noise power, that is effectively at low SNR, the influence of the channel in the denominator in Equation (2.16) becomes negligible and the problem reduces essentially to the case of constant amplification factor. In the other asymptotic regime of very low noise power, that is effectively at high SNR, the influence of the noise in the denominator in Equation (2.16) becomes negligible and the problem here reduces to the case of a single channel. Problems arise in the intermediate regions where at the time of writing of this book, little is known about this type of channel. The only available results [140] deal with a transparent two-hop relaying scenario for which the amplification factor given in Equation (2.16) has been assumed. Indeed, it was observed that the power of the end-to-end transparent relay channel with variable amplification [140]

$$g = \frac{P_2}{P_1} \cdot \frac{g_1 g_2}{g_1 + \sigma_1^2/P_1} \tag{2.133}$$

is in structure the same as the end-to-end SNR in transparent relay channels with constant amplification. This allows one to reuse tools and results developed for said channels where one literally needs to replace the SNR by the channel power, replace occurring constants by σ_1^2/P_1 and apply a scaling by P_2/P_1; the envelope is then obtained by simply invoking the PDF transformation given in Equation (2.23). First results for the dual hop Rayleigh channel [165, 167, 271, 272] can be used. Generalizations to more hops, different amplification factors and other channel statistics can be found [168, 270, 273–276]. Of particular interest here are the results [168, 275] that allow one to obtain some bounds on the expressions for the end-to-end fading statistics for fairly arbitrary fading statistics in each relaying segment. We will briefly summarize the available findings for some cascaded distributions:

- **Cascaded Rayleigh Fading.** For a simple dual hop transparent relaying system we follow the above insights [140] which are based on Hasna and Alouini, [167] to obtain the PDFs of the channel envelope and powers:

$$p_a(a) = \frac{4a P_1}{P_2 \bar{g}_2} e^{-\frac{P_1 a^2}{P_2 \bar{g}_2}}$$

$$\times \left(\sqrt{\frac{\sigma_1^2 a^2}{P_2 \bar{g}_1 \bar{g}_2}} K_1 \left[2\sqrt{\frac{\sigma_1^2 a^2}{P_2 \bar{g}_1 \bar{g}_2}} \right] + \frac{\sigma_1^2}{P_1 \bar{g}_1} K_0 \left[2\sqrt{\frac{\sigma_1^2 a^2}{P_2 \bar{g}_1 \bar{g}_2}} \right] \right) \tag{2.134a}$$

$$p_g(g) = \frac{2 P_1}{P_2 \bar{g}_2} e^{-\frac{P_1 g}{P_2 \bar{g}_2}}$$

$$\times \left(\sqrt{\frac{\sigma_1^2 g}{P_2 \bar{g}_1 \bar{g}_2}} K_1 \left[2\sqrt{\frac{\sigma_1^2 g}{P_2 \bar{g}_1 \bar{g}_2}} \right] + \frac{\sigma_1^2}{P_1 \bar{g}_1} K_0 \left[2\sqrt{\frac{\sigma_1^2 g}{P_2 \bar{g}_1 \bar{g}_2}} \right] \right). \tag{2.134b}$$

The CDFs of the channel envelope and powers respectively read:

$$F_a(a) = 1 - 2\sqrt{\frac{\sigma_1^2 a^2}{P_2 \bar{g}_1 \bar{g}_2}} e^{-\frac{P_1 a^2}{P_2 \bar{g}_2}} K_1 \left[2\sqrt{\frac{\sigma_1^2 a^2}{P_2 \bar{g}_1 \bar{g}_2}} \right] \tag{2.135a}$$

$$F_g(g) = 1 - 2\sqrt{\frac{\sigma_1^2 g}{P_2 \bar{g}_1 \bar{g}_2}} e^{-\frac{P_1 g}{P_2 \bar{g}_2}} K_1 \left[2\sqrt{\frac{\sigma_1^2 g}{P_2 \bar{g}_1 \bar{g}_2}} \right] \tag{2.135b}$$

Closed form solutions for more general fading scenarios are not available at the time of writing of this book. However, some bounds and approximations for a multihop network are provided [168,

276]. These are given below for the Nakagami fading case, where one only needs to set $m_i = 1$ for $\forall i$.

- **Cascaded Ricean Fading.** Similarly to the case of constant gain, an integral expression could again be derived that has eluded a closed form expression to date. However, the author of [168] has observed that above end-to-end channel power expression can be expressed in form of a harmonic mean which in turn is upper bounded by a geometric mean. He then derives the statistical moments of this upper bound. These moments can then be used for performance analysis [277] or to reconstruct the approximate PDF of the end-to-end fading channel via a series expansion of the MGF [278]. Following the analysis by Karagiannidis [168], the bounded moments for the Ricean case can be expressed as:

$$\overline{a_{\text{bound}}^k} = \mathcal{Z}_N^k \prod_{i=1}^{N} \left(\frac{\overline{g_i}}{K_i + 1} \right)^{k(N+1-i)/N} \Gamma\left(1 + \frac{k(n+1-i)}{N} \right)$$

$$\times {}_1F_1\left(-\frac{k(N+1-i)}{N}, 1; -K_i \right), \qquad (2.136)$$

where $\mathcal{Z}_N = \frac{1}{N} \prod_{i=1}^{N} C_i^{-\frac{N-i}{N}}$, $C_i = \sigma_i^2 / P_i$ and ${}_1F_1(\cdot, \cdot; \cdot)$ is the Kummer confluent hypergeometric function.

Note that for sufficiently large Ricean fading factors K_i in each relaying segment, the resulting distribution can again be approximated by another Rice fading distribution with

$$K \approx \frac{\prod_{i=1}^{N} K_i}{\sum_{i=1}^{N} \prod_{j=1, j \neq i}^{N} K_j}. \qquad (2.137)$$

For the double cascaded case this reduces to $K = K_1 K_2 / (K_1 + K_2)$. This allows one to use the PDF and CDF of the Ricean channel discussed before.

- **Cascaded Nakagami-m Fading.** Closed form PDF expressions for the envelope and power of the dual hop Nakagami fading case can be derived by following, for example, the approach outlined by Tsiftsis *et al.* [279]:

$$p_a(a) = 2a \sum_{i=0}^{m_1-1} \sum_{j=0}^{i} \Xi[i, j] e^{-\frac{m_1 a^2}{\overline{g_1}}} \left[\frac{m_1}{\overline{g_1}} (a^2)^{(2i+m_2-j)/2} K_{m_2-j}(2\sqrt{\tau a^2}) \right.$$

$$+ \frac{\sqrt{\tau}}{2} (a^2)^{(2i+m_2-j-1)/2} K_{m_2-j-1}(2\sqrt{\tau a^2})$$

$$+ \frac{\sqrt{\tau}}{2} (a^2)^{(2i+m_2-j-1)/2} K_{m_2-j+1}(2\sqrt{\tau a^2})$$

$$\left. - \frac{1}{2} (2i + m_2 - j)(a^2)^{(2i+m_2-j-2)/2} K_{m_2-j}(2\sqrt{\tau a^2}) \right], \qquad (2.138a)$$

$$p_g(g) = \sum_{i=0}^{m_1-1} \sum_{j=0}^{i} \Xi[i, j] e^{-\frac{m_1 g}{\overline{g_1}}} \left[\frac{m_1}{\overline{g_1}} (g)^{(2i+m_2-j)/2} K_{m_2-j}(2\sqrt{\tau g}) \right.$$

$$+ \frac{\sqrt{\tau}}{2} (g)^{(2i+m_2-j-1)/2} K_{m_2-j-1}(2\sqrt{\tau g})$$

$$+ \frac{\sqrt{\tau}}{2} (g)^{(2i+m_2-j-1)/2} K_{m_2-j+1}(2\sqrt{\tau g})$$

$$\left. - \frac{1}{2} (2i + m_2 - j)(g)^{(2i+m_2-j-2)/2} K_{m_2-j}(2\sqrt{\tau g}) \right], \qquad (2.138b)$$

where

$$\Xi[i,j] = \frac{2\binom{i}{j}}{\Gamma(m_2)i!} \left(\frac{m_1}{\gamma_1}\right)^{\frac{2i+m_2-j}{2}} \left(\frac{Cm_2}{\gamma_2}\right)^{\frac{m_2+j}{2}}$$

(2.139)

and $\tau = \frac{m_1 m_2 C}{\bar{g}_1 \bar{g}_2}$ and $C = \frac{\sigma_1^2}{P_1}$. The respective CDFs read:

$$F_a(a) = 1 - \sum_{i=0}^{m_1-1} \sum_{j=0}^{i} \Xi[i,j] e^{-\frac{m_1 a^2}{\bar{g}_1}} a^{(2i+m_2-j)} K_{m_2-j}(2\sqrt{\tau a^2}).$$

(2.140a)

$$F_g(g) = 1 - \sum_{i=0}^{m_1-1} \sum_{j=0}^{i} \Xi[i,j] e^{-\frac{m_1 g}{\bar{g}_1}} g^{(2i+m_2-j)/2} K_{m_2-j}(2\sqrt{\tau g}).$$

(2.140b)

For a system with N relaying segments, an upper bound on the moments of the channel power can be obtained [168] as:

$$\overline{a_{\text{bound}}^k} = \mathcal{Z}_N^k \prod_{i=1}^{N} \left(\frac{\bar{g}_i}{m_i}\right)^{k(N+1-i)/N} \frac{\Gamma\left(m_i + \frac{k(n+1-i)}{N}\right)}{\Gamma(m_i)},$$

(2.141)

where $\mathcal{Z}_N = \frac{1}{N} \prod_{i=1}^{N} C_i^{-\frac{N-i}{N}}$ and $C_i = \sigma_i^2/P_i$. Note that another bound on the end-to-end channel PDF based on the assumption that the relay only inverts the instantaneous channel fade of the previous hop has been derived [276], which is not further explored here.

It can hence be reiterated, the generalization to higher order cascaded channels is possible but leads to expressions that are often only tractable numerically.

2.4.2.3 Relationship With SNR

Due to the importance of the end-to-end SNR in system performance, we shall briefly discuss the relationship between channel power and SNR statistics. In the case of transparent relaying, this relationship is unfortunately not as simple as with the regenerative relaying case. A general expression for the end-to-end SNR γ has been derived [273]:

$$\gamma = \frac{\prod_{i=1}^{N} \gamma_n}{\sum_{i=1}^{N} \frac{\prod_{t=i+1}^{N} \gamma_t}{\prod_{t=1}^{i-1} A_t^2 \sigma_t^2}}$$

(2.142)

where $\gamma_i = a_i^2 E_s/N_0$, E_s is the symbol energy and N_0 the power spectral density. For the case of constant amplification factor $A_i^2 = 1/(C\sigma_i^2)$, above expression simplifies to:

$$\gamma = \frac{\prod_{i=1}^{N} \gamma_i}{\sum_{i=1}^{N} C^{-i+1} \prod_{t=i+1}^{N} \gamma_t}.$$

(2.143)

For the case of variable amplification factor $A_i^2 = 1/(a_i^2 + \sigma_i^2)$, above expression simplifies to:

$$\gamma = \left[\prod_{i=1}^{N} \left(1 + \frac{1}{\gamma_i}\right) - 1\right]^{-1}.$$

(2.144)

Note that the PDF $p_\gamma(\gamma)$ of the end-to-end SNR is generally very difficult to obtain and either numerical or moment-based methods need to be invoked.

2.4.3   Temporal Fading Characteristics

We will subsequently quantify the temporal behavior of the transparent relaying channel. Similarly to the previous section on regenerative relaying channels, albeit in much more condensed form, we will commence with the system assumptions and then move to the canonical as well as more sophisticated propagation scenarios.

2.4.3.1   System Assumptions

To characterize the small scale behavior of the transparent relaying channel, we shall again commence with the geometrical two-ring scattering model of the canonical transparent relaying scenario. It should be noted though that no explicit geometrical channel model has been presented so far. However, modifications of previously discussed models as well as latest results emerging from [280], allow us to proceed. Assuming a two-hop transparent relaying system, the following assumptions are taken as a basis for modeling:

- a wave emitted by the transmitter will impinge upon several clutter surfaces that are closest to the transmitter;
- each of these reflections off clutter surfaces yields a new wave that impinges upon clutter surfaces closest to the relay;
- a wave re-emitted by the relay will impinge upon several clutter surfaces closest to the relay;
- each of these reflections off clutter surfaces yields a new wave which impinges upon clutter surfaces closest to the receiver;
- clutter surrounds transmitter, relay and receiver circularly and uniformly, forming a ring of scatterers;
- waves that are reflected off clutter surfaces that are further away are negligible because they suffer from increased pathloss;
- waves that experience more reflections are negligible because each reflection leads to significant power losses;
- the ring of scatterers is fixed so that the mobile environment can be regarded as quasi-static for sufficiently short periods;
- the number of scatterers at transmitter, relay and receiver tends towards infinity, requiring that the power of each wave is negligible compared to the total mean power.

This scenario is depicted in Figure 2.32 and formalized as follows where the subsequent notations resemble those of the SISO regenerative relaying channel of Section 2.3.3. To simplify notation, as already explained at the beginning of this chapter, we introduce a subscript i for the i-transparent relaying segment. Each segment is composed of a transmitter and receiver; in the two-hop case, for example, the first segment's transmitter is the source, the first segment's receiver is the receiving relay, the second segment's transmitter is the transmitting relay and the second segment's receiver is the destination.

We fix a two-dimensional coordinate system with the origin in the transmitter. The transmitter in the ith relaying stage is moving with speed $v_{t,i}$ at an angle $\gamma_{t,i}$. The receiver is moving with speed $v_{r,i}$ at an angle $\gamma_{r,i}$. The distance between transmitter and receiver is D_i. The clutter rings at transmitter and receiver have a radius $R_{t,i}$ and R_r, respectively. There are M_i and N_i scatterers surrounding the transmitter and receiver respectively. The transmitting wave impinging upon the m_ith clutter object surrounding the transmitter has an angle of departure (AOD) of $\alpha_{m,i}$. Similarly, the receiving wave having been reflected off the nth clutter object surrounding the receiver has an angle of arrival (AOA) of $\alpha_{n,i}$.

The baseband narrowband channel realization in each transparent relaying segment can therefore be written as a superposition of any LOS, single bounced (SB) components that have only been bounced at the transmitter (SBT) and only bounced at the receiver (SBR), and the double bounced (DB) components. For the channel of the i-th relaying segment this reads [172, 173]:

$$h_i(t) = h_i^{\text{LOS}}(t) + h_i^{\text{SBT}}(t) + h_i^{\text{SBR}}(t) + h_i^{\text{DB}}(t). \tag{2.145}$$

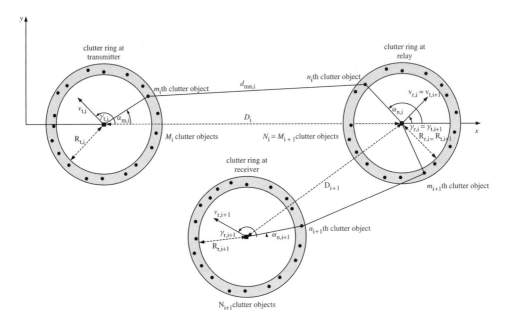

Figure 2.32   Double-bounced two-ring model for SISO transparent relaying channel

Similarly to Section 2.3.3, we will only deal with the double bounced component in this section so as to ease understanding; an extension to the other components is straightforward and illustrated in Section 2.4.4. The individual's relay segment double bounced component hence reads [172]:

$$h_i^{\text{DB}}(t) = \lim_{M_i, N_i \to \infty} \sum_{m_i=1}^{M_i} \sum_{n_i=1}^{N_i} a_{mn,i} \cdot e^{j\left(2\pi f_{mn,i} t - k d_{mn,i} + \phi_{mn,i}\right)}, \tag{2.146}$$

where $a_{mn,i}$ and $\phi_{mn,i}$ are the joint channel gain and phase shift caused by the interaction of m_ith scatterers at the transmitter and n_ith scatterer at the receiver in the ith relaying segment, respectively. Furthermore, $f_{mn,i}$ is the Doppler shift induced by moving transmitter and receiver experienced by the wave reflected off the m_ith scatterer and n_ith scatterer. Finally, $k d_{mn,i}$ is the phase shift induced due to the wave traversing a distance $d_{mn,i}$ whilst traveling from transmitter to receiver. These quantities can be calculated as follows:

- **Amplitude $a_{mn,i}$.** The amplitude introduced by the m_ith clutter at the transmitter is of the same relative magnitude but independent of the amplitude due to the n_ith clutter at the receiver in each relaying segment:

$$a_{mn,i} = a_{m,i} \cdot a_{n,i} = \frac{1}{\sqrt{M_i N_i}}, \tag{2.147}$$

where the last equation results from normalizing conditions which require that the power of Equation (2.146) remains bounded and equal to unity as $M_i, N_i \to \infty$.

- **Phase $\phi_{mn,i}$.** The phase shift introduced by the m_ith clutter at the transmitter is independent of the phase shift introduced by the n_ith clutter at the receiver in each relaying segment:

$$\phi_{mn,i} = \phi_{m,i} + \phi_{n,i}. \tag{2.148}$$

These phase shifts are random and, since the number of scatterers at either end is approaching infinity, these discrete random variables become continuous with PDFs $p_{\phi_{t,i}}(\phi_{t,i})$ and $p_{\phi_{r,i}}(\phi_{r,i})$, respectively. It can generally be assumed that they are uniformly distributed in $[0, 2\pi)$.

- **Doppler Shift $f_{mn,i}$.** The Doppler shifts depend on the geometrical relationship between direction of movement of transmitter or receiver and direction of departure or arrival of the wave. The Doppler shift introduced by the m_ith clutter at the transmitter is generally independent of the Doppler shift introduced by the n_ith clutter at the receiver in each relaying segment:

$$f_{mn,i} = f_{m,i} + f_{n,i} \tag{2.149a}$$

$$f_{m,i} = \hat{f}_{t,i} \cdot \cos\left(\alpha_{m,i} - \gamma_{t,i}\right), \tag{2.149b}$$

$$f_{n,i} = \hat{f}_{r,i} \cdot \cos\left(\alpha_{n,i} - \gamma_{r,i}\right), \tag{2.149c}$$

where $\hat{f}_{t,i}$ and $\hat{f}_{r,i}$ are respectively the maximum Doppler shifts experienced at the transmitter and receiver and which are given as $\hat{f}_{t,i} = f_{c,i} \cdot v_{t,i}/c = v_{t,i}/\lambda_i$ and $\hat{f}_{r,i} = f_{c,i} \cdot v_{r,i}/c = v_{r,i}/\lambda_i$ with $f_{c,i}$ being the carrier frequency and λ_i the wavelength in the ith relaying segment. Note that these carrier frequencies and wavelengths are distinct if the transparent relaying system operates in half duplex mode, and are the same if it operates in full duplex mode. Note further that since the location of the scatterers is not known *a priori*, all AODs and AOAs are discrete random variables. However, since the number of scatterers at source, relays and destination is approaching infinity, these discrete random variables become continuous with given PDFs.

- **Path Length $d_{mn,i}$.** The path length is impacted by the geometrical arrangement of clutter w.r.t. transmitter and receiver in each relaying segment. We will now distinguish two cases, that is D_i is not significantly larger than $\max(R_{t,i}, R_{r,i})$ for all i and D_i is significantly larger than $\max(R_{t,i}, R_{r,i})$ for all i. The insights of these two extreme cases can then be used to derive the behavior for cases where, in some relaying stages, the former relation holds and in others the remaining. Note that the former case is typically encountered indoors and requires the exact expression for $d_{mn,i}$ to be used:

$$d_{mn,i} = R_{t,i} + R_{r,i}$$
$$+ \sqrt{\left(R_{t,i} \sin\alpha_{m,i} - R_{r,i} \sin\alpha_{n,i}\right)^2 + \left(D_i - R_{t,i} \cos\alpha_{m,i} + R_{r,i} \cos\alpha_{n,i}\right)^2}, \tag{2.150}$$

which essentially prevents the decoupling of the two sums in Equation (2.146) into two separate sums since the exact expression for $d_{mn,i}$ can not be separated into two terms of the form $d_{mn,i} = d_{m,i} + d_{n,i}$. The CLT therefore applies to the entire expression and the envelope $a_i = |h_i|$ is Rayleigh distributed as per Section 2.3.2. Thus, the envelope of the end-to-end transparent relaying channel $a = |h| = \prod_{i=1}^{N} |h_i| = \prod_{i=1}^{N} a_i$ is $N * $ Rayleigh distributed.

The latter case, that is $D_i \gg \max(R_{t,i}, R_{r,i})$, is typically encountered outdoors and allows the following simplification to be used [172]:

$$d_{mn,i} \approx D_i + R_{t,i} \cdot \left(1 - \cos\alpha_{m,i}\right) + R_{r,i} \cdot \left(1 + \cos\alpha_{n,i}\right), \tag{2.151}$$

which allows the decoupling of the two sums in Equation (2.146) into two separate sums as per Equation (2.151). Therefore, the CLT applies to each of the sums separately and the envelope $a_i = |h_i|$ is double-Rayleigh distributed as per Section 2.3.2. Thus, the envelope of the end-to-end transparent relaying channel $a = |h| = \prod_{i=1}^{N} |h_i| = \prod_{i=1}^{N} a_i$ is $2N * $ Rayleigh distributed.

Under these assumptions, Equation (2.146) can be rewritten as:

$$h_i^{DB}(t) = \lim_{M_i, N_i \to \infty} \sum_{m_i=1}^{M_i} \sum_{n_i=1}^{N_i} \frac{1}{\sqrt{M_i N_i}} \cdot e^{j\left(2\pi(f_{m,i}+f_{n,i})t - kd_{mn,i} + \phi_{m,i} + \phi_{n,i}\right)} \tag{2.152}$$

and the transparent end-to-end channel reads:

$$h(t) = \prod_{i=1}^{N} h_i = \prod_{i=1}^{N} h_i^{\text{DB}} \tag{2.153a}$$

$$= \prod_{i=1}^{N} \lim_{M_i, N_i \to \infty} \sum_{m_i=1}^{M_i} \sum_{n_i=1}^{N_i} \frac{1}{\sqrt{M_i N_i}} \cdot e^{j(2\pi(f_{m,i}+f_{n,i})t - kd_{mn,i} + \phi_{m,i} + \phi_{n,i})} \tag{2.153b}$$

The similarities of the above formulation in Equation (2.37) with the one given in Equation (2.152) indicates that very similar mathematical approaches for the transparent relaying channel can be invoked as has already been done for the regenerative channel.

2.4.3.2 Canonical Scenario

Without going into the same modeling details as for the transparent relaying channel, we now state several temporal key characteristics related to the transparent narrowband relaying fading channel. We will generally concentrate on the two-hop relaying case since an extension to more relaying segments is straightforward.

- **Temporal Autocorrelation Function.** The first to report on the ACF of the transparent relaying channel were Patel *et al.* [140], who used the fact that Equation (2.153) is of the form $h = h_1 h_2$ where h_1 and h_2 are respectively the channels of the first and second relaying segments.

  Assuming first a constant amplification factor A, the normalized ACF of Equation (2.153) can be calculated as [140]:

$$R(\Delta t) = \mathbb{E}\left\{h(t + \Delta t) \cdot h^*(t)\right\} \tag{2.154a}$$

$$= \mathbb{E}\left\{h_1(t + \Delta t)h_1^*(t)h_2(t + \Delta t)h_2^*(t)\right\} \tag{2.154b}$$

$$= R_1(\Delta t) \cdot R_2(\Delta t) \tag{2.154c}$$

$$= \prod_{i=1}^{2} J_0\left(2\pi \hat{f}_{t,i} \Delta t\right) \cdot J_0\left(2\pi \hat{f}_{r,i} \Delta t\right). \tag{2.154d}$$

Assuming now a variable amplification factor A and a high operational SNR that allows the end-to-end channel to be approximated by $h(t) \approx \sqrt{P_2/P_1} \exp[j\theta_1(t)] \cdot h_2(t)$, the normalized ACF of Equation (2.153) can be calculated as [140]:

$$R(\Delta t) \approx \mathbb{E}\left\{\exp[j\theta_1(t + \Delta t)]\exp[-j\theta_1(t)]h_2(t + \Delta t)h_2^*(t)\right\} \tag{2.155a}$$

$$= J_0\left(2\pi \hat{f}_{t,1}\Delta t\right)\left[1 - J_0^2\left(2\pi \hat{f}_{t,1}\Delta t\right)\right]$$

$$\times {}_2F_1\left(\frac{3}{2}, \frac{3}{2}, 2, J_0^2\left(2\pi \hat{f}_{t,1}\Delta t\right)\right)$$

$$J_0\left(2\pi \hat{f}_{r,1}\Delta t\right)\left[1 - J_0^2\left(2\pi \hat{f}_{r,1}\Delta t\right)\right]$$

$$\times {}_2F_1\left(\frac{3}{2}, \frac{3}{2}, 2, J_0^2\left(2\pi \hat{f}_{r,1}\Delta t\right)\right)$$

$$\times J_0\left(2\pi \hat{f}_{t,2}\Delta t\right) \cdot J_0\left(2\pi \hat{f}_{r,2}\Delta t\right). \tag{2.155b}$$

A closed form solution for arbitrary SNRs has not yet been derived. The ACFs are depicted in Figure 2.33, where we compare the ACFs of the base-to-mobile channel, regenerative mobile-to-mobile channel and transparent mobile relaying channel with constant and variable amplification.

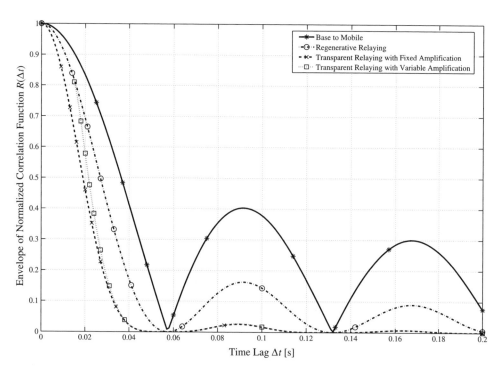

Figure 2.33 Envelop of the temporal autocorrelation function of the complex fading channel comparing the cases of base-to-mobile, regenerative relaying and transparent relaying with constant and variable amplification factor

All mobile terminal speeds have been fixed to 1 m/s. In the case of transparent relaying, the carrier frequency in the first relaying segment is $f_{c,1} = 2.0$ GHz and in the second relaying segment $f_{c,2} = 2.1$ GHz. Again, mobility causes a quicker decorrelation. Notably, there is little difference between the cases of constant and variable amplification factor.

- **Doppler Power Spectrum.** The Doppler power spectrum of the complex channel $h(t)$ is determined from the Fourier transform of its temporal autocorrelation function $R(\Delta t)$ as per Equation (2.40). To date, no closed form solution has been obtained. Some numerical results however have been reported [140].
- **Level Crossing Rate.** Again, Patel *et al.* were the first to report on the LCR of the transparent relaying channel [140], who developed integral expressions for the cases of fixed and variable amplification factors.

  The LCR of the channel envelope $a = |h|$ for a fixed amplification factor can be derived as [140]:

$$N(a_{\text{thr}}) = \frac{4\sqrt{\pi} a_{\text{thr}}^2}{\sqrt{2\bar{g}_1 \bar{g}_2}} \int_0^\infty \frac{1}{y^2} \exp\left(-\frac{\bar{g}_2 a_{\text{thr}}^2 + \bar{g}_1 y^4}{\bar{g}_1 \bar{g}_2 y^2}\right)$$

$$\times \sqrt{\bar{g}_1 \left(\hat{f}_{t,1}^2 + \hat{f}_{r,1}^2\right) y^4 + \bar{g}_2 \left(\hat{f}_{t,2}^2 + \hat{f}_{r,2}^2\right) a_{\text{thr}}^2} \, dy \qquad (2.156)$$

for which no closed form expression is known but for which approximate expressions are available [281].

The LCR of the channel envelope $a = |h|$ for a variable amplification factor was not directly developed [140] but hinted at when deriving the LCR for the end-to-end SNR with constant amplification. Performing some basic manipulations, one can show that it can be derived as:

$$
N(a_{\text{thr}}) = \frac{4\sqrt{\pi}a_{\text{thr}}}{\sqrt{2}\bar{g}_1\bar{g}_2} \exp\left(-a_{\text{thr}}^2/\bar{g}_1\right) \int_0^\infty \frac{1}{y^2} \exp\left(-\frac{C\bar{g}_2 a_{\text{thr}}^2 + \bar{g}_1 y^4}{\bar{g}_1\bar{g}_2 y^2}\right)
$$
$$
\times \sqrt{\bar{g}_1(\hat{f}_{t,1}^2 + \hat{f}_{r,1}^2)y^4(y^2 + C) + C^2\bar{g}_2(\hat{f}_{t,2}^2 + \hat{f}_{r,2}^2)a_{\text{thr}}^2}\, dy \qquad (2.157)
$$

where $C = \sigma_1^2/P_1$. Again, no closed form expression is known but approximate expressions can be developed following the insights of Hadzi-Velkov et al. [281].

- **Average Fade Duration (AFD).** The AFD of the channel envelope $T(a_{\text{thr}})$ can be derived using the definition given in Equation (2.11), where for the case of constant amplification factor the CDF $F_a(a)$ is given by Equation (2.124) and $N(a_{\text{thr}})$ by Equation (2.156), and for the case of variable amplification factor $F_a(a)$ is given by Equation (2.135) and $N(a_{\text{thr}})$ by Equation (2.157). Again, no closed form expressions are known but can be approximated using the analysis of Hadzi-Velkov et al. [281].

With these insights to hand, we are now able to move to more advanced narrowband SISO propagation scenarios, such as nonisotropic clutter distributions and Nakagami fading conditions.

2.4.3.3 Nonisotropic Scattering Scenario

We consider here scatter distributions that differ from the prior assumed isotropic circular distribution. Similarly to the section on regenerative channels, we will henceforth use the von Mises/Tikhonov distribution because it can be used to approximate the other distributions and has the neat property of facilitating closed form expressions for the ACF. Using the PDF of the von Mises/Tikhonov distribution with mean μ and concentration κ given in Equation (2.44), we are able to derive the transparent relaying channel's ACF.

- **Temporal Autocorrelation Function.** Applying this PDF to the AODs and AOAs and plugging it into Equation (2.39d), one obtains after some algebraic manipulations the ACF for the case with constant amplification factor:

$$
R(\Delta t) = \prod_{i=1}^{2} R_i(\Delta t) \qquad (2.158a)
$$

$$
R_i(\Delta t) = \frac{I_0\left(\sqrt{\kappa_{t,i}^2 - 4\pi^2 \hat{f}_{t,i}^2 \Delta t^2 + j4\pi\kappa_{t,i}\hat{f}_{t,i}\Delta t \cos\mu_{t,i}}\right)}{I_0(\kappa_{t,i})}
$$

$$
\times \frac{I_0\left(\sqrt{\kappa_{r,i}^2 - 4\pi^2 \hat{f}_{r,i}^2 \Delta t^2 + j4\pi\kappa_{r,i}\hat{f}_{r,i}\Delta t \cos\mu_{r,i}}\right)}{I_0(\kappa_{r,i})}, \qquad (2.158b)
$$

where the various μ are the mean and κ the concentration factors of the nonisotropic PDFs of the AOD at the transmitter and AOA at the receiver in each relaying segment. Certain geometrical arrangements of clutter may lead to a deterministic relationship between the AOA and AOD at the relay(s).

Assuming now a variable amplification factor A and a high operational SNR, which allows the normalized ACF to be calculated as:

$$
R(\Delta t) \approx \mathbb{E}\left\{\exp[j\theta_1(t + \Delta t)]\exp[-j\theta_1(t)]h_2(t + \Delta t)h_2^*(t)\right\} \qquad (2.159a)
$$

$$= J_0\left(2\pi \hat{f}_{t,1}\Delta t\right)\left[1 - J_0^2\left(2\pi \hat{f}_{t,1}\Delta t\right)\right]$$

$$\times {}_2F_1\left(\frac{3}{2}, \frac{3}{2}; 2; J_0^2\left(2\pi \hat{f}_{t,1}\Delta t\right)\right)$$

$$J_0\left(2\pi \hat{f}_{r,1}\Delta t\right)\left[1 - J_0^2\left(2\pi \hat{f}_{r,1}\Delta t\right)\right]$$

$$\times {}_2F_1\left(\frac{3}{2}, \frac{3}{2}; 2; J_0^2\left(2\pi \hat{f}_{r,1}\Delta t\right)\right)$$

$$\times \frac{I_0\left(\sqrt{\kappa_{t,2}^2 - 4\pi^2 \hat{f}_{t,2}^2\Delta t^2 + j4\pi \kappa_{t,2}\hat{f}_{t,2}\Delta t \cos \mu_{t,2}}\right)}{I_0(\kappa_{t,2})}$$

$$\times \frac{I_0\left(\sqrt{\kappa_{r,2}^2 - 4\pi^2 \hat{f}_{r,2}^2\Delta t^2 + j4\pi \kappa_{r,2}\hat{f}_{r,2}\Delta t \cos \mu_{r,2}}\right)}{I_0(\kappa_{r,2})}. \qquad (2.159\text{b})$$

Similarly, the PSD, ACF and LCR can be calculated but in most cases leads to integral expressions that require numerical evaluation.

2.4.3.4 Nakagami Fading Scenario

The correlation properties of channels other than the dual-hop Rayleigh channel have also been investigated. Notably, Talha and Patzhold [267] have investigated the LCR and AFD of a double Rice fading channel; all expressions, however, remain in integral form. An interesting step forward has been taken [282–284], which quantified the LCR and AFD for end-to-end channels that are composed of one direct Rayleigh link and one transparently relayed double Rayleigh link; all expressions, however, are fairly complex integrals. Hadzi-Velkov *et al.*, [281] have developed expressions for the LCR and AFD of a $N * $ Rayleigh transparent relay channel with constant amplification factor; note that whilst the exact expressions are involved integrals, a viable approximation in closed form has also been developed. We will only briefly highlight the LCR and AFD results developed by Hadzi-Velkov and coworkers [285] for the dual hop transparent relaying Nakagami fading channel with constant amplification.

- **Temporal Autocorrelation Function.** Following Filko and Yakoub [286] and Patel *et al.*, [140], which essentially assumes that the clutter on the transmitting and receiving ends is uncorrelated, the following ACF for the case with constant amplification factor can be derived:

$$R(\Delta t) = \prod_{i=1}^{2} R_i(\Delta t) \qquad (2.160\text{a})$$

$$R_i(\Delta t) = \frac{\Gamma^2\left(m_i + \frac{1}{2}\right)}{m_i \Gamma^2(m_i)} {}_2F_1\left(-\frac{1}{2}, -\frac{1}{2}; m_i; J_0^2\left(2\pi \hat{f}_{t,i}\Delta t\right)\right)$$

$$\times \frac{\Gamma^2\left(m_i + \frac{1}{2}\right)}{m_i \Gamma^2(m_i)} {}_2F_1\left(-\frac{1}{2}, -\frac{1}{2}; m_i; J_0^2\left(2\pi \hat{f}_{r,i}\Delta t\right)\right), \qquad (2.160\text{b})$$

where $\Gamma(\cdot)$ is the complete gamma function and ${}_2F_1(\cdot, \cdot; \cdot; \cdot)$ is the Gauss hypergeometric function.
- **Doppler Power Spectrum.** This requires taking the Fourier transform of Equation (2.160) w.r.t. Δt. However, no closed form expressions have been reported so far.

- **Level Crossing Rate.** An exact expression for the LCR of the envelope with constant amplification has been developed [285], which reads:

$$N(a_{\text{thr}}) = \frac{1}{\sqrt{2\pi}} \frac{4a_{\text{thr}}^{2m_2-1} \sqrt{\bar{\bar{g}}_2}}{\Gamma(m_1)\Gamma(m_2)} \left(\frac{m_1}{\bar{g}_1}\right)^{m_1} \left(\frac{m_2}{\bar{g}_2}\right)^{m_2}$$

$$\times \int_0^\infty \sqrt{1 + \frac{a_{\text{thr}}^2}{x^4}\frac{\bar{\bar{g}}_1}{\bar{\bar{g}}_2}} x^{2(m_1-m_2)} e^{-\left(\frac{m_1 x^2}{\bar{g}_1} + \frac{m_2 a_{\text{thr}}^2}{\bar{g}_2 x^2}\right)} dx, \tag{2.161}$$

where $\bar{\bar{g}}_1 = \pi^2(\hat{f}_{t,1}^2 + \hat{f}_{r,1}^2)\bar{g}_1/m_1$ and $\bar{\bar{g}}_2 = \pi^2(\hat{f}_{t,2}^2 + \hat{f}_{r,2}^2)\bar{g}_2/m_2$. An approximation has also been developed [285]:

$$N(a_{\text{thr}}) = \frac{4a_{\text{thr}}^{2m_2-1} \sqrt{\bar{\bar{g}}_2}}{\Gamma(m_1)\Gamma(m_2)} \left(\frac{m_1}{\bar{g}_1}\right)^{m_1} \left(\frac{m_2}{\bar{g}_2}\right)^{m_2}$$

$$\times \frac{g(x_0, a_{\text{thr}})}{\sqrt{f''(x_0, a_{\text{thr}})}} e^{-f(x_0, a_{\text{thr}})}, \tag{2.162}$$

where

$$f(x, a_{\text{thr}}) = \frac{m_1 x^2}{\bar{g}_1} + \frac{m_2}{\bar{g}_1}\left(\frac{a_{\text{thr}}}{x}\right)^2 - \ln\left(x^{2(m_1-m_2)}\right) \tag{2.163a}$$

$$g(x, a_{\text{thr}}) = \sqrt{1 + \frac{a_{\text{thr}}^2}{x^4}\frac{\bar{\bar{g}}_1}{\bar{\bar{g}}_2}} \tag{2.163b}$$

$$f''(x, a_{\text{thr}}) = \frac{2m_1}{\bar{g}_1} + \frac{6m_2 a_{\text{thr}}^2}{\bar{g}_2 x^4} + \frac{2(m_1 - m_2)}{x^2} \tag{2.163c}$$

$$x_0(a_{\text{thr}}) = \sqrt{\frac{\bar{g}_1\bar{g}_2(m_1 - m_2) + \sqrt{\bar{g}_1^2\bar{g}_2^2(m_1 - m_2)^2 + 4m_1 m_2 \bar{g}_1\bar{g}_2 a_{\text{thr}}^2}}{2m_1\bar{g}_2}} \tag{2.163d}$$

which is obtained by invoking the theory of Laplace approximation.
- **Average Fade Duration (AFD).** Using again the definition of the AFD (Equation 2.11), one only needs to know the CDF of the end-to-end fading channel, which has been derived [270] as:

$$T(a_{\text{thr}}) = \frac{1}{\Gamma(m_1)\Gamma(m_2)} G_{1,3}^{2,1}\left[a_{\text{thr}}^2 \frac{m_1 m_2}{\bar{g}_1\bar{g}_2} \bigg|_{m_1, m_1, 0}^{1}\right], \tag{2.164}$$

where $G[\cdot]$ is the Meijer's G function.

The insights gained in the above discussed canonical and advanced narrowband propagation scenarios can be used to derive the temporal characteristics of any type of propagation scenarios.

2.4.4 Spatial–Temporal Fading Characteristics

We now extend the temporal insights to the spatial domain where we assume that the transmitter, relay and receiver are equipped with multiple antennas. We will first dwell on the underlying system assumptions that allow us to obtain key results for the chosen canonical configuration. We finally discuss some case studies originating from the canonical scenario.

2.4.4.1 System Assumptions

So far, Talha and Patzold [280] have reported on a geometrical channel model for the transparent relaying case with constant amplification, for which the space–time correlation function have been obtained. The geometrical model there is based on a three-ring scatter model, where transmitter, relay and destination are each surrounded by isotropic clutter. We will however formulate the problem in equivalence to the regenerative relaying model developed in Section 2.3.4 which is based on Zajic and Stubber [173] and later develop a few correlation functions following the insights of Talha and Patzold [280]. These models follow the same assumptions as already outlined in Section 2.4.3, as well as the following set of assumptions:

- a LOS component exists between MIMO source and relay as well as between MIMO relay and destination;
- there is no direct LOS component between MIMO source and MIMO destination;
- the existence of a LOS component increases the likelihood of having also single-bounced components;
- the resultant channel is hence modeled as a superposition of LOS, single and double bounced rays.

This scenario is similar to the one already depicted in previous sections and hence not depicted further. It has been formalized [173, 280] as follows where the remaining notations resemble those of the SISO case. To simplify notation, as already explained at the beginning of this chapter, we introduce a subscript i for the ith transparent relaying segment. Each segment is composed of a transmitter and receiver; in the two-hop case for example, the first segment's transmitter is the source, the first segment's receiver is the receiving relay, the second segment's transmitter is the transmitting relay and the second segment's receiver is the destination.

We commence by fixing again a 2D coordinate system with the origin at the transmitter. We will again assume a configuration where the transmitter is equipped with $n_{t,i}$ transmit and the receiver with $n_{r,i}$ receive antennas. The distance between the antenna elements is respectively denoted by $\Delta d_{t,i}$ and $\Delta d_{r,i}$. With this choice of notation, clearly, $\Delta d_{r,i} = \Delta d_{t,i+1}$.

The transmitter is moving with speed $v_{t,i}$ at an angle $\gamma_{t,i}$. The receiver is moving with speed $v_{r,i}$ at an angle $\gamma_{r,i}$. There are M_i and N_i scatterers surrounding the transmitter and receiver respectively. The transmitted wave departed from transmit antenna p_i impinging upon the m_ith clutter object surrounding the transmitter has an angle of departure (AOD) of $\alpha_{pm,i}$. The transmitted wave departed from transmit antenna p_i impinging upon the n_ith clutter object surrounding the receiver has an angle of departure (AOD) of $\alpha_{pn,i}$. The receiving wave having been reflected off the m_ith clutter object surrounding the transmitter and impinging onto receive antenna q_i has an angle of arrival (AOA) of $\alpha_{mq,i}$. The receiving wave having been reflected off the n_ith clutter object surrounding the receiver and impinging onto receive antenna q_i has an angle of arrival (AOA) of $\alpha_{nq,i}$. The angle of the LOS component between transmit antenna p_i and receive antenna q_i is denoted as $\alpha_{pq,i}$.

The distance between transmitter and receiver is D_i. The clutter rings at transmitter and receiver have a radius $R_{t,i}$ and $R_{r,i}$, respectively. The distance $d_{pm,i}$ is distance between the p_ith transmit antenna and the m_ith clutter surrounding the transmitter, $d_{mq,i}$ between the m_ith clutter surrounding the transmitter and the q_ith receive antenna, $d_{pn,i}$ between the p_ith transmit antenna and the n_ith clutter surrounding the receiver, $d_{nq,i}$ between the n_ith clutter surrounding the receiver and the q_ith receive antenna, $d_{mn,i}$ between the m_ith clutter surrounding the transmitter and the n_ith clutter surrounding the receiver, and finally $d_{pq,i}$ is the distance between the p_ith transmit and q_ith receive antenna.

The baseband narrowband channel realization per relaying segment can therefore be written as a superposition of the LOS, single bounced (SB) components that have been bounced at the transmitter (SBT) and receiver (SBR), and the double bounced (DB) components. For the channel between the p_ith transmit and q_ith receive antenna this reads:

$$h_{pq,i}(t) = h_{pq,i}^{\text{LOS}}(t) + h_{pq,i}^{\text{SBT}}(t) + h_{pq,i}^{\text{SBR}}(t) + h_{pq,i}^{\text{DB}}(t), \tag{2.165}$$

where the respective components read

$$h_{pq,i}^{LOS}(t) = a_{pq,i}^{LOS} e^{j\left(2\pi f_{pq,i}^{LOS}t - kd_{pq,i}^{LOS}\right)} \tag{2.166a}$$

$$h_{pq,i}^{SBT}(t) = \lim_{M_i \to \infty} \sum_{m_i=1}^{M_i} a_{pq,m,i}^{SBT} \cdot e^{j\left(2\pi f_{pq,i}^{SBT}t - kd_{pq,i}^{SBT} + \phi_{m,i}^{SBT}\right)} \tag{2.166b}$$

$$h_{pq,i}^{SBR}(t) = \lim_{N_i \to \infty} \sum_{n_i=1}^{N_i} a_{pq,n,i}^{SBR} \cdot e^{j\left(2\pi f_{pq,i}^{SBR}t - kd_{pq,i}^{SBR} + \phi_{n,i}^{SBR}\right)} \tag{2.166c}$$

$$h_{pq,i}^{DB}(t) = \lim_{M_i,N_i \to \infty} \sum_{m_i=1}^{M_i} \sum_{n_i=1}^{N_i} a_{pq,mn,i}^{DB} \cdot e^{j\left(2\pi f_{pq,i}^{DB}t - kd_{pq,i}^{DB} + \phi_{mn,i}^{DB}\right)}. \tag{2.166d}$$

Omitting the sub- and superscripts, a and ϕ are the channel gains and phase shifts caused by the interaction of waves and scatterers. Furthermore, f are the Doppler shifts experienced by the waves due to the moving transmitter and receiver. Finally, kd is the phase shift induced due to the wave traversing a distance d whilst traveling from transmitter to receiver. These quantities can be calculated in the very same manner as has already been done in Section 2.3.4 with each variable indexed by the relaying segment i. These equations are omitted here for the sake of brevity. Above equations allow us to build the end-to-end MIMO channel, which for two antenna elements at source, relay and destination reads:

$$\mathbf{H}(t) = \mathbf{H}_2(t)\mathbf{H}_1(t) \tag{2.167a}$$

$$\begin{pmatrix} h_{11}(t) & h_{12}(t) \\ h_{21}(t) & h_{22}(t) \end{pmatrix} = \prod_{i=2}^{1} \begin{pmatrix} h_{11,i}(t) & h_{12,i}(t) \\ h_{21,i}(t) & h_{22,i}(t) \end{pmatrix}. \tag{2.167b}$$

These equations only hold for constant amplification factor. Any theoretical approach assuming a variable amplification factor is highly complex and is therefore omitted in subsequent derivations.

2.4.4.2 Canonical Scenario

We are now in a position to derive temporal and spatial key quantities related to the transparent MIMO relay fading channel. Following Talha and Patzold [280], the space–time correlation function (STCF) can be calculated as follows:

• **Space–Time Correlation Function.** The same approach as in previous sections can be used here but leads to very different results. As such, the STCF of the complex fading channel between any transmit–receive antenna pair at source and destination $p_1 q_2$ and any other pair $\tilde{p}_1 \tilde{q}_2$ is composed of a sum of products between the LOS, SBT, SBR and DB components in each relaying segment. The derivations are possible but quite involved. For the sake of illustration, we will only calculate the STFC between the first and second transmit-receive pair for the DB components [280]:

$$R_{11,22}^{DB}(\Delta \mathbf{d}, \Delta t) = \mathbb{E}_{\mathbf{v}}\left\{h_{11}^{DB}(t + \tau) \cdot (h_{22}^{DB}(t))^*\right\} \tag{2.168}$$

where $\Delta \mathbf{d} = (\Delta d_{t,1}, \Delta d_{r,1}, \Delta d_{t,2}, \Delta d_{r,2})$ and the expectation is taken w.r.t. all involved random variable. Furthermore, from Equations (2.165) and (2.167), we have:

$$h_{11}^{DB}(t) = h_{11,2}^{DB}h_{11,1}^{DB} + h_{12,2}^{DB}h_{21,1}^{DB} \tag{2.169a}$$

$$h_{22}^{DB}(t) = h_{21,2}^{DB}h_{12,1}^{DB} + h_{22,2}^{DB}h_{22,1}^{DB} \tag{2.169b}$$

Plugging in Equation (2.166d) and following the procedure outlined in previous sections, one obtains for isotropic scattering [280]:

$$R_{11,22}^{\mathrm{DB}}(\Delta\mathbf{d}, \Delta t) = R_{\mathrm{t},1}^{\mathrm{DB}}(\Delta d_{\mathrm{t},1}, \Delta t) R_{\mathrm{r}1,t2}^{\mathrm{DB}}(\Delta d_{\mathrm{r},1}, \Delta d_{\mathrm{t},2}, \Delta t) R_{\mathrm{r},2}^{\mathrm{DB}}(\Delta d_{\mathrm{r},2}, \Delta t), \qquad (2.170)$$

where

$$R_{\mathrm{t},1}^{\mathrm{DB}}(\Delta d_{\mathrm{t},1}, \Delta t) =$$

$$J_0\left(2\pi\sqrt{\left(\frac{\Delta d_{\mathrm{t},1}}{\lambda_1}\right)^2 + \left(\hat{f}_{\mathrm{t},1}\Delta t\right)^2 - 2\frac{\Delta d_{\mathrm{t},1}}{\lambda_1}\hat{f}_{\mathrm{t},1}\Delta t \cos(\gamma_{\mathrm{t},1} - \theta_{\mathrm{t},1})}\right) \qquad (2.171a)$$

$$R_{\mathrm{r}1,t2}^{\mathrm{DB}}(\Delta d_{\mathrm{r},1}, \Delta d_{\mathrm{t},2}, \Delta t) = \frac{1}{3} J_0\left(2\pi\,\hat{f}_{\mathrm{r},1}\Delta t\right) J_0\left(2\pi\,\hat{f}_{\mathrm{t},2}\Delta t\right)$$

$$+ \frac{1}{3} J_0\left(2\pi\sqrt{\left(\frac{\Delta d_{\mathrm{r},1}}{\lambda_1}\right)^2 + \left(\hat{f}_{\mathrm{r},1}\Delta t\right)^2 - 2\frac{\Delta d_{\mathrm{r},1}}{\lambda_1}\hat{f}_{\mathrm{r},1}\Delta t \cos(\gamma_{\mathrm{r},1} - \theta_{\mathrm{r},1})}\right)$$

$$\times J_0\left(2\pi\sqrt{\left(\frac{\Delta d_{\mathrm{t},2}}{\lambda_2}\right)^2 + \left(\hat{f}_{\mathrm{t},2}\Delta t\right)^2 - 2\frac{\Delta d_{\mathrm{t},2}}{\lambda_2}\hat{f}_{\mathrm{t},2}\Delta t \cos(\gamma_{\mathrm{t},2} - \theta_{\mathrm{t},2})}\right)$$

$$+ \frac{1}{3} J_0\left(2\pi\sqrt{\left(\frac{\Delta d_{\mathrm{r},1}}{\lambda_1}\right)^2 + \left(\hat{f}_{\mathrm{r},1}\Delta t\right)^2 + 2\frac{\Delta d_{\mathrm{r},1}}{\lambda_1}\hat{f}_{\mathrm{r},1}\Delta t \cos(\gamma_{\mathrm{r},1} - \theta_{\mathrm{r},1})}\right)$$

$$\times J_0\left(2\pi\sqrt{\left(\frac{\Delta d_{\mathrm{t},2}}{\lambda_2}\right)^2 + (\hat{f}_{\mathrm{t},2}\Delta t)^2 + 2\frac{\Delta d_{\mathrm{t},2}}{\lambda_2}\hat{f}_{\mathrm{t},2}\Delta t \cos(\gamma_{\mathrm{t},2} - \theta_{\mathrm{t},2})}\right) (2.171b)$$

$$R_{\mathrm{r},2}^{\mathrm{DB}}(\Delta d_{\mathrm{r},2}, \Delta t) =$$

$$J_0\left(2\pi\sqrt{\left(\frac{\Delta d_{\mathrm{r},2}}{\lambda_2}\right)^2 + (\hat{f}_{\mathrm{r},2}\Delta t)^2 - 2\frac{\Delta d_{\mathrm{r},2}}{\lambda_2}\hat{f}_{\mathrm{r},2}\Delta t \cos(\gamma_{\mathrm{r},2} - \theta_{\mathrm{r},2})}\right). \qquad (2.171c)$$

We have assumed here that the wavelength in each relaying segment is different due to the half-duplex transparent relaying constraint. Note further that changing above equations to the more general case of nonisotropic scattering conditions is straightforward.

Expressions for the PSD, LCR, AFD, etc., can be similarly obtained following the approach outlined in previous sections.

2.4.4.3 Case Studies

An exemplification of Equation (2.170) has been depicted in Figure 2.34, where the following parameterization has been assumed: All mobile terminals move at a speed of 1 m/s. Furthermore, $f_{\mathrm{c},1} = 2.0\,\mathrm{GHz}$, $f_{\mathrm{c},2} = 2.1\,\mathrm{GHz}$, $\theta_{\mathrm{t},1} = \theta_{\mathrm{r},1} = \theta_{\mathrm{t},2} = \theta_{\mathrm{r},1} = \pi/2$, $\gamma_{\mathrm{t},1} = \pi$, $\gamma_{\mathrm{r},1} = \gamma_{\mathrm{t},2} = \pi/2$ and $\gamma_{\mathrm{r},2} = 0$. Similar insights as already discussed in the context of regenerative relays hold here.

2.4.5 Spectral–Spatial–Temporal Fading Characteristics

We finally extend the spatio-temporal insights to the spectral domain where we assume that the source, relay and destination are equipped with multiple antennas and the signal components are resolvable

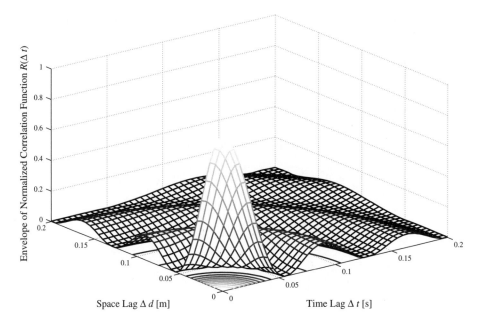

Figure 2.34 Spatial and temporal domains decorrelate similarly; $\Delta d = \Delta d_{\mathrm{t},1} = \Delta d_{\mathrm{r},1} = \Delta d_{\mathrm{t},2} = \Delta d_{\mathrm{r},2}$

in time. We will only briefly dwell on the underlying system assumptions since at the time of writing of this book, no contributions were available that relate to the transparent wideband MIMO relaying channel. We thus briefly hint at how to extend prior models to this case. With respect to the model assumptions of Sections 2.4.3 and 2.4.4, the following assumption holds:

- scatterers are distributed in 3D within concentric cylinders around source, relay and destination, which allows the PDP to be modeled more precisely in urban environments.

The parameterization of this model follows the one outlined in Section 2.3.5 with the sub-indexing approach of Sections 2.4.3 and 2.4.4. Omitting the notational details, we thus arrive at the time-variant frequency transfer function of each relaying segment:

$$T_{pq,i}(t, f) = T_{pq,i}^{\mathrm{LOS}}(t, f) + T_{pq,i}^{\mathrm{SBT}}(t, f) + T_{pq,i}^{\mathrm{SBR}}(t, f) + T_{pq,i}^{\mathrm{DB}}(t, f), \tag{2.172}$$

where the respective components read

$$T_{pq,i}^{\mathrm{LOS}}(t, f) = a_{pq,i}^{\mathrm{LOS}} e^{j\left(2\pi f_{pq,i}^{\mathrm{LOS}} t - 2\pi f \tau_{pq,i}^{\mathrm{LOS}} - k d_{pq,i}^{\mathrm{LOS}}\right)} \tag{2.173a}$$

$$T_{pq,i}^{\mathrm{SBT}}(t, f) = \lim_{M_i \to \infty} \sum_{l_i=1}^{L_i} \sum_{m_i=1}^{M_i^{(l_i)}} a_{pq,ml,i}^{\mathrm{SBT}} \cdot e^{j\left(2\pi f_{pq,ml,i}^{\mathrm{SBT}} t\right)}$$

$$\times e^{j\left(-2\pi f \tau_{pq,ml,i}^{\mathrm{SBT}} - k d_{pq,ml,i}^{\mathrm{SBT}} + \phi_{ml,i}^{\mathrm{SBT}}\right)} \tag{2.173b}$$

$$T_{pq,i}^{\mathrm{SBR}}(t, f) = \lim_{N_i \to \infty} \sum_{k_i=1}^{P_i} \sum_{n_i=1}^{N_i^{(k_i)}} a_{pq,nk,i}^{\mathrm{SBR}} \cdot e^{j\left(2\pi f_{pq,nk,i}^{\mathrm{SBR}} t\right)}$$

$$\times e^{j\left(-2\pi f \tau_{pq,nk,i}^{\mathrm{SBR}} - k d_{pq,nk,i}^{\mathrm{SBR}} + \phi_{nk,i}^{\mathrm{SBR}}\right)} \tag{2.173c}$$

$$T_{pq,i}^{\mathrm{DB}}(t,f) = \lim_{M_i,N_i \to \infty} \sum_{l_i,m_i=1}^{L_i,M_i^{(l_i)}} \sum_{k_i,n_i=1}^{P_i,N_i^{(k_i)}} a_{pq,ml,nk,i}^{\mathrm{DB}} \cdot e^{j\left(2\pi f_{pq,ml,nk,i}^{\mathrm{DB}}t\right)}$$

$$\times\, e^{j\left(-2\pi f \tau_{pq,ml,nk,i}^{\mathrm{DB}} - k d_{pq,ml,nk,i}^{\mathrm{DB}} + \phi_{ml,nk,i}^{\mathrm{DB}}\right)}. \tag{2.173d}$$

Omitting the sub- and superscripts, a and ϕ are the channel gains and phase shifts caused by the interaction of waves and scatterers. Furthermore, f are the Doppler shifts experienced by the waves due to the moving transmitter and receiver. Then, τ are the delays of the MPCs induced by the 3D clutter. Finally, kd is the phase shift induced due to the wave traversing a distance d whilst traveling from transmitter to receiver. These quantities can again be calculated in the very same manner as already done in Section 2.3.5 with each variable indexed by the relaying segment i. These equations are omitted here for the sake of brevity. The above equations allow us to build the end-to-end MIMO channel, which for two antenna elements at source, relay and destination reads:

$$\mathbf{H}(t,f) = \mathbf{H}_2(t,f) * \mathbf{H}_1(t,f), \tag{2.174}$$

where $*$ denoted convolution, which is performed here over frequency. The convolution is due to the multiplicative nature of the transparent relaying channel.

The temporal, spatial and spectral key quantities related to the transparent MIMO relay fading channel can thus be obtained by following exactly the same procedure as outlined in previous sections. Results can readily be obtained in integral form and may require some approximations to be cast in closed form. These derivations are omitted here for the sake of brevity.

2.4.6 Simulating Transparent Fading Channels

Simulating the transparent relay channel is crucial in characterizing the performance of such system by means of link layer simulations. Ideally, any simulation should mimic the statistical behavior of the channel as precisely as possible. Since the above calculations assumed an infinite number of scatterers, an identical reproduction of the channel behavior is prohibitive in terms of computational complexity.

Comparing Sections 2.3.3, 2.3.4 and 2.3.5 with Sections 2.4.3, 2.4.4 and 2.4.5 reveals that the formulation of the channel is essentially equivalent. Therefore, the simulators developed in Section 2.3.6 for regenerative relays can readily be applied to the transparent relaying channel. This procedure is omitted here for the sake of brevity.

2.4.7 Measurements and Empirical Models

As of today, no measurements dedicated to the transparent relaying channel have been reported. Undoubtedly, many but not all insights can be obtained by concatenating in a transparent fashion the channel models of the regenerative relaying measurement campaigns. This procedure is again omitted here for the sake of brevity and left to the reader.

2.4.8 Estimating Transparent Fading Channels

Whether analyzed or simulated by means of theoretical or empirical models, the estimation of the relaying channel has a profound impact onto the performance of transparent relaying protocols [250, 287–296]. The aim of this section is again not to go into great detail but rather to highlight the few available contributions in this area.

The milestone contribution in the area of channel estimation and pilot designs for transparent relaying channels has been made by Patel and Stuber [297], who developed pilot-aided estimation schemes. The resultant estimator has been quantified and verified by means of performance analysis.

Kim *et al.* [298] derive the Cramer–Rao lower bound (CRLB) of the transparent relay estimator. They then move on to the design of an optimal preamble that minimizes the CRB under several power constraints. Moreover, a minimum variance unbiased (MVU) estimator that achieves the CRB is introduced for the cooperative channel.

Cui and coworkers [299, 300] studied a training based channel estimation for transparent relay networks. They first pointed out that separately estimating the channel from source to relay and the channel from relay to destination incurs several problems. They thereupon propose a new estimation scheme that directly estimates the overall channel from source to destination, where both linear least-square estimator and minimum mean-square-error estimators are studied. The corresponding optimal training sequences and the optimal precoding matrices are also derived.

Herdin and Auer [301] propose inserting a new pilot grid at the relay station in addition to the existing pilot grid to aid the channel estimation process. At the destination, channel estimation is performed in two steps, where the second and first hop channels are estimated separately. The proposed pilot and channel estimation scheme allows the gains through chunk reordering at the relay station to be fully exploited, a scheme previously proposed by the same authors, with only a small increase in complexity.

Woo and coworkers [302–304] propose a MMSE channel estimation technique for a transparent OFDM relaying system. The estimator takes the propagation delays due to relaying into account. The authors also demonstrate the superiority of the proposed algorithm in terms of the mean square error (MSE) and BER.

Gao *et al.* [305, 306] deal with the channel estimation problem for the two-way transparent relay network. They derive the maximum likelihood (ML) channel estimator as well as a new estimator called the linear maximum signal-to-noise ratio (LMSNR) estimator. It is shown that the proposed methods give superior performance compared with the common channel estimators like the least-square (LS) and the linear minimum-mean-squared-error (LMMSE) under the given assumptions scenario. The provided study is based on any given training sequence, that is no optimization has been performed.

Zhang *et al.* [254, 255] deal with channel estimation in a transparent but also regenerative cooperative orthogonal frequency-division multiplexing (OFDM) network in the presence of frequency offsets. In order to eliminate the multiple access interference occurring in such systems, the maximum number of active regenerative relays is $\lceil N/2L - 1 \rceil$, where N is the total number of subcarriers and L is the channel order. Various tradeoffs are then formulated in the paper.

Lalos *et al.* [307] present efficient channel estimation algorithms for wideband transparent relay networks. They show that all channels in the network from the source through the relays to the destination node can be blindly estimated up to a phase ambiguities vector that contains the phases of the direct source to destination channel frequency response.

Liu *et al.* [308] provide three channel estimation techniques, that is LS (least square) estimation, LMMSE (linear minimum mean square error) estimation and LR-MMSE (low rank minimum mean square error) estimation utilized at the relay channel to approximate the optimal solution in OFDM systems. It is shown that LMMSE estimation outperforms the other two estimations at the price of being very complex. The performance of LR-MMSE estimation approaches that of the LMMSE estimation, but LR-MMSE exhibits less complexity in terms of number of multiplications.

Gedik and Uysal [309] investigated the performance of amplify-and-forward relaying with two different pilot-symbol-assisted channel estimation methods. In the first estimation method, the cascaded channel consisting of source-to-relay and relay-to-destination links was estimated at the destination terminal. In the second estimation method, the estimation of cascaded channel was disintegrated into separate estimations of source-to-relay and relay-to-destination links, which are carried out at the relay and destination terminals, respectively. The latter method involves feed-forwarding a quantized version of the source-to-relay channel estimate to the destination terminal. Results indicate that cascaded channel estimation outperforms the other approach with a small number of quantization bits. As the number of employed quantization bits increases, the performance of disintegrated channel estimation approaches the cascaded one until eventually outperforming it.

Behbahani and Eltawil [310] designed and analyzed a training based LMMSE channel estimator for transparent relaying networks. Various scenarios were considered and compared with each other.

Pham *et al.* [311, 312] designed a training sequence for bi-directional transparent relaying networks using single carrier cyclic prefix (SCCP). They proposed a design for training sequences from two nodes to minimize the MSE of the channel estimation according to the zero forcing (ZF) criterion.

Firag and Garth [313] developed a practical adaptive equalizer structure for time reversal space time block-coded transparent relaying systems. This essentially eliminated the need for a separate decoding block requiring explicit channel estimation. The adaptive equalizer length is also independent of the data block length, making it particularly suitable for frequency selective fading channels with short delay spreads. The authors derive the MMSE solution and, using the resulting structure, develop a block recursive least squares (RLS) adaptive algorithm. Applying this algorithm to previously developed AF protocols shows the viability and superiority of the developed solution.

Yan *et al.* [314] focus on LMMSE channel estimation algorithms tailored to transparent OFDM-based relaying systems with multiple deployed AF relays. The thread of the analysis is around complexity, which has been shown to be sufficiently low.

Hu *et al.* [315] focused on channel estimation for bidirectional relaying applied to cellular systems. Based on some analytical insights, they proposed a novel pilot-aided transmission strategy based on a semiorthogonal pilot structure. The main idea is to preequalize the signals at the BS in dependency of the characteristics of the BS-RS link. At the RS, part of the CSI is estimated and used to equalize the relayed signals. With the thus proposed transmission scheme and pilot structure, the number of pilots is essentially halved and the channel estimation at the MS has much lower complexity.

Roemer and Haardt [316] also focused on transparent two-way relaying where the simple AF relays do not have any CSI available. The reliable estimation of all relevant channel parameters is performed by means of a novel tensor-based channel estimation algorithm, referred to as TENCE, which provides both terminals with full knowledge of all channel parameters involved in the transmission. The solution is algebraic, that is, it does not require any iterative procedures and updates as many of related blind algorithms. Applicable to arbitrary antenna configurations, criteria for the design of the pilot symbols and the corresponding relay amplification matrices are derived.

Thiagarajan *et al.* [317] exploit a simple training scheme available for estimating space–time channels is exploited to estimate the channel frequency offset (CFO) at the relay for OFDM-based transparent relaying systems. ML and LS based joint CFO and channel estimators are constructed for estimating the CFO and the end-to-end product channel at the destination. They then show that the LS based estimator is equivalent to the ML based estimator.

Rajan and Rajan [318] utilized insights from Dayal–Brehler–Varanasi who have proved that training codes in a MIMO setting achieve the same diversity order as that of the underlying coherent STBC if a simple minimum mean squared error estimate of the channel formed using the training part is employed for coherent detection of the underlying STBC. They extend this approach to transparent relay networks using coherent, distributed STBCs by means of a combination of training, channel estimation and detection. They also extend these results to asynchronous relay networks using orthogonal frequency division multiplexing.

From this brief survey it becomes apparent that, compared with the regenerative relaying case, more contributions are available is due to the transparent relaying channel exhibiting entirely novel features. Nonetheless, developments related to channel estimation in transparent relaying channels are very recent with many open issues pertaining, mainly in the context of the two-way relaying system as well as phase estimation and channel prediction.

2.5 Distributed MIMO Channel

We will only briefly dwell on the distributed MIMO channel since most problems can be reduced to a superposition of already discussed problems.

2.5.1   Problem Reduction

As per Figure 2.35, the distributed MIMO channel can be assembled from elements of the regenerative and transparent relaying channel, as has been discussed throughout Sections 2.3 and 2.4. Some differences, however, need to be taken into account:

- **Independent Antenna Elements.** The main difference is that each antenna element needs to be modeled with independent speed, direction of movement, clutter environment, K-factor, etc. This complicates formulation a little, in particular if each node itself is equipped with multiple antennas; however, since the derivations are straightforward, they are not further investigated here.
- **Coverage Limited System.** Another difference is that when the system is range limited, the distances between source and first stage relays forming a virtual antenna array (VAA) as well as the destination VAA and the actual destination are typically very small compared with the distances formed between VAAs. This implies that the performance is dictated by the channel between the source and destination VAA, that is the end-to-end distributed MIMO channel.
- **Capacity Limited System.** As already alluded to in Chapter 1, when the system is capacity limited, there does usually exist a direct link between source and destination, in addition to the link via the relay. This implies that for all transparent relaying protocols with a full-duplex relay and some transparent relaying protocols with a half-duplex relay, the channel as seen by the destination is a superposition of the direct and the transparently relayed channel. This changes the channel's behavior, which has been analyzed for the LCR and AFD [282–284]. The topic, however, remains largely unexplored.

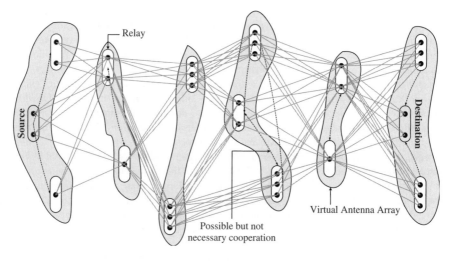

Figure 2.35   Assembly of spatially distributed and possibly cooperating terminals forming a distributed MIMO channel by means of VAAs

At high level, such an end-to-end MIMO system is characterized by its channel matrix $\mathbf{H}$:

$$\mathbf{H} = \begin{pmatrix} h_{11} & h_{12} & \cdots & h_{1,n_t} \\ h_{21} & h_{22} & \cdots & h_{2,n_t} \\ \vdots & \vdots & \ddots & \vdots \\ h_{n_r,1} & h_{n_r,2} & \cdots & h_{n_r,n_t}, \end{pmatrix}$$

where n_t and n_r are respectively the number of VAA transmit and receive antennas, h_{kl} is the channel from kth transmit antenna of the source VAA to the lth receive antenna of the destination VAA. Each entry h_{kl} can itself be wideband in nature and is characterized by the theoretical body exposed in previous sections.

2.5.2 Main Design Criteria

The above representation allows one to gain insights into important distributed MIMO design parameters, that is the rank, diversity gain and power gain of the channel:

- **Rank.** As for the rank, also referred to as the degree-of-freedom, this corresponds to the minimum number of nonzero rows and nonzero columns in **H**. It depends on the amount of clutter in the channel and the antenna separation. It generally determines the data multiplexing capabilities of the channel.
- **Diversity Gain.** As for the diversity gain, it corresponds to the number of nonzero entries in **H**. It depends on the connectivity of the channel and the antenna separation and generally determines the reliability of the channel. In addition to the traditional diversity offered at small scale fading, distributed MIMO channels also enjoy macro diversity gains, which we will briefly allude to below.
- **Power Gain.** Finally, as for the power gain, this is characterized by the strongest eigenvalue of **H** w.r.t. the weakest eigenvalue, that is the condition number $\max \lambda_i / \min \lambda_i$. It essentially determines the beamforming capabilities of the distributed MIMO system.

In the context of distributed cooperative communication systems, as depicted in Figure 2.35, such distributed topologies are usually submerged in a rich clutter environment, resulting in a full-rank channel with maximum degrees of freedom (that is high data throughput), a fully connected channel with maximum diversity gain (that is high reliability), and a well conditioned channel with little beamforming gain (that is limited range).

2.5.3 Macro Diversity Gains

As mentioned above, distributed MIMO systems not only enjoy micro diversity gains for fading but also macro diversity gain for shadowing. With reference to Figure 2.35, the receiver enjoys macro diversity in that the probability of all channels being in a poor shadowing state diminishes with an increasing number of channels. One way to exploit such macro diversity gain is to communicate only over the strongest channel, that is the one that is least shadowed. If, for instance, the best of N available channels is chosen for communication then the effective PDF of the shadowing process in decibels as seen by the receiver can be calculated from order statistics [319] as:

$$p\left(S^{(N)}\right) = N \cdot F^{N-1}(S) \cdot p(S) \tag{2.175}$$

$$= N \cdot \left(1 - Q\left(\frac{S}{\sigma_{dB}}\right)\right)^{N-1} \cdot \frac{1}{\sqrt{2\pi}\,\sigma_{dB}} e^{-\frac{S^2}{2\sigma_{dB}^2}}, \tag{2.176}$$

where $Q(\cdot)$ is the Q-function. The PDF is visualized in Figure 2.36 for different N. Clearly, macro diversity significantly improves the performance as the low power tail is shifted to the right for an increasing number of available channels. From this, an improvement in outage and error performance can be observed (see for example, [320] and references therein).

2.6 Concluding Remarks

This chapter serves to introduce channel modeling issues related to cooperative relaying systems. To this end, we first dealt with key modeling issues of a general point-to-point fading channel. We showed

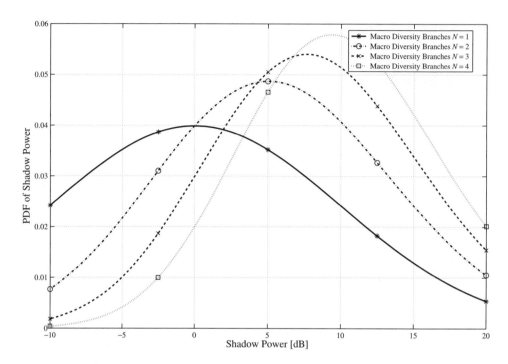

Figure 2.36 Resultant PDF of the shadow process in decibels when macro diversity with N diversity branches is employed; $\sigma_{dB} = 10\,dB$

that the complicated propagation behavior can be broken down into the three underlying effects of pathloss, shadowing and fading. We then discussed their characteristics and statistical behavior.

We then moved on to more detailed modeling issues of the regenerative and transparent wireless channel. For each case, we summarized the pathloss and shadowing modeling approaches before treating fading in greater detail. We dealt with the fading amplitude distributions and the respective temporal, spatial and spectral correlation functions. We then also summarized the latest contributions pertaining to channel simulation and estimation as well as measurements and the resulting empirical models. We then showed that the distributed cooperative relaying channel can be broken down into transparent and regenerative channels as dealt with earlier.

Throughout we observed that the biggest difference between the cooperative relaying and traditional channels is the difference in height of transmit and/or receive antenna where usually both ends are of low height and thus well submerged into clutter. Another important difference is the movement of both transmitter and receiver, which greatly affected temporal correlation behavior. Another major difference is the reduced communication distance in cooperative systems, which results in communication happening within breakpoint distance and shadowing variations reduced.

With respect to each other, regenerative and transparent relaying as well as distributed systems exhibit very different behavior:

- **Regenerative Relaying Channel.** First, the reduced communication distances yield, generally, less aggregated pathloss, less aggregated shadowing variations and shorter delay spreads, and hence a significant reduction in frequency selectivity. Second, due to a mobile transmitter and mobile receiver, the temporal correlation is greatly impacted. Third, due to the regenerative nature of the system, each relaying segment is decoupled leading to the observation of known amplitude statistics.

The major part of the performance gains in regenerative relaying systems originates indisputably from pathloss and to some extent shadowing gains but not from fading gains.

- **Transparent Relaying Channel.** First, despite the reduced communication distances and hence generally less pathloss and shadowing per relaying stage, the transparent nature of the system amplifies the breakpoint behavior, augments the shadowing variations and typically increases the end-to-end delay spread. Second, due to mobile transmitter and mobile receiver, the temporal correlation is again greatly impacted. Third, due to the transparent nature of the system, all relaying segments are coupled making fading behavior dependent on the system topology and leading to novel amplitude statistics of more severe behavior. The major part of the performance loss in transparent relaying systems originates indisputably from the aggravation of the statistical variations of each relaying segment, be they due to shadowing or fading.

- **Distributed Relaying Channel.** Depending on whether regenerative or transparent relaying systems are used to realize the distributed system, the respective insights from above hold. One further major advantage of distributed relaying systems is the ability to capitalize on macro and micro diversity gains in that the best from the available distributed channels is used for communication.

Quite a few problems remain open at the time of writing of this book, more of which will be discussed in Chapter 6. For instance, there are no channel measurements or empirical models available that capture the behavior of the transparent as well as distributed relaying channel. Furthermore, most of the analytical results for the transparent relaying channel remain in integral form or have not even been obtained. Finally, the temporal, spatial and spectral characteristics and characterization of the eigenvalues of the MIMO regenerative and transparent relaying as well as distributed channel are largely an open problem.

3

Transparent Relaying Techniques

3.1 Introductory Note

3.1.1 Chapter Contents

Analyzing the physical layer performance of a wireless system is vital in understanding, optimizing and synthesizing system parameters. As of 2009, there are several hundreds of highly complex contributions on transparent PHY layer analysis and design available. This requires us to concentrate on some canonical techniques that can then be applied to other scenarios. For this reason, we proceed in this chapter with the following topics:

- transparent relaying protocols;
- transparent space–time processing protocols;
- distributed system optimization.

Transparent relaying protocols are realized by relays that transparently relay the received signal, either by simply amplifying it or performing some linear (analog) operations. For more general topologies, such as with several serial relaying stages or multiple parallel relaying branches, we will generally assume that these do not interfere. Distributed space–time processing protocols, on the other hand, facilitate distributed space–time transmission with the aid of one or several transparent relays. Note that for the below topologies we assume a sufficient degree of synchronization between nodes, since asynchronous transparent space–time processing protocols are not dealt with in this chapter. Finally, the system optimization includes issues pertaining to distributed power allocation as well as distributed relay selection.

This chapter is fairly short compared with the subsequent chapter on regenerative protocols. The prime reason is because we feel that the applicability of stand-alone transparent relaying techniques is limited in real-world applications. This observation is rooted in the very poor wireless channel, as already discussed in Chapter 2, and the only slightly inferior complexity/cost w.r.t. regenerative solutions, to be discussed in Chapter 5. The aim of this chapter is hence to offer some basic tools in analysis and design in the context of some canonical relaying topologies that can easily be extended to more complex scenarios if the need arises. Also, some hybrid protocols based on transparent and regenerative approaches will be dealt with in Chapter 4, for which below theory forms a sound basis.

3.1.2 Choice of Notation

Concerning the notation, we will often, but not always, adopt the notation used in the open literature. For two-hop topologies, we will often subscript the source–destination, source–relay and

Cooperative Communications Mischa Dohler and Yonghui Li
© 2010 John Wiley & Sons, Ltd

relay–destination links in an explicit manner by, respectively, SD, SR and RD. If the need arises, we will add further subscripts to the number of relays.

For general topologies, however, we will stick to the notation used in previous chapters, that is subscript zero is the source–destination link and numbers respectively the relaying links. For instance, the number of transmit antennas at the transmitter of the ith stage becomes $n_{t,i}$, the transmit power into the ith stage is P_i, the channel of the ith stage is h_i, the number of receive antennas at the receiver of the ith stage is $n_{r,i}$, the noise at the receiver of the ith relaying stage is σ_i^2, etc.

If more subscripts are required, then they simply follow the one on the relaying stage separated by a comma. For instance, if the ith relaying stage is spanned by a MIMO channel with $n_{t,i}$ transmit and $n_{r,i}$ receive antennas, then the channel spanned by the kth transmit and lth receive antenna element is denoted as $h_{i,kl}$, and so on.

With this choice of notation, it could be that the parameters at the receiver of the ith relaying stage is the same as the parameters at the transmitter of the $(i+1)$th stage. For instance, the number of receiver antennas is typically the same as the number of transmit antennas in a relaying node; however, they do not have to be the same, which is why the above notation is not only convenient but also more general.

The most important variables, symbols and functions used throughout this chapter are tabulated at the beginning of this book.

3.2 Transparent Relaying Protocols

We will here deal with some canonical transparent relaying protocols that can be used to construct more complicated relaying topologies, such as depicted in Figure 3.1. This general topology essentially realizes a general transparent relaying topology with parallel noninterfering information flows. Each node may have multiple antennas. Compared with the transparent space–time processing case to be discussed in Section 3.3, there is no formation of virtual antenna arrays (VAA); compared with the regenerative case to be discussed in Chapter 4, there is no first VAA and also no cooperation between nodes of the same VAA. A large amount of literature has been devoted to these topologies [88, 167, 168, 247, 255, 276, 321–381]. This generally analyzes end-to-end error rates in closed form for various fading channels or gives some simple upper bounds. Other transparent relaying protocols,

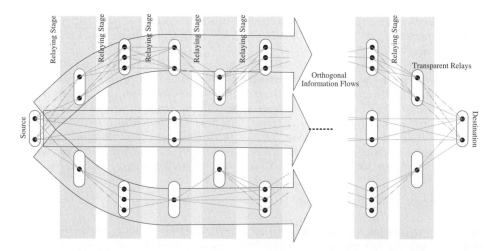

Figure 3.1 General transparent relaying topology with parallel noninterfering information flows, realizing a multibranch multihop relaying topology

such as those dealing with interference and/or doing some linear processing, can also be constructed from these canonical approaches.

3.2.1 Single-Branch Dual-Hop AF

The canonical of all canonical relaying topologies is arguably the single-branch dual-hop relay case. Whilst a large number of techniques from the above publication list is applicable in one form or another to this canonical topology, we will restrict our attention those [167, 346] that developed expressions for the Rayleigh and Nakagami fading channels, respectively. Both approaches also use two different theoretical bodies, where the former relies on the moment generating function (MGF) and the latter on an alternative representation of the error integral through the channel's cumulative distribution function (CDF).

3.2.1.1 System Assumptions

The topology is depicted in Figure 3.2. Its core characteristics can be summarized as follows:

- dual-hop single-branch topology;
- absence of source–destination link;
- fixed gain amplification (semi-blind);
- flat Nakagami-m fading channels without shadowing.

We thus consider the case where only a single relay is present, which essentially leads to a system with two relaying segments. Due to the absence of a direct link between source and destination, this additionally leads to a single-branch topology. Furthermore, we will assume that the relay only has knowledge on the average fading power of the first relaying segment. It can hence only utilize a constant amplification, which is also sometimes referred to as semi-blind relay. We will then deal with the Rayleigh fading case, that is $m = 1$, and with the general Nakagami fading case.

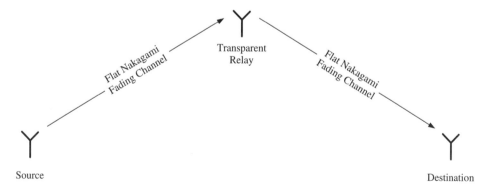

Figure 3.2 Canonical single-branch dual-hop topology without direct link operating over flat Rayleigh fading channels without shadowing

3.2.1.2 Rayleigh Fading Channels

Hasna and Alouini [167] and some earlier papers were the first to derive error rate expressions for the AF relaying protocol over flat Rayleigh fading channels. They essentially derived the MGF in closed-form, which is most useful in quantifying the performance of virtually any transceiver design [91]. Note, however, that most error rate expressions still remain in integral form, which is generally no problem since most expressions reduce to a simple single integral that can be evaluated using standard numerical packages.

Following the analysis outlined [167], the end-to-end SNR at the receiver with fixed or constant amplification factor and receiving channel state information (CSI) at the relay can be written as:

$$\gamma = \frac{\gamma_1\gamma_2}{C+\gamma_2}, \tag{3.1}$$

where $\gamma_i = P_i|h_i|^2/\sigma_i^2 = P_i a_i^2/\sigma_i^2 = P_i g_i/\sigma_i^2$, P_i is the transmission power and σ_i^2 the noise power. Here, h_i is a zero-mean complex Gaussian random variable, a_i its envelope which is Rayleigh distributed, and g_i its gain/power, which is central chi-square distributed − all of these variables represent the flat Rayleigh fading channel in the ith relaying segment. Furthermore, for the sake of power normalization, the constant power amplification in the relay is such that $C = \overline{\gamma}_1 \left[e^{1/\overline{\gamma}_1} \mathrm{Ei}(1/\overline{\gamma}_1) \right]^{-1}$, where $\mathrm{Ei}(\cdot)$ is the exponential integral function and $\overline{\gamma}_1 = \mathbb{E}\{\gamma_1\}$.

To characterize many types of coded and uncoded communication systems, it is advisable to derive the system's MGF [91]. One would typically perform such derivations via the outage probability, which we perform in this chapter once for illustrative purposes but omit these derivations for subsequently discussed topologies. Using the definition of the outage probability of the instantaneous SNR, we obtain:

$$P_{\mathrm{out}} = P[\gamma < \gamma_{\mathrm{thr}}] = \int_0^\infty P\left[\frac{\gamma_1\gamma_2}{\gamma_2+C} < \gamma_{\mathrm{thr}}|\gamma_2\right] p_{\gamma_2}(\gamma_2)d\gamma_2 \tag{3.2a}$$

$$= \int_0^\infty \frac{1}{\overline{\gamma}_2}\left[1 - e^{-(\gamma_{\mathrm{thr}}/\overline{\gamma}_1)(1+C/\gamma_2)}\right]e^{-\gamma_2/\overline{\gamma}_2}d\gamma_2 \tag{3.2b}$$

$$= 1 - 2\sqrt{\frac{C\gamma_{\mathrm{thr}}}{\overline{\gamma}_1\overline{\gamma}_2}}e^{-\gamma_{\mathrm{thr}}/\overline{\gamma}_1}K_1\left(2\sqrt{\frac{C\gamma_{\mathrm{thr}}}{\overline{\gamma}_1\overline{\gamma}_2}}\right). \tag{3.2c}$$

By taking the derivative w.r.t. γ_{thr} and replacing γ_{thr} by γ, we obtain the PDF of γ as:

$$p_\gamma(\gamma) = \frac{2e^{-\gamma/\overline{\gamma}_1}}{\overline{\gamma}_1}\left[\sqrt{\frac{C\gamma}{\overline{\gamma}_1\overline{\gamma}_2}}K_1\left(2\sqrt{\frac{C\gamma}{\overline{\gamma}_1\overline{\gamma}_2}}\right) + \frac{C}{\overline{\gamma}_2}K_0\left(2\sqrt{\frac{C\gamma}{\overline{\gamma}_1\overline{\gamma}_2}}\right)\right]. \tag{3.3}$$

Here, $K_\nu(\cdot)$ is the νth order modified Bessel function of the second kind. The MGF is finally found by evaluating $M(s) = \int_0^\infty p_\gamma(\gamma)e^{-s\gamma}d\gamma$, which leads to:

$$M(s) = \frac{1}{\overline{\gamma}_1 s+1} + \frac{C\overline{\gamma}_1 s e^{C/\overline{\gamma}_2(\overline{\gamma}_1 s+1)}}{\overline{\gamma}_2(\overline{\gamma}_1 s+1)^2}\mathrm{Ei}\left(\frac{C}{\overline{\gamma}_2(\overline{\gamma}_1 s+1)^2}\right). \tag{3.4}$$

From Equation (3.4), the performance of many coherent and differential modulation schemes can be either derived in closed form or evaluated by some standard numerical simulation packages.

For instance, the bit error probability (BEP) of differential binary phase shift keying (D-BPSK) is simply given as $P_b(e) = \frac{1}{2}M(1)$. The performance of D-BPSK over dual-hop and traditional single-hop networks is depicted in Figure 3.3. We have assumed a pathloss coefficient of $n = 4$ and that half power is allocated to each relaying segment, which, as already demonstrated in Section 1.2, yields a SNR gain of 16 w.r.t. the case of direct single-hop transmission. Note that the MGF of the Rayleigh fading channel is simply $M(s) = 1/(\overline{\gamma}s+1)$. The slopes at high SNRs for both cases are the same, indicating that the diversity order is simply unity. The dual-hop case, however, yields performance gains that are solely due to the power gains. In the absence of the power gain of 16 due to a reduced aggregated pathloss, direct communication would outperform the multihop case.

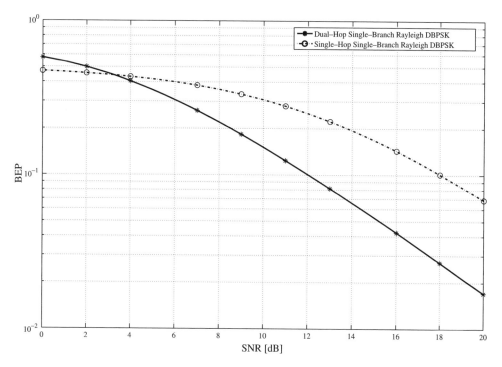

Figure 3.3 BEP of D-BPSK over flat single and dual-hop Rayleigh fading channel with $\mathrm{SNR} = 16 \cdot \overline{\gamma} = \overline{\gamma}_1 = \overline{\gamma}_2$

3.2.1.3 Nakagami Fading Channels

The Nakagami-m fading channel is of practical interest, as has already been discussed in Chapter 2. Contributions pertaining to dual-hop single-branch topologies over said channels emerged as early as 2004 [167, 274] with further results available [168, 276, 345]. However, all these contributions derive bounds – some tighter than others – that are of great use in understanding the performance trends but lack the generality of closed-form expressions.

Recently, some closed form expressions were developed [346], which starts with the end-to-end SNR expression given in Equation (3.1) and the PDF of the SNR obtained through the PDF of the Nakagami fading channel in each relaying segment given in Equation (2.28b). Subsequent derivations then are based on an alternative representation of the error integral through the end-to-end channel's CDF. As such, the average BEP over fading channels is calculated as:

$$P_{\mathrm{b}}(e) = \int_0^\infty P_{\mathrm{b}}(e|\gamma) p_\gamma(\gamma) \, d\gamma, \tag{3.5}$$

where $P_{\mathrm{b}}(e|\gamma)$ can be expressed as a finite weighted sum of $Q(\sqrt{b\gamma})$ [382], with $Q(\cdot)$ being the Gaussian Q-function and b some modulation dependent constant. For instance, for BPSK $P_{\mathrm{b}}(e|\gamma) = Q(\sqrt{2\gamma})$ and for QPSK $P_{\mathrm{b}}(e|\gamma) = Q(\sqrt{\gamma})$. With the aid of the theory exposed by Tsiftsis et $al.$ [279], above average BEP can be recast as:

$$P_{\mathrm{b}}(e) = \frac{1}{\sqrt{2\pi}} \int_0^\infty F_\gamma\left(t^2/c\right) e^{-t^2/2} \, dt. \tag{3.6}$$

The CDF of the end-to-end transparent relaying channel with constant amplification has been derived [279] as:

$$F_\gamma(\gamma) = 1 - \sum_{i=1}^{m_1-1} \sum_{j=0}^{i} \Xi[i,j] e^{-m_1\gamma/\overline{\gamma}_1}$$

$$\times \gamma^{\frac{2i+m_2-j}{2}} K_{m_2-j}\left(2\sqrt{\frac{m_1 m_2 C\gamma}{\overline{\gamma}_1\overline{\gamma}_2}}\right), \quad (3.7)$$

where

$$\Xi[i,j] = \frac{2\binom{i}{j}}{\Gamma(m_2)i!}\left(\frac{m_1}{\overline{\gamma}_1}\right)^{\frac{2i+m_2-j}{2}}\left(\frac{Cm_2}{\overline{\gamma}_2}\right)^{\frac{m_2+j}{2}}. \quad (3.8)$$

Performing some algebraic manipulations on Equation (3.6) with Equation (3.7), one can obtain the BEP in closed form as [346]:

$$P_b(e) = \frac{1}{2} - \frac{1}{\sqrt{2\pi}} \sum_{i=1}^{m_1-1}\sum_{j=0}^{i} \Xi[i,j]\frac{1}{4} \frac{\vartheta^{\lambda_1}}{\left(b\left(\frac{1}{2}+\frac{m_1}{\overline{\gamma}_1 b}\right)\right)^{\lambda_2}} \frac{\Gamma\left(i+\frac{1}{2}\right)\Gamma\left(2\lambda_2-i+\frac{1}{2}\right)}{\sqrt{\frac{m_1 m_2 C}{\overline{\gamma}_1\overline{\gamma}_2 b}}}$$

$$\times \Psi\left(2\lambda_2-i+\frac{1}{2}, 2\lambda_1; \vartheta\right), \quad (3.9)$$

where $\lambda_1 = (m_2 - j + 1)/2$, $\lambda_2 = (2i + m_2 - j)/2$, $\vartheta = 2m_1 m_2 C/((2m_1 + \overline{\gamma}_1 b)\overline{\gamma}_2)$ and $\Psi(\cdot, \cdot; \cdot)$ is the Tricomi confluent hypergeometric function. Suraweera and Karagiannidis [346] also developed an asymptotic expression that holds at sufficiently high SNRs but at error rates of interest, that is:

$$P_b(e) \approx \Lambda(m_1; m_2)_2F_2\left(m_1, 0.5+m_1; 1+m_1, 1+m_1-m_2; \frac{2m_1 m_2}{\overline{\gamma}_2 b}\right)$$

$$+ \Lambda(m_2; m_1)_2F_2\left(m_2, 0.5+m_2; 1+m_2, 1+m_2-m_1; \frac{2m_1 m_2}{\overline{\gamma}_2 b}\right), \quad (3.10)$$

where

$$\Lambda(m_i; m_j) = \frac{2^{m_i}}{\sqrt{\pi}} \frac{\Gamma(m_j - m_i)\Gamma(0.5+m_i)}{\Gamma(m_i)\Gamma(m_j)} \frac{\left(\frac{m_i m_j}{\overline{\gamma}_2 b}\right)^{m_i}}{m_i}. \quad (3.11)$$

A generalization for the case of different fading statistics and amplification factors has been advocated [380, 381] following the above procedures; it features fairly complex expressions, however, most of them in closed form or easy to evaluate numerically. Finally note that the MGF of this Nakagami fading case can be equally derived, leading to the following result [360]:

$$M_\gamma(s) = \sum_{i=0}^{m_1} A_i(s) \cdot G_{1,2}^{2,1}\left(B(s)\Big|_{0,m_1-m_2-i}^{1-m_2-i}\right) \quad (3.12a)$$

$$A_i(s) = \frac{m_1^{m_1} m_2^{m_2}}{\overline{\gamma}_1^{m_1}\overline{\gamma}_2^{m_2}\Gamma(m_1)\Gamma(m_2)}\binom{m_1}{i}\frac{\left(\frac{m_1}{\overline{\gamma}_1}\right)^{m_2-m_1+i-2} C^{m_2}}{\left(\frac{m_1}{\overline{\gamma}_1}+s\right)^{m_2+i}} \quad (3.12b)$$

$$B(s) = \frac{\frac{m_1}{\overline{\gamma}_1}\frac{m_2}{\overline{\gamma}_2}C}{\frac{m_1}{\overline{\gamma}_1}+s}. \quad (3.12c)$$

3.2.2 Single-Branch Multihop AF

Above canonical topology is now extended to the multihop case where several relays are connected serially. Whilst not a lot of literature is devoted to this case, the techniques developed for above canonical case as well as other topologies can easily be adopted to the single-branch multihop amplify-and-forward case. A very elegant method deriving bounds on the end-to-end SNR and error rate has been developed in [168, 276, 326, 327] and essentially forms the basis for below treatment.

3.2.2.1 System Assumptions

The topology is depicted in Figure 3.4. Its core characteristics can be summarized as follows:

- N-hop single-branch topology;
- absence of source–destination link;
- fixed gain amplification (semiblind);
- flat Nakagami-m fading channels without shadowing.

We thus consider the case where N relaying segments are present. Due to the absence of a direct link between source and destination, this leads again to a single-branch topology. Furthermore, we will generally assume that the relays only have knowledge of the average fading power of the previous relaying segment, that is semiblind relays. Note that the below results can be extended to CSI-assisted as well as blind relays. We will later deal with the Rayleigh fading case, that is $m = 1$, and then with the general Nakagami fading case.

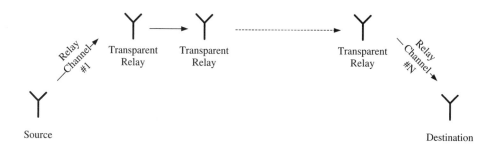

Figure 3.4 Canonical single-branch multi-hop topology without direct link operating over flat fading channels without shadowing

3.2.2.2 Rayleigh Fading Channels

Following the analysis outlined [168, 273, 276], the end-to-end SNR at the receiver with semiblind amplification factor at the relay can be written as:

$$\gamma = \left[\sum_{n=1}^{N} \prod_{j=1}^{n} \frac{C_{j-1}}{\gamma_j} \right]^{-1}$$

(3.13)

where for the Rayleigh case [168]:

$$C_i = \left[-\frac{1}{\overline{\gamma}_i} e^{1/\overline{\gamma}_i} \mathrm{Ei} \left(\frac{1}{\overline{\gamma}_i} \right) \right]^{-1}$$

(3.14)

for $i \geq 1$ and $C_0 = 1$. Since treatment Equation (3.13) has eluded a closed form expression to date, a viable upper bound [168, 274, 276] has been proposed. This is based on the insight that Equation (3.13) essentially represents the harmonic mean, which is known to be upper bounded by its geometric mean. This allows Equation (3.13) to be simplified:

$$\gamma \leq \ulcorner \gamma \urcorner = \frac{1}{N} \prod_{i=1}^{N} \frac{\gamma_i^{(N+1-i)/N}}{C_i^{(N-i)/N}}. \tag{3.15}$$

Using the corollaries [276], one can derive the PDF and CDF of the upper bounded end-to-end SNR. For instance, the CDF reads:

$$F_{\ulcorner\gamma\urcorner}(\gamma) = A \cdot G_{1,\rho+1}^{\rho,1} \left(B\gamma^N \Big|_{\Lambda_1,\Lambda_2,\ldots,\Lambda_N,0}^{1} \right), \tag{3.16}$$

where

$$\rho = \frac{N(N+1)}{2} \tag{3.17a}$$

$$A = \frac{\prod_{i=1}^{N} \sqrt{N+1-i}}{\left(\sqrt{2\pi}\right)^{N(N-1)/2}} \tag{3.17b}$$

$$B = N^N \prod_{i=1}^{N} C_i^{N-i} \prod_{i=1}^{N} \left[\frac{1}{\overline{\gamma}_i (N+1-i)} \right]^{N+1-i} \tag{3.17c}$$

$$\Lambda_i = \Delta(N+1-i, 1) \tag{3.17d}$$

$$\Delta(k, u) = u/k, (u+1)/k, \ldots, (u+k-1)/k. \tag{3.17e}$$

When inserted into Equation (3.6), one obtains a lower bound on the desired error rates [276]:

$$P_b(e) \geq \frac{G_{2N,2N}^{N,2N} \left(\left(\frac{N^2}{b}\right)^N \prod_{i=1}^{N} \frac{1}{\overline{\gamma}_i} \Big|_{1,1,\ldots,1,\Delta(N,0)}^{\Delta(N,1),\Delta(N,1/2)} \right)}{\sqrt{2} \left(\sqrt{2\pi}\right)^N}. \tag{3.18}$$

Performance results shown in Ref. [276] indicate that the bound is not asymptotically tight but yields sufficiently precise results in the error region of interest. Note finally that Karagiannidis [168] develops higher order moments for the upper bounded end-to-end SNR which, when used in conjunction with the theory of Padé approximation, also allows one to obtain error rates of sufficient precision.

3.2.2.3 Nakagami Fading Channels

The Nakagami-m fading channel follows essentially the same theory as the Rayleigh channel, which is why we only state the end results for the PDF, CDF and BEP:

$$p_{\ulcorner\gamma\urcorner}(\gamma) = \frac{NA}{\gamma} \cdot G_{0,\rho}^{\rho,0} \left(B\gamma^N \Big|_{\Lambda_1,\Lambda_2,\ldots,\Lambda_N,0}^{-} \right) \tag{3.19a}$$

$$F_{\ulcorner\gamma\urcorner}(\gamma) = A \cdot G_{1,\rho+1}^{\rho,1} \left(B\gamma^N \Big|_{\Lambda_1,\Lambda_2,\ldots,\Lambda_N,0}^{1} \right) \tag{3.19b}$$

$$P_b(e) \geq \frac{G_{2N,2N}^{N,2N} \left(\left(\frac{N^2}{b}\right)^N \prod_{i=1}^{N} \frac{m_i}{\overline{\gamma}_i} \Big|_{m_1,m_2,\ldots,m_N,\Delta(N,0)}^{\Delta(N,1),\Delta(N,1/2)} \right)}{\sqrt{2} \left(\sqrt{2\pi}\right)^N \prod_{i=1}^{N} \Gamma(m_i)}, \tag{3.19c}$$

where now

$$A = \frac{\prod_{i=1}^{N}(N+1-i)^{m_i-1/2}}{\left(\sqrt{2\pi}\right)^{N(N-1)/2}\prod_{i=1}^{N}\Gamma(m_i)} \quad (3.20a)$$

$$B = N^N \prod_{i=1}^{N} C_i^{N-i} \prod_{i=1}^{N}\left[\frac{m_i}{\bar{\gamma}_i(N+1-i)}\right]^{N+1-i} \quad (3.20b)$$

$$\Lambda_i = \Delta(N+1-i, m_i). \quad (3.20c)$$

3.2.3 Multibranch Dual-Hop AF

Again, multiple contributions have emerged in the context of dual-hop multi-branch AF protocols [for example, 272, 325, 333, 360, 373]. We will concentrate here on one [360] that deals with blind and semiblind relays as well as some [272, 373] that deal with CSI-assisted relays.

3.2.3.1 System Assumptions

The topology is depicted in Figure 3.5. Its core characteristics can be summarized as follows:

- dual-hop L-branch topology;
- presence of source-destination link;
- fixed and variable gain amplification factors;
- flat Nakagami-m fading channels without shadowing.

The prime assumption in the context of these topologies is that the L relaying branches do not interfere with each other. This can be achieved by allocating different resources in time, frequency, code, etc., so that transmissions remain orthogonal.

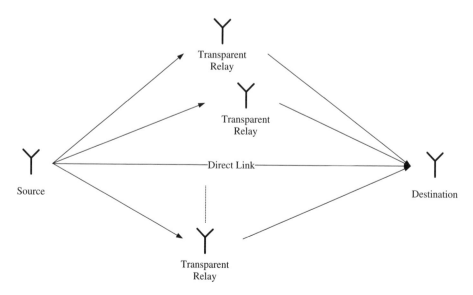

Figure 3.5 Canonical multibranch dual-hop topology with direct link operating over flat fading channels without shadowing

Furthermore, the multibranch topology allows the destination receiver to perform maximum ratio combining (MRC) with all received diversity branches or selection combining (SC) of the best branch, where the respective end-to-end SNRs read:

$$\gamma_{\text{MRC}} = \gamma_0 + \sum_{l=1}^{L} \gamma_l \tag{3.21a}$$

$$\gamma_{\text{SC}} = \max\{\gamma_0, \gamma_1, \ldots, \gamma_L\}. \tag{3.21b}$$

Here, γ_0 is the SNR of the direct link and γ_l is the SNR of the lth relaying branch.

We will generally deal with the MRC case only since the SC is dealt with in a similar fashion. We will first deal with the case where each relay does some fixed gain amplification, that is by means of a blind or semiblind relaying procedure, and then with the case where each relay amplifies the signal such that it aims at compensating for the fading effects of the preceding segment, that is by means of a CSI-assisted relaying procedure.

3.2.3.2 Blind and Semiblind Relays

In the case of blind and semiblind relays, γ_l is given in Equation (3.1) subscripted on branch l, that is:

$$\gamma_l = \frac{\gamma_{l1}\gamma_{l2}}{C_l + \gamma_{l2}}. \tag{3.22}$$

Therefore, the PDFs, CDFs and MGFs derived in previous sections for the dual and multihop case can be reused here on a per-branch basis, such as, for example, Equation (3.12). They are not repeated here for brevity. Assuming independent relaying branches, the MGF of the end-to-end channel is given as $M(s) = \prod_{l=0}^{L} M_{\gamma_l}(s)$, which allows us to obtain various performance measures in closed or simple integral form [91]. Note finally that Di Renzo et al. [360] introduce some bounds on above expressions, allowing essentially noninteger Nakagami-m fading parameters as well as Weibull fading distributions to be accounted for.

3.2.3.3 CSI-Assisted Relays

Based on Refs [276] and [373], we will now deal with the case of CSI-assisted relays. With a variable amplification factor, the end-to-end SNR in the l-diversity branch can be written as:

$$\gamma_l = \frac{\gamma_{l1}\gamma_{l2}}{\gamma_{l1} + \gamma_{l2} + 1}. \tag{3.23}$$

The PDF and CDF for this case have been derived [279] and also the MGF [373], which read per relaying branch as follows:

$$F_{\gamma_l}(\gamma) = 1 - \sum_{i=0}^{m_{l1}-1} \sum_{j=0}^{i} \sum_{k=0}^{m_{l2}-1} \left\{ \Theta(i,j,k)\gamma^{(2m_{l2}+2i-j-k-1)/2} \right.$$

$$\times (1+\gamma)^{(j+k+1)/2} \exp(-(\Omega_{l1}+\Omega_{l2})\gamma)$$

$$\left. \times K_{k-j+1}\left(2\sqrt{(\Omega_{l1}\Omega_{l2})\gamma(1+\gamma)}\right) \right\} \tag{3.24a}$$

$$M_{\gamma_l}(s) = 1 - \sum_{i=0}^{m_{l1}-1} \sum_{j=0}^{i} \sum_{k=0}^{m_{l2}-1} \left\{ \Theta(i,j,k) \sum_{t=0}^{k+j+2} \binom{k+j+2}{t} \right.$$

$$\left. (-1)^{m_{l2}+i+t-j-k-1} s \frac{\partial^{m_{l2}+i+t-j-k-1}\Upsilon(s)}{\partial s^{m_{l2}+i+t-j-k-1}} \right\} \tag{3.24b}$$

where

$$\Theta(i, j, k) = \frac{2}{i!(m_{l2} - 1)!} \binom{i}{j} \binom{m_{l2} - 1}{k} \Omega_{l1}^{(2i-j+k+1)/2} \Omega_{l2}^{(2m_{l2}+j-k-1)/2} \quad (3.25a)$$

$$\Omega_{li} = \frac{m_{li}}{\overline{\gamma}_{li}} \quad (3.25b)$$

$$\Upsilon(s) = \frac{(k+1)!j!}{2 (\Omega_{l1}\Omega_{l2})^{j-k-1}} \Psi(k+2; k-j+2; 0.5z_+)\Psi(k+2; k-j+2; 0.5z_-) \quad (3.25c)$$

$$z_\pm = s + \Omega_{l1} + \Omega_{l2} \pm \sqrt{(s + \Omega_{l1} + \Omega_{l2})^2 - 4\Omega_{l1}\Omega_{l2}} \quad (3.25d)$$

and $K_\nu(\cdot)$ is the νth order modified Bessel function of the second kind, $\Psi(\cdot, \cdot; \cdot)$ is the Tricomi confluent hypergeometric function and $\frac{\partial^n f(x)}{\partial x^n}$ denotes the nth order derivative of $f(x)$ w.r.t. x; note that $\frac{\partial^0 f(x)}{\partial x^0} = f(x)$. The derivations can be performed symbolically using standard numerical evaluation packages.

To facilitate analysis, Anghel and Kaveh [272] have derived some upper and lower bounds for the Rayleigh fading case. These bounds are invoked at an asymptotically high SNR at which the noise term in the relays can be neglected, essentially simplifying Equation (3.23) to:

$$\gamma_l = \frac{\gamma_{l1}\gamma_{l2}}{\gamma_{l1} + \gamma_{l2}}. \quad (3.26)$$

Again, the MGF could be calculated via the derivative of the outage probability; however, this leads to an intractable integral for the SER. Anghel and Kaveh [272] hence proposed an asymptotically tight lower and upper bound for the SEP of MPSK:

$$P_s(e) \geq \left[(1 + \overline{\gamma}_0 g_{PSK}) \prod_{l=1}^{L} \left(1 + g_{PSK} \frac{\overline{\gamma}_{p,l}}{\overline{\gamma}_{\sigma,l}} \right) \right]^{-1} \cdot W(L, M) \quad (3.27a)$$

$$P_s(e) \leq \frac{2^L}{g_{PSK}^{L+1}} \left[\overline{\gamma}_0 \prod_{l=1}^{L} \frac{\overline{\gamma}_{p,l}}{\overline{\gamma}_{\sigma,l}} \right]^{-1} \cdot W(L, M), \quad (3.27b)$$

where $g_{PSK} = \sin^2(\pi/M)$, $\overline{\gamma}_{p,l} = \overline{\gamma}_{l1}\overline{\gamma}_{l2}$ and $\overline{\gamma}_{\sigma,l} = \overline{\gamma}_{l1} + \overline{\gamma}_{l2}$. Furthermore,

$$W(L, M) = \frac{1}{\pi} \int_0^\xi \sin^\nu \phi d\phi$$

$$= -\frac{1}{\pi} \cos(\xi) \cdot {}_2F_1 \left(\frac{1}{2}, \frac{1-\nu}{2}; \frac{3}{2}; \cos^2(\xi) \right), \quad (3.28)$$

where $\xi = \pi(M - 1)/M$, $\nu = 2L + 2$ and ${}_2F_1(\cdot, \cdot; \cdot; \cdot)$ is the Gauss hypergeometric function. The tightness of the proposed bound is investigated in Figure 3.6. It can be observed that the bounds are sufficiently tight at the SERs of interest for a fairly low number of relay branches but increasingly diverges for an increasing number of branches. Whilst numerically attractive, it is use is hence restricted to a few relaying branches.

3.2.4 Multibranch Multihop AF

The literature on multi-hop multibranch AF protocols is scarce; however, most of above techniques are readily extendable to such generic relaying topologies. Example publications specifically dedicated to said topologies exist [275, 333, 355].

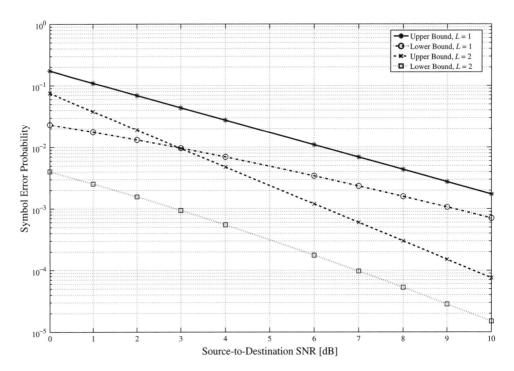

Figure 3.6 Upper and lower symbol error probability bounds for QPSK, a pathloss coefficient of $n = 4$ and the relays placed exactly midway between source and destination

3.2.4.1 System Assumptions

The topology is depicted in Figure 3.7. Its core characteristics can be summarized as follows:

- N-hop L-branch topology;
- presence of source-destination link;
- fixed and variable gain amplification factors;
- flat Nakagami-m fading channels without shadowing.

The prime assumption in the context of these topologies is again that the N relaying hops and L relaying branches do not interfere with each other. This can again be achieved by allocating different resources in time, frequency, code, etc., so that transmissions remain orthogonal.

We will first deal with a general framework, which was originally developed for topologies with one antenna element per terminal. This was then extended to the case where the terminals are allowed to have multiple antenna elements.

3.2.4.2 SISO Topologies

Ribeiro *et al.* [275] have produced a remarkable observation in that the integrant in Equation (3.5) drops very quickly to zero. Indeed, Figure 3.8 compares the values of $p_\gamma(\gamma)$ for a normalized Rayleigh distribution with $P_b(e|\gamma) = Q(\sqrt{\gamma})$ for various average SNRs $\bar{\gamma}$. Clearly, the contributing part of the Q-function tends to zero for an increasing average SNR. This means that the majority of the captured area under the curve is in the asymptotic region of low γ, which allows one to invoke a McLaurin

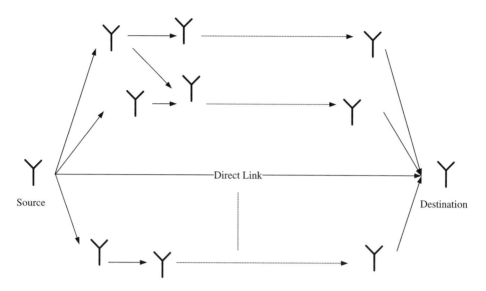

Figure 3.7 Canonical multibranch multihop topology with direct link operating over flat fading channels without shadowing

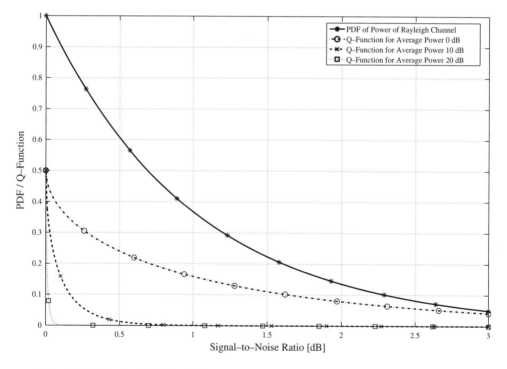

Figure 3.8 Behavior of the Rayleigh channel's PDF, $p_\gamma(\gamma)$, and various realizations of $Q(\sqrt{\gamma})$

series expansion and hence significantly simplify the integrant in Equation (3.5) and thus its solution. For an arbitrarily large $\overline{\gamma}$, the PDF is henceforth approximated arbitrarily well by the first term of the McLaurin series. This has been rigorously formalized [383] to yield the asymptotic expression for the PDF as $p_\gamma(\gamma) \to a\gamma^t + o(\gamma)$. This, in turn, leads to the following simple asymptotic SEP expression:

$$P_s(e) \to \frac{2^t a \Gamma(t + 3/2)}{\sqrt{\pi}(t + 1)} (b\overline{\gamma})^{-(t+1)}, \qquad (3.29)$$

where a and t depend on the channel. For example, for the Nakagami-m channel, $a = m^m / \Gamma(m)$ and $t = m - 1$; for the Rice-K channel, $a = (1 + K)\exp(-K)$ and $t = 0$; etc. This allows us to obtain the asymptotic end-to-end SEPs for generic relaying topologies in closed form for a variety of general fading channels.

For instance, we assume now a two-hop single-branch topology with a direct link between source and destination communicating at sufficiently high SNR so that (i) above expressions can be used, and (ii) the asymptotic SNR expression Equation (3.26) for the relying link can be used. Working on Equation (3.29) as outlined by Ribeiro *et al.* [275], leads to a generic but very simple asymptotic end-to-end SEP expression for this topology in the form of:

$$P_s(e) \to \frac{3}{4b^2} \left[p_{\gamma_1}(0) + p_{\gamma_2}(0) \right] p_{\gamma_0}(0), \qquad (3.30)$$

which, in the case of a Rician-K channel with the PDF given in Chapter 2, becomes:

$$P_s(e) \to \frac{3(K + 1)^2}{4b^2} \left(\frac{1}{\overline{\gamma}_1} + \frac{1}{\overline{\gamma}_2} \right) \frac{1}{\overline{\gamma}_0}. \qquad (3.31)$$

In the case of a generic relaying topology as shown in Figure 3.7, the SER expression can be shown to be [275]:

$$P_s(e) \to \frac{C(L)}{b^{L+1}} p_0(0) \prod_{i=1}^{L} \sum_{j=0}^{N_i} p_{ij}(0), \qquad (3.32)$$

where $C(L) = (\prod_{l=1}^{L+1}(2l - 1))/(2(L + 1)!l^{(L+1)})$, $p_0(0)$ and $p_{ij}(0)$ are respectively the channel's power PDF of the direct and of the ith relay branch and jth relay hop evaluated at zero, N_i is the number of relay hops in the ith relay branch, and b is a modulation dependent constant. It has been [275], shown that these asymptotic expressions hold with high precision for arbitrary topologies in the error region of interest.

3.2.4.3 MIMO Topologies

Extending above SISO topologies to the MIMO case where each terminal is equipped with multiple antenna elements, has some very interesting implications. For the sake of illustration, we will only briefly dwell on the single-branch dual-hop SIMO topology, depicted in Figure 3.9. Following the analysis outlined by Mheidat and Uysal [333], we will proceed with the cases of constant as well as variable amplification. We restrict the analysis to the case where the noise as well as assigned transmit power in both relaying stages is the same, leading to $\gamma_1 = \gamma_2 = \gamma$.

In the case of constant amplification through a blind or semiblind relay, an exact expression for the SEP of MPSK as well as an upper bound are given as [333]:

$$P_s(e) = \frac{1}{\pi} \int_0^{\pi(M-1)/M} \frac{(-\xi)^{n_r-1} e^\xi \Gamma(0, \xi) - \sum_{j=1}^{n_r-1}(j - 1)!(-\xi)^{n_r-1-j}}{\Gamma(n_r)(1 + 1/\xi)^{n_r} 1/\xi} d\theta$$

$$\leq \frac{M - 1}{M} \frac{\Gamma(n_r - 1)}{\Gamma(n_r)} \left(\frac{\gamma}{2} \right)^{-(n_r+1)}, \qquad (3.33)$$

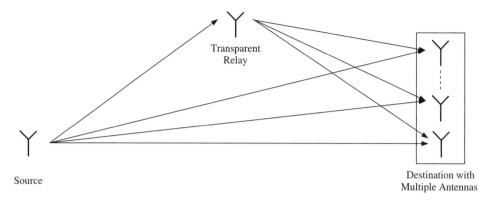

Figure 3.9 Canonical single-branch two-hop SIMO topology with direct link operating over flat fading channels without shadowing

where n_r is the number of receiver antennas at the destination, $\xi = \sin^2\theta/(\sin^2(\pi/M)\gamma)$, $\Gamma(\cdot)$ is the gamma function and $\Gamma(\cdot,\cdot)$ is the upper incomplete gamma function. From the bound it can be concluded that the diversity order with constant amplification is n_{r+1}.

In the case of variable amplification through a CSI-assisted relay, an exact expression for the SEP of MPSK as well as an upper bound are given as [333]:

$$
P_s(e) = \frac{1}{\pi}\int_0^{\pi(M-1)/M}\left(1 + \frac{\sin^2(\pi/M)}{\sin^2\theta}\gamma\right)^{-2n_r}d\theta
$$

$$
\leq \frac{M-1}{M}\left(\frac{\gamma}{2}\right)^{-2n_r}, \tag{3.34}
$$

Here, a diversity order of $2n_r$ is achieved.

Comparing both cases clearly shows that full diversity of $2n_r$ is only achieved when the relay deploys variable gain amplification. Similar insights hold for general MIMO relaying topologies.

3.3 Transparent Space–Time Processing

Recent key milestones in telecommunications are indisputably: (i) the extension of Shannon's information-theoretical channel capacity limits to the spatial domain [117, 118]; and (ii) the construction of space–time codes (STCs) that operate asymptotically close to this limit [121, 122]. With STC, extraordinarily high spectral efficiencies can be achieved; however, at the expense of more complex transceiver structures, where complexity increases linearly with the number of transmit and receive antennas and exponentially with the constellation size and trellis depth. To accomplish a MIMO channel, additional transmit and receive RF blocks have to be provided that allow the formation of multiple-input–multiple-output (MIMO) channels. Additionally, the created MIMO channel has to be suitable for space–time encoded transmission, that is it ought not to be singular or close to it. This requires the channel to be uncorrelated to a certain extent at the transmitting and receiving ends. Decorrelation is achieved by placing the antenna elements sufficiently far apart, where the distance is dictated by the operating wavelength and the surrounding environment.

A logical step forward is thus to deploy distributed and cooperative space–time encoded relaying systems as per Figure 3.10, where we concentrate on transparent relaying architectures. The space–time processing approaches at hand to date are space–time block coding (STBC), space–time trellis coding (STTC), spatial multiplexing (SM) and beamforming (BF). This general topology essentially realizes a

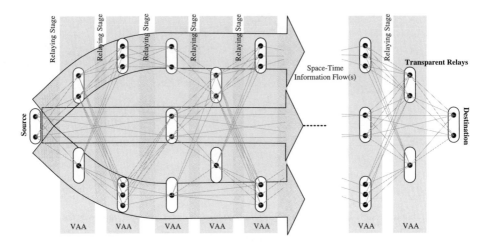

Figure 3.10 General transparent space–time processing topology with (interfering) information flow(s), realizing a multihop distributed space–time topology

general transparent space–time processing architecture with one (or generally several but interfering) information flow(s). Again, each node may have multiple antennas. Compared with the transparent relaying case discussed in Section 3.2, there is a formation of VAAs; compared with the regenerative case to be discussed in Chapter 4, there is again no first VAA and also no cooperation between nodes of the same VAA. The application of some distributed space–time processing techniques to a subset of said architectures is now investigated in subsequent sections.

3.3.1 Distributed Space–Time Block Codes

In a general wireless relay network, different relays receive different noisy copies of the same information symbols. The relays process these received signals and forward them to the destination. The distributed processing at the different relay nodes thus forms a virtual antenna array, as alluded to before. Therefore, conventional space–time block coding schemes can be applied to relay networks to achieve the cooperative diversity and coding gain. The distributed space–time block coding can be constructed by using either transparent relay protocols or regenerative relay protocols. The design approaches in these two coding types are quite different.

In this section, we focus on the design of distributed space–time block codes based on the transparent relay protocol and will discuss the design of distributed space–time block codes using regenerative relay protocols in the next chapter. The amount of literature on transparent space-time block codes is surprisingly rich, with a nonexhaustive set of publications [295, 318, 336, 338, 342, 367, 384–401]. Due to their superior performance, however, we will mainly concentrate on the design proposed by Hassibi and coworkers [402, 403].

3.3.1.1 Distributed Linear Dispersion Space–Time Codes

We will here study distributed space–time codes based on a linear dispersion code (DLD-STC) proposed by Jing and Hassibi [403]. In the DLD-STC, upon receiving signals from the source, each relay performs a linear transformation of the previously received T signals. All the relays then forward the linearly transformed signals to the destination at the same time. The signals received at the destination then form a DLD-STC. There are two major differences between the LD codes in MIMO systems with collocated antennas [402] and DLD-STC for relay networks [403]. In MIMO systems, since all antennas are collocated, the LD codes can be designed in a centralized manner. In wireless relay networks, however, relay nodes are spatially distributed so that the DLD-STC has to be implemented in a distributed manner. Secondly, in MIMO systems, all transmit antennas have perfect knowledge of the information symbols. However, in relay networks, each relay only has access to a noisy copy of the original information symbols. These differences essentially result in the different designs between DLD-STC and traditional LD codes.

We consider a general two-hop relay network, consisting of one source, n relays and one destination. For simplicity, we assume that there is no direct link from the source to the destination. Figure 3.11 depicts the block diagram of the DLD-STC system. Let $\mathbf{S} = (s(1), \ldots, s(T))^T$ be a vector of source information symbols, where $s(k)$ represents the kth source information symbol and we assume $\mathbb{E}\{|s(k)|^2\} = 1$. The overall transmission is divided into two phases. The source first broadcasts $\mathbf{S}$ with power p_S to all the relays in the first T time slots. The received signals at the relay i, denoted by $\mathbf{y}_{SR}^i$, can be expressed as:

$$\mathbf{y}_{SR}^i = \sqrt{p_S T} h_{SR}^i \mathbf{S} + \mathbf{n}_{SR}^i, \tag{3.35}$$

where h_{SR}^i is the fading coefficient from the source to the relay i. We assume that the fading coefficients keep constant during these T time slots. Furthermore, $\mathbf{n}_{SR}^i$ is the noise vector at the relay and each element in $\mathbf{n}_{SR}^i$ is a zero-mean complex Gaussian random variable with two sided power spectral density of $N_0/2$ per dimension.

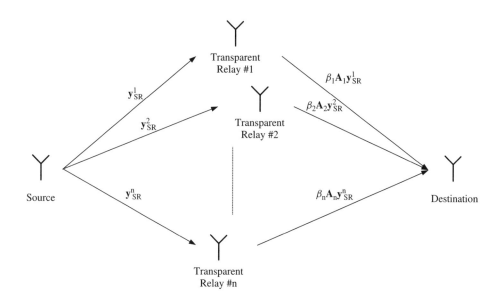

Figure 3.11 The block diagram of distributed linear dispersion space time codes

In DLD-STC, upon receiving signals from the source, each relay performs a linear transformation of its received signal, that is:

$$\mathbf{X}^i = \mathbf{A}_i \mathbf{y}_{SR}^i, \tag{3.36}$$

where $\mathbf{A}_i$ is a $T \times T$ linear transformation matrix at the relay i given by:

$$\mathbf{A}_i = (\mathbf{A}_{i,1}, \ldots, \mathbf{A}_{i,T}), \tag{3.37}$$

and $\mathbf{A}_{i,j}$ is the jth column of $\mathbf{A}_i$:

$$\mathbf{A}_{i,j} = (a_{i,1j}, \ldots, a_{i,Tj})^T . \tag{3.38}$$

In general, the transformation matrix $\mathbf{A}_i$ could be arbitrary. To simplify the analysis, we assume that $\mathbf{A}_i$ is a unitary matrix, that is $(\mathbf{A}_i)^H \mathbf{A}_i = \mathbf{I}$. The relay then scales the transformed $\mathbf{X}^i$ by a scaler β_i and forwards to the destination $\mathbf{X}_R^i$, given by:

$$\mathbf{X}_R^i = \beta_i \mathbf{X}^i . \tag{3.39}$$

Here, β_i is an amplification factor such that:

$$\mathbb{E}\{|\mathbf{X}_R^i|^2\} = p_R T, \tag{3.40}$$

where p_R is the relay transmission power. For the tractability of analysis, we consider a long-term power constraint at the relay, which is different from other sections, where we consider a short-term power constraint with the expectation in Equation (3.40) done with respect to the noise only. With the long-term power constraint, Equation (3.40) is averaged with respect to both the noise and the fading coefficients and can be further written as:

$$\mathbb{E}\{|\mathbf{X}_R^i|^2\} = \beta_i^2 \mathbb{E}\{(\mathbf{X}^i)^H \mathbf{X}^i\} = \beta_i^2 \mathbb{E}\{(\mathbf{A}_i \mathbf{y}_{SR}^i)^H (\mathbf{A}_i \mathbf{y}_{SR}^i)\}$$
$$= \beta_i^2 \mathbb{E}\{(\mathbf{y}_{SR}^i)^H (\mathbf{A}_i)^H \mathbf{A}_i \mathbf{y}_{SR}^i\} = \beta_i^2 \mathbb{E}\{(\mathbf{y}_{SR}^i)^H \mathbf{y}_{SR}^i\} = p_R T. \tag{3.41}$$

Substituting Equation (3.35) now into $\mathbb{E}\{(\mathbf{y}_{SR}^i)^H \mathbf{y}_{SR}^i\}$, we have:

$$\mathbb{E}\{(\mathbf{y}_{SR}^i)^H \mathbf{y}_{SR}^i\} = (p_S + N_0)T. \tag{3.42}$$

Inserting the above equation into Equation (3.41), we get

$$\mathbb{E}\{\mathbf{X}_R^i\} = \beta_i^2 (p_S + N_0)T = p_R T, \tag{3.43}$$

and β_i can be calculated from the above equation as:

$$\beta_i = \sqrt{\frac{p_R}{(p_S + N_0)}}. \tag{3.44}$$

The signals received at the destination are a superposition of signals transmitted from all relays, that is:

$$\mathbf{y}_D = \sum_{i=1}^{n} h_{RD}^i \mathbf{X}_R^i + \mathbf{n}_D = \sum_{i=1}^{n} h_{RD}^i \left(\beta_i \mathbf{A}_i \mathbf{y}_{SR}^i\right) + \mathbf{n}_D$$
$$= \sum_{i=1}^{n} \beta_i \sqrt{p_S T} h_{SR}^i h_{RD}^i \mathbf{A}_i \mathbf{S} + \tilde{\mathbf{n}}_D \tag{3.45}$$

where $\mathbf{n}_D$ is a destination noise vector and

$$\tilde{\mathbf{n}}_D = \sum_{i=1}^{n} h_{RD}^i \beta_i \mathbf{A}_i \mathbf{n}_{SR}^i + \mathbf{n}_D \tag{3.46}$$

is the equivalent noise vector with variance

$$\sigma_{\tilde{n}_D}^2 = \left(\sum_{i=1}^{n} |h_{RD}^i \beta_i|^2 + 1\right) N_0 = \left(\frac{p_R}{p_S + N_0} \sum_{i=1}^{n} |h_{RD}^i|^2 + 1\right) N_0. \tag{3.47}$$

Equation (3.45) can be further written in matrix form as

$$\mathbf{y}_D = \sqrt{\frac{p_S p_R T}{p_S + N_0}} \mathbf{A}_S \mathbf{H} + \tilde{\mathbf{n}}_D = \alpha_D \mathbf{A}_S \mathbf{H} + \tilde{\mathbf{n}}_D, \tag{3.48}$$

where $\alpha_D = \sqrt{\frac{p_S p_R T}{p_S + N_0}}$,

$$\mathbf{A}_S = [\mathbf{A}_1 \mathbf{S}, \ldots, \mathbf{A}_n \mathbf{S}] \tag{3.49}$$

and

$$\mathbf{H} = (h_{SR}^1 h_{RD}^1, \ldots, h_{SR}^n h_{RD}^n)^T. \tag{3.50}$$

As is evident from Equation (3.46), $\tilde{\mathbf{n}}_D$ is a linear combination of n Gaussian distributed random vectors, making $\tilde{\mathbf{n}}_D$ still Gaussian distributed. Thus,

$$P(\mathbf{y}_D|\mathbf{S}) = \frac{\exp\left(-\frac{\|\mathbf{y}_D - \alpha_D \mathbf{A}_S \mathbf{H}\|^2}{\sigma_{\tilde{n}_D}^2}\right)}{\pi^T \left(\sigma_{\tilde{n}_D}^2\right)^T}, \tag{3.51}$$

where $\|\mathbf{x}\|$ represents the Frobenius norm of vector $\mathbf{x}$. The maximum likelihood decoding of the DLD-STC is now obtained by maximizing $P(\mathbf{y}_D|\mathbf{S})$ and can be formulated as

$$\hat{\mathbf{S}} = \arg\max_{\mathbf{S}} P(\mathbf{y}_D|\mathbf{S}) = \arg\min_{\mathbf{S}} \|\mathbf{y}_D - \alpha_D \mathbf{A}_S \mathbf{H}\|^2. \tag{3.52}$$

In the subsequent sections, we present the pairwise error probability (PEP) of the DLD-STC scheme.

3.3.1.2   Chernoff Bound of General Communication System

We first briefly revisit the calculation of the Chernoff bound for a general communication system, from which the PEP upper bound of the DLD-STC can then be derived. For a given random variable z and a function $f(z)$, which satisfies

$$f(z) \geq \begin{cases} 1 & z \geq 0 \\ 0 & z < 0, \end{cases} \tag{3.53}$$

if the various order statistics of z always exists, the Chernoff bound says that the following inequality always exists:

$$P(z > 0) \leq \mathbb{E}\{f(z)\}. \tag{3.54}$$

For example, if we choose $f(z) = exp(\lambda z)$, then the Chernoff bound becomes:

$$P(z > 0) \leq \mathbb{E}\{\exp(\lambda z)\}. \tag{3.55}$$

We consider a general point-to-point single antenna communication system. The received signal of this system is expressed as

$$y = hs + n, \tag{3.56}$$

where s is the transmitted signal, h is the fading coefficient and n is a Gaussian random noise with the spectrum density of N_0 per dimension. The maximum likelihood (ML) decoding of such a system is given by:

$$\hat{s} = \arg\max_s P(y|s) = \arg\min_s |y - hs|^2. \tag{3.57}$$

That is, in the case of ML decoding, the decoder always selects the symbol that has the minimum distance to the received signal. Thus, the probability that the decoder decides in favor of a wrong symbol $\bar{s}$ when s is transmitted, denoted by $P(s \to \bar{s}|y, h)$, is given by:

$$P(s \to \bar{s}|y, h) = P(|y - hs|^2 > |y - h\bar{s}|^2) = P((|y - hs|^2 - |y - h\bar{s}|^2) > 0). \tag{3.58}$$

Applying the Chernoff bound of Equation (3.55) to above equation, we get:

$$P(s \to \bar{s}|y, h) \leq \mathbb{E}_n \left\{ \exp(\lambda(|y - hs|^2 - |y - h\bar{s}|^2)) \right\}, \tag{3.59}$$

where $\mathbb{E}_n\{\cdot\}$ represents the expectation of the argument with respect to the random variable n. After some algebraic manipulation, Equation (3.59) can be calculated as

$$P(s \to \bar{s}|y, h) \leq \mathbb{E}_n \left\{ \exp(\lambda(|y - hs|^2 - |y - h\bar{s}|^2)) \right\} \tag{3.60}$$

$$= \exp\left(-\lambda h^2(1 - N_0\lambda)|s - \bar{s}|^2\right).$$

The above equation is minimized when $\lambda = \frac{1}{2N_0}$ and Equation (3.60) then becomes

$$P(s \to \bar{s}|y, h) \leq \exp\left(-\frac{1}{4N_0}h^2|s - \bar{s}|^2\right). \tag{3.61}$$

Similarly, for a multiple antenna space–time coded system, the transmitted codeword becomes a matrix, denoted by $\mathbf{S}$, and the received signal matrix can be expressed as:

$$\mathbf{Y} = \mathbf{HS} + \mathbf{N}, \tag{3.62}$$

where $\mathbf{H} = \{h_{ij}\}$ is the channel matrix with h_{ij} representing the fading channel coefficient from the transmit antenna i to the receive antenna j. Furthermore $\mathbf{N}$ is a random noise vector and each element is Gaussian distributed with the spectrum density of N_0 per dimension.

Following the same calculation as in the single antenna system, the PEP of the decoder deciding in favor of a another codeword $\bar{\mathbf{S}}$ when actually $\mathbf{S}$ has been transmitted, can be upper bounded as

$$P(\mathbf{S} \to \bar{\mathbf{S}}|\mathbf{Y}, \mathbf{H}) \leq \exp\left(-\frac{1}{4N_0}\mathbf{H}^H(\mathbf{S} - \bar{\mathbf{S}})^H(\mathbf{S} - \bar{\mathbf{S}})\mathbf{H}\right), \tag{3.63}$$

where $\mathbf{H}^H$ represents the Hermitian of matrix $\mathbf{H}$.

3.3.1.3 PEP Upper Bound of the DLD-STC Scheme

In this section, we use the above introduced Chernoff bound to derive the PEP upper bound for the DLD-STC system. As shown before, $\tilde{n}_D$ also follows the Gaussian distribution with variance of $\sigma_{\tilde{n}_D}^2 = \left(\frac{P_R}{P_S + N_0}\sum_{i=1}^n |h_{RD}^i|^2 + 1\right)N_0$. By substituting $N_0 = \sigma_{\tilde{n}_D}^2$ and $\mathbf{S} = \sqrt{\frac{P_S P_R T}{P_S + N_0}}\mathbf{A}_S$ into Equation (3.63), the PEP of the DLD-STC is upper bounded as

$$P(\mathbf{S} \to \bar{\mathbf{S}}|\mathbf{Y}, \mathbf{H}) \tag{3.64}$$

$$\leq \exp\left(-\frac{P_S P_R T}{4(1 + P_S + P_R \sum_{i=1}^n |h_{RD}^i|^2)}\mathbf{H}^H(\mathbf{A}_S - \bar{\mathbf{A}}_S)^H(\mathbf{A}_S - \bar{\mathbf{A}}_S)\mathbf{H}\right)$$

Averaging the above equation with respect to the $|h^i_{SR}|^2$, we have [403]:

$$P(\mathbf{S} \rightarrow \overline{\mathbf{S}}|\mathbf{Y}, h^i_{RD}, i = 1, \ldots, n) \leq \det^{-1}\left[\mathbf{I}_n + \frac{P_S P_R T}{4(1 + P_S + P_R \sum_{i=1}^n |h^i_{RD}|^2)}\mathbf{M}\mathbf{G}_{RD}\right], \quad (3.65)$$

where $\mathbf{M} = (\mathbf{A}_S - \overline{\mathbf{A}}_S)^H(\mathbf{A}_S - \overline{\mathbf{A}}_S)$ and $\mathbf{G}_{RD} = \text{diag}\{h^1_{RD}, \ldots, h^n_{RD}\}$. Here, the notation $\text{diag}\{x^1, \ldots, x^n\}$ represents a diagonal matrix with the diagonal elements of $x^1, \ldots, x^n$.

If we compare the above expression with the PEP Chernoff bound of a multiple antenna system, we can see that their expressions are very similar with the only major difference that in the above equation we still have the random variables $|h^i_{RD}|^2, i = 1, \ldots, n$. In order to derive the final PEP upper bound, we need to average Equation (3.65) over $|h^i_{RD}|^2, i = 1, \ldots, n$. Unfortunately, the expectations over all $|h^i_{RD}|^2$ are very difficult to obtain in closed-form, which forces us to resort to some approximations. Let $g = \sum_{i=1}^n |h^i_{RD}|^2$, leading to the following gamma distribution [403]:

$$f_g(x) = \frac{x^{n-1}e^{-x}}{(n-1)!}. \quad (3.66)$$

Both mean and variance of g are n. For large n, g can be approximated by its mean, that is $g \approx n$ [403]. Thus, Equation (3.65) can be approximated as:

$$P(\mathbf{S} \rightarrow \overline{\mathbf{S}}|\mathbf{Y}, h^i_{RD}, i = 1, \ldots, n) \leq \det^{-1}\left[\mathbf{I}_n + \frac{P_S P_R T}{4(1 + P_S + P_R n)}\mathbf{M}\mathbf{G}_{RD}\right]. \quad (3.67)$$

The above equation is minimized when $\frac{P_S P_R T}{4(1 + P_S + P_R n)}$ is maximized. Let $P = P_S + P_R n$ represents the total transmission power used at the source and all n relays. Then we have:

$$\frac{P_S P_R T}{4(1 + P_S + P_R n)} = \frac{P_S(P - P_S)T}{4n(1 + P)} \quad (3.68)$$

and the above equation achieves its maximum value $\frac{P^2 T}{16n(1+P)}$ when

$$P_S = P/2 \quad P_R = P/2n. \quad (3.69)$$

That is, the optimal power allocation strategy allocates half of the total power to the source node and the remaining half equally to all the relays. Now, substituting Equation (3.69) into Equation (3.67), we get:

$$P(\mathbf{S} \rightarrow \overline{\mathbf{S}}|\mathbf{Y}, h^i_{RD}, i = 1, \ldots, n) \leq \det^{-1}\left[\mathbf{I}_n + \frac{PT}{16n}\mathbf{M}\mathbf{G}_{RD}\right]. \quad (3.70)$$

Integrating the above equation with respect to $|h^i_{RD}|^2$ and assuming that $\mathbf{M}$ is a full rank matrix and $T \geq n$, the average PEP of the DLD-STC can be further approximated as [403]:

$$P(\mathbf{S} \rightarrow \overline{\mathbf{S}}|\mathbf{Y}) \leq \det^{-1}[\mathbf{M}]\left(\frac{8n}{T}\right)^n P^{-n\left(1 - \frac{\log\log P}{\log P}\right)}. \quad (3.71)$$

From Equation (3.71) we can see that when $\mathbf{M}$ is full rank and $T \geq n$, then the diversity order achieved by the DLD-STC is $n\left(1 - \frac{\log\log P}{\log P}\right)$. When P is very large, $\frac{\log\log P}{\log P} \rightarrow 0$, and the asymptotic diversity order at high SNRs is thus n. In general, however, the diversity order of the DLD-STC system depends on the total transmission power P. In the above equation, we have assumed that $T \geq n$. In the general case, the diversity order achieved by the DLD-STC is $\min\{n, T\}\left(1 - \frac{\log\log P}{\log P}\right)$. In order to maximize the coding gain of the DLD-STC, the determinant of the matrix $\mathbf{M}$ should thus be maximized.

Let us compare the PEP expression of the DLD-STC with the STTC in multiple antenna system. As has been shown [122] and also in Section 3.3.2, the PEP upper bound at high SNR for STTC systems can be approximated as:

$$\det{}^{-1}[\mathbf{M}]\left(\frac{4n}{PT}\right)^{n}. \tag{3.72}$$

If we compare the above expression with the PEP of DLD-STC given in Equation (3.71), we can see that under the same BER and power condition, the DLD-STC is $(3 + 10\log_{10}\log P)$ dB worse than the STTC. The 3-dB SNR loss is due to the fact that the relay only uses half of total power for transmission and the rest of power is used by the source node. The second term, $10\log_{10}\log P$ actually represents the diversity loss of DLD-STC compared to the STTC. This is due to the distributed construction of the DLD-STC and the noise residing in the received signals in the relay networks. Note however that, as already discussed throughout Chapters 1 and 2, this power loss is in most cases compensated for by the aggregated pathloss and/or shadowing gains.

Similarly, at the low SNR, $P \ll 1$, the PEP upper bound can be approximated as [403]:

$$P(\mathbf{S} \to \overline{\mathbf{S}}|\mathbf{Y}) \lesssim \left(1 - \frac{P^2 T}{16n} TR\{\mathbf{M}\}\right) + o(P^2), \tag{3.73}$$

where $TR\{\mathbf{M}\}$ represents the trace of the matrix $\mathbf{M}$. From the above equation, we can see that the coding gain at low transmission power is dominated by the trace of code matrix $\mathbf{M}$ and the design criterion is thus to maximize $TR\{\mathbf{M}\}$.

Based on the above discussion, we can summarize the following design criteria for the DLD-STC system:

Theorem 3.3.1 *For a DLD-STC system with n relay nodes, let $\mathbf{A}_i$ represents the linear transformation matrix performed at the relay i. Then the maximum diversity order achieved by such a system is $\min\{n, T\}\left(1 - \frac{\log\log P}{\log P}\right)$. The coding gain at high SNR is dominated by the determinant of code difference matrix $\mathbf{M} = (\mathbf{A}_S - \overline{\mathbf{A}}_S)^H(\mathbf{A}_S - \overline{\mathbf{A}}_S)$, where $\mathbf{A}_S = [\mathbf{A}_1\mathbf{S}, \dots, \mathbf{A}_n\mathbf{S}]$. In order to optimize the system performance at high SNR, the determinant of the matrix $\mathbf{M}$ should be maximized. Furthermore, the coding gain at low SNR is determined by the trace of code difference matrix $\mathbf{M}$ and thus to achieve the best performance at the low SNR, the trace of $\mathbf{M}$ should be maximized.*

3.3.2 Distributed Space–Time Trellis Codes

Whilst the open literature is rich in contributions pertaining to regenerative distributed space-time trellis codes, it is fairly scarce in the case of transparent systems. Some example publications exist [404–407], which we will later make use of.

3.3.2.1 System Assumptions

The generic topology under consideration is depicted in Figure 3.10. Its core characteristics can be summarized as follows:

- N-hop single ST-flow topology;
- presence of source–destination link;
- fixed gain amplification factors (semiblind);
- arbitrary flat fading channels without shadowing.

The prime assumption in the context of these topologies is that there is a single flow composed of N transparent relaying segments where transmitted symbols within this flow interfere. Suitable space–time trellis processing techniques are therefore needed to circumvent and possibly use this interference.

The foundations of such techniques was laid Tarokh *et al.* [122], who derived the design criteria for STTCs over slow flat fading channels. The authors showed that in the case of independent Rayleigh fading, the average PEP can be upper-bounded by:

$$P(\mathbf{c} \rightarrow e) \leq \left(\prod_{i=1}^{r} \lambda_i \right)^{-m} \cdot \left(\frac{\gamma}{4} \right)^{-rm} \tag{3.74}$$

where $\mathbf{c}$ and e are the originally sent and erroneously received codewords, respectively; γ the effective SNR; m, n the number of receive and transmit antennas, respectively; and λ_i the ith eigenvalue of the distance matrix $\mathbf{A}(\mathbf{c}, e) = \mathbf{B}(\mathbf{c}, e)\mathbf{B}^{\mathrm{H}}(\mathbf{c}, e)$, where $\mathbf{B}^{\mathrm{H}}$ denotes the Hermitian of $\mathbf{B}$, and r its rank. The difference matrix $\mathbf{B}$ for codewords of length l is given as [122]:

$$\mathbf{B}(\mathbf{c}, e) = \begin{pmatrix} e_1^1 - c_1^1 & \cdots & e_l^1 - c_l^1 \\ \vdots & \ddots & \vdots \\ e_1^n - c_1^n & \cdots & e_l^n - c_l^n \end{pmatrix} \tag{3.75}$$

Tarokh *et al.* concluded that the utmost important design goal is to maximize the diversity gains first as this will dictate the gradient of the error curve; therefore, the sought generator matrix of the STTC has to produce solely full-rank codeword combinations. Over all full-rank generator matrices that one has to be chosen which maximizes the determinant and, thus, the coding gain.

Vucetic *et al.* then derived the design criteria for STTCs over slow flat fading when the diversity figure, defined as the product of transmit and receive antennas, is greater than three [408]. The derivation is based on the observation that for increasing diversity figures the influence of the channel coefficients vanishes. The final expression for the pairwise error probability is given as [408]:

$$\lim_{m \rightarrow \infty} P(\mathbf{c} \rightarrow e) \leq \frac{1}{2} \cdot e^{-m \cdot \frac{\gamma}{4} \cdot \sum_{i=1}^{n} \lambda_i} \tag{3.76}$$

From this upper bound it is clear that the trace of the distance matrix $\mathbf{A}(\mathbf{c}, e)$ has to be maximized.

Moving now to the distributed framework, we will now first deal with a general distributed space–time trellis architecture and show that above original STTC design criteria suffice for such topologies. We then quantify the design of distributed STTCs for some chosen topologies.

3.3.2.2 Generic Design Criteria

Subsequent developments generally follow the insights given by Dohler *et al.* [404]. We assume the generic topology of Figure 3.10 where the source is equipped with n transmit and the destination with m receive antennas. Furthermore, we assume that the source space–time trellis encodes the incoming symbol stream over its n collocated transmit antennas, as outlined by Tarokh *et al.* [122]. This stream is then broadcast over a generic fading channel to the surrounding relays who may perform any form of amplification. The destination receives the information stream on its m receive antennas and then uses the detector described by Tarokh *et al.* [122] to recover the data stream.

The upper bound on the pairwise error probability in dependency of the observed channel fading coefficients $\beta_{i,j}$, each of which is a combination of the fading coefficients $a_{i,j}$ as well as amplification factors $A_{i,j}$ encountered on the relaying path, can thus be calculated as [122, 404]:

$$P(\mathbf{c} \rightarrow e | (\beta_{1,1}, \ldots, \beta_{n,m})) \leq \prod_{j=1}^{m} \prod_{i=1}^{n} e^{-\frac{1}{4}\gamma \lambda_i |\beta_{i,j}|^2}. \tag{3.77}$$

The average error probability $\langle P(\mathbf{c} \to e) \rangle$ is then obtained via

$$\langle P(\mathbf{c} \to e) \rangle \leq \mathrm{E}_{(\beta_{1,1},\dots,\beta_{n,m})} \left\{ P(\mathbf{c} \to e | (\beta_{1,1},\dots,\beta_{n,m})) \right\}$$

$$= \int_{\beta \in \mathcal{R}^{n \times m}} d\beta \cdot p(\beta) \prod_{j=1}^{m} \prod_{i=1}^{n} e^{-\frac{1}{4}\gamma \lambda_i |\beta_{i,j}|^2} \tag{3.78}$$

where $p(\beta)$ is the PDF of the $n \times m$ dimensional random vector $\beta = (\beta_{1,1},\dots,\beta_{n,m})$. To upper bound this expression further, we assuming independent channel realizations which allows the bound on the PEP to be simplified to:

$$\langle P(\mathbf{c} \to e) \rangle \leq \prod_{j=1}^{m} \prod_{i=1}^{n} \left[\int_{\beta_{i,j}} p(\beta_{i,j}) \cdot e^{-\frac{1}{4}\gamma \lambda_i |\beta_{i,j}|^2} d\beta_{i,j} \right]. \tag{3.79}$$

Defining $g(\beta_{i,j}, \lambda_i) = e^{-\frac{1}{4}\gamma \lambda_i |\beta_{i,j}|^2}$, $x = \beta_{i,j}$ and using Schwarz' integral inequality, it can be shown that:

$$\int p(x) \cdot g(x, \lambda_i) dx \leq \sqrt{\int p^2(x) dx} \cdot \sqrt{\int g^2(x, \lambda_i) dx}$$

$$= C \cdot \sqrt{\int g^2(x, \lambda_i) dx} \tag{3.80}$$

under the conditions that $p(x)$ is integrable and finite over its range of definition. One finally obtains for the PEP:

$$\langle P(\mathbf{c} \to e) \rangle \leq \prod_{j=1}^{m} \prod_{i=1}^{n} \left[\left(\frac{1}{\frac{2\gamma}{\pi}} \right)^{\frac{1}{4}} \cdot \left(\frac{1}{\lambda_i} \right)^{\frac{1}{4}} \right]$$

$$= \left(\prod_{i=1}^{r} \lambda_i \right)^{-\frac{m}{4}} \cdot \left(\frac{2\gamma}{\pi} \right)^{-\frac{mr}{4}} \tag{3.81}$$

which yields again that the minimum determinant of all codeword difference matrices needs to be maximized. The derivation thus proves that, generally, the determinant criterion constitutes a sufficient upper bound for channels with any PDF. Note that it does not prove that the determinant criterion guarantees optimum codes.

As for the rank and trace criterion, since the actual channel coefficients (including their statistics and thus amplitude) play a minor role in the derivation of the design criteria, the same rules hold as derived by Yuan *et al.* [408] and given by Equation (3.76). Note that Equation (3.76) only holds for large SNR and $n \to \infty$. However, with the additional noise introduced in each relaying link, the convergence of the criteria is slower and thus the determinant criteria will apply for diversity figures, which were prior covered by the trace criterion.

The performance of the traditional space–time trellis codes deploying VAA is depicted in Figures 3.12 and 3.13. The transceiver configuration resembles that of Tarokh *et al.* [122], that is 130 symbols per frame, QPSK modulation, no outer channel code, STTC generator matrix derived by means of computer search and listed by Tarokh *et al.* [122], quasi-static fading realization during one frame, etc. The case where the main links were Rayleigh distributed and the relaying links Ricean were simulated with $K = 4\,$dB. Further, the relay introduced additional noise leading to the reduction of the SNR of the relayed stream at the destination by a factor of two; nonetheless, the

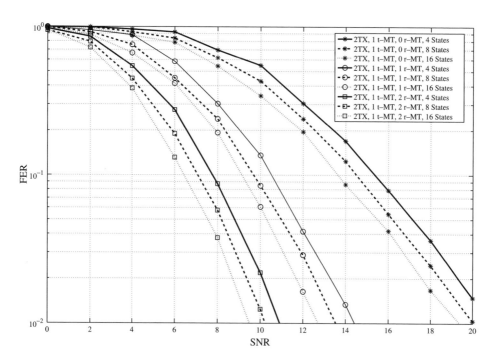

Figure 3.12 FER versus SNR for STTCs deployed with VAA: 2 source, 1 relay and a varying number of destination antennas and STTC states

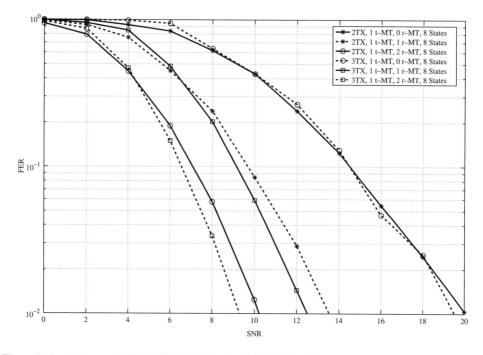

Figure 3.13 FER versus SNR for STTCs deployed with VAA: 2 and 3 source, 1 relay and a varying number of destination antennas for 8-state STTC

STTCs perform sufficiently well. It should also be noted that an extensive search over all possible codeword combinations yields many optimum codes under the given design criteria. Dohler *et al.* [404] discovered codes which slightly outperform the codes given Yuan *et al.* [408]; however, the performance differences are minimal and thus not worth considering.

As such, Figure 3.12 depicts the case of a transmit antenna array with two elements, a destination with one element and a varying number of relays with one antenna element each and STTC states. Obviously, increasing the number of relays acting as virtual receive antennas m has a stronger influence than achieving maximum coding gain; however, the difference between codes of different strength is notable at a FER of 1% and mounts to approximately 0.7 dB. Further, a gain of 10 dB is achieved when comparing the performance of a system with and without VAA, operating with an eight-state STTC. That is quite considerable but needs to be weighed against the shadowing losses discussed in Chapter 2.

Figure 3.13 depicts the performance of the STTCs in VAA for the number of trellis states fixed at eight. It is notable that increasing the number of transmit antennas from two to three, yields additional gains of approximately 1 dB.

In conclusion, it could be proved that the sufficient upper bound on the pairwise error probability, governed by the determinant of the codeword difference matrix, does not depend on the actual channel fading statistics. Therefore, a STTC optimized for Rayleigh channel will be sufficient for Ricean, double Rayleigh, log-normal and any other channel PDF. Note, however, that this bound is not tight; better codes can be derived using other criteria and approaches, such as discussed subsequently. Note finally that Nabar *et al.* [166] have later established similar insights.

3.3.2.3 Protocol-Specific Design Criteria

Subsequent developments generally follow the insights given by Canpolat *et al.* [405]. The subset topology of the generic topology of Figure 3.10 is shown in Figure 3.14, where we assume for the below protocol that the source is equipped with one transmit antenna, each of the R relays is equipped with one antenna and the destination is equipped with n_r receive antennas. Canpolat *et al.* [405] analyzed three different relay protocols, all of which originated Nabar *et al.* [166]. Note

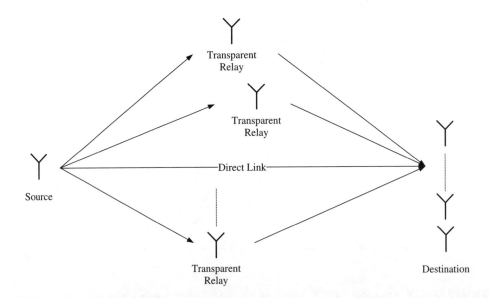

Figure 3.14 Transparent space–time processing topology of [405]

that none of the three relaying protocols is truly adapted to transparent operation as all three protocols require the relay to store the analog signal prior to relaying it. Having said this, the below design criteria derived in the temporal domain, which essentially reduce the conditions to derive the STTC generator matrix coefficients to the traditional approach by invoking some asymptotics, are equally applicable in the spectral domain. We shall now briefly revisit the three transparent relaying protocols:

- **Protocol I.** In this protocol, the source broadcasts the first $M - 1$ modulated symbols $x_1, x_2, \ldots, x_{M-1}$ to the R relays in a sequential manner such that only one relay is served at each slot. Meanwhile, the destination receives each of the above sent symbols. In the Mth slot, the source sends the Mth symbol and all R relays their symbols simultaneously to the destination. This implies that the system is mainly capacity limited as a direct link exists between source and destination. Furthermore, whilst the silent periods for the relays may seem suboptimum, it can be utilized by the relays to engage in other information flows/exchanges.
- **Protocol II.** This protocol is the same as above Protocol I with the only difference that now every relay repeats its transmission of the received symbol in each following time slot.
- **Protocol III.** This protocol is the same as above Protocol I with the only difference being that the source communicates with the destination in the last slot only. Again, this liberates the destination to be engaged in other communications whilst the source communicates with the R relays.

Space time trellis encoding is essentially performed by using the source and relays as $n_t = R + 1$ transmit antennas and the n_r antennas at the destination as receive antennas. Let x_m^l be the modulated space time trellis encoded signal transmitted by the source to the mth relay in the lth signaling frame, where $l = 1, 2, \ldots, L'$. This yields a total transmission length of $L = L' \times M$. The received signal at the nth receive antenna at the destination can hence be cast into matrix form as follows [405]:

$$\mathbf{R}_n = \mathbf{H}_n \mathbf{X} + \mathbf{N}_n, \tag{3.82}$$

where the received matrix $\mathbf{R}_n = [\mathbf{r}_n^1, \mathbf{r}_n^2, \ldots, \mathbf{r}_n^{L'}]$, space-time matrix $\mathbf{X} = [\mathbf{x}^1, \mathbf{x}^2, \ldots, \mathbf{r}^{L'}]$ and the AWGN matrix $\mathbf{N}_n = [\mathbf{n}_n^1, \mathbf{n}_n^2, \ldots, \mathbf{n}_n^{L'}]$, with each entry of each noise vector having an equal noise power, are of size $M \times L'$. The $M \times M$ channel matrix is protocol specific and reads for the respective transparent STTC relaying protocols as follows [405]:

$$\mathbf{H}_n^{\mathrm{I}} = \begin{bmatrix} h_0 & 0 & \cdots & 0 & 0 \\ 0 & h_0 & \cdots & 0 & 0 \\ \vdots & & \ddots & & \vdots \\ 0 & 0 & \cdots & h_0 & 0 \\ a^{\mathrm{I}} \tilde{h}_{1n} & a^{\mathrm{I}} \tilde{h}_{2n} & \cdots & a^{\mathrm{I}} \tilde{h}_{(M-1)n} & a^{\mathrm{I}} h_0 \end{bmatrix} \tag{3.83a}$$

$$a^{\mathrm{I}} = \left(1 + \sum_{m=1}^{M-1} \frac{\gamma_{2,m}}{1 + \gamma_{1,m}} |h_{2,mn}|^2 \right)^{-1/2} \tag{3.83b}$$

and

$$\mathbf{H}_n^{\mathrm{II}} = \begin{bmatrix} a_0^{\mathrm{II}} h_0 & 0 & \cdots & 0 & 0 \\ a_1^{\mathrm{II}} \tilde{h}_{1n} & a_1^{\mathrm{II}} h_0 & \cdots & 0 & 0 \\ \vdots & & \ddots & & \vdots \\ a_{M-2}^{\mathrm{II}} \tilde{h}_{1n} & a_{M-2}^{\mathrm{II}} \tilde{h}_{2n} & \cdots & a_{M-2}^{\mathrm{II}} h_0 & 0 \\ a_{M-1}^{\mathrm{II}} \tilde{h}_{1n} & a_{M-1}^{\mathrm{II}} \tilde{h}_{2n} & \cdots & a_{M-1}^{\mathrm{II}} \tilde{h}_{(M-2)n} & a_{M-1}^{\mathrm{II}} h_0 \end{bmatrix} \tag{3.84a}$$

$$a_i^{II} = \left(1 + \sum_{m=1}^{i} \frac{\gamma_{2,m}}{1 + \gamma_{1,m}} |h_{2,mn}|^2\right)^{-1/2} \qquad i = 1, \cdots, M-1 \tag{3.84b}$$

$$a_0^{II} = 1 \tag{3.84c}$$

and

$$\mathbf{H}_n^{III} = \begin{bmatrix} 0 & 0 & \cdots & 0 & 0 \\ 0 & 0 & \cdots & 0 & 0 \\ \vdots & & \ddots & & \vdots \\ 0 & 0 & \cdots & 0 & 0 \\ a^{III}\tilde{h}_{1n} & a^{III}\tilde{h}_{2n} & \cdots & a^{III}\tilde{h}_{(M-1)n} & a^{III}h_0 \end{bmatrix} \tag{3.85a}$$

$$a^{III} = \left(1 + \sum_{m=1}^{M-1} \frac{\gamma_{2,m}}{1 + \gamma_{1,m}} |h_{2,mn}|^2\right)^{-1/2} \tag{3.85b}$$

with

$$\tilde{h}_{mn} = \sqrt{\frac{\gamma_{1,m}}{1 + \gamma_{1,m}}} h_{1,m} h_{2,mn}. \tag{3.86}$$

Here, the various γ's are the respective SNRs and, due to similar shadowing and pathloss, we have assumed that $\gamma_{2,mn} = \gamma_{2,m}$. Assuming maximum likelihood (ML) decoding at the receiver, the average PEP can be upper bounded by [405]:

$$P(\mathbf{c} \to e) \le e^{-\frac{1}{4}\gamma_0 \cdot d^2}, \tag{3.87}$$

where the distance metric d^2 is again different for each relaying protocol and, assuming Rayleigh fading, reads respectively [405]:

$$d_I^2 = n_r \left(\Lambda_{L'}^M + \sum_{m=1}^{M-1} \Lambda_{L'}^m\right) + \frac{4}{\gamma_0} \sum_{m=1}^{M-1} \ln\left(1 + n_r \Lambda_{L'}^m \frac{\gamma_{2,m}}{4}\right) \tag{3.88a}$$

$$d_{II}^2 = n_r \sum_{m=1}^{M} \Lambda_{L'}^m + \frac{4}{\gamma_0} \sum_{m=1}^{M-1} \ln\left(1 + n_r(M-m)\Lambda_{L'}^m \frac{\gamma_{2,m}}{4}\right) \tag{3.88b}$$

$$d_{III}^2 = n_r \sum_{m=1}^{M} \Lambda_{L'}^m + \frac{4}{\gamma_0} \sum_{m=1}^{M-1} \ln\left(1 + n_r \Lambda_{L'}^m \frac{\gamma_{2,m}}{4}\right), \tag{3.88c}$$

where $\Lambda_{L'}^m = \sum_{l=1}^{L'} |x_m^l - e_m^l|^2$. The criteria simplify significantly if deterministic fading is assumed, which has been the basis for finding suitable STTCs [405]. Their superiority to prior known STTCs has also been demonstrated in the paper and is omitted here for the sake of brevity. Note finally that an improved design criterium has been derived [406] which offers slight performance advantages.

3.3.3 Distributed Spatial Multiplexing

In a network with multiple source, relay and destination nodes, it is desirable for the source nodes to send multiple data streams to the destination nodes without causing interference to the other data streams. We refer to such transmission scheme as the distributed spatial multiplexing (DSM).

In DSM, when multiple independent data streams are being sent from the source nodes, they interfere with each other at the relay and destination nodes. Therefore, the major challenge is to properly design the system to obtain interference free signals at the destination nodes. This can be done by interference cancelation techniques, which can be performed at the source, relay and destination nodes or any combination thereof. The level of processing at the source, relay and destination nodes is often dependent on several factors, such as whether there is cooperation between nodes, or if nodes have stringent energy requirements and thus cannot perform much processing. A number of processing techniques have previously been proposed in various combinations at the source, relay or destination nodes [296, 409–416].

In this section, we consider two-hop AF relay networks for four different system setups based on zero forcing design strategies developed by Louie & Li & Vucetic [416]. In particular, we consider the use of zero forcing precoders and/or receivers for processing at the (i) source and relay, (ii) relay, (iii) relay and destination and (iv) destination nodes. We consider a zero forcing approach because it is a practical and simple scheme often considered for interference cancelation. For these systems, we will study the outage probability derived by Louie & Li & Vucetic [416].

3.3.3.1 System Model

We consider a wireless communications system where n_s source nodes transmit to n_d destination nodes through n_r relay nodes, as shown in Figure 3.15. Let P_s and P_r represent the total source and relay transmission power, respectively. Let $\mathbf{H}_{SR} = \{h_{SR}^{ij}\}$ and $\mathbf{H}_{RD} = \{h_{RD}^{ki}\}$ represent the $n_r \times n_s$ source–relay channel and $n_d \times n_r$ relay–destination channel, where h_{SR}^{ij} and h_{RD}^{ki} are the fading channel coefficients from the source antenna j to the relay antenna i and that from the the relay antenna i to the destination antenna k, respectively. Let $\mathbf{W}_s, \mathbf{W}_r, \mathbf{W}_d$ denote the $n_s \times n_s$ source weight matrix, $n_r \times n_r$ relay weight matrix, $n_d \times n_d$ destination weight matrix, respectively. Furthermore, let $\mathbf{n}_{SR}$ and $\mathbf{n}_{RD}$ be the noise vectors at the relays and destinations, respectively. Each noise element is a Gaussian distributed variable with spectral density of $N_0/2$ per dimension.

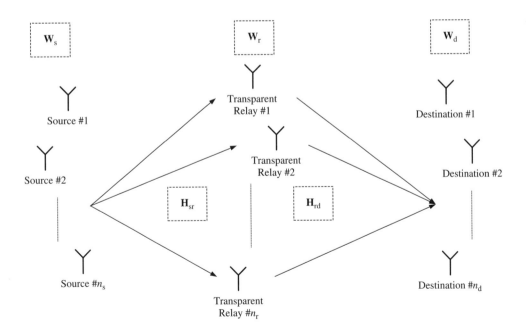

Figure 3.15 A distributed spatial multiplexing system with n_s source nodes, n_r relay nodestransmit and n_d destination nodes

The source nodes first broadcast n_s data streams $\mathbf{X}$ to n_r relay nodes where $E(\mathbf{X}^H\mathbf{X}) = 1$. The received signal vectors at the relay nodes can be expressed as:

$$\mathbf{Y}_{SR} = Q_s\mathbf{H}_{SR}\mathbf{W}_s\mathbf{X} + \mathbf{n}_{SR}, \tag{3.89}$$

where Q_s is the normalization factor designed to ensure that the long term total transmit power at the source nodes,

$$Q_s = \sqrt{\frac{P_s}{\alpha_1}} \tag{3.90}$$

where

$$\alpha_1 = Tr\left\{E(\mathbf{W}_s(\mathbf{W}_s)^H)\right\}. \tag{3.91}$$

Upon receiving signals from the source nodes, the relay nodes apply the weighting matrix $Q_r\mathbf{W}_r$ on $\mathbf{Y}_{SR}$. The received signals at the destination nodes can be expressed as:

$$\mathbf{Y}_{RD} = Q_r\mathbf{H}_{RD}\mathbf{W}_r\mathbf{Y}_{SR} + \mathbf{n}_{RD}$$
$$= Q_rQ_s\mathbf{H}_{RD}\mathbf{W}_r\mathbf{H}_{SR}\mathbf{W}_s\mathbf{X} + Q_r\mathbf{H}_{RD}\mathbf{W}_r\mathbf{n}_{SR} + \mathbf{n}_{RD}, \tag{3.92}$$

where Q_r is the normalization constant designed to ensure that the long term total transmit power at the relay nodes is constrained, and is given by:

$$Q_r = \sqrt{\frac{\overline{\gamma}_{RD}\alpha_1}{\overline{\gamma}_{SR}\alpha_2 + \alpha_1\alpha_3}}, \tag{3.93}$$

where

$$\alpha_2 = Tr\{E(\mathbf{W}_r\mathbf{H}_{SR}\mathbf{W}_s(\mathbf{W}_r\mathbf{H}_{SR}\mathbf{W}_s)^H)\} \tag{3.94a}$$

$$\alpha_3 = Tr\{E(\mathbf{W}_r(\mathbf{W}_r)^H)\}. \tag{3.94b}$$

and $\overline{\gamma}_{SR} = \frac{P_s}{N_0}$ and $\overline{\gamma}_{RD} = \frac{P_r}{N_0}$. We shall note that we only consider the long term power constraint here as the analysis for the short term power is not tractable. We then apply a destination weight matrix $\mathbf{W}_d$ to the received signal vector resulting in the following data estimate:

$$\tilde{\mathbf{X}} = \mathbf{W}_d\mathbf{Y}_{RD}^H$$
$$= Q_rQ_s\mathbf{W}_d\mathbf{H}_{RD}\mathbf{W}_r\mathbf{H}_{SR}\mathbf{W}_s\mathbf{X} + Q_r\mathbf{W}_d\mathbf{H}_{RD}\mathbf{W}_r\mathbf{n}_{SR} + \mathbf{W}_d\mathbf{n}_{RD}. \tag{3.95}$$

The destination received SNR for an arbitrary transmitted kth stream can be written as:

$$\gamma = \frac{\overline{\gamma}_{SR}\overline{\gamma}_{RD}\left[\mathbf{W}_d\mathbf{H}_{RD}\mathbf{W}_r\mathbf{H}_{SR}\mathbf{W}_s(\mathbf{W}_d\mathbf{H}_{RD}\mathbf{W}_r\mathbf{H}_{SR}\mathbf{W}_s^H)\right]_{k,k}}{\left[\overline{\gamma}_{RD}\alpha_1\mathbf{W}_d\mathbf{H}_{RD}\mathbf{W}_r(\mathbf{W}_d\mathbf{H}_{RD}\mathbf{W}_r)^H + (\overline{\gamma}_{SR}\alpha_2 + \alpha_1\alpha_3)\mathbf{W}_d(\mathbf{W}_d)^H\right]_{k,k}}, \tag{3.96}$$

where $[\mathbf{A}]_{k,k}$ represent the kkth element of matrix $\mathbf{A}$.

To obtain interference free signals at each destination node, we use the principles of zero forcing to design the weight matrices $\mathbf{W}_s$, $\mathbf{W}_r$ and $\mathbf{W}_d$. The weight matrices are designed depending on the amount of processing allowed at the source, relay and destination nodes. This is often dependent on the level of cooperation, energy requirements and the feedback capabilities at each source, relay and destination nodes.

Note that we assume node cooperation by means of which the zero forcing processing is facilitated. For example, if we consider the use of zero forcing at the source and relay, we assume that the source nodes can cooperate, and the relay nodes can cooperate. In this section, cooperation refers to the cooperating nodes being able to exchange channel information and transmitted symbols with each other, which essentially facilitates the zero forcing processing. Our results also apply to the specific case when the cooperating nodes are antennas belonging to a single entity, such as a base station. As this scenario may be more common in practice, we will often use this scenario to motivate the different schemes in the following sections. We note, however, that our results also apply to the general case of multiple source, relay and destination nodes.

Finally, we make the common assumption that the number of nodes where the processing is performed is greater than or equal to the number of nodes at the other side of the link. For example, if zero forcing is performed at the source and relay, we have $n_s \geq n_r$ and $n_r \geq n_d$.

Next, we will study the performance of four different system setups based on the zero forcing design strategies described in this section. To derive the outage probability, we will make use of the following lemma [417]:

Lemma 3.3.2 *Let* **V** *be a* $n \times m$ *complex normal gaussian variable with mean* **0** *and variance* $\mathbf{I}_n \otimes \mathbf{I}_m$ *with* $m > n$. *Then*

$$Tr\{E(\mathbf{V}\mathbf{V}^H)^{-1}\} = \frac{n}{m-n}. \tag{3.97}$$

3.3.3.2 Zero Forcing at the Source and Relay Nodes

In this section, we consider the DSM with zero forcing at the source and relay nodes. This corresponds to many practical scenarios including those where there is cooperation between the source nodes and cooperation between the relay nodes, but no cooperation between the destination nodes. An example includes a base station transmitting to multiple destination mobiles via a base-station relay, where the mobiles are power limited and thus are unable to perform much processing, and the base station and base-station relay are equipped with multiple antennas. In this scenario, the source and relay nodes correspond to individual antennas at the source and relay base stations, respectively.

As the zero forcing processing is performed at the source and relay, we assume that $n_s \geq n_r$ and $n_r \geq n_d$. The principles of zero forcing results in setting

$$\mathbf{W}_s = (\mathbf{H}_{SR})^H (\mathbf{H}_{SR}(\mathbf{H}_{SR})^H)^{-1}\mathbf{P}_1 \tag{3.98a}$$

$$\mathbf{W}_r = (\mathbf{H}_{RD})^H (\mathbf{H}_{RD}(\mathbf{H}_{RD})^H)^{-1}\mathbf{P}_2 \tag{3.98b}$$

$$\mathbf{W}_d = \mathbf{I}_{n_d}, \tag{3.98c}$$

where $\mathbf{P}_1$ is a $n_r \times n_s$ matrix designed to ensure that only n_r out of n_s data streams are transmitted at the source, and $\mathbf{P}_2$ is a $n_d \times n_r$ matrix designed to ensure that only n_d out of n_r data streams are transmitted at the relay. We construct $\mathbf{P}_1$ by horizontally concatenating a $n_r \times n_r$ permutation matrix and a $n_r \times (n_s - n_r)$ zero matrix while we construct $\mathbf{P}_2$ by horizontally concatenating a $n_d \times n_d$ permutation matrix and a $n_d \times (n_r - n_d)$ zero matrix. For example, if we assume $n_s = 4$, $n_r = 3$ and $n_d = 2$, then a possible choice of $\mathbf{P}_1$ and $\mathbf{P}_2$ is given by

$$\mathbf{P}_1 = \begin{pmatrix} 1 & 0 & 0 & 0 \\ 0 & 1 & 0 & 0 \\ 0 & 0 & 1 & 0 \end{pmatrix} \tag{3.99a}$$

$$\mathbf{P}_2 = \begin{pmatrix} 1 & 0 & 0 \\ 0 & 1 & 0 \end{pmatrix} \tag{3.99b}$$

Note that the permutation matrices can be chosen to prioritize certain data streams, or allocate equal time periods to each data stream to maintain fairness. For the former case, this could be achieved by choosing those data streams that give the best performance for each channel realization, while for the latter case, this could be achieved by designing the permutation matrices to transmit data streams in a round-robin manner. In this section, however, we consider choosing the permutation matrices to maintain fairness. The permutation matrices are therefore chosen randomly for each channel realization.

Now, substituting Equation (3.98) into Equation (3.96), the signal-to-noise-and-interference-ratio (SINR) for an arbitrary kth stream approaches a constant, and is given by:

$$\gamma_{S/R} = \frac{\overline{\gamma}_{SR}\overline{\gamma}_{RD}}{\overline{\gamma}_{RD}\alpha_1 + \overline{\gamma}_{SR}\alpha_2 + \alpha_1\alpha_3}, \tag{3.100}$$

where

$$\alpha_1 = Tr\{E(\mathbf{W}_s(\mathbf{W}_s)^H)\} = Tr\{E(\mathbf{H}_{SR}(\mathbf{H}_{SR})^H)^{-1}\}$$

$$= \frac{n_r}{n_s - n_r} \tag{3.101a}$$

$$\alpha_2 = \alpha_3 = Tr\{E(\mathbf{H}_{RD}(\mathbf{H}_{RD})^H)^{-1}\}$$

$$= \frac{n_d}{n_r - n_d}. \tag{3.101b}$$

Substituting these equations into Equation (3.90) and Equation (3.93), we have:

$$Q_s = \sqrt{\frac{P_s(n_s - n_r)}{n_r}} \tag{3.102a}$$

$$Q_r = \sqrt{\frac{\overline{\gamma}_{RD}n_r(n_r - n_d)}{\overline{\gamma}_{SR}n_d(n_s - n_r) + n_r n_d}}. \tag{3.102b}$$

Substitute now the above equations into Equation (3.100), and the SINR can be rewritten as:

$$\gamma_{S/R} = \frac{\overline{\gamma}_{SR}\overline{\gamma}_{RD}(n_s - n_r)(n_r - n_d)}{\overline{\gamma}_{RD}n_r(n_r - n_d) + \overline{\gamma}_{SR}n_d(n_s - n_r) + n_r n_d}. \tag{3.103}$$

By taking the derivative of $\gamma_{S/R}$ w.r.t. n_s, it is easy to show that the SINR is always an increasing function of n_s. Note that we see in Equations (3.102a) and (3.102b) that increasing n_s increases the normalization constant at the source, and thus decreases the received SINR at the relay. However, from Equation (3.102b), we see that increasing n_s decreases the normalization constant at the relay. This indicates that the detrimental effect of the normalization constant at the relay, due to increasing n_s, dominates the beneficial effect of the normalization constant at the source.

Further, by taking the derivative w.r.t. n_d, it is easy to show that the SINR is always an increasing function of n_d. This is because increasing the number of destination nodes decreases the normalization constant at the relay and hence the received SINR is increased.

By taking the derivative of $\gamma_{S/R}$ w.r.t. n_r, we can obtain an expression such that the SINR is an increasing function of n_r. However, the resulting expression is complicated and is not amenable to useful insights. We thus consider the case when $\overline{\gamma} = \overline{\gamma}_{SR} = \overline{\gamma}_{RD}$, from which we see that the SINR increases with n_r if and only if:

$$\overline{\gamma} > \frac{n_d}{n_s - n_d} \quad \text{and} \quad n_s n_d > n_r^2 \quad \text{or}$$

$$\overline{\gamma} < \frac{n_d}{n_s - n_d} \quad \text{and} \quad n_s n_d < n_r^2. \tag{3.104}$$

We see from Equation (3.104) the interesting result that the SINR increases with n_r under certain power and antenna constraints. Note that increasing the relay nodes has the dual effect of decreasing the normalization constant at the source, and also increasing the normalization constant at the relay. The condition in Equation (3.104) characterizes this tradeoff by explicitly showing the exact conditions under which the SINR is an increasing function of the relay nodes. This is thus very useful in designing the number of relays which will result in a positive increase in the SINR.

Figure 3.16 shows the SINR versus the number of relay nodes for a fixed number of source and destination nodes. The 'analytical' curve is obtained from Equation (3.100) and clearly agrees with the curve obtained by Monte Carlo simulations. We see that the SINR increases when the number of relay nodes increases from three to four, but decreases when additional relays nodes are added. This agrees with our previous analysis where we observed that adding relay nodes can have the dual effect of decreasing the normalization constant at the source, and increasing the normalization constant at the relay. We also see in Figure 3.16 that for a low number of relay nodes, the positive impact of the decreased normalization constant at the source dominates the negative impact of the increased normalization constant at the relay, but vice versa for a larger number of relay nodes.

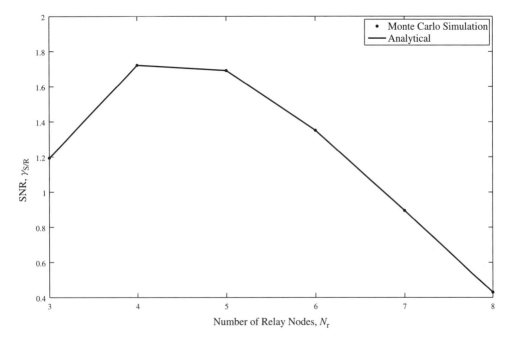

Figure 3.16   SINR versus number of source nodes for zero forcing at the source and relay nodes with $n_d = 2$, $n_s = 9$, $\overline{\gamma}_{SR} = 6\,\text{dB}$ and $\overline{\gamma}_{RD} = 5\,\text{dB}$. (Reproduced by permission of © IEEE)

3.3.3.3   Zero Forcing at the Relay Nodes Only

In this section, we consider the DSM with zero forcing at the relay nodes only. This corresponds to practical scenarios where there is cooperation at the relay nodes but no cooperation at the source and destination nodes. An example includes mobile nodes transmitting to other mobile nodes via a base station relay equipped with multiple antennas, where the mobiles nodes are power limited and thus are unable to perform much processing. In this scenario, the relay nodes correspond to individual antennas at a single relay base station. We consider using zero forcing precoders and receivers at the relay nodes. As the zero forcing processing is performed at the relay only, we assume $n_r \geq n_s$ and $n_r \geq n_d$, and as such, the number of end-to-end data streams, or end-to-end spatial multiplexing gain is $\min(n_s; n_d)$. However, due to mathematical intractability, our results are only valid when $n_r > n_d$

and $N = n_s = n_d$. By applying the principles of zero forcing in this setting, we have:

$$\mathbf{W}_s = \mathbf{I}_{n_s} \tag{3.105a}$$

$$\mathbf{W}_r = (\mathbf{H}_{RD})^H (\mathbf{H}_{RD}\mathbf{H}_{RD}^H)^{-1}((\mathbf{H}_{SR})^H \mathbf{H}_{SR})^{-1}(\mathbf{H}_{SR})^H \tag{3.105b}$$

$$\mathbf{W}_d = \mathbf{I}_{n_d}. \tag{3.105c}$$

Substituting Equation (3.105) into Equation (3.96), the SINR for an arbitrary k–th stream is derived as:

$$\gamma_R = \frac{\overline{\gamma}_{SR}\overline{\gamma}_{RD}}{\overline{\gamma}_{RD}\alpha_1 \left[(\mathbf{H}_{SR}^H \mathbf{H}_{SR})^{-1} \right]_{k,k} + \overline{\gamma}_{SR}\alpha_2 + \alpha_1\alpha_3}, \tag{3.106}$$

where

$$\alpha_1 = Tr\{E(\mathbf{W}_s(\mathbf{W}_s)^H)\} = n_s \tag{3.107a}$$

$$\alpha_2 = Tr\{E(\mathbf{W}_r\mathbf{H}_{SR}\mathbf{W}_s(\mathbf{W}_r\mathbf{H}_{SR}\mathbf{W}_s)^H)\} = Tr\{E(\mathbf{H}_{RD}(\mathbf{H}_{RD})^H)^{-1}\}$$
$$= \frac{n_d}{n_r - n_d} \tag{3.107b}$$

$$\alpha_3 = Tr\{E((\mathbf{H}_{SR})^H\mathbf{H}_{SR})\mathbf{H}_{RD}(\mathbf{H}_{RD})^H)\}$$
$$= \frac{\sum_{k=1}^n \det(\mathbf{C}^k)}{\prod_{l=1}^n (n_r - l)!(N - l)!(n_r - l)!} \tag{3.107c}$$

with $N = n_s = n_d$ and $\mathbf{C}^k$ being a $N \times N$ matrix with ijth entry given by:

$$\mathbf{C}_{ij}^k \geq \begin{cases} (n_r - N + i - 2)!(n_r - N + i + j - 3)! & \text{for } j = k \\ (n_r - N + i - 1)!(n_r - N + i + j - 2)! & \text{for } j \neq k \end{cases}. \tag{3.108}$$

Substituting these equations into Equations (3.90) and (3.93), we have:

$$Q_s = \sqrt{\frac{P_s}{n_s}} \tag{3.109a}$$

$$Q_r = \sqrt{\frac{\overline{\gamma}_{RD}n_s(n_r - n_d) \prod_{l=1}^N (n_r - l)!(N - l)!(n_r - l)!}{\overline{\gamma}_{SR}n_d \prod_{l=1}^N (n_r - l)!(N - l)!(n_r - l)! + n_s(n_r - n_d) \sum_{k=1}^N \det(\mathbf{C}^k)}}. \tag{3.109b}$$

At high SNR, the latter reduces to

$$Q_r = \sqrt{\frac{\overline{\gamma}_{RD}n_s(n_r - n_d)}{\overline{\gamma}_{SR}n_d}}. \tag{3.110}$$

By using the above equations, we can derive the outage probability given by the following theorem.

Theorem 3.3.3 *The outage probability for the two hop AF relay network using zero forcing at the relay nodes is given by [416]:*

$$F_{\gamma_R}(\gamma_{th}) = 1 - \frac{\Gamma\left(n_r - n_s + 1, \frac{\overline{\gamma}_{RD}\alpha_1\gamma_{th}}{\overline{\gamma}_{SR}\overline{\gamma}_{RD}-\gamma_{th}\Upsilon}\right)}{\Gamma(n_r - n_s + 1)}\mathbf{1}_{\gamma_{th}\leq\frac{\overline{\gamma}_{SR}\overline{\gamma}_{RD}}{\Upsilon}} \tag{3.111}$$

where $\Upsilon = \overline{\gamma}_{SR}\alpha_2 + \alpha_1\alpha_3$ and $\Gamma(\cdot)$ is the gamma function and $\Gamma(\cdot, \cdot)$ is the upper incomplete gamma function.

By taking a Taylor series expansion of the above equation around $\gamma_{th} = 0$, we can obtain the asymptotic outage probability at high SNR, high $\overline{\gamma}_{SR}$ and $\overline{\gamma}_{RD}$ values, given by [416]:

$$F_{\gamma_R}(\gamma_{th}) \approx \left(\frac{n_s \gamma_{th}}{\overline{\gamma}_{SR}}\right)^{n_r - n_s + 1} \frac{1}{(n_r - n_s + 1)!}. \qquad (3.112)$$

It has been shown [383], and frequently used throughout this book, that if the first order expansion of the CDF of γ can be written in the form

$$F_\gamma(\gamma) = \frac{C_0 \gamma^{N+1}}{\overline{\gamma}^{N+1}(N+1)} + o\left(\gamma^{N+1+\varepsilon}\right), \varepsilon > 0 \qquad (3.113)$$

then the asymptotic BER at high SNR can be approximated as [383]:

$$P_b = E\left[aQ\left(\sqrt{2b\gamma}\right)\right] = \frac{2^N a C_0 \Gamma\left(N + \frac{3}{2}\right)}{\sqrt{\pi}(N+1)} (b\overline{\gamma})^{-(N+1)} + o\left(\overline{\gamma}^{-(N+1)}\right), \qquad (3.114)$$

where $\overline{\gamma}$ is the average transmit SNR. The diversity order achieved by this system is thus $N + 1$. Based on Equations (3.113) and (3.114), we can easily derive the asymptotic BER expressions and we can see that the diversity order achieved by the DSM with zero forcing at the relay nodes only can be shown to be $n_r - n_s + 1$. Equation (3.112) allows us to obtain useful insights into the performance of the system. We see that the asymptotic outage probability is independent of the number of destination nodes. Further, for a fixed $n_r - n_s$, we see that increasing n_s results in an increase in outage probability, hence is detrimental to system performance.

Figure 3.17 shows the outage probability for various numbers of relay nodes and fixed numbers of source and destination nodes. The "analytical" curves are obtained from (3.111), and clearly agree with the curves obtained by Monte Carlo simulations. We see that increasing the number of relay nodes leads to a substantial decrease in the outage probability. This is because increasing the

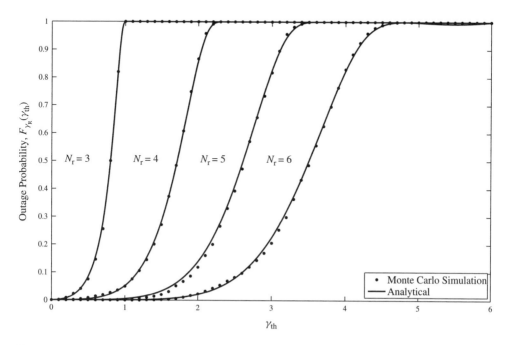

Figure 3.17 Outage probability versus SINR threshold for zero forcing at the relay nodes with $n_s = n_d = 2$, $\overline{\gamma}_{SR} = 6 \, \text{dB}$ and $\overline{\gamma}_{RD} = 5 \, \text{dB}$ [416]. (Reproduced by permission of © IEEE)

number of relay nodes increases the diversity order from the source to relay link, and decreases the
normalization constant at the relay.

3.3.3.4   Zero Forcing at the Relay and Destination Nodes

In this section, we consider a DSM with zero forcing at the relay and destination nodes. This corresponds to practical scenarios where there is cooperation between relay nodes and cooperation between
destination nodes. An example includes mobiles transmitting to a base station via a base-station relay,
where the mobiles are power limited and thus are unable to perform much processing, and the base stations are equipped with multiple antennas. In this scenario, the relay and destination nodes correspond
to individual antennas at the relay and destination base stations respectively.

As the zero forcing processing is performed at the relay and destination nodes, we assume $n_r \geq n_s$
and $n_d \geq n_r$, and as such, the number of end-to-end data streams, or end-to-end spatial multiplexing
gain is n_s. However, due to mathematical intractability, our results are only valid when $n_r > n_s$. By
applying the principles of zero forcing, we get:

$$\mathbf{W}_s = \mathbf{I}_{n_s} \tag{3.115a}$$

$$\mathbf{W}_r = \mathbf{P}((\mathbf{H}_{SR})^H \mathbf{H}_{SR})^{-1}(\mathbf{H}_{SR})^H \tag{3.115b}$$

$$\mathbf{W}_d = ((\mathbf{H}_{RD})^H \mathbf{H}_{RD})^{-1}(\mathbf{H}_{RD})^H, \tag{3.115c}$$

where $\mathbf{P}$ is a matrix designed to ensure that n_r data streams are sent at the relay. Note that after applying
the zero forcing matrix at the relay, there are n_s data streams to be sent out by $n_r \geq n_s$ relays. There
are thus an additional $n_r - n_s$ relay nodes that are not being utilized. Sophisticated signal processing
techniques can be used to make use of the $n_r - n_s$ nodes, such as diversity techniques [418], which
will result in a better error performance. In this section, however, we consider the analysis of an
arbitrary transmitted stream at the relay that is not subject to any sophisticated signal processing
techniques. Note that this will result in a worse performance when compared to the case when the
data stream(s) are subject to the sophisticated signal processing technique.

Now, substituting Equation (3.115) into Equation (3.96), the SINR for an arbitrary kth stream is
given by:

$$\gamma_{R/D} = \frac{\overline{\gamma}_{SR}\overline{\gamma}_{RD}}{\overline{\gamma}_{RD}\alpha_1 \left[(\mathbf{H}_{SR}^H \mathbf{H}_{SR})^{-1}\right]_{k,k} + (\overline{\gamma}_{SR}\alpha_2 + \alpha_1\alpha_3)\left[(\mathbf{H}_{RD}^H \mathbf{H}_{RD})^{-1}\right]_{k,k}}, \tag{3.116}$$

where

$$\alpha_1 = Tr\{E(\mathbf{W}_s(\mathbf{W}_s)^H)\} = n_s \tag{3.117a}$$

$$\alpha_2 = Tr\{E(\mathbf{W}_r\mathbf{H}_{SR}\mathbf{W}_s(\mathbf{W}_r\mathbf{H}_{SR}\mathbf{W}_s)^H)\} = n_r \tag{3.117b}$$

$$\alpha_3 = Tr\{E(\mathbf{W}_r(\mathbf{W}_r)^H)\} = Tr\{E[((\mathbf{H}_{SR})^H\mathbf{H}_{SR})^{-1}]\} = \frac{n_s}{n_r - n_s} \tag{3.117c}$$

The above equation is only valid for $n_r > n_s$. Substituting these equations into Equations (3.90) and
(3.93), we obtain:

$$Q_s = \sqrt{\frac{P_s}{n_s}} \tag{3.118a}$$

$$Q_r = \sqrt{\frac{\overline{\gamma}_{RD}n_s(n_r - n_s)}{\overline{\gamma}_{SR}n_r(n_r - n_s) + n_s^2}}. \tag{3.118b}$$

From the above SINR expression, we can derive the outage probability given by Theorem 3.3.4.

Theorem 3.3.4 *The outage probability for the two hop AF relay network using zero forcing at the relay and destination nodes is given by [416]:*

$$F_{\gamma_{R/D}}(z) = 1$$

$$- \frac{2 \left(\frac{z(\bar{\gamma}_{SR}\alpha_2 + \alpha_1\alpha_3)}{\bar{\gamma}_{SR}\bar{\gamma}_{RD}} \right)^{n_d - n_r + 1} e^{-\frac{z}{\bar{\gamma}_{SR}\bar{\gamma}_{RD}}(\bar{\gamma}_{SR}\alpha_2 + \alpha_1\alpha_3 + \bar{\gamma}_{RD}\alpha_1)}}{\Gamma(n_d - n_r + 1)}$$

$$\times \sum_{p=0}^{n_r - n_s} \frac{\left(\frac{z\bar{\gamma}_{RD}\alpha_1}{\bar{\gamma}_{SR}\bar{\gamma}_{RD}} \right)^p}{p!} \sum_{q=0}^{n_d - n_r + p} \binom{n_d - n_r + p}{q} \left(\frac{\bar{\gamma}_{RD}\alpha_1}{\bar{\gamma}_{SR}\alpha_2 + \alpha_1\alpha_3} \right)^{\frac{q - p + 1}{2}}$$

$$\times K_{p-q-1} \left(\frac{2z\sqrt{\bar{\gamma}_{RD}\alpha_1(\bar{\gamma}_{SR}\alpha_2 + \alpha_1\alpha_3)}}{\bar{\gamma}_{SR}\bar{\gamma}_{RD}} \right) \qquad (3.119)$$

where $K.(\cdot)$ is the modified Bessel function of the second kind.

It can be further approximated at high SNR as [416]:

$$F_{\gamma_{R/D}}(\gamma_{th}) \approx \begin{cases} \frac{n_s^{n_r - n_s + 1}}{(n_r - n_s + 1)!} \left(\frac{\gamma_{th}}{\bar{\gamma}_{SR}} \right)^{n_r - n_s + 1} & \text{for } n_r - n_s < n_d - n_r \\ \frac{n_r^{n_d - n_r + 1}}{(n_d - n_r + 1)!} \left(\frac{\gamma_{th}}{\bar{\gamma}_{RD}} \right)^{n_d - n_r + 1} & \text{for } n_r - n_s > n_d - n_r \\ \frac{N^N}{N!} (\frac{1}{\kappa^N} + 1)^N \left(\frac{\gamma_{th}}{\bar{\gamma}_{SR}} \right)^N & \text{for } N = n_r - n_s + 1 = n_d - n_r + 1 \end{cases} \qquad (3.120)$$

where $\kappa = \frac{\bar{\gamma}_{RD} n_s}{\bar{\gamma}_{SR} n_r}$ Similarly, based on Equations (3.113) and (3.114), the diversity order of the system can be shown as [416]:

$$G_d = \begin{cases} n_r - n_s + 1 & \text{for } n_r - n_s < n_d - n_r \\ n_d - n_r + 1 & \text{for } n_r - n_s > n_d - n_r \\ N & \text{for } N = n_r - n_s + 1 = n_d - n_r + 1 \end{cases} \qquad (3.121)$$

Equations (3.120) and (3.121) allow us to obtain useful insights into the performance of the system. We see that the diversity order is dependent on the particular source, relay and destination node configuration. In particular, we see in Equation (3.121) that for a fixed n_s and n_d, a diversity order of $n_d - n_r + 1$ will be achieved for low n_r, and that increasing n_r is beneficial for system performance. However, as n_r increases, we see that there is a particular number of relay nodes where the diversity order will shift to $n_d - n_r + 1$, and adding more relays is detrimental to system performance. We further see that for a fixed $n_r - n_s$ when $n_r - n_s < n_d - n_r$, increasing n_s results in a increase in outage probability, hence is detrimental to system performance. In addition, we see that for a fixed $n_d - n_r$ when $n_r - n_s > n_d - n_r$, increasing n_r results in a increase in outage probability, hence is detrimental to system performance.

Figure 3.18 shows the outage probability for various numbers of destination nodes and fixed numbers of source and relay nodes. The 'analytical' curves are obtained from Equation (3.119), and clearly agree with the curves obtained by Monte Carlo simulations. We see that increasing the number of destination nodes leads to a substantial decrease in the outage probability. This is because increasing n_d results in an increase in diversity order from the relay to destination nodes link, hence is beneficial for system performance.

Figure 3.19 shows the outage probability for various numbers of relay nodes and fixed numbers of source and destination nodes. Interestingly, we see that the outage probability decreases when going from four to five relay nodes. However, adding more relay nodes results in an increase in the outage probability. This agrees with our previous analysis where we observed that adding relay nodes is beneficial for a low number of relay nodes, but detrimental for a higher number of relay nodes.

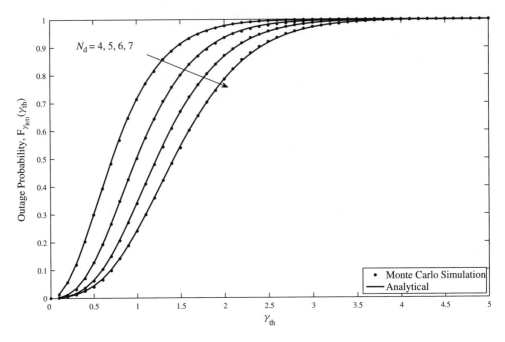

Figure 3.18 Outage probability versus SINR threshold for zero forcing at the relay and destination nodes with $n_s = 2$, $n_r = 3$, $\overline{\gamma}_{SR} = 8\,\text{dB}$ and $\overline{\gamma}_{RD} = 9\,\text{dB}$ [416]. (Reproduced by permission of © IEEE)

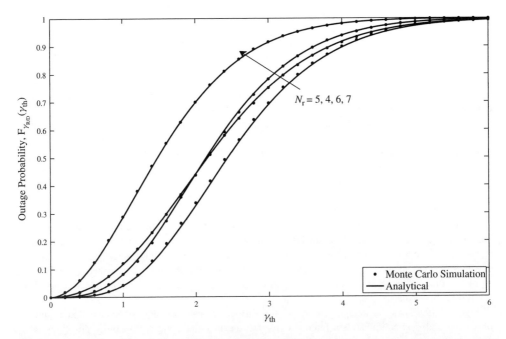

Figure 3.19 Outage probability versus SINR threshold for zero forcing at the relay and destination nodes with $n_s = 3$, $n_d = 8$, $\overline{\gamma}_{SR} = 6\,\text{dB}$ and $\overline{\gamma}_{RD} = 5\,\text{dB}$ [416] (Reproduced by permission of © IEEE)

Although our analysis was for the high SNR regime, Figure 3.19 indicates that our analysis also holds for all SNR values.

3.3.3.5 Zero Forcing at the Destination Node Only

In this section, we finally consider a DSM with zero forcing at the destination only. This corresponds to many practical scenarios, including those where there is cooperation at the destination nodes. In addition, the source nodes send independent data streams to the relay nodes, which then forward them to the destination nodes. The destination nodes then collaboratively process the signal to recover the desired data. An example is mobiles transmitting to a base station via mobile relay nodes, where the mobiles are power limited and thus are unable to perform much processing, and the base station is equipped with multiple antennas. In this scenario, the destination nodes correspond to individual antennas at the destination base station.

As the zero forcing processing is performed at the relay only, we assume $n_d \geq n_s$ and $n_d \geq n_r$, and as such, the number of end-to-end data streams, or end-to-end spatial multiplexing gain is $\min\{n_s; n_r\}$. By applying the principles of zero forcing, we get:

$$\mathbf{W}_s = \mathbf{I}_{n_s} \tag{3.122a}$$

$$\mathbf{W}_r = \mathbf{I}_{n_r} \tag{3.122b}$$

$$\mathbf{W}_d = ((\mathbf{H}_{RD}\mathbf{H}_{SR})^H \mathbf{H}_{RD}\mathbf{H}_{SR})^{-1}(\mathbf{H}_{RD}\mathbf{H}_{SR})^H. \tag{3.122c}$$

Substituting Equation (3.122a) into Equation (3.96), the SINR for an arbitrary kth stream is given by:

$$\gamma_D = \frac{\overline{\gamma}_{SR}\overline{\gamma}_{RD}}{\overline{\gamma}_{RD}\alpha_1\left[(\mathbf{H}_{SR}^H\mathbf{H}_{SR})^{-1}\right]_{k,k} + (\overline{\gamma}_{SR}\alpha_2 + \alpha_1\alpha_3)\left[((\mathbf{H}_{RD}\mathbf{H}_{SR})^H\mathbf{H}_{RD}\mathbf{H}_{SR})^{-1}\right]_{k,k}} \tag{3.123}$$

where

$$\alpha_1 = Tr\{E(\mathbf{W}_s(\mathbf{W}_s)^H)\} = n_s \tag{3.124a}$$

$$\alpha_2 = Tr\{E(\mathbf{W}_r\mathbf{H}_{SR}\mathbf{W}_s(\mathbf{W}_r\mathbf{H}_{SR}\mathbf{W}_s)^H)\} = Tr\{E(\mathbf{H}_{SR}\mathbf{H}_{SR}^H)\} \tag{3.124b}$$

$$= N_s N_r$$

$$\alpha_3 = Tr\{E(\mathbf{W}_r(\mathbf{W}_r)^H)\} = n_r. \tag{3.124c}$$

Substituting these equations into Equations (3.90) and (3.93), we have:

$$Q_s = \sqrt{\frac{P_s}{n_s}} \tag{3.125a}$$

$$Q_r = \sqrt{\frac{\overline{\gamma}_{RD}}{\overline{\gamma}_{SR}n_r + n_r}}. \tag{3.125b}$$

From the above SINR expression, we can derive the outage probability, given in Theorem 3.3.5.

Theorem 3.3.5 *The outage probability for the two-hop AF relay network using zero forcing at the destination nodes is given by:*

$$F_{\gamma_D}(\gamma_{th}) = 1 - 2e^{-\frac{\gamma_{th}\alpha_1}{\overline{\gamma}_{SR}}}\sum_{p=0}^{n_d-n_r}\frac{\left(\frac{\gamma_{th}(\overline{\gamma}_{SR}\alpha_2 + \alpha_1\alpha_3)}{\overline{\gamma}_{SR}\overline{\gamma}_{RD}}\right)^{\frac{p}{2}+\frac{1}{2}}}{p!}$$

$$\times K_{1-p}\left(2\sqrt{\frac{(\overline{\gamma}_{SR}\alpha_2 + \alpha_1\alpha_3)\gamma_{th}}{\overline{\gamma}_{SR}\overline{\gamma}_{RD}}}\right). \tag{3.126}$$

At high SNR, that is high $\overline{\gamma}_{SR}$ and $\overline{\gamma}_{RD}$ values, the above outage probability can be further approximated as [416]:

$$F_{\gamma_D}(\gamma_{\text{th}}) \approx \left(\frac{N_s N_r}{\overline{\gamma}_{RD}(n_d - n_r)} + \frac{n_s}{\overline{\gamma}_{SR}} \right) \gamma_{\text{th}}. \tag{3.127}$$

From the above equation, we can see that the diversity order of the system is only unity. From Equation (3.127), we also see that the outage probability is always an increasing function of n_s and n_r, and thus increasing the number of source and relay nodes is detrimental to system performance. Further, we see from Equation (3.127) that increasing the number of destination nodes is beneficial to system performance.

Figure 3.20 shows the outage probability for various numbers of destination nodes and fixed numbers of source and relay nodes. The 'analytical' curves are obtained from Equation (3.126), and clearly agree with the curves obtained by Monte Carlo simulations. We see that increasing the number of destination nodes leads to a substantial decrease in the outage probability. This agrees with our previous analysis at high SNR where we showed that adding destination nodes is beneficial to system performance. Figure 3.20 indicates that this also applies to all SNR values.

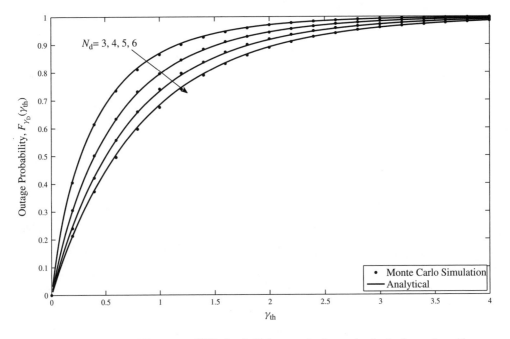

Figure 3.20 Outage probability versus SINR threshold for zero forcing at the destination nodes with $n_s = n_r = 2$, $\overline{\gamma}_{SR} = 6\,\text{dB}$ and $\overline{\gamma}_{RD} = 5\,\text{dB}$ [416]. (Reproduced by permission of © IEEE)

Figure 3.21 shows the outage probability for various numbers of source, relay and destination nodes, when the differences between the number of source, relay and destination nodes are fixed. In this figure, we set $n_s = n_r = n_d - 1$. We see that unlike the system considered in Figure 3.20, there is a substantial increase in the outage probability when more destination nodes are added. This can be explained because in this scenario, $n_d - n_r$ is held fixed, and it can easily be shown from Equation (3.127) that for a fixed $n_d - n_r$, increasing n_s and n_r is detrimental to the outage probability at high SNR. Clearly, Figure 3.21 indicates that this also applies for arbitrary SNR values.

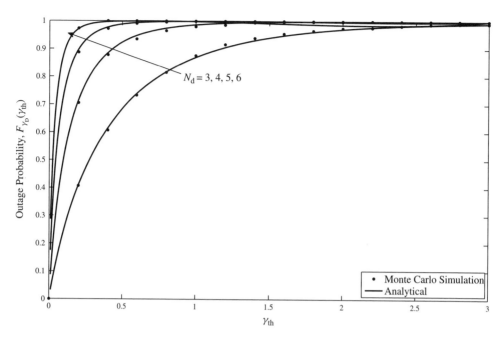

Figure 3.21 Outage probability versus SINR threshold for zero forcing at the destination nodes with $n_s = n_r = n_d - 1$, $n_d = 8$, $\overline{\gamma}_{SR} = 6\,\text{dB}$ and $\overline{\gamma}_{RD} = 5\,\text{dB}$ [416]. (Reproduced by permission of © IEEE)

3.3.4 Distributed Beamforming

It is well known that for a MIMO system without channel state information (CSI) at the transmitter, space–time coding can achieve the full spatial diversity and considerable coding gains. When full CSI is available at the transmitter, the system performance can be further improved, particularly, by controlling the phase and relative amplitude of the signal transmitted through each antenna so as to form a strongest possible beam in the direction of the receiver. The realized transmit beamforming can provide significant array gains and is proven to be the optimal transmission scheme under these conditions [419].

Similarly, in cooperative wireless networks, distributed space–time coding can achieve the full spatial diversity order as has been observed in Section 3.3.1. Similarly to the MIMO system, the performance of cooperative transmission can be further improved when the full CSI at the relay is exploited, which allows for distributed linear processing. Examples of this are distributed beamforming [124, 420–429] and distributed MMSE [430–434], where significant performance improvements over the distributed STBC have been observed when using these distributed linear signal processing techniques. There are however some major differences between the conventional beamforming in MIMO systems and the distributed beamforming in wireless relaying networks:

- **Collocated versus Distributed Antennas.** The first difference is that in conventional beamforming, all antennas are collocated at one transmitter, which allows the optimal beamforming vectors to be designed jointly; whilst in distributed beamforming systems, antennas/relays are located in different places and each relay has to process its own received signal based on its own CSI and some common feedback information from the destination receiver. That is, the beamforming design in cooperative wireless networks has to be done in a distributed way.

- **Perfect versus Partial Knowledge.** The second difference is that in the collocated MIMO system, each antenna has perfect knowledge on the information symbol, whilst in cooperative networks, the received signals at each relay are a noise-corrupted version of original source information and each relay can only get an estimation of the source information. Therefore, the noise effects and estimation errors at the relays have to been taken into account in the design of distributed beamforming vectors in cooperative wireless networks.

These differences make the distributed beamforming design more complex than the conventional beamforming design. In this section, we will therefore introduce the design of distributed beamformers relying on the AF relaying protocol. We will discuss the beamforming designs under two power constraints, that is (i) a global sum power constraint for which the total power consumption of all relays is not greater than a given power level, and (ii) an individual relay power constraint for which each relay has its own transmission power limit.

3.3.4.1 System Model

We consider again a general two-hop relay network, consisting of one source, n relays and one destination. We assume that there is no direct link from the source to the destination. Figure 3.22 shows the block diagram of this two-hop distributed beamforming system.

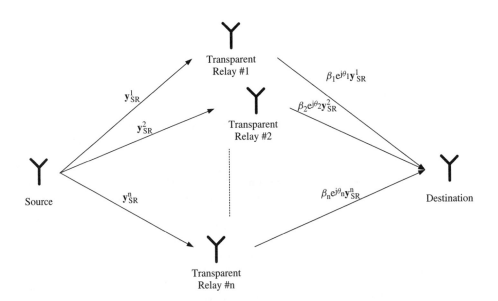

Figure 3.22 A general two-hop distributed beamforming system

In this system, the source first broadcasts the information symbol x to all the relays. The corresponding received signal at the relay i, denoted by y^i_{SR}, can be expressed as:

$$y^i_{SR} = \sqrt{p^i_{SR}} h^i_{SR} x + n^i_{SR}, \qquad (3.128)$$

where $p^i_{SR} = p_s \cdot (G^i_{SR})^2$ is the received signal power at the relay i, p_s is the source transmit power, G^i_{SR} is the channel gain between the source and relay i. Also h^i_{SR} is the fading coefficient between the source and the relay i. Furthermore, n_{SR}^i is a noise vector that is a zero mean complex Gaussian random variable with a two-sided power spectral density of $N_0/2$ per dimension.

After receiving the signal from the source, the relay i scales the received signal by an (analog) factor $w_i = \beta_i \exp(-\theta_i)$ and forwards to the destination the signal $\tilde{x}_R^i$:

$$\tilde{x}_R^i = w_i y_{SR}^i = \beta_i \exp(-\theta_i) \left(\sqrt{p_{SR}^i} h_{SR}^i x + n_{SR}^i \right), \tag{3.129}$$

where β_i is a power allocation factor and can be calculated from the transmission power constraint at the relay i, that is:

$$\beta_i = \sqrt{\frac{p_{R,i}}{p_{SR}^i |h_{SR}^i|^2 + N_0}}. \tag{3.130}$$

Here, $p_{R,i}$ is the power allocated to the relay i. The corresponding received signal at the destination is finally given by:

$$\begin{aligned} y_{RD} &= \sum_{i=1}^{n} \left[G_{RD}^i h_{RD}^i \tilde{x}_R^i \right] + n_D \\ &= \sum_{i=1}^{n} \left[G_{RD}^i h_{RD}^i \left(\beta_i \exp(-\theta_i) \left(\sqrt{p_{SR}^i} h_{SR}^i x + n_{SR}^i \right) \right) \right] + n_D \\ &= \sum_{i=1}^{n} \left[\beta_i \exp(-\theta_i) G_{RD}^i \sqrt{p_{SR}^i} h_{SR}^i h_{RD}^i x \right] + \tilde{n}_{SRD}, \end{aligned} \tag{3.131}$$

where $\tilde{n}_{SRD} = \sum_{i=1}^{n} G_{RD}^i h_{RD}^i \beta_i \exp(-\theta_i) n_{SR}^i + n_D$ is the equivalent noise at the destination with the variance of

$$\sigma_{n,SRD}^2 = \sum_{i=1}^{n} |G_{RD}^i h_{RD}^i \beta_i|^2 N_0 + N_0. \tag{3.132}$$

3.3.4.2 Design under Global Sum Power Constraint

Now let us design the optimal beamforming coefficients $w_i = \beta_i \exp(-\theta_i)$, $i = 1, \ldots, n$ under the global sum power constraint [124]:

$$\sum_{i=1}^{n} p_{R,i} = p_R. \tag{3.133}$$

The optimal beamforming vectors are obtained by maximizing the SNR at the destination. From Equation (3.131), the received SNR at the destination can be calculated as:

$$\begin{aligned} \gamma_{\text{des}} &= \frac{\left| \sum_{i=1}^{n} \beta_i \exp(-\theta_i) G_{RD}^i \sqrt{P_{SR}^i} h_{SR}^i h_{RD}^i \right|^2}{\sum_{i=1}^{n} |G_{RD}^i h_{RD}^i \beta_i|^2 N_0 + N_0} \\ &= \frac{\left| \sum_{i=1}^{n} \beta_i \exp(-\theta_i + \arg(h_{SR}^i) + \arg(h_{RD}^i)) G_{RD}^i \sqrt{P_{SR}^i} |h_{SR}^i h_{RD}^i| \right|^2}{\sum_{i=1}^{n} |G_{RD}^i h_{RD}^i \beta_i|^2 N_0 + N_0} \end{aligned} \tag{3.134}$$

From the above equation, we can see that the destination SNR is maximized when

$$-\theta_i + \arg(h_{SR}^i) + \arg(h_{RD}^i) = 0 \tag{3.135}$$

that is,

$$\theta_i = \arg(h_{SR}^i) + \arg(h_{RD}^i), i = 1, \ldots, n \tag{3.136}$$

is the optimal choice of phase for the beamforming coefficient. Essentially, matched filters should be used at the relays to cancel the phases of channels. By substituting these values and Equation (3.130) into Equation (3.134), we get:

$$
\begin{aligned}
\gamma_{\text{des}} &= \frac{\left| \sum_{i=1}^n \beta_i G_{RD}^i \sqrt{p_{SR}^i} |h_{SR}^i h_{RD}^i| \right|^2}{\sum_{i=1}^n \left| G_{RD}^i h_{RD}^i \beta_i \right|^2 N_0 + N_0} \\[2mm]
&= \frac{\left| \sum_{i=1}^n \sqrt{p_{R,i}} \sqrt{\frac{\gamma_{SR}^i \gamma_{RD}^i}{\gamma_{SR}^i + 1}} \right|^2}{\sum_{i=1}^n p_{R,i} \frac{\gamma_{RD}^i}{\gamma_{SR}^i + 1} + p_R} \\[2mm]
&= \frac{\left| \sum_{i=1}^n \sqrt{p_{R,i}} \sqrt{\frac{\gamma_{SR}^i \gamma_{RD}^i}{\gamma_{SR}^i + 1}} \right|^2}{\sum_{i=1}^n p_{R,i} \frac{\gamma_{RD}^i}{\gamma_{SR}^i + 1} + \sum_{i=1}^n p_{R,i}} \\[2mm]
&= \frac{\left| \sum_{i=1}^n \sqrt{p_{R,i}} \sqrt{\frac{\gamma_{SR}^i \gamma_{RD}^i}{\gamma_{SR}^i + 1}} \right|^2}{\sum_{i=1}^n p_{R,i} \frac{\gamma_{RD}^i + \gamma_{SR}^i + 1}{\gamma_{SR}^i + 1}},
\end{aligned}
\tag{3.137}
$$

where $\gamma_{SR}^i = \frac{p_{SR}^i |h_{SR}^i|^2}{N_0}$ and $\gamma_{RD}^i = \frac{p_R |G_{RD}^i h_{RD}^i|^2}{N_0}$. Let

$$a_i = \sqrt{p_{R,i} \frac{\gamma_{RD}^i + \gamma_{SR}^i + 1}{\gamma_{SR}^i + 1}}$$

$$b_i = p_{R,i} \frac{\gamma_{SR}^i \gamma_{RD}^i}{\gamma_{SR}^i + 1}$$

$$c_i = b_i / |a_i|^2 = \frac{\gamma_{SR}^i \gamma_{RD}^i}{\gamma_{RD}^i + \gamma_{SR}^i + 1}. \tag{3.138}$$

Equation (3.137) can then be further written as:

$$\gamma_{\text{des}} = \frac{\left| \sum_{i=1}^n |a_i| \sqrt{\frac{b_i}{|a_i|^2}} \right|^2}{\sum_{i=1}^n |a_i|^2} = \frac{\left| \sum_{i=1}^n |a_i| \sqrt{c_i} \right|^2}{\sum_{i=1}^n |a_i|^2}. \tag{3.139}$$

By using the Cauchy–Schwarz inequality on the nominator [124], we have:

$$\left| \sum_{i=1}^n |a_i| \sqrt{c_i} \right|^2 \le \left(\sum_{i=1}^n |a_i|^2 \right) \left(\sum_{i=1}^n |c_i| \right). \tag{3.140}$$

Furthermore, by substituting Equation (3.140) into Equation (3.139), we get:

$$\gamma_{\text{des}} \le \sum_{i=1}^n |c_i|. \tag{3.141}$$

The equality, which corresponds to the maximum SNR, is achieved if and only if:

$$|a_i|^2 = C_0 c_i, \tag{3.142}$$

where C_0 is constant. From Equation (3.142), we can obtain the optimum power allocation factor $p_{R,i}$ [124]:

$$p_{R,i} = C_0 \frac{\gamma_{SR}^i \gamma_{RD}^i (\gamma_{SR}^i + 1)}{(\gamma_{RD}^i + \gamma_{SR}^i + 1)^2}. \tag{3.143}$$

By substituting Equation (3.143) into Equation (3.133), we can obtain C_0 as:

$$C_0 = \frac{p_R}{\sum_{i=1}^n \frac{\gamma_{SR}^i \gamma_{RD}^i (\gamma_{SR}^i + 1)}{(\gamma_{RD}^i + \gamma_{SR}^i + 1)^2}}. \tag{3.144}$$

Substituting Equations (3.143) and (3.144) into Equation (3.130), we can obtain the optimum beamforming coefficient at relay i, given by:

$$w_i = \beta_i \exp(-\theta_i) = \sqrt{\frac{C_0}{N_0} \frac{\gamma_{SR}^i \gamma_{RD}^i (\gamma_{SR}^i + 1)}{(\gamma_{RD}^i + \gamma_{SR}^i + 1)^2}} \sqrt{\frac{1}{\gamma_{SR}^i + 1}} \exp(-(\arg(h_{SR}^i) + \arg(h_{RD}^i))). \tag{3.145}$$

As we can observe from Equation (3.145), in order to perform a distributed beamforming, each relay node can decide its transmission power and phase based on knowledge of its incoming and outgoing channel information γ_{SR}^i and γ_{RD}^i, as well as a common constant C_0, which is the same for all the relays and can thus be periodically obtained from the destination node via a broadcast channel.

The maximum SNR achieved by the distributed beamforming under the global sum power constraint is finally given by:

$$\gamma_{des} = \sum_{i=1}^n |c_i| = \sum_{i=1}^n \frac{\gamma_{SR}^i \gamma_{RD}^i}{\gamma_{RD}^i + \gamma_{SR}^i + 1}. \tag{3.146}$$

From Equation (3.146), we can see that the expression of the destination SNR in the distributed beamforming scheme is very similar to the destination SNR expression of the conventional AF system with maximum ratio combining (MRC), where multiple relays transmit on the orthogonal channels and received signals are combined using the MRC. Therefore, its capacity, BER and outage probability will have very similar expressions to the analysis of AF in Section 3.2.

3.3.4.3 Design under Individual Relay Power Constraint

In the previous section, we discussed the optimal beamforming design under a global sum power constraint where each relay controls its own transmitted power to optimize the received SNR. This guarantees that the total power consumption of all relays does not exceed a certain power level. This clearly does not pose any restrictions on the individual power at each relay. In practical systems, however, each relay is equipped with its own battery and thus rather suffers from individual power limits. It is therefore more practical to consider the individual relay power constraint. Jing and Jafarkhani [427] considered the beamforming design under such an individual power constraint. When changing to the individual power constraint, the power control algorithm at each relay has to be modified accordingly to ensure that its own transmission power limits are met. It has been shown that under the individual relay power constraint, some relays may not use their maximum power in order for the destination to achieve the optimal SNR. In this section, we will follow the results obtained by Jing and Jafarkhani [423, 427] to discuss the beamforming design under individual power constraints.

We consider the same system as discussed in Section 3.3.4. Following the same calculations as in Section 3.3.4, the SNR of the signal received at the destination is given by:

$$\gamma_{\text{des}} = \frac{\left|\sum_{i=1}^{n} \beta_i \exp(-\theta_i + \arg(h_{SR}^i) + \arg(h_{RD}^i)) G_{RD}^i \sqrt{p_{SR}^i} |h_{SR}^i h_{RD}^i|\right|^2}{\sum_{i=1}^{n} |G_{RD}^i h_{RD}^i \beta_i|^2 N_0 + N_0} \tag{3.147}$$

where $\beta_i = \sqrt{\frac{p_{R,i}}{p_{SR}^i |h_{SR}^i|^2 + N_0}}$ is the amplification factor at the relay i, $p_{R,i} = p_i \alpha_i^2$, p_i is the transmit power limit of the relay i, $0 \le \alpha_i \le 1$ is introduced in this model for the purpose of power control and is referred to as the power control factor of relay i.

Similarly to the discussions in Section 3.3.4, the destination SNR is maximized when $-\theta_i + \arg(h_{SR}^i) + \arg(h_{RD}^i) = 0$. By substituting this equation into Equation (3.147), the destination SNR can be rewritten as:

$$\gamma_{\text{des}} = \frac{\left(\sum_{i=1}^{n} \alpha_i \sqrt{p_i} \sqrt{\frac{\gamma_{SR}^i \gamma_{RD}^i}{\gamma_{SR}^i + 1}}\right)^2}{1 + \sum_{i=1}^{n} \alpha_i^2 p_i \frac{\gamma_{RD}^i}{\gamma_{SR}^i + 1}} = \frac{\left(\sum_{i=1}^{n} \alpha_i \sqrt{p_i} \mu_i\right)^2}{1 + \sum_{i=1}^{n} \alpha_i^2 p_i \nu_i}, \tag{3.148}$$

where $\gamma_{SR}^i = \frac{p_{SR}^i |h_{SR}^i|^2}{N_0}$, $\gamma_{RD}^i = \frac{|G_{RD}^i h_{RD}^i|^2}{N_0}$ (note that the definition of γ_{RD}^i here is slightly different from that defined in Equation (3.137)), and

$$\mu_i = \sqrt{\frac{\gamma_{SR}^i \gamma_{RD}^i}{\gamma_{SR}^i + 1}} \tag{3.149}$$

$$\nu_i = \frac{\gamma_{RD}^i}{\gamma_{SR}^i + 1}. \tag{3.150}$$

The beamforming design under individual relay power constraint is then to find the optimal power control factors $0 \le \alpha_i \le 1, i = 1, \ldots, n$, so that the destination SNR in Equation (3.148) is maximized. Rewriting the destination SNR γ_{des} as a function of $\alpha_i, i = 1, \ldots, n$, the optimization problem can be formulated as follows:

$$\max_{\alpha_1, \ldots, \alpha_n} \gamma_{\text{des}}(\alpha_1, \ldots, \alpha_n) = \max_{\alpha_1, \ldots, \alpha_n} \frac{\left(\sum_{i=1}^{n} \alpha_i \sqrt{p_i} \mu_i\right)^2}{1 + \sum_{i=1}^{n} \alpha_i^2 p_i \nu_i},$$

$$\text{subject to } 0 \le \alpha_i \le 1, i = 1, \ldots, n. \tag{3.151}$$

To find the optimal power control factors $0 \le \alpha_i \le 1, i = 1, \ldots, n$, let us first take the derivative of Equation (3.151) with respect to α_i, to obtain:

$$\frac{\partial \gamma_{\text{des}}(\alpha_1, \ldots, \alpha_n)}{\partial \alpha_i} = \frac{2\sqrt{p_i} \mu_i \left(\sum_{i=1}^{n} \alpha_i \sqrt{p_i} \mu_i\right) \left(1 + \sum_{i=1}^{n} \alpha_i^2 p_i \nu_i\right) - 2\alpha_i p_i \nu_i \left(\sum_{i=1}^{n} \alpha_i \sqrt{p_i} \mu_i\right)^2}{\left(1 + \sum_{i=1}^{n} \alpha_i^2 p_i \nu_i\right)^2}$$

$$= \frac{2\sqrt{p_i} \left(\mu_i^2 \alpha_i \sqrt{p_i} + \mu_i \sum_{j \ne i} \alpha_j \sqrt{p_j} \mu_j\right) \left(1 + \alpha_i^2 p_i \nu_i + \sum_{j \ne i} \alpha_j^2 p_j \nu_j\right)}{\left(1 + \sum_{i=1}^{n} \alpha_i^2 p_i \nu_i\right)^2}$$

$$- \frac{2\alpha_i p_i \nu_i \left(\alpha_i \sqrt{p_i} \mu_i + \sum_{j \ne i} \alpha_j \sqrt{p_j} \mu_j\right)^2}{\left(1 + \sum_{i=1}^{n} \alpha_i^2 p_i \nu_i\right)^2}$$

$$= \frac{2\sqrt{p_i} \left(\sum_{i=1}^{n} \alpha_i \sqrt{p_i} \mu_i\right) \left(\mu_i \left(1 + \sum_{j \ne i} \alpha_j^2 p_j \nu_j\right) - \nu_i \alpha_i \sqrt{p_i} \sum_{j \ne i} \alpha_j \sqrt{p_j} \mu_j\right)}{\left(1 + \sum_{i=1}^{n} \alpha_i^2 p_i \nu_i\right)^2}. \tag{3.152}$$

For ease of subsequent exposure, we define:

$$\phi_i = \frac{\mu_i \left(1 + \sum_{j \neq i} \alpha_j^2 p_j v_j\right)}{v_i \sqrt{p_i} \sum_{j \neq i} \alpha_j \sqrt{p_j} \mu_j}. \tag{3.153}$$

From the above equation, we can see that

$$\frac{\partial \gamma_{\text{des}}(\alpha_1, \cdots, \alpha_n)}{\partial \alpha_i} > 0 \text{ if only if } \alpha_i < \phi_i \tag{3.154a}$$

$$\frac{\partial \gamma_{\text{des}}(\alpha_1, \cdots, \alpha_n)}{\partial \alpha_i} < 0 \text{ if only if } \alpha_i > \phi_i \tag{3.154b}$$

and

$$\frac{\partial \gamma_{\text{des}}(\alpha_1, \ldots, \alpha_n)}{\partial \alpha_i} = 0 \text{ if only if } \alpha_i = \phi_i = \frac{\mu_i \left(1 + \sum_{j \neq i} \alpha_j^2 p_j v_j\right)}{v_i \sqrt{p_i} \sum_{j \neq i} \alpha_j \sqrt{p_j} \mu_j}. \tag{3.155}$$

We can see from the above equations that the power control factors $\alpha_1, \ldots, \alpha_n$ are related to each other. Finding the optimal values of power control factors is thus quite complex. Let us first get some inspiration by solving the above problem for the case of two relays, for which

$$\phi_1 = \frac{\mu_1 \left(1 + \alpha_2^2 p_2 v_2\right)}{v_1 \sqrt{p_1} \alpha_2 \sqrt{p_2} \mu_2} \tag{3.156}$$

$$\phi_2 = \frac{\mu_2 \left(1 + \alpha_1^2 p_1 v_1\right)}{v_2 \sqrt{p_2} \alpha_1 \sqrt{p_1} \mu_1} \tag{3.157}$$

and let α_i^{opt} be the optimal power control factors. We then we have the following proposition [423]:

Proposition 3.3.6 *For a relay network with two relay nodes, let $\gamma_{\text{des}}(\alpha_1, \alpha_2)$ represent the destination SNR when employing the distributed beamforming under individual relay power constraint. Then the optimal beamforming power control factors have the following properties [423]:*

(i) $\gamma_{\text{des}}(\alpha_1, \alpha_2)$ achieves its maximum value when either $\alpha_1^{\text{opt}} = 1$ or $\alpha_2^{\text{opt}} = 1$;

(ii) When $\alpha_1^{\text{opt}} = 1$, $\gamma_{\text{des}}(\alpha_1, \alpha_2)$ is maximal at $\alpha_2^{\text{opt}} = \min\{1, \phi_2|_{\alpha_1=1}\}$, where $\phi_2|_{\alpha_1=1}$ is obtained by substituting $\alpha_1 = 1$ into ϕ_2, that is, $\phi_2|_{\alpha_1=1} = \frac{\mu_2(1+p_1 v_1)}{v_2 \sqrt{p_2} \sqrt{p_1} \mu_1}$;

(iii) When $\alpha_2^{\text{opt}} = 1$, $\gamma_{\text{des}}(\alpha_1, \alpha_2)$ is maximal at $\alpha_1^{\text{opt}} = \min\{1, \phi_1|_{\alpha_2=1}\}$, where $\phi_1|_{\alpha_2=1} = \frac{\mu_1(1+p_2 v_2)}{v_1 \sqrt{p_1} \sqrt{p_2} \mu_2}$.

(iv) Therefore, if $\gamma_{\text{des}}(1, \min\{1, \phi_2\}) > \gamma_{\text{des}}(\min\{1, \phi_1\}, 1)$, the optimal power control solution is $\alpha_1^{\text{opt}} = 1$ and $\alpha_2^{\text{opt}} = \min\{1, \phi_2\}$; otherwise, the optimal power control solution is $\alpha_1^{\text{opt}} = \min\{1, \phi_1\}$ and $\alpha_2^{\text{opt}} = 1$.

Proof: The proof of above propositions is given elsewhere [423] and briefly summarized below. Let us first prove item (i). We define:

$$\omega(\beta, \alpha_1, \alpha_2) = \gamma_{\text{des}}(\sqrt{\beta} \alpha_1, \sqrt{\beta} \alpha_2) = \frac{\beta \left(\sum_{i=1}^{n} \alpha_i \sqrt{p_i} \mu_i\right)^2}{1 + \beta \sum_{i=1}^{n} \alpha_i^2 p_i v_i}. \tag{3.158}$$

Then

$$\frac{\partial \omega(\beta, \alpha_1, \alpha_2)}{\partial \beta} = \frac{\left(\sum_{i=1}^{n} \alpha_i \sqrt{p_i} \mu_i\right)^2}{\left(1 + \beta \sum_{i=1}^{n} \alpha_i^2 p_i v_i\right)^2} > 0. \tag{3.159}$$

Therefore $\omega(\beta, \alpha_1, \alpha_2) = \gamma_{\text{des}}(\sqrt{\beta}\alpha_1, \sqrt{\beta}\alpha_2)$ is an increasing function of β. We assume that $0 < \alpha_1^{\text{opt}} < 1$ and $0 < \alpha_2^{\text{opt}} < 1$. Without loss of generality, we further assume that $\alpha_1^{\text{opt}} \geq \alpha_2^{\text{opt}}$. Let $\beta_0 = 1/(\alpha_1^{\text{opt}})^2$. We then have $\beta_0 > 1$. Thus,

$$\gamma_{\text{des}}\left(1, \sqrt{\beta_0}\alpha_2^{\text{opt}}\right) = \gamma_{\text{des}}\left((1/\alpha_1^{\text{opt}})\alpha_1^{\text{opt}}, \left(1/\alpha_1^{\text{opt}}\right)\alpha_2^{\text{opt}}\right) = \omega\left(\beta_0, \alpha_1^{\text{opt}}, \alpha_2^{\text{opt}}\right)$$

$$> \omega\left(1, \alpha_1^{\text{opt}}, \alpha_2^{\text{opt}}\right) = \gamma_{\text{des}}\left(\alpha_1^{\text{opt}}, \alpha_2^{\text{opt}}\right) \qquad (3.160)$$

This however contradicts the assumption of α_1^{opt} and α_2^{opt} are the optimal solutions. This concludes the proof of item (i).

Items (ii) and (iii) are symmetrical, and we therefore only need to prove one of them. Let us hence concentrate on item (ii). As shown in Equations (3.154a) and (3.154b), $\frac{\partial \gamma_{\text{des}}(1,\alpha_2)}{\partial \alpha_2} > 0$ when $\alpha_2 < \phi_2$ and $\frac{\partial \gamma_{\text{des}}(1,\alpha_2)}{\partial \alpha_2} < 0$ when $\alpha_2 > \phi_2$. Therefore, $\gamma_{\text{des}}(1, \alpha_2)$ achieves its maximum value at ϕ_2. However, power allocation factors should be in the range of [0,1], that is, $0 \leq \alpha_1 \leq 1$ and $0 \leq \alpha_2 \leq 1$. If, on one hand, $\phi_2 \leq 1$, the optimal choice of α_2 is $\alpha_2^{\text{opt}} = \phi_2$. If, on the other hand, $\phi_2 > 1$, we have $\alpha_2 \leq 1 < \phi_2$. From Equation (3.154a) we therefore have $\frac{\partial \gamma_{\text{des}}(1,\alpha_2)}{\partial \alpha_2} > 0$. $\gamma_{\text{des}}(1, \alpha_2)$ is a monotonically increasing function of α_2 in the range of $0 \leq \alpha_2 \leq 1$ and achieves its maximum value at $\alpha_2 = 1$. This concludes the proof of item (ii). Similarly, we can prove item (iii). Based on the items (ii) and (iii), the proof of item (iv) is straightforward.

As shown in the above proposition for the case of two relays, in order to achieve the optimal destination SNR under the individual relay power constraint, at least one relay has to use its maximum power and the rest of the relays may only use a part of their maximum transmission power, the value of which is determined by not only their own channel and power but also the other relays' channels as well.

Let us next discuss the general case with an arbitrary number of relays. The calculation of optimal power control factors for the network with more than two relays is however very complex. Here we give directly the final solutions and skip all the proofs. The detailed proof can be found elsewhere [427]. Define

$$\eta_i = \frac{\mu_i}{\nu_i \sqrt{p_i}} = \sqrt{\frac{\gamma_{SR}^i(1 + \gamma_{SR}^i)}{p_i \gamma_{RD}^i}}, i = 1, \ldots, n \qquad (3.161)$$

and

$$\eta_{n+1} = 0. \qquad (3.162)$$

We reorder η_i in a descending order as:

$$\eta_{\kappa_1} \geq \eta_{\kappa_2} \geq \cdots \geq \eta_{\kappa_n} \geq \eta_{\kappa_{n+1}} \qquad (3.163)$$

where $\eta_{\kappa_{n+1}} = \eta_{n+1} = 0$. Let us also define

$$\lambda_i = \frac{\left(1 + \sum_{m=1}^{i} p_{\kappa_m} \nu_{\kappa_m}\right)}{\sum_{m=1}^{i} \sqrt{p_{\kappa_m}} \mu_{\kappa_m}}. \qquad (3.164)$$

Then the solutions to the optimization problem of finding the optimal power control factors can be summarized in the following theorem [427]:

Theorem 3.3.7 *Define* $\alpha^i = (\alpha^i(1), \alpha^i(2), \ldots, \alpha^i(n))$ *as*

$$\alpha^i(m) = \begin{cases} 1 & m = \kappa_1, \ldots, \kappa_i \\ \lambda_i \eta_m & m = \kappa_{i+1}, \ldots, \kappa_n \end{cases} \qquad (3.165)$$

Then the optimal power control factors under individual relay power constraint is then given by $\alpha^{\text{opt}} = \alpha^{i_0}$, where i_0 is the smallest i such that $\lambda_i \eta_{\kappa_{i+1}} < 1$.

From the above theorem, we can see a similar phenomenon as in the case of two relays. In order to achieve the maximum destination SNR under the individual power constraint, not all the relays should use their maximum transmission power. Only i_0 relays whose η values are the largest use their maximum power and the remaining $n - i_0$ relays will only use part of their powers. The reason why not all the relays should use their maximum power is that each relay's transmission power has two conflicting effects on the overall destination SNR. On the one hand, increasing the transmission power at the relays increases the signal power at the destination. On the other hand, increasing the relay power also increases the noise power at the destination as the noise power at the relay is amplified as well.

Therefore, the optimal power used at each relay also depends on the power and channels of the other relays. Since the destination has the knowledge of all relay channels and their maximum transmission power, power control can be done by the destination in a centralized way. From (3.161), (3.164) and (3.165), we can see that, in order for each relay to calculate the optimal power allocation factor, each relay also needs to know a common factor λ_{i_0} in addition to its own maximum power and channel coefficients. Therefore, the destination can broadcast the common factor λ_{i_0} and indexes those relays that use their full powers. Once a relay hears its own index from the destination, it will use its maximum power to forward the received signal. Otherwise, if it does not hear its index, it will use the common factor λ_{i_0} and its own channel and power information to calculate the power used for transmission based on (3.161), (3.164) and (3.165). Figure 3.23 shows the block diagram of a distributed beamforming system designed under individual relay power constraint.

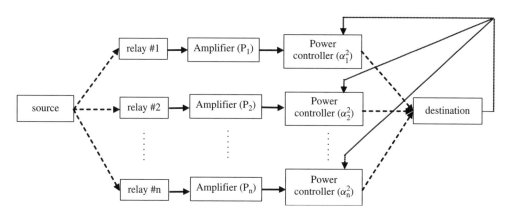

Figure 3.23 Distributed beamforming system under individual relay power constraint

3.3.4.4 Simulation Results

In this section, we provide the simulation results for two beamforming schemes under the global sum power constraint and the individual relay power constraint. We define $\overline{\gamma}^i_{SR} = p^i_{SR}/N_0$ and $\overline{\gamma}^i_{RD} = p_R|G^i_{RD}|^2/N_0$. We assume that $\overline{\gamma}^i_{SR} = \overline{\gamma}_{SR}$ and $\overline{\gamma}^i_{RD} = \overline{\gamma}_{RD}$ for $i = 1, \ldots, n$. Figures 3.24–3.26 show the BER performance of distributed beamforming under individual and global sum constraint for various number of relays in various SNR configurations, where the horizontal axis is $\overline{\gamma}_{RD}$. From these figures we infer that distributed beamforming under individual power constraint is worse by around 1 dB than under the total power constraint as the the former is just a special case of the latter. Both schemes can achieve the full diversity order, which is equal to the number of relays.

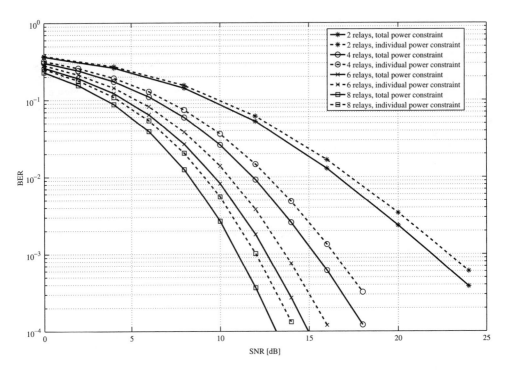

Figure 3.24   BER performance comparison for various numbers of relays, $\overline{\gamma}_{SR} = \overline{\gamma}_{RD} - 10\,\text{dB}$

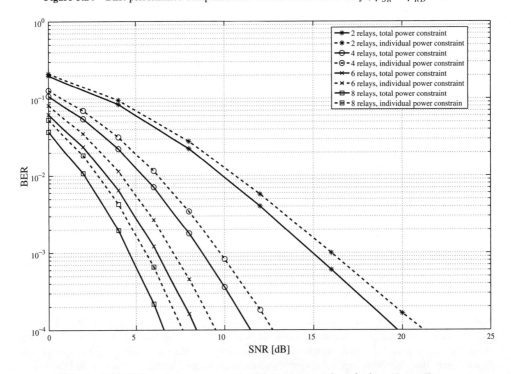

Figure 3.25   BER performance comparison for various number of relays, $\overline{\gamma}_{SR} = \overline{\gamma}_{RD}$

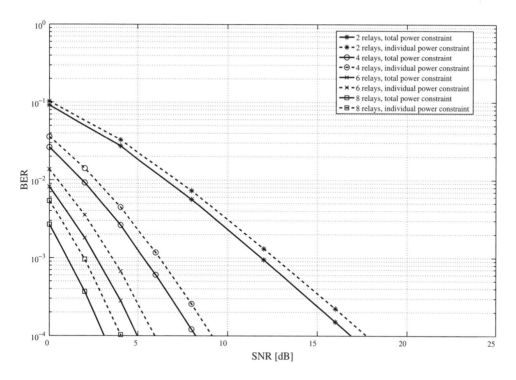

Figure 3.26 BER performance comparison for various number of relays, $\overline{\gamma}_{SR} = \overline{\gamma}_{RD} + 10\,\mathrm{dB}$

3.4 Distributed System Optimization

A crucial aspect of system design is the optimization of all or at least the most influential parameters. Already a challenge for point-to-point systems, such optimization is far from trivial for distributed cooperative systems. We will therefore address various facets of allocating transmission power in a distributed fashion so that given performance metrics are optimized under various constraints. We will also deal with relay selection in distributed transparent communication systems, where we will introduce a simple relay selection algorithm. This in turn facilitates the system design for nonorthogonal two-hop multiple relay networks as it considerably improves the system performance and capacity compared with the all-participation orthogonal AF relay schemes.

3.4.1 Distributed Adaptive Power Allocation

In the previous section, we designed optimal distributed beamforming under the global sum power constraint and individual relay power constraint. As we can see, the key to designing an optimal distributed beamforming scheme is to find the optimal power allocation or power control factors. How to distributively allocate the total transmission power among all relays or distributively control the transmission power of each relay to satisfy the individual relay power constraint is one of the most important issues in the system design. In this optimization process, we have not so far considered the source transmission power. In a practical wireless sensor network, for instance, in order to prolong the lifetime of each sensor node, it is very important to minimize the overall network consumption power, which is the overall transmission power of all transmission terminals including the source terminal and all intermediate terminals.

In this section, we will therefore concentrate on such an optimization problem for distributed power allocation between the source and relaying terminals. We will find the power allocation solutions to maximize the overall received SNR. Since the bit error rate can be described as a monotonic decreasing function of the destination SNR for a diversity system, the maximization of the overall received SNR is also equivalent to the minimization of the bit error rate. We find that the exact solutions are not tractable, so we consider a received SNR upper bound, from which we calculate the power allocation coefficients. Simulation results demonstrate that the proposed power allocation scheme can bring to a two-hop diversity relay system a considerable SNR gain compared with equal power allocation. This gain is monotonically increased as the number of relays increases. This demonstrates that the distributed adaptive power allocation is very important in optimizing system performance and reducing the overall consuming power of a wireless relay network.

In a practical system, the power allocation factors can be calculated at the destination, and then the reverse link channels can be used to feedback/broadcast these factors to the source and relays. The source and relays then adjust their transmit power based on these feedback values. There are various papers addressing the issues relevant to the power adjustment/control in radio systems [435, 436]. In adapted form, the same method can also be used in distributed networks.

3.4.1.1 System Model

We study a general two-hop wireless diversity relay network with multiple relaying terminals in parallel. As shown in Figure 3.5, it consists of one source, n parallel relays and one destination. Let h^i_{SR}, h^i_{RD} and h_{SD} represent the fading coefficients from the source to the ith relay, from the ith relay to the destination and from the source to destination, respectively. They are modeled as zero-mean, independent complex Gaussian random variables. We consider a quasi-static fading channel, for which the fading coefficients are constant within one transmission block, but change independently from one frame to another. We also assume that the fading channels between the source and destination, between the source and each relay, and between each relay and the destination are independent. We further assume that the transmission channels for all relays are orthogonal, such that the transmitted signals from each relay can be separated at the destination without any interference from other relays.

The source first broadcasts the information symbol x to both the destination and relays. The received signals at the relay i and the destination, denoted by y^i_{SR} and y_{SD}, can be expressed as:

$$y^i_{SR} = \sqrt{p_S} G^i_{SR} h^i_{SR} x + n^i_{SR} \tag{3.166a}$$

$$y_{SD} = \sqrt{p_S} G_{SD} h_{SD} x + n_{SD}, \tag{3.166b}$$

where $\mathbb{E}\{|x|^2\} = 1$, p_S is the source transmission power, and

$$G^i_{SR} = G_0 \left(\frac{d_{SR,i}}{d_0} \right)^{-\kappa/2} \tag{3.167a}$$

$$G_{SD} = G_0 \left(\frac{d_{SD}}{d_0} \right)^{-\kappa/2} \tag{3.167b}$$

are the channel gains between the source and relay i and that between the source and destination, respectively; $d_{SR,i}$ and d_{SD} are the distances between the source and relay i, and that between the source and destination, respectively; d_0 is a reference distance, G_0 is a constant depending on the carrier wavelength, and κ is a pathloss factor with values typically in the range $2 \leq \kappa \leq 6$. n^i_{SR} and n_{SD} are zero mean complex Gaussian random variables with the two sided spectrum density of $N_0/2$ per dimension. In this section, we assume that all noise processes are of the same variance without loss of generality. Different noise variances can be taken into account by appropriately adjusting the channel gains.

Upon receiving signals from the source, each relay amplifies its received signals as follows:

$$x_R^i = \mu_i y_{SR}^i, \tag{3.168}$$

where μ_i is an amplification factor such that:

$$\mathbb{E}\left\{|\mu_i y_{SR}^i|^2\right\} = \mu_i^2\left(|G_{SR}^i h_{SR}^i|^2 p_S + N_0\right) \le \alpha_{R,i} P. \tag{3.169}$$

Here, $\alpha_{R,i} P$, $i = 1, 2, \ldots, n$, are the powers allocated to the ith relay, and the overall network power should not exceed the overall network power limit P:

$$p_S + \sum_{i=1}^{n} \alpha_{R,i} P \le P. \tag{3.170}$$

Let $p_S = \alpha_S P$, then Equation (3.170) can be rewritten as:

$$\alpha_S + \sum_{i=1}^{n} \alpha_{R,i} \le 1. \tag{3.171}$$

From Equation (3.169), the amplification factor μ_i can be calculated as:

$$\mu_i \le \sqrt{\frac{\alpha_{R,i} P}{|G_{SR}^i h_{SR}^i|^2 \alpha_S P + N_0}}. \tag{3.172}$$

The corresponding received signal at the destination is then given by:

$$\begin{aligned} y_{RD}^i &= G_{RD}^i h_{RD}^i x_R^i + n_{RD}^i \\ &= G_{RD}^i h_{RD}^i \mu_i \left(G_{SR}^i h_{SR}^i x + n_{SR}^i\right) + n_{RD}^i, \end{aligned} \tag{3.173}$$

where $G_{RD}^i = G_0\left(d_{RD,i}/d_0\right)^{-\kappa/2}$ is the channel gain between the source and relay i, $d_{RD,i}$ is the distance between relay i and the destination, and n_{RD}^i is a zero mean complex Gaussian random variable with the two-sided spectrum density of $N_0/2$ per dimension.

The signals originating from the source and relays and received at the destination, given in Equations (3.166b) and (3.173), are then combined as follows:

$$w_S y_{SD} + \sum_{i=1}^{n} w_{R,i} y_{RD}^i$$

$$= w_S(G_{SD} h_{SD} x + n_{SD}) + \sum_{i=1}^{n} w_{R,i} \left(G_{RD}^i h_{RD}^i \mu_i \left(G_{SR}^i h_{SR}^i x + n_{SR}^i\right) + n_{RD}^i\right), \tag{3.174}$$

where w_S and $w_{R,i}$, $i = 1, 2, \ldots, n$ are the combination coefficients. The corresponding destination SNR, denoted by γ_{AF}, can be calculated from Equation (3.174) as:

$$\gamma_{AF} = \frac{\alpha_S P \left|G_{SD} h_{SD} w_S + \sum_{i=1}^{n} \mu_i G_{RD}^i G_{SR}^i h_{RD}^i h_{SR}^i w_{R,i}\right|^2}{N_0\left(|w_S|^2 + \sum_{i=1}^{n} |w_{R,i}|^2 \left(|\mu_i G_{RD}^i h_{RD}^i|^2 + 1\right)\right)} \tag{3.175}$$

The above SNR can be maximized by taking partial derivatives w.r.t. w_S and $w_{R,i}$, $i = 1, \ldots, n$ and their optimal values can be calculated as:

$$w_S = \frac{\sqrt{\alpha_S P} G_{SD} h_{SD}^*}{N_0}, \quad w_{R,i} = \frac{\mu_i \sqrt{\alpha_S P} G_{RD}^i (h_{RD}^i)^* G_{SR}^i (h_{SR}^i)^*}{(\mu_i^2 |G_{RD}^i h_{RD}^i|^2 + 1) N_0}, i = 1, \ldots, n. \tag{3.176}$$

By substituting Equation (3.176) into Equation (3.175), γ_{AF} can be expressed as:

$$\gamma_{AF} = \left(|G_{SD}h_{SD}|^2 + \sum_{i=1}^{n} \frac{\mu_i^2 |G_{SR}^i G_{RD}^i h_{SR}^i h_{RD}^i|^2}{\mu_i^2 |G_{RD}^i h_{RD}^i|^2 + 1} \right) \frac{\alpha_S P}{N_0}. \tag{3.177}$$

By substituting Equation (3.172) into Equation (3.177), we finally have:

$$\gamma_{AF} \leq \left(|G_{SD}h_{SD}|^2 + \sum_{i=1}^{n} \frac{|G_{SR}^i G_{RD}^i h_{SR}^i h_{RD}^i|^2 \alpha_{R,i} P}{\alpha_{R,i} P |G_{RD}^i h_{RD}^i|^2 + \alpha_S P |G_{SR}^i h_{SR}^i|^2 + N_0} \right) \frac{\alpha_S P}{N_0}$$

$$= \left(a + \sum_{i=1}^{n} \frac{b_i c_i \alpha_{R,i}}{c_i \alpha_{R,i} + b_i \alpha_S + N_0/P} \right) \frac{\alpha_S P}{N_0}, \tag{3.178}$$

where $a = |G_{SD}h_{SD}|^2$, $b_i = |G_{SR}^i h_{SR}^i|^2$, and $c_i = |G_{RD}^i h_{RD}^i|^2$. The above bound is achieved when each relay transmits signals with its maximum transmission power $\alpha_{R,i} P$, that is:

$$\mu_i = \sqrt{\frac{\alpha_{R,i} P}{\alpha_S P |h_{SR}^i|^2 + N_0}}. \tag{3.179}$$

The optimization problem now consists in calculating the maximum γ_{AF} under the condition:

$$\alpha_S + \sum_{i=1}^{n} \alpha_{R,i} = 1, \, 0 < \alpha_S \leq 1 \text{ and } 0 \leq \alpha_{R,i} < 1. \tag{3.180}$$

Following the method of Lagrange multipliers, we take the partial derivative of Equation (3.178) to arrive at:

$$\frac{\partial}{\partial \alpha_{R,i}} \left[\left(a + \sum_{i=1}^{n} \frac{b_i c_i \alpha_{R,i}}{c_i \alpha_{R,i} + b_i \alpha_S + N_0/P} \right) \frac{\alpha_S P}{N_0} - \lambda \left(\alpha_S + \sum_{i=1}^{n} \alpha_{R,i} - 1 \right) \right] = 0 \tag{3.181a}$$

$$\frac{\partial}{\partial \alpha_S} \left[\left(a + \sum_{i=1}^{n} \frac{b_i c_i \alpha_{R,i}}{c_i \alpha_{R,i} + b_i \alpha_S + N_0/P} \right) \frac{\alpha_S P}{N_0} - \lambda \left(\alpha_S + \sum_{i=1}^{n} \alpha_{R,i} - 1 \right) \right] = 0, \tag{3.181b}$$

where λ is the Lagrange multiplier. The values of α_S and $\alpha_{R,i}$, $i = 1, \ldots, n$ can therefore be derived from Equations (3.180), (3.181a) and (3.181b).

3.4.1.2 Distributed Adaptive Power Allocation

The general closed-form power allocation solutions based on the exact SNR expression are not tractable. In this section we therefore calculate a received SNR upper bound and derive the power allocation solutions based on this bound approximation.

At high SNR, the destination received SNR in Equation (3.178) can be further approximated as:

$$\gamma_{AF} \leq \left(a + \sum_{i=1}^{n} \frac{b_i c_i \alpha_{R,i}}{c_i \alpha_{R,i} + b_i \alpha_S} \right) \frac{\alpha_S P}{N_0}$$

$$= \left(a \alpha_S + \sum_{i=1}^{n} \left((c_i \alpha_{R,i})^{-1} + (b_i \alpha_S)^{-1} \right)^{-1} \right) \frac{P}{N_0}$$

$$\leq \left(a \alpha_S + \frac{1}{2} \sum_{i=1}^{n} \sqrt{b_i c_i \alpha_{R,i} \alpha_S} \right) \frac{P}{N_0}, \tag{3.182}$$

where we use the inequality $(x^{-1} + y^{-1})^{-1} \leq \frac{1}{2}\sqrt{xy}$ for $x, y \geq 0$ and the bound in Equation (3.182) is achieved when $c_i\alpha_{R,i} = b_i\alpha_S$ for $i = 1, \ldots, n$. Now we use the above bound to calculate the optimum power allocation solutions. By using the the method of Lagrange multipliers, we can derive the following equations:

$$\alpha_S + \sum_{i=1}^{n} \alpha_{R,i} = 1, 0 < \alpha_S \leq 1, 0 \leq \alpha_{R,i} < 1 \tag{3.183}$$

$$\left(a + \frac{1}{4}\sum_{i=1}^{n} \sqrt{b_i c_i}\sqrt{\frac{\alpha_{R,i}}{\alpha_S}}\right)\frac{P}{N_0} - \lambda = 0 \tag{3.184}$$

$$\frac{1}{4}\sqrt{b_i c_i}\sqrt{\frac{\alpha_S}{\alpha_{R,i}}}\frac{P}{N_0} - \lambda = 0, i = 1, \ldots, n \tag{3.185}$$

From Equation (3.185), we have

$$\frac{\alpha_{R,i}}{b_i c_i} = \frac{\alpha_{R,j}}{b_j c_j}, i = 1, \ldots, n. \tag{3.186}$$

Substituting Equation (3.186) into Equation (3.184), we obtain:

$$a + \frac{1}{4}\sqrt{\frac{\alpha_{R,j}}{\alpha_S}}\frac{1}{\sqrt{b_j c_j}}\sum_{i=1}^{n} b_i c_i - \frac{1}{4}\sqrt{b_j c_j}\sqrt{\frac{\alpha_S}{\alpha_{R,j}}} = 0, \ j = 1, \ldots, n. \tag{3.187}$$

Let now $z_j = \sqrt{\frac{\alpha_S}{\alpha_{R,j}}} > 0$. Then Equation (3.187) can be written as:

$$b_j c_j z_j^2 - 4a\sqrt{b_j c_j}z_j - \sum_{i=1}^{n} b_i c_i = 0, \ j = 1, \ldots, n. \tag{3.188}$$

Solving the above equation, we get:

$$z_j = \frac{2a \pm \sqrt{4a^2 + \sum_{i=1}^{n} b_i c_i}}{\sqrt{b_j c_j}}. \tag{3.189}$$

Since $z_j = \sqrt{\frac{\alpha_S}{\alpha_{R,j}}} > 0$, only $z_j = \frac{2a+\sqrt{4a^2+\sum_{i=1}^{n} b_i c_i}}{\sqrt{b_j c_j}}$ is a solution. Therefore, we have:

$$\frac{\alpha_S}{\alpha_{R,j}} = \frac{8a^2 + \sum_{i=1}^{n} b_i c_i + 8a\sqrt{a^2 + \frac{1}{4}\sum_{i=1}^{n} b_i c_i}}{b_j c_j}. \tag{3.190}$$

From the above equation, we can get the following power allocation solutions:

$$\alpha_{R,j}^{PA} = \frac{b_j c_j}{8a^2 + 2\sum_{i=1}^{n} b_i c_i + 8a\sqrt{a^2 + \frac{1}{4}\sum_{i=1}^{n} b_i c_i}} \tag{3.191a}$$

$$\alpha_S^{PA} = \frac{8a^2 + \sum_{i=1}^{n} b_i c_i + 8a\sqrt{a^2 + \frac{1}{4}\sum_{i=1}^{n} b_i c_i}}{8a^2 + 2\sum_{i=1}^{n} b_i c_i + 8a\sqrt{a^2 + \frac{1}{4}\sum_{i=1}^{n} b_i c_i}}, \tag{3.191b}$$

From Equation (3.191a) and (3.191b), we note that the condition of $0 < \alpha_S \leq 1$ and $0 \leq \alpha_{R,i} < 1$ always holds for any values of a, b_i and c_i. It can also be observed from Equation (3.191a) that the fraction of power allocated to the relay i is proportional to $b_i c_i = |h_{SR}^i|^2|h_{RD}^i|^2$, which actually represents the equivalent channel gain in the link from source $\rightarrow$ relay-i $\rightarrow$ destination. We refer to this link as ith relay link. The power allocation factor for the ith relaying node is proportional to the channel gain of the i-relay link. More power is thus allocated to the good relay links and less power to the worse relay link. This principle is the same as for a general communications system.

Finally, substituting Equations (3.191b) and (3.191a) into Equation (3.178), the maximum received SNR can then be calculated as:

$$\gamma_{AF}^{\max} = \left(a + \sum_{i=1}^{n} \frac{b_i c_i \alpha_{R,i}^{PA}}{c_i \alpha_{R,i}^{PA} + b_i \alpha_S^{PA} + N_0/P} \right) \frac{\alpha_S^{PA} P}{N_0}. \tag{3.192}$$

3.4.1.3 Simulation Results

In this section, we evaluate the performance of the AF system with adaptive power allocation. For simplicity, we assume that all relay nodes are the same distance from the source and also the same distance from the destination node, that is, $d_{SR,i} = d_{SR}$ and $d_{RD,i} = d_{RD}$, for all $i = 1, 2, \ldots, n$. Therefore, we have $G_{SR}^i = G_{SR}$ and $G_{RD}^i = G_{RD}$, for all $i = 1, 2, \ldots, n$. We assume that the channels are normalized with respect to d_{sd}, such that $G_{SD} = 1$. We further assume a pathloss factor $\kappa = 2$. We consider the following four different configurations:

- **Case 1:** $G_{SR}^2 = G_{RD}^2 = G_{SD}^2 = 1$. In this case, all the channels have the same path attenuation.
- **Case 2:** $G_{SR}^2 = 4G_{SD}^2$, and $G_{RD}^2 = G_{SD}^2 = 1$. In this case, the relays are closer to the source.
- **Case 3:** $G_{RD}^2 = 4G_{SD}^2$, and $G_{SR}^2 = G_{SD}^2 = 1$. In this case, the relays are closer to the destination.
- **Case 4:** $G_{SR}^2 = G_{RD}^2 = 4G_{SD}^2$. In this case, the relays are located at the central point of the line between the source and the destination.

Figure 3.27 illustrates the average received SNR gain of the distributed adaptive power allocation solutions relative to the equal power allocation with an AF relay protocol for various numbers of relays. We employ the power allocation solutions derived in Equations (3.191a) and (3.191b). It can be observed from the figure that the received SNR gain due to adaptive power allocation monotonically increases with the number of relays. For example, in Cases 1 and 3, the SNR gain for $n = 2$ is about 2.5–3 dB, while it increases to about 5 dB for $n = 10$. This implies that the adaptive power allocation is very effective, especially in a network with a large number of relays.

3.4.2 Distributed Relay Selection

In the previous sections, dealing with transparent relay processing, all relays participate in the relayed transmission. For some of them it was assumed that they transmit in orthogonal channels. Although the orthogonality assumption reduces the system implementation complexity, it limits the system throughput. Some of transparent relay processing protocols relax the orthogonality constraint, such as the distributed beamforming. It can lead to a further capacity increase, but system implementation complexity is also increased as a considerable amount of information has to be fed back from the destination to all the relays, such as some common factors and the CSI of the forward channels from the relays to the destination. To overcome these problems, in this section, we introduce a simple relay selection algorithm based on an AF relay protocol to facilitate the system design for nonorthogonal two-hop multiple relay networks. The relay selection considerably improves the system performance and capacity compared to the all-participation orthogonal AF relay schemes.

3.4.2.1 System Model

We consider the same two-hop relay network dealt with in Section 3.3.4, which consists of one source, n relays and one destination, and we assume that there is no direct link. Let h_{SR}^i and h_{RD}^i,

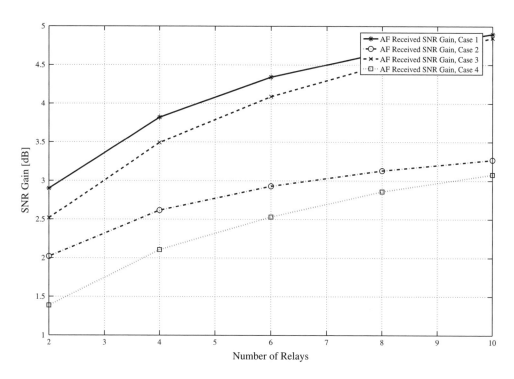

Figure 3.27   SNR gains due to adaptive power allocation in AF relay networks

$i = 1, 2, \cdots, n$ denote the channel fading coefficient for the links between the source and the ith relay and that between the ith relay and the destination. Let G^i_{SR} and G^i_{RD}, $i = 1, 2, \cdots, n$ represent the channel gain (pathloss) for the links between the source and the ith relay and that between the ith relay and the destination. We assume that the variance of h^i_{SR} and h^i_{RD} are same and equal to unity. We consider quasi-static fading channels, for which the fading coefficients are constant within one frame and change independently from one frame to another. We also assume that the amplitudes of the channel fading coefficients are Rayleigh distributed. Let $\gamma^i_{SR} = \left| h^i_{SR} \right|^2 \overline{\gamma}^i_{SR}$ and $\gamma^i_{RD} = \left| h^i_{RD} \right|^2 \overline{\gamma}^i_{RD}$ denote the instantaneously received SNR in the link from the source to the relay i and that from the relay i to the destination, where $\overline{\gamma}^i_{SR} = \frac{p_S (G^i_{SR})^2}{N_0}$ and $\overline{\gamma}^i_{RD} = \frac{p^i_R (G^i_{RD})^2}{N_0}$ are the average SNRs in the link from the source to the relay i and that from the relay i to the destination, p_S is the source transmission power and p^i_R is relay transmission power. For simplicity, we assume that $\overline{\gamma}^i_{SR} = \overline{\gamma}_{SR}$ and $\overline{\gamma}^i_{RD} = \overline{\gamma}_{RD}$ for all $i = 1, \ldots, n$. Let L_i represent the the overall link from the source to the destination via the relay i and we refer to it as the relay link i. Let γ^i_{SRD} be the destination received SNR in the relay link i. Then, following the same calculation as for the AF protocol, γ^i_{SRD} can be calculated as:

$$\gamma^i_{SRD} = \frac{\gamma^i_{SR} \gamma^i_{RD}}{\gamma^i_{RD} + \gamma^i_{SR} + 1}. \tag{3.193}$$

In the AF relay selection strategy, among all the relays, a single relay, which has the maximum source–relay–destination link SNR γ^i_{SRD} is selected to forward the information symbols from the source to the destination, whilst the other unselected relays will remain idle. The single selected relay, denoted by S, is determined by:

$$S = \underset{1 \leq i \leq n}{\arg\max} \{\gamma^i_{SRD}\} \tag{3.194}$$

which is the strongest link among all n relay links. Similarly to Bletsas *et al.* [437], to perform selection at the relay nodes, each relay needs to listen to the request-to-send (RTS) and the clear-to-send (CTS) packets from the source and the destination, respectively. Based on that, each relay node estimates its channel power gains from the source and the destination. Then, a back-off timer is set to be inversely proportional to the relaying channel quality, γ^i_{SRD}. The best relay with the largest γ^i_{SRD} and the smallest back-off timer can occupy the channel first. In this way, the selection is performed at the relay in a decentralized manner and each relay node needs to estimate only the channel amplitudes from the source RTS packet and the destination CTS packet.

3.4.2.2 Performance Analysis

In this section, we analyze the performance of AF relay selection, which we henceforth refer to as AF-RS.

General Method for Calculating the Asymptotic BER In Fading Channel. To facilitate the BER analysis of the AF-RS, we first introduce the general method for calculating the asymptotic BER in fading channels. This brief summary follows the methodology of Wang and Giannakis [383] as already discussed in Section 3.2.4. The BER for a variety of modulation schemes can be calculated as:

$$P_b = \mathbb{E}\left\{ aQ\left(\sqrt{2b\gamma}\right) \right\} \tag{3.195}$$

where γ is the instantaneously received SNR, $Q(x) = \frac{1}{\sqrt{2\pi}} \int_x^\infty e^{-\frac{y^2}{2}} dy$ is the Q-function, a and b are modulation specific parameters. Table 3.1 lists the values of a and b for various modulation schemes. After some manipulation and integration by parts, Equation (3.195) can be further written as:

$$P_b = E\left[aQ\left(\sqrt{2b\gamma}\right) \right] = \frac{a\sqrt{b}}{2\sqrt{\pi}} \int_0^\infty \frac{e^{-b\gamma} F_\gamma(\gamma)}{\sqrt{\gamma}} d\gamma \tag{3.196}$$

Table 3.1   Values of a and b for various modulation schemes

Modulation scheme	a	b
BPSK	1	1
QPSK	1	0.5
M-ary PAM	$2(M-1)/M$	$3/(M^2-1)$
Minimum M-ary PSK	2	$\sin^2(\pi/M)$
BFSK with orthogonal signaling	1	0.5
BFSK with minimum correlation	1	0.715

where $F_\gamma(\gamma)$ is the cumulative distribution function (CDF) of γ. Equation (3.196) is very useful as it allows us to obtain the BER directly by using the CDF of γ. The CDF of γ can also be used to obtain the asymptotic BER at high SNR. In particular, by using a first order expansion of the CDF of γ, we can get the approximated BER in the high SNR regime. If the first order expansion of the CDF of γ can be written in the form:

$$F_\gamma(\gamma) = \frac{C_0 \gamma^{N+1}}{\bar{\gamma}^{N+1}(N+1)} + o\left(\gamma^{N+1+\varepsilon}\right), \varepsilon > 0, \tag{3.197}$$

or the PDF of γ can be written as:

$$f_\gamma(\gamma) = \frac{C_0 \gamma^N}{\overline{\gamma}^N} + o\left(\gamma^{N+\varepsilon}\right), \varepsilon > 0, \tag{3.198}$$

then the asymptotic BER at high SNR can be approximated as [383]:

$$P_b = \mathbb{E}\left\{a Q\left(\sqrt{2b\gamma}\right)\right\} = \frac{2^N a C_0 \Gamma(N + \frac{3}{2})}{\sqrt{\pi}(N+1)} (b\overline{\gamma})^{-(N+1)} + o\left(\overline{\gamma}^{-(N+1)}\right), \tag{3.199}$$

where $\overline{\gamma}$ is the average transmit SNR.

Order Statistics of Multiple Variables. To analyze the performance of AF-RS schemes, we also need to use the knowledge of order statistics. In this section, we briefly introduce some useful basic order statistical tools, which will be used later for the performance analysis. For a comprehensive introduction of the theory of order statistics, the reader can refer to the literature [319].

We consider n independent and identically distributed random variables, $X_1, \ldots, X_n$, each with a CDF $F(x)$. If we arrange the variables $X_1, \ldots, X_n$ in an increasing order of magnitude, denoted by:

$$X_{(1)} \leq X_{(2)} \leq \cdots \leq X_{(n)}, \tag{3.200}$$

we call $X_{(i)}$ the ith order statistic ($i = 1, \cdots, n$). Then the CDF of the ith order statistic X_i, denoted by $F_{(i)}(x)$, is given by [319]:

$$\begin{aligned} F_{(i)}(x) &= \text{Prob}\{X_{(i)} \leq x\} \\ &= \text{Prob}\{\text{at least } i \text{ of the } X_i \text{ are less than or equal to } x\} \\ &= \sum_{k=i}^{n} \binom{n}{i} F^i(x)[1 - F(x)]^{n-i}. \end{aligned} \tag{3.201}$$

Now let us consider two special cases of $X_{(1)}$ and $X_{(n)}$. Substituting $i = 1$ and $i = n$ into the above equations, we have:

$$F_{(n)}(x) = F^n(x) \tag{3.202}$$

$$F_{(1)}(x) = 1 - [1 - F(x)]^n. \tag{3.203}$$

We can also get the PDFs of $X_{(1)}$ and $X_{(n)}$ by differentiating their respective CDFs to get:

$$f_{(n)}(x) = \frac{\partial F_{(n)}(x)}{\partial x} = nf(x)F^{n-1}(x) \tag{3.204}$$

$$f_{(1)}(x) = \frac{\partial F_{(1)}(x)}{\partial x} = nf(x)[1 - F(x)]^{n-1}. \tag{3.205}$$

Similarly, we can get the joint PDF of $X_{(n_1)}, \ldots, X_{(n_k)}(1 \leq n_1 < n_2 < \cdots < n_k < n)$ for $x_1 \leq x_2 \leq \cdots \leq x_k$:

$$\begin{aligned} f_{(n_1),\cdots,(n_k)}(x_1, \cdots, x_k) = & \frac{n!}{(n_1 - 1)!(n_2 - n_1 - 1)! \cdots (n - n_k)!} \\ & F^{n_1-1}(x_1)f(x_1)\left[F(x_2) - F(x_1)\right]^{n_2-n_1-1} f(x_2) \cdots [1 - F(x_k)]^{n-n_k} f(x_k). \end{aligned}$$

In particular, the joint PDF of all n order statistics is given by:

$$n! f(x_1)f(x_2) \cdots f(x_n), x_1 \leq \ldots \leq x_n. \tag{3.206}$$

Harmonic Mean of Two Random Variables. In this section, we study the statistics of the harmonic mean and the modified harmonic mean of two i.i.d. gamma random variables X and Y. We will see in the next section that the destination received SNR in the relay link i, denoted by γ^i_{SRD}, can actually be represented by the the modified harmonic mean of two i.i.d. gamma random variables. We have already alluded to this in Section 3.2.

The harmonic mean of two random variables X and Y is defined as [274]:

$$\mu_H(X, Y) = \frac{2XY}{X + Y},\tag{3.207}$$

and the modified harmonic mean of two random variables X and Y is defined as [438–440]:

$$\mu_{MH}(X, Y, C) = \frac{2XY}{X + Y + C}.\tag{3.208}$$

When $C = 0$, it reduces to the conventional harmonic mean [274, 441]. We use the notation $X \sim g(k, a)$ to denote that X is distributed according to a gamma distribution with PDF given by:

$$f_X(x) = \frac{x^{k-1}e^{-x/a}}{(k-1)!a^k}\tag{3.209}$$

and CDF given by:

$$F_X(x) = 1 - e^{-x/a} \sum_{p=0}^{k} \frac{(x/a)^p}{p!}.\tag{3.210}$$

The following theorem presents the CDF of the modified harmonic mean of two i.i.d. gamma random variables:

Theorem 3.4.1 *Let X and Y be two i.i.d. gamma random variables with $X \sim g(k_1, a_1)$ and $Y \sim g(k_2, a_2)$. Then the CDF of the modified harmonic mean of X and Y, denoted by $Z_C = \mu_{MH}(X, Y, C)$, is given by [438]:*

$$F_{Z_c}(z) = 1 - \frac{e^{-\frac{z}{2}(1/a_1 + 1/a_2)}\sqrt{a_2/a_1}z^{k_2 - 1/2}\sqrt{z + 2C}}{(k_2 - 1)!2^{k_2 - 1}a_2^{k_2}} \sum_{p=0}^{k_1-1} \left(\frac{z^2 + 2Cz}{a_1 a_2}\right)^{p/2} \frac{1}{p!2^p}$$

$$\times \sum_{j=0}^{p} \binom{p}{j}\left(\frac{a_2 z}{a_1(z + 2C)}\right)^{j/2} \sum_{q=0}^{k_2-1} \binom{k_2 - 1}{q}\left(\frac{a_2(z + 2C)}{a_1 z}\right)^{q/2}$$

$$\times K_{q+j-p+1}\left(\sqrt{\frac{z^2 + 2Cz}{a_1 a_2}}\right)\tag{3.211}$$

where $K_v(\cdot)$ is the vth order modified Bessel function of the second kind.

By substituting $C = 0$ into the above equation, we can derive the distribution of the harmonic mean of two i.i.d. gamma random variables in the following corollary [438].

Corollary 3.4.2 *Let X and Y be two i.i.d. gamma random variables with $X \sim g(k_1, a_1)$ and $Y \sim g(k_2, a_2)$. Then the CDF of the harmonic mean of X and Y, denoted by $Z_H = \mu_H(X, Y)$, is given by:*

$$F_{Z_H}(z) = 1 - \frac{e^{-\frac{z}{2}(1/a_1 + 1/a_2)}\sqrt{a_2/a_1}z^{k_2}}{(k_2 - 1)!2^{k_2 - 1}a_2^{k_2}} \sum_{p=0}^{k_1-1} \frac{z^p}{(a_1 a_2)^{p/2}p!2^p}$$

$$\times \sum_{j=0}^{k_2+p-1} \binom{k_2 + p - 1}{j}\left(\frac{a_2}{a_1}\right)^{j/2} K_{p-j+1}\left(\frac{z}{\sqrt{a_1 a_2}}\right).\tag{3.212}$$

Based on the above theorem and corollary, we can obtain the distribution of the modified harmonic mean of two exponential random variables as per the following corollary [438].

Corollary 3.4.3 *Let X and Y be two i.i.d. exponential random variables X_e and Y_e with the respective PDF of $f_{X_e}(x_e) = b_1 e^{-b_1 x_e}$ and $f_{Y_e}(y_e) = b_2 e^{-b_2 y_e}$. Then the CDF of the modified harmonic mean of X_e and Y_e, denoted by $Z_{MH,e} = \mu_{MH}(X_e, Y_e, C)$, is given by:*

$$F_{Z_{MH,e}}(z) = 1 - e^{-\frac{z}{2}(b_1+b_2)}\sqrt{b_1 b_2 (z^2 + 2CZ)}K_1\left(\sqrt{b_1 b_2 (z^2 + 2CZ)}\right). \tag{3.213}$$

The proof of the above corollary follows by substituting $k_1 = k_2 = 1$, $b_1 = 1/a_1$ and $b_2 = 1/a_2$. Similarly, by substituting $C = 0$ into the above equation, we can get the distribution of the harmonic mean of two exponential random variables.

Corollary 3.4.4 *Let X and Y be two i.i.d. exponential random variables X_e and Y_e with the respective PDF of $f_{X_e}(x_e) = b_1 e^{-b_1 x_e}$ and $f_{Y_e}(y_e) = b_2 e^{-b_2 y_e}$. Then the CDF of the harmonic mean of X_e and Y_e, denoted by $Z_{H,e} = \mu_H(X_e, Y_e)$, is given by*

$$F_{Z_{H,e}}(z) = 1 - ze^{-\frac{z}{2}(b_1+b_2)}\sqrt{b_1 b_2}K_1\left(z\sqrt{b_1 b_2}\right). \tag{3.214}$$

Following corollaries presents the PDF of the modified harmonic mean and the harmonic mean of two i.i.d. exponential variables X_e and Y_e.

Corollary 3.4.5 *Let X and Y be two i.i.d. exponential random variables X_e and Y_e with the respective PDF of $f_{X_e}(x_e) = b_1 e^{-b_1 x_e}$ and $f_{Y_e}(y_e) = b_2 e^{-b_2 y_e}$. Then the PDF of the modified harmonic mean of X_e and Y_e, denoted by $Z_{MH,e} = \mu_{MH}(X_e, Y_e, C)$, is given by [438]:*

$$f_{Z_{MH,e}}(z) = \frac{1}{2}e^{-\frac{z}{2}(b_1+b_2)}(b_1 b_2)^{\frac{3}{2}}\left(\frac{2}{\sqrt{b_1 b_2}}(C+z)K_0\left(\sqrt{zb_1 b_2 (2c+z)}\right)\right.$$
$$\left. + \left(\frac{1}{b_1} + \frac{1}{b_2}\right)\sqrt{z(2C+z)}K_1\left(\sqrt{zb_1 b_2 (2C+z)}\right)\right). \tag{3.215}$$

Corollary 3.4.6 *Let X and Y be two i.i.d. exponential random variables X_e and Y_e with the respective PDF of $f_{X_e}(x_e) = b_1 e^{-b_1 x_e}$ and $f_{Y_e}(y_e) = b_2 e^{-b_2 y_e}$. Then the PDF of harmonic mean of X_e and Y_e, denoted by $Z_{H,e} = \mu_H(X_e, Y_e)$, is given by [441]:*

$$f_{Z_{H,e}}(z) = \frac{1}{2}e^{-\frac{z}{2}(b_1+b_2)}(b_1 b_2)^{\frac{3}{2}}\left(\frac{2z}{\sqrt{b_1 b_2}}K_0\left(z\sqrt{b_1 b_2}\right)\right.$$
$$\left. + \left(\frac{1}{b_1} + \frac{1}{b_2}\right)zK_1\left(z\sqrt{b_1 b_2}\right)\right). \tag{3.216}$$

BER Performance Analysis of AF Relay Selection in Rayleigh Fading Channels. In this section, we use the general tools introduced in the previous sections to analyze the performance of AF relay selection. For simplicity, in the following, we consider BPSK modulation.

Let $f_{SR,i}(x)$ and $f_{RD,i}(x)$ represent the PDF of γ_{SR}^i and γ_{RD}^i, respectively. They both follow exponential distributions and their PDFs are given by:

$$f_{SR,i}(x) = \frac{1}{\bar{\gamma}_{SR}}e^{-x/\bar{\gamma}_{SR}} \tag{3.217a}$$

$$f_{RD,i}(x) = \frac{1}{\bar{\gamma}_{RD}}e^{-x/\bar{\gamma}_{RD}} \tag{3.217b}$$

Let us first calculate the distribution of the destination SNR of the relay link i, denoted by γ^i_{SRD}, which is given by:

$$\gamma^i_{SRD} = \frac{\gamma^i_{SR}\gamma^i_{RD}}{\gamma^i_{RD} + \gamma^i_{SR} + 1} = \frac{1}{2}\mu_{MH}(\gamma^i_{SR}, \gamma^i_{RD}, 1). \tag{3.218}$$

Let $f_R(x)$ and $F_R(x)$ denote the PDF and CDF of γ^i_{SRD}, then its CDF can be obtained by substituting $x = z/2$, $b_1 = \frac{1}{\overline{\gamma}_{SR}}$ and $b_2 = \frac{1}{\overline{\gamma}_{RDR}}$ into Corollary 3.4.3:

$$F_R(x) = 1 - e^{-x\left(\frac{1}{\overline{\gamma}_{SR}} + \frac{1}{\overline{\gamma}_{RD}}\right)}\sqrt{\frac{4(x^2 + Cx)}{\overline{\gamma}_{SR}\overline{\gamma}_{RD}}}K_1\left(\sqrt{\frac{4(x^2 + Cx)}{\overline{\gamma}_{SR}\overline{\gamma}_{RD}}}\right). \tag{3.219}$$

Similarly, its PDF can be derived from Corollary 3.4.5 as:

$$f_R(x) = \frac{e^{-x\left(\frac{1}{\overline{\gamma}_{SR}} + \frac{1}{\overline{\gamma}_{RD}}\right)}}{(\overline{\gamma}_{SR}\overline{\gamma}_{RD})^{\frac{3}{2}}}\left(2\sqrt{\overline{\gamma}_{SR}\overline{\gamma}_{RD}}(C + 2x)K_0\left(2\sqrt{\frac{x(C + x)}{\overline{\gamma}_{SR}\overline{\gamma}_{RD}}}\right)\right.$$
$$\left. + 2\left(\overline{\gamma}_{SR} + \overline{\gamma}_{RD}\right)\sqrt{x(C + x)}K_1\left(2\sqrt{\frac{x(C + x)}{\overline{\gamma}_{SR}\overline{\gamma}_{RD}}}\right)\right). \tag{3.220}$$

In the AF relay selection strategy, among all the relays, a single relay, denoted by S, which has the maximum source–relay–destination link SNR γ^S_{SRD}, is selected to forward the information symbols from the source to the destination; the remaining relays remain idle. That is,

$$S = \underset{1 \leq i \leq n}{\text{argmax}}\{\gamma^i_{SRD}\}. \tag{3.221}$$

In other words, if we arrange the variables $\gamma^1_{SRD}, \ldots, \gamma^n_{SRD}$ in an increasing order of magnitude, denoted by:

$$\gamma^{(1)}_{SRD} \leq \gamma^{(2)}_{SRD} \leq \cdots \leq \gamma^{(n)}_{SRD}, \tag{3.222}$$

then we have:

$$\gamma^S_{SRD} = \gamma^{(n)}_{SRD}. \tag{3.223}$$

Then, according to the order statistics theory introduced in Section 3.4.2, the CDF and PDF of γ^S_{SRD}, denoted by $F_{(S)}(x)$ and $f_S(x)$, are given by:

$$F_{(S)}(x) = (F_R(x))^n \tag{3.224a}$$

$$f_{(S)}(x) = nf_R(x)(F_R(x))^{n-1}. \tag{3.224b}$$

As $F_R(x)$ and $f_R(x)$ are already very complex, the calculation of the BER using the exact CDF and PDF of γ^S_{SRD} is fairly involved. To simplify the analysis, we do some high SNR approximations on the destination SNR as follows:

$$\gamma^i_{SRD} = \frac{\gamma^i_{SR}\gamma^i_{RD}}{\gamma^i_{RD} + \gamma^i_{SR} + 1} \approx \frac{\gamma^i_{SR}\gamma^i_{RD}}{\gamma^i_{RD} + \gamma^i_{SR}} = \frac{1}{2}\mu_H(\gamma^i_{SR}, \gamma^i_{RD}) \tag{3.225}$$

At high SNR regime, γ^i_{SRD} can thus be approximated by the harmonic mean of γ^i_{SR} and γ^i_{RD}. Similarly, by using Corollaries 3.4.4 and 3.4.6, the CDF and PDF of the approximate γ^i_{SRD} are given as:

$$F_R(x) = 1 - \frac{2xe^{-x\left(\frac{1}{\overline{\gamma}_{SR}} + \frac{1}{\overline{\gamma}_{RD}}\right)}}{\sqrt{\overline{\gamma}_{SR}\overline{\gamma}_{RD}}}K_1\left(\frac{2x}{\sqrt{\overline{\gamma}_{SR}\overline{\gamma}_{RD}}}\right) \tag{3.226a}$$

$$f_R(x) = \frac{e^{-x\left(\frac{1}{\bar{\gamma}_{SR}} + \frac{1}{\bar{\gamma}_{RD}}\right)}}{(\bar{\gamma}_{SR}\bar{\gamma}_{RD})^{\frac{3}{2}}} \left(4x\sqrt{\bar{\gamma}_{SR}\bar{\gamma}_{RD}}K_0\left(\frac{2x}{\sqrt{\bar{\gamma}_{SR}\bar{\gamma}_{RD}}}\right)\right.$$

$$\left. + 2x\left(\bar{\gamma}_{SR} + \bar{\gamma}_{RD}\right)K_1\left(\frac{2x}{\sqrt{\bar{\gamma}_{SR}\bar{\gamma}_{RD}}}\right)\right) \tag{3.226b}$$

Equations (3.226a) and (3.226b) are still too complex to be evaluated as they contains the modified Bessel function, requiring us to invoke some further approximations. For the modified Bessel function, we have the following properties: $K_0(x) \to 0$ and $K_1(x) \to 1/x$ as $x \to 0$. By using this approximation, Equations (3.226a) and (3.226b) can be further approximated at high SNR as:

$$F_R(x) \approx 1 - e^{-x\left(\frac{1}{\bar{\gamma}_{SR}} + \frac{1}{\bar{\gamma}_{RD}}\right)} \tag{3.227a}$$

$$f_R(x) \approx \left(\frac{1}{\bar{\gamma}_{SR}} + \frac{1}{\bar{\gamma}_{RD}}\right)e^{-x\left(\frac{1}{\bar{\gamma}_{SR}} + \frac{1}{\bar{\gamma}_{RD}}\right)}. \tag{3.227b}$$

By substituting the approximated CDF and PDF of γ_{SRD}^i into Equations (3.224a) and (3.224b), the destination SNR for the selected relay S can be approximated as:

$$F_{(S)}(x) \approx \left(1 - e^{-x\left(\frac{1}{\bar{\gamma}_{SR}} + \frac{1}{\bar{\gamma}_{RD}}\right)}\right)^n \tag{3.228a}$$

$$f_{(S)}(x) \approx n\left(\frac{1}{\bar{\gamma}_{SR}} + \frac{1}{\bar{\gamma}_{RD}}\right)e^{-x\left(\frac{1}{\bar{\gamma}_{SR}} + \frac{1}{\bar{\gamma}_{RD}}\right)}\left(1 - e^{-x\left(\frac{1}{\bar{\gamma}_{SR}} + \frac{1}{\bar{\gamma}_{RD}}\right)}\right)^{n-1} \tag{3.228b}$$

Substituting Equation (3.228a) into Equation (3.196) and setting $a = 1, b = 1$ for BPSK modulation, the average BER of AF-RS, denoted by $P_{b,AF-RS}$ is given as:

$$P_{b,AF-RS} = \mathbb{E}\left\{Q\left(\sqrt{2\gamma_{SRD}^S}\right)\right\} = \frac{1}{2\sqrt{\pi}}\int_0^\infty \frac{e^{-x}}{\sqrt{x}}F_{(S)}(x)dx$$

$$= \frac{1}{2\sqrt{\pi}}\int_0^\infty \frac{e^{-x}}{\sqrt{x}}\left(1 - e^{-x\left(\frac{1}{\bar{\gamma}_{SR}} + \frac{1}{\bar{\gamma}_{RD}}\right)}\right)^n dx$$

$$= \frac{1}{2\sqrt{\pi}}\sum_{p=0}^n \binom{n}{p}(-1)^p \int_0^\infty \frac{e^{-x}}{\sqrt{x}}e^{-px\left(\frac{1}{\bar{\gamma}_{SR}} + \frac{1}{\bar{\gamma}_{RD}}\right)}dx$$

$$= \frac{1}{2}\sum_{p=0}^n \binom{n}{p}(-1)^p \frac{1}{\sqrt{1 + p\left(\frac{1}{\bar{\gamma}_{SR}} + \frac{1}{\bar{\gamma}_{RD}}\right)}} \tag{3.229}$$

where $F_\gamma(\gamma)$ is the CDF of γ. At high SNR, the first order expansion of $F_{(S)}(x)$ is given by:

$$F_{(S)}(x) = \left(1 - e^{-x\left(\frac{1}{\bar{\gamma}_{SR}} + \frac{1}{\bar{\gamma}_{RD}}\right)}\right)^n = \left(x\left(\frac{1}{\bar{\gamma}_{SR}} + \frac{1}{\bar{\gamma}_{RD}}\right)\right)^n + o(x^{n+\varepsilon}), 0 < \varepsilon < 1 \tag{3.230}$$

Finally, the asymptotic BER of the AF-RS at a high SNR can be approximated as:

$$P_{b,AF-RS} \approx \frac{2^{n-1}\Gamma(n + \frac{1}{2})}{\sqrt{\pi}}\left(\frac{1}{\bar{\gamma}_{SR}} + \frac{1}{\bar{\gamma}_{RD}}\right)^n = \frac{(2n-1)!!}{2}\left(\frac{1}{\bar{\gamma}_{SR}} + \frac{1}{\bar{\gamma}_{RD}}\right)^n \tag{3.231}$$

From the above equation, we can see that also the AF relay selection scheme can achieve the full diversity order of n.

3.4.2.3 Simulation Results

In this section, we provide some simulation results for the AF relay selection scheme and compare it with the analytical results. We consider a BPSK modulation and uncoded system. Figures 3.28–3.30 compare the BER simulation results with the analytical results obtained in the previous section for various numbers of relays. The x-axis is $\overline{\gamma}_{RD}$. From these figures, we can see that the diversity of the system increases as the number of relay increases. Also the asymptotic analytical results approach the simulation results as the SNR increases.

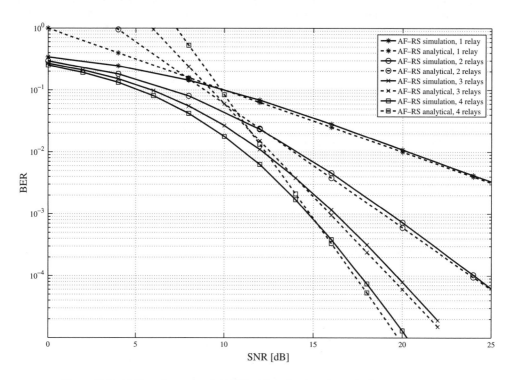

Figure 3.28   BER performance of AF-RS with various numbers of relays, $\overline{\gamma}_{SR} = \overline{\gamma}_{RD}$

3.5   Concluding Remarks

In this chapter we have introduced some transparent physical layer protocols facilitating the deployment of transparent cooperative relaying systems. It requires a relay to receive a signal, transparently process it and then retransmit it. Transparent processing essentially implies that the relay is not accessing the information contents but only performs analog changes to the waveform by means of, for example, analog amplification, analog phase shifting, etc.

We first dealt with canonical relaying systems where information is relayed by several relays in a noninterfering manner. We then moved on to distributed space–time processing architectures where the relays retransmit their information in a controlled manner so that known space–time processing mechanisms can be used in an adapted manner. We finally looked at ways of optimizing the transparent relaying system in a distributed fashion. Observations pertaining to these parts are summarized below:

- **Transparent Relaying Protocols.** We have analyzed four canonical topologies that differ in the number of parallel relaying branches and serial relaying segments, that is single-branch dual-hop,

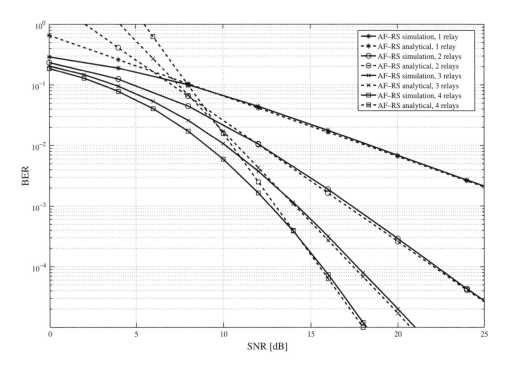

Figure 3.29　BER performance of AF-RS with various numbers of relays, $\overline{\gamma}_{SR} = \overline{\gamma}_{RD} + 5\,\mathrm{dB}$

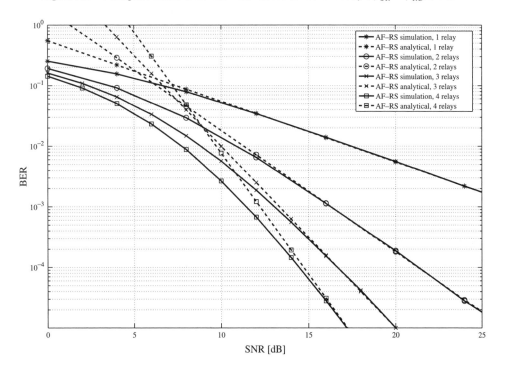

Figure 3.30　BER performance of AF-RS with various numbers of relays, $\overline{\gamma}_{SR} = \overline{\gamma}_{RD} + 10\,\mathrm{dB}$

single-branch multihop, multibranch single-hop and multibranch multihop. These can be used to construct any type of transparent relaying topology. The exposed mathematical tools can be used to analyze and synthesize more complex system settings. For example, one would typically use the CDF analysis to obtain the CDF and thus PDF and MGF of the end-to-end SNR of serially concatenated transparent multihop relaying channels. Parallel transparent multibranch relaying systems, on the other hand, are directly dealt with by using the MGF approach. A generic framework for any type of relaying topology and arbitrary fading channels (obeying some loose requirements) has also been discussed as it only involves the knowledge of the channel PDF at zero SNR. From a performance point of view, it could generally be observed that the multihop fading channel performed worse than the direct channel unless the fading (and shadowing) gains are taken into account. The multibranch channel, on the other hand, enjoys a diversity gain already at fading level and hence outperforms systems without relays.

- **Transparent Distributed Space–Time Processing.** We have shown how the traditional space–time processing techniques originally developed for spatially collocated antenna elements can be adapted to topologies with spatially distributed antenna elements. We have thus dealt with canonical algorithms pertaining to distributed space–time block coding, space–time trellis coding, spatial multiplexing and beamforming. A general observation is that, if synchronization and other system parameters are perfect or at least as good as for collocated MIMO antenna arrays, then originally developed space–time techniques work just fine. Having said this, the imperfections and distributed use leave a lot of room for improvements, which most of recent literature in this area has capitalized on. This has led to improved space time block and trellis codes that sometimes yield a performance superior by an order of magnitude when compared with original designs. We have, for example, dealt with space–time codes based on linear dispersion codes where each relay performs a linear transformation of the previously received signals and then forwards them simultaneously to the destination. We then also discussed the design of distributed space–time trellis codes where the search for the optimum generator matrix has been based on design criteria adapted to the cascaded transparent relaying channel. We have also dwelt on distributed multiplexing schemes, where we observed that the majority is based on some simple zero-forcing principles. The analysis has revealed some surprising insights, such as on the number of nodes to be used at the source, relay and destination so that the overall system capacity is not jeopardized. Finally, we have investigated beamforming designs under global sum and individual relay power constraints. A challenge for such beamforming systems is the optimum phase but also power allocation, which generally requires a very high degree of synchronism in the system.
- **Transparent Distributed System Optimization.** A cornerstone in system design is the optimization of all or at least the most influential parameters. From the above analysis, we have seen that a key parameter is the allocated transmission power. One of the challenges we hence addressed was how to allocate distributively the total transmission power among all relays or control distributively the transmission power of each relay to satisfy the individual relay power constraint. This is of paramount importance in, for instance, wireless sensor networks so as to prolong the lifetime of sensor nodes. Another very important issue we touched upon related to relay selection in distributed transparent communication systems. Among the numerous ways of carrying out relay selection to the benefit of the overall system performance, we introduced a simple relay selection algorithm based on an AF relay protocol. This facilitates the system design for nonorthogonal two-hop multiple relay networks as it considerably improves the system performance and capacity compared with the all-participation orthogonal AF relay schemes.

Most of the analysis carried out in this chapter was without consideration to outer channel codes, the effect of interference in the system, clock and general timing asynchronisms, the impact of wideband and shadow fading, etc. These are important issues arising in today's modern communication systems and generally impact the behavior of transparent relaying architectures. We will discuss in Chapter 6 how techniques exposed in this chapter can be used in conjunction with aforementioned impairments.

Whilst transparent relaying protocols have been explored in great detail, quite a few problems mainly related to distributed space–time processing and optimization remain open at the time of writing of this book. For instance, the design of space–time block and trellis code matrixes, which are applicable to truly general topologies and which are robust to topological changes as well as general system impairments, remains largely an open issue. Furthermore, the entire area of system synthesis in which important parameters are automatically tuned by the cooperative system is in its infancy. Again, we will highlight some more open issues in Chapter 6.

4

Regenerative Relaying Techniques

4.1 Introductory Note

4.1.1 Chapter Contents

Analyzing the physical layer performance of a wireless system is vital in understanding, optimizing and synthesizing system parameters. As of 2009, there are several hundreds of highly complex contributions on regenerative PHY layer analysis and design available. This requires us to concentrate on some canonical techniques that can then be applied to other scenarios. For this reason, we proceed in this chapter with the following topics:

- regenerative relaying protocols;
- distributed space–time coding protocols;
- distributed network coding protocols.

Regenerative relaying protocols are realized by relays that regenerate the received signal prior to retransmission. Such regeneration can include sample, demodulate, deccode, etc., as well as any joint combination thereof. Distributed space–time coding protocols, on the other hand, facilitate distributed space–time transmission with the aid of one or several regenerative relays where we assume again that sufficient synchronization is guaranteed. In this book, we will dwell on distributed space–time block, space–time trellis and turbo codes. Finally, we also touch upon the recently emerged advanced topic of distributed network coding, where we will deal with distributed network-channel coding and network coding division multiplexing.

4.1.2 Choice of Notation

Concerning the notation, we will almost exclusively adopt the notation used in the open literature, which may sometimes lead to different notations between sections. The prime reason for doing this is because we had to sub-sample a virtually infinite amount of material, which inevitably requires the reader to cross-refer to cited and related publications. Keeping the original notation thus eases such a cross-referencing process.

Furthermore, since we often deal with a two-hop topology, we will use subscripts for the source–destination, source–relay and relay–destination links in an explicit manner by, respectively, SD, SR and RD. If the need arises, we will use further subscripts on the number of relays.

Cooperative Communications   Mischa Dohler and Yonghui Li
© 2010 John Wiley & Sons, Ltd

For some protocols over general topologies, such as the regenerative distributed space–time block coding approach, we will stick to the notation used in previous chapters, that is numerical subscripts for the relaying stages. For instance, the number of transmit antennas at the transmitter of the ith stage becomes $n_{t,i}$, the transmit power into the ith stage is P_i, the channel of the ith stage is h_i, the number of receive antennas at the receiver of the ith stage is $n_{r,i}$, the noise at the receiver of the ith relaying stage is σ_i^2, etc.

Note that the most important variables, symbols and functions used throughout this chapter are tabulated at the beginning of this book.

4.2   Regenerative Relay Protocols

A general topology of a regenerative relaying system is depicted in Figure 4.1. This general topology essentially realizes a general regenerative relaying topology with parallel (non) interfering information flows. Each node may have multiple antennas. Compared with the regenerative space–time processing case to be discussed in Section 4.3, there is no formation of virtual antenna arrays (VAA) but only of groups composed of relays belonging to the same relaying stage; compared with the transparent case already discussed in Chapter 3, there is also no source/destination VAA but now there is potential cooperation between nodes of the same relaying stage. Since the multibranch multihop topology can be constructed by suitably combining the contributions of a single-branch two-hop relaying system with a potentially direct link between source and destination, we will restrict our attention mainly to this canonical topology.

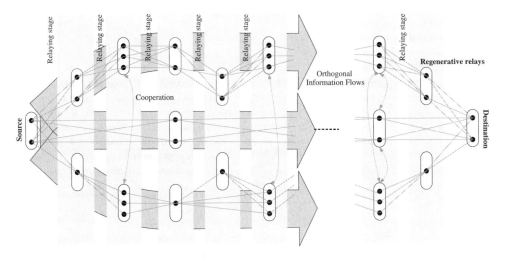

Figure 4.1   General regenerative relaying topology with parallel (non-)interfering information flows, realizing a multibranch multihop relaying topology

We will then deal with the decode-and-forward (DF) protocol where the relay detects, demodulates and, if applicable, decodes the received signal before either re-encoding, remodulating and retransmitting or simply remodulating and retransmitting it. This is in contrast to the estimate-and-forward (EF) protocol, which is essentially the same as the DF but without the coding procedure. We will then also deal with the compress-and-forward (CF) protocol where the relay retransmits a sampled and compressed version of the received signal. We then finally deal with more advanced relaying protocols, such as the soft information, adaptive and selective DF relaying protocols.

4.2.1 Decode and Forward

In this section, we study the commonly used decode-and-forward (DF) regenerative relay protocol. Most of distributed coding schemes are based on the DF protocol. In the conventional DF protocol, the relay decodes the received signals and forwards it either as is or re-encoded to the destination regardless whether the relay can decode correctly or not. The relay could also only forward the correctly decoded messages, which is referred to as the selective DF (S-DF) protocol and dealt with in Section 4.2.5. In this section, however, we will focus on the conventional DF protocol. We first introduce the system model. We then introduce an equivalent one-hop channel model of source-relay-destination link and the near optimal combining technique proposed in [442]. The performance of DF is evaluated through simulations and compared to other schemes.

4.2.1.1 System Model

We consider a single relay two-hop network consisting of one source, one relay and one destination. For simplicity, we consider again BPSK modulation and an uncoded system. An extension to higher modulation orders and/or outer channel codes is possible but generally cumbersome.

The overall transmission of the DF is divided into two phases. The source first broadcasts the modulated signal x_k to both the relay and destination. The corresponding received signals at the relay and destination are:

$$y_{SR,k} = \sqrt{p_{SR}}h_{SR}x_k + n_{SR,k} \tag{4.1a}$$

$$y_{SD,k} = \sqrt{p_{SD}}h_{SD}x_k + n_{SD,k}, \tag{4.1b}$$

where p_{SR}, p_{SD} are the received signal power at the relay and destination, respectively. Also h_{SR} and h_{SD} are the fading coefficients between the source and relay and between the source and destination, respectively. They are modeled as zero-mean, independent circular symmetric complex Gaussian random variables. We consider a quasi-static fading channel, for which the fading coefficients are constant within one frame and change independently from one frame to another. Furthermore, $n_{SR,k}$ and $n_{SD,k}$ are noise terms. Here we assume that all noise terms are zero mean random variables with two sided power spectral density of $N_0/2$ per dimension.

In the uncoded DF protocol, upon receiving signals from the source, the relay estimates x_k by using maximum likelihood detection:

$$\hat{x}_k = \arg\min_x |y_{SR,k} - \sqrt{p_{SR}}h_{SR}x|^2, \tag{4.2}$$

where $\hat{x}_k$ is the estimated symbol of x_k at the relay. The relay then forwards $\hat{x}_k$ with power p_R to the destination. The received signal at the destination is thus given by:

$$y_{RD,k} = G_{RD}h_{RD}\sqrt{p_R}\hat{x}_k + n_{RD,k} = \sqrt{p_{RD}}h_{RD}\hat{x}_k + n_{RD,k}, \tag{4.3}$$

where $p_{RD} = p_R G_{RD}^2$, G_{RD} and h_{RD} are the channel gain and fading coefficient between the relay and destination, and $n_{RD,k}$ is the destination noise.

After receiving the signals transmitted from the source $y_{SD,k}$ and from the relay $y_{RD,k}$, the destination jointly processes these signals to obtain the estimation of x_k. The optimum decoder that minimizes the probability of error is the maximum likelihood (ML) decoder. It picks a symbol that maximizes the likelihood function $p(y_{SD,k}, y_{RD,k}|x)$:

$$\tilde{x}_k = \arg\max_x p(y_{SD,k}, y_{RD,k}|x), \tag{4.4}$$

where $\tilde{x}_k$ is the ML decoded symbol at the destination. Since $y_{SD,k}$ and $y_{RD,k}$ are independent, Equation (4.4) can be further expressed as:

$$\tilde{x}_k = \arg\max_x p(y_{SD,k}, y_{RD,k}|x) = \arg\max_x p(y_{SD,k}|x)p(y_{RD,k}|x), \tag{4.5}$$

where $p(y_{SD,k}|x)$ is given by:

$$p(y_{SD,k}|x) = \frac{1}{\sqrt{2\pi N_0}} \exp\left(-\frac{|y_{SD,k} - \sqrt{P_{SD}}h_{SD}x|^2}{2N_0}\right). \tag{4.6}$$

Let us now calculate $p(y_{RD,k}|x)$. Due to the decoding errors at the relay, the calculation of $p(y_{RD,k}|x)$ is quite different from Equation (4.6). Let $P_{SR}(\gamma_{SR})$ be the conditional bit error probability in the link from the source to the relay, where $\gamma_{SR} = \overline{\gamma}_{SR}|h_{SR}|^2$ and $\overline{\gamma}_{SR} = \frac{P_{SR}}{N_0}$. Then we have:

$$\hat{x}_k = -x_k \text{ with probability of } P_{SR}(\gamma_{SR}) \tag{4.7a}$$

$$\hat{x}_k = x_k \text{ with probability of } 1 - P_{SR}(\gamma_{SR}), \tag{4.7b}$$

where $\hat{x}_k$ is the estimated symbol of x_k at the relay. From Equation (4.7a), $p(y_{RD,k}|x)$ can be calculated as follows:

$$\tilde{x}_k = (1 - P_{SR}(\gamma_{SR}))\frac{1}{\sqrt{2\pi N_0}} \exp\left[\frac{|y_{RD,k} - \sqrt{P_{RD}}h_{RD}x|^2}{2N_0}\right]$$

$$+ P_{SR}(\gamma_{SR})\frac{1}{\sqrt{2\pi N_0}} \exp\left[-\frac{|y_{RD,k} + \sqrt{P_{RD}}h_{RD}x|^2}{2N_0}\right]. \tag{4.8}$$

By substituting now Equations (4.6) and (4.8) into Equation (4.5), the ML decoding can be formulated as follows:

$$\tilde{x}_k = \arg\max_x \left\{ (1 - P_{SR}(\gamma_{SR}))\frac{\exp\left[-\frac{|y_{SD,k}-\sqrt{P_{SD}}h_{SD}x|^2+|y_{RD,k}-\sqrt{P_{RD}}h_{RD}x|^2}{2N_0}\right]}{2\pi N_0} \right.$$

$$\left. + P_{SR}(\gamma_{SR})\frac{\exp\left[-\frac{|y_{SD,k}-\sqrt{P_{SD}}h_{SD}x|^2+|y_{RD,k}+\sqrt{P_{RD}}h_{RD}x|^2}{2N_0}\right]}{2\pi N_0} \right\}. \tag{4.9}$$

The ML decoding is clearly quite complex as the destination has to calculate the error probability in the link from the source to relay $P_{SR}(\gamma_{SR})$). In the next section, we introduce a simple linear complexity decoding method [442].

4.2.1.2 Equivalent Model of S-R-D Link

As discussed in the previous section, the ML decoding is rather complex. In this section, we introduce a simple combining method with near ML performance. It is known that maximum ratio combining (MRC) is the optimal combining technique with only linear complexity. To reduce the complexity of the ML decoder at the destination in DF relay systems, we borrow the concept of MRC and apply it here to combine the received signals $y_{SD,k}$ and $y_{RD,k}$ as follows:

$$y_{D,k} = w_{SD}y_{SD,k} + w_{RD}y_{RD,k} \tag{4.10}$$

and the estimated symbol $\tilde{x}_k$ can be obtained as:

$$\tilde{x}_k = \arg\min_{x_k} \left|(w_{SD}y_{SD,k} + w_{RD}y_{RD,k}) - (w_{SD}\sqrt{P_{SD}}h_{SD} + w_{RD}\sqrt{P_{RD}}h_{RD})x_k\right|^2 \tag{4.11}$$

where w_{SD} and w_{RD} are the combining coefficients. The conventional MRC employs the weights $w_{SD} = \sqrt{P_{SD}}h_{SD}^*$ and $w_{RD} = \sqrt{P_{RD}}h_{RD}^*$. These weights are optimal when there are no decoding errors at the relay. However, when we consider the detection errors at the relay in the practical relay networks, such weights are far from being optimal. Finding the optimal weights is a challenging problem. To simplify the optimization problem, an equivalent one-hop link model has been proposed [442] to represent the source–relay–destination (S-R-D) two-hop link, such that the equivalent one-hop link has the same BEP as the S-R-D link. We now use this equivalent link to design the optimal combining strategy for the DF protocol.

Let γ_{eq} represent the equivalent instantaneous SNR in the S-R-D link. We then can represent the destination received signal transmitted from the relay in S-R-D link as follows:

$$y_{SRD,k} = h_{RD}\sqrt{p_{RD}}x_k + \alpha_E \hat{n}_{SD,k}, \tag{4.12}$$

where $\alpha_E > 1$ is the noise variance amplification factor used to take into account the detection errors at the relay and $\hat{n}_{SD,k}$ is the equivalent destination noise with two-sided power spectral density of $N_0/2$ per dimension. From Equation (4.12), γ_{eq} can be calculated as:

$$\gamma_{eq} = \frac{|h_{RD}|^2}{\alpha_E^2}\frac{p_{RD}}{N_0} = \frac{\gamma_{RD}}{\alpha_E^2}, \tag{4.13}$$

from which we can obtain:

$$\alpha_E = \sqrt{\frac{\gamma_{RD}}{\gamma_{eq}}} \tag{4.14}$$

with $\gamma_{RD} = \overline{\gamma}_{RD}|h_{RD}|^2$ and $\overline{\gamma}_{RD} = p_{RD}/N_0$. Figure 4.2 shows the system model for the one-hop equivalent link in the S-R-D link. By using the above one-hop equivalent model of the S-R-D link, the combined signal in Equation (4.10) can be rewritten as:

$$y_{D,k} = w_{SD}y_{SD,k} + w_{RD}y_{SRD,k}. \tag{4.15}$$

The optimal weights can now be calculated by maximizing the SNR of the above combined signal:

$$w_{SD} = \sqrt{p_{SD}}h_{SD}^*$$

$$w_{RD} = \frac{\sqrt{p_{RD}}h_{RD}^*}{\alpha_E^2} = \sqrt{p_{RD}}h_{RD}^*\frac{\gamma_{eq}}{\gamma_{RD}} = h_{RD}^*\frac{\gamma_{eq}}{\sqrt{p_{RD}}|h_{RD}|^2}. \tag{4.16}$$

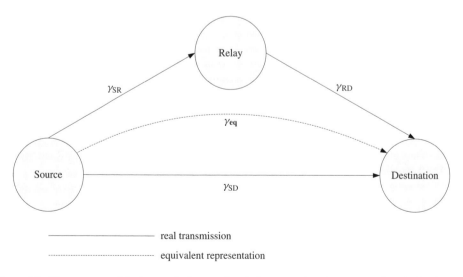

Figure 4.2 The system model of an equivalent one-hop link representing the source–relay–destination two-hop links in a DF relaying protocol

The weights shown in Equation (4.16) are the optimal weights for DF when taking into account the detection errors at the relay. The MRC combining using the above equivalent one-hop link is referred to as the cooperative MRC (C-MRC) [442].

The overall received SNR of the combined signal at the destination is thus given by:

$$\gamma_{\text{all}} = \gamma_{SD} + \gamma_{\text{eq}}, \tag{4.17}$$

where $\gamma_{SD} = \overline{\gamma}_{SD}|h_{SD}|^2$ and $\overline{\gamma}_{SD} = p_{SD}/N_0$. After taking into account the detection errors at the relay, the decision metrics in Equation (4.11) are modified to:

$$\tilde{x}_k = \arg\min_{x_k} \left| \left(w_{SD} y_{SD,k} + w_{RD} y_{RD,k} \right) \right.$$

$$\left. - \left(w_{SD}\sqrt{p_{SD}} h_{SD} + w_{RD} h_{RD} \sqrt{p_{RD}} \right) x_k \right|^2. \tag{4.18}$$

From the above equations, we infer that the simple decoding at the destination is possible in the DF protocol by using the equivalent one-hop link model, as long as γ_{eq} is known.

In the next step, we calculate the output SNR γ_{eq} of the equivalent one-hop link. Let P_{SRD} $(\gamma_{SR}, \gamma_{RD}|h_{SR}, h_{RD})$ represent the overall conditional BEP in the S-R-D link. Let $P_{SR}(\gamma_{SR}|h_{SR})$ and $P_{RD}(\gamma_{RD}|h_{RD})$ denote the conditional BEP in the link from the source to the relay and that from the relay to the destination. Notice that for BPSK modulation the signal at the destination in the S-R-D link is received in error only either when the source–relay transmission is received correctly and relay–destination transmission is received in error, or the source–relay transmission is received in error and the relay–destination transmission is received correctly. Therefore the conditional end-to-end BEP of this two-hop source–relay–destination link can be calculated as [442]:

$$P_{SRD}(\gamma_{SR}, \gamma_{RD}|h_{SR}, h_{RD}) = P_{SR}(\gamma_{SR}|h_{SR}) * (1 - P_{RD}(\gamma_{RD}|h_{RD}))$$

$$+ (1 - P_{SR}(\gamma_{SR}|h_{SR})) P_{RD}(\gamma_{RD}|h_{RD}). \tag{4.19}$$

The output SNR of the equivalent one-hop link γ_{eq} can be calculated as:

$$\gamma_{\text{eq}} := \frac{1}{\alpha} \left\{ Q^{-1} \left[P_{SRD}(\gamma_{SR}, \gamma_{RD}|h_{SR}, h_{RD}) \right] \right\}^2 \tag{4.20}$$

where $Q(x)$ is the Q-function, $Q^{-1}(x)$ is the inverse function of $Q(x)$, and α is a constant depending on the signal constellation; for example, for the BPSK modulation, $\alpha = 2$. Defining $\gamma_{\min} :=$ $\min\{\gamma_{RD}, \gamma_{SR}\}$, it has been proved [442] that $\gamma_{\min}$ and γ_{eq} have the following relationship:

$$\gamma_{\min} - \frac{3.24}{\alpha} < \gamma_{\text{eq}} \leq \gamma_{\min}. \tag{4.21}$$

This equation indicates that $\gamma_{\min}$ actually is a very tight approximation of γ_{eq} at high SNR as the factor $3.24/\alpha$ is very small compared with $\gamma_{\min}$ at high SNR and thus negligible. That is, at high SNR, we have:

$$\gamma_{\text{eq}} \approx \gamma_{\min}. \tag{4.22}$$

Therefore, in the DF protocol, we can use $\gamma_{\min}$ to approximate the overall received SNR of the S-R-D link. This approximation can significantly simplify the destination detector design.

4.2.1.3   Simulation Results

In this section, we evaluate the performance of the DF with the cooperative MRC and also compare it to the amplify-and-forward (AF) protocol. All simulations are performed for uncoded BPSK modulation and a frame size of 100 symbols over quasi-static fading channels. We have evaluated three scenarios, in which the relay node is located either equidistant from both the source and the destination node, or it is closer to the destination node, or closer to the source node.

Let $\overline{\gamma}_{SR}$, $\overline{\gamma}_{RD}$ and $\overline{\gamma}_{SD}$ represent the average output SNRs in the S-R, R-D and S-D links. For the convenience of evaluating various scenarios, we represent these SNRs in the following way:

$$SNR_{SR} = 10\log(\overline{\gamma}_{SR}) = SNR + Gap_{SR} \qquad (4.23a)$$

$$SNR_{RD} = 10\log(\overline{\gamma}_{RD}) = SNR + Gap_{RD} \qquad (4.23b)$$

$$SNR_{SD} = 10\log(\overline{\gamma}_{SD}) = SNR + Gap_{SD}. \qquad (4.23c)$$

Figure 4.3 compares the performance of DF with C-MRC for various scenarios with output SNRs of $(Gap_{SR}, Gap_{RD}, Gap_{SD}) = (0, 0, 0)$, $(Gap_{SR}, Gap_{RD}, Gap_{SD}) = (0, 30, 0)$ and $(Gap_{SR}, Gap_{RD}, Gap_{SD}) = (30, 0, 0)$ respectively with all numbers given in dB. It can be seen from the figure that the first scenario has the worst performance, because both S-R and R-D links have a low SNR at the same time. The third scenario $(Gap_{SR}, Gap_{RD}, Gap_{SD}) = (30, 0, 0)$ outperforms other scenarios because the S-R link has a higher SNR, therefore less error propagation to the destination. From the figure we can also see that the performance of DF with C-MRC is slightly worse than that of the AF protocol, but both schemes achieve the full diversity order of 2.

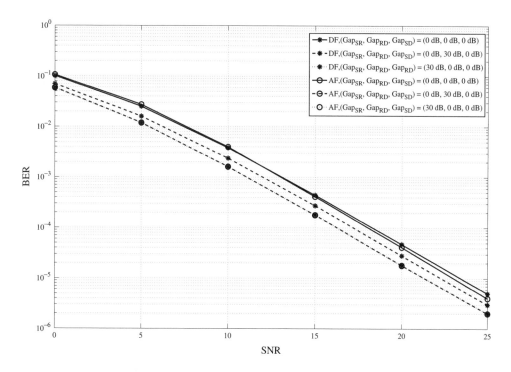

Figure 4.3 Performance of DF with cooperative MRC versus AF

4.2.2   Compress and Forward

In wireless relay networks, the signals received at the relay and the destination are different noisy versions of essentially the same source signal and they are hence correlated. The relay can thus use this correlation to compress the signals received at the relay. Protocols exploiting such correlation to compress the signal are referred to as compress-and-forward (CF) protocols. In CF, the relay quantizes the received signals, compresses the quantized signals and forwards them to the destination. CF is another important regenerative relay protocol, but has not received as much attention as the AF and DF relaying protocols.

The concept of CF was originally proposed by Cover and Gamal [96]. Recently some practical CF schemes have been developed, including CF based on quantization-and-forward [443, 444], CF based on Wyner–Ziv Coding (CF-WZC) [445–447] and CF based on Slepian–Wolf Coding (CF-SWC) [448, 449]. In this section, we will mainly focus on the CF-WZC and CF-SWC schemes. At this stage, we shall note that in some papers compress-and-forward is also referred to as quantize-and-forward or even estimate-and-forward [450]. In the more general sense, amplifying and decoding can be viewed as a means of estimation allowing AF and DF to be regarded as a special case of CF.

4.2.2.1 CF based on Wyner–Ziv Coding

In this section, we introduce a CF scheme based on the Wyner–Ziv Coding scheme [451]. The Wyner–Ziv encoder usually consists of a quantizer followed by an index encoder. The quantizer converts the input analog signals into digital signals. The digital signals are then processed by the succeeding index encoder to get further compression. For Wyner–Ziv coding, the quantizer design has to take into account the type of index encoder. The commonly used index encoders include Slepian–Wolf coding, single source entropy coding, etc. In this section, we will focus on the quantizer design. We will discuss the Slepian–Wolf coding in the next section. Here, for simplicity, we consider a fixed rate scalar quantizer as proposed by Hu and Li [445].

Before quantization, the relay first needs to check the quality of the received signals and thereupon determines whether the signals can be decoded correctly or not. In practical systems, cyclic redundancy check (CRC) bits are usually appended to each information block and it is therefore easy to check if the decoding is successful or not. The relay can also estimate the decoding outcomes by simply measuring the received SNR. It has been shown [452] that for a given channel coding scheme, there exists a threshold calculated on the basis of the Bhattacharyya code parameter. When the received SNR is greater than that threshold, the received signals can be decoded asymptotically correctly when the block length goes to infinity. Therefore, each relay can simply compare its received SNR with this threshold to estimate if the decoding is successful or not.

If the relay decodes the received signals successfully, it then resorts to the DF protocol to forward the signals to the destination. In this case, no quantization is needed. If the relay decoding fails, the compress-and-forward protocol can then be used at the relay. In the following, we consider the scenario where the received signals are not decoded correctly and thus CF protocol is used by the relay to help forward the signals.

Let $y_{SR,k}$ represent the received signal at the relay at time k. The relay can directly quantize $y_{SR,k}$. The relay can also first decode $y_{SR,k}$, then calculate the corresponding soft information, such as soft symbol estimate or log-likelihood ratio (LLR), and quantize the soft information. Here we will not discuss the calculation and modeling of the soft information. There is a separate protocol, called the soft information relay, which will implement the calculation and relaying of soft information. We will discuss this protocol separately in Section 4.2.3. Here, we introduce a quantizer design approach for general input signals. That is, the input to the quantizer could be either the relay received signal, or the signal after being processed by the relay receiver, such as soft information, etc.

Let r_k denote the input to the quantizer. The quantizer will quantize the analog input r_k into a binary sequence $\mathbf{Q}_k = (q_{k,1}, \cdots, q_{k,M})$, where 2^M is the total number of quantization levels. The quantizer design here is to find the boundaries of the bins $B_0, \ldots, B_i, \ldots, B_{2^M}$ so as to optimize a certain criterion. When $B_{i-1} < r_k < B_i$, r_k is mapped to a binary sequence $\mathbf{L}_i$. That is,

$$\mathbf{Q}_k = Q(r_k) = \mathbf{L}_i \quad B_{i-1} < r_k < B_i, \tag{4.24}$$

where $Q(\cdot)$ here represents the quantization function. We consider here the criterion of maximizing the information of the message contained by the quantizer sequence. This is equivalent to minimizing the entropy of the source signal x_k conditioned on the quantization output $\mathbf{Q}_k$ as follows:

$$\arg \min H(x_k|\mathbf{Q}_k), \tag{4.25}$$

where $H(x_k|\mathbf{Q}_k)$ is the conditional entropy defined as:

$$H(x_k|\mathbf{Q}_k) = -\sum_{x_k, \mathbf{Q}_k} p(x_k, \mathbf{Q}_k) \log(p(x_k|\mathbf{Q}_k)). \tag{4.26}$$

Here,

$$p(x_k, \mathbf{Q}_k) = p(\mathbf{Q}_k|x_k)p(x_k) \tag{4.27a}$$

$$p(x_k|\mathbf{Q}_k) = p(x_k, \mathbf{Q}_k)/p(\mathbf{Q}_k) \tag{4.27b}$$

and

$$p(\mathbf{Q}_k = \mathbf{L}_i|x_k) = \int_{B_{i-1}}^{B_i} p(r_k|x_k)dr_k \tag{4.28a}$$

$$p(\mathbf{Q}_k = \mathbf{L}_i) = \int_{B_{i-1}}^{B_i} p(r_k)dr_k \tag{4.28b}$$

where $p(r_k|x_k)$ is the conditional probability of r_k given x_k and

$$p(r_k) = \frac{1}{2}\left(p(r_k|x_k = 1) + p(r_k|x_k = -1)\right). \tag{4.29}$$

As we will see in Section 4.2.3, no matter whether r_k is the relay received signal or the soft information, we can represent r_k in the following format:

$$r_k = \mu_r x_k + n_k, \tag{4.30}$$

where μ_r denotes the fading coefficient when r_k is the relay received signal, whereas it is some scalar produced by the soft information when r_k is the soft information (refer to Section 4.2.3 for details). Similarly, n_k represents the equivalent noise, which is the receiver noise when r_k is the relay received signal and is the estimation errors contained in the soft information when r_k is the soft information.

In either case, when we assume that the noise term n_k follows a Gaussian distribution, $p(r_k|x_k)$ can be calculated as follows:

$$p(r_k|x_k) = \frac{1}{\sqrt{2\pi\sigma_r^2}} \exp\left(-\frac{(r_k - \mu_r x_k)^2}{2\sigma_r^2}\right) \tag{4.31}$$

where σ_r is the variance of n_k. Assume that $p(r_k)$, the probability density of r_k, is symmetrical with respect to the origin. This is true when r_k is the received signal or soft information. Then, the bin boundaries should be symmetrical with respect to the origin as well. That is,

$$B_i = -B_{2M-i}, \tag{4.32}$$

where $B_0 = -\infty$ and $B_{2M} = \infty$. As an example, we consider a four level quantizer design [445] for which the bin boundaries are B_0, B_1, B_2, B_3, B_4. By using the relationship in Equation (4.33), we have:

$$B_0 = -\infty, B_4 = \infty, B_3 = -B_1 = B_d, B_2 = 0. \tag{4.33}$$

We thus only have one unknown variable, that is B_d. Substituting Equations (4.27)–(4.31) into Equation (4.26), we can represent $H(x_k|\mathbf{Q}_k)$ as a function of B_d, denoted by $G(B_d)$. The optimization problem in Equation (4.25) can therefore be reformulated as:

$$\arg \min_{B_d} G(B_d).$$

(4.34)

The optimal solution can then be obtained by solving the following equations:

$$\frac{\partial G(B_d)}{\partial B_d} = 0$$

(4.35a)

$$\frac{\partial^2 G(B_d)}{\partial^2 B_d} > 0$$

(4.35b)

The closed-form solution of above equations is very difficult to derive. In a practical system, we can derive the solutions numerically.

After finding the optimal quantizer, r_k is then quantized into $\mathbf{Q}_k$; thereupon it is modulated and forwarded to the destination. Let $\mathbf{y}_{RD,k}$ be the corresponding signal received at the destination. At the destination, in order to estimate the transmitted source symbol x_k, the receiver needs to calculate the a posterior probability of x_k. Assume that we use maximum *a posteriori* (MAP) decoding (refer to Section 4.2.3 for the introduction of the MAP decoder), the branch metric associated with x_k in $\gamma_k(m, m')$ should be modified as follows:

$$\gamma_k(m, m') = \exp\left[-\frac{|y_{SD,k} - \sqrt{p_{SD}}h_{SD}x_k|^2}{N_0} \right] p(\mathbf{y}_{RD,k}|x_k),$$

(4.36)

where m and m' are a pair of states connected with x_k in the trellis, $y_{SD,k}$, p_{SD} and h_{SD} are respectively the received signal directly transmitted from the source, the source transmission power and the fading coefficient from the source to destination. The quantity $p(\mathbf{y}_{RD,k}|x_k)$ can be calculated by using the Markov chain $x_k \rightarrow r_k \rightarrow \mathbf{Q}_k \rightarrow \mathbf{y}_{RD,k}$ as follows:

$$p(\mathbf{y}_{RD,k}|x_k) = \sum_{\mathbf{Q}_k} p(\mathbf{y}_{RD,k}|\mathbf{Q}_k) p(\mathbf{Q}_k|x_k),$$

(4.37)

where $p(\mathbf{Q}_k|x_k)$ can be calculated (Equation 4.28a). The calculation of $p(\mathbf{y}_{RD,k}|\mathbf{Q}_k)$ depends on how $\mathbf{Q}_k$ is modulated. If the whole bit vector $\mathbf{Q}_k$ is modulated into a single symbol S_k, where $E(|S_k|^2) = 1$, then $\mathbf{y}_{RD,k}$ becomes $y_{RD,k}$ and $p(y_{RD,k}|\mathbf{Q}_k)$ is given by:

$$p(y_{RD,k}|\mathbf{Q}_k) = \frac{1}{\sqrt{2\pi N_0}} \exp\left(-\frac{(y_{RD,k} - \sqrt{p_R}G_{RD}h_{RD}S_k)^2}{2N_0} \right),$$

(4.38)

where p_r, G_{RD} and h_{SD} represent the relay transmission power, the channel gain and fading coefficient from the relay to the destination. If each bit $q_{k,i}$ in $\mathbf{Q}_k = (q_{k,1}, \cdots, q_{k,M})$ is modulated onto a symbol $s_{k,i}$, then we have:

$$p(\mathbf{y}_{RD,k}|\mathbf{Q}_k) = \prod_{i=1}^{M} p(y_{RD,k,i}|q_{k,i}),$$

(4.39)

where $y_{RD,k,i}$ is the received signal corresponding to $s_{k,i}$ and

$$p(y_{RD,k,i}|q_{k,i}) = \frac{1}{\sqrt{2\pi N_0}} \exp\left(-\frac{(y_{RD,k,i} - \sqrt{p_R}G_{RD}h_{RD}s_{k,i})^2}{2N_0} \right).$$

(4.40)

4.2.2.2   CF based on Slepian–Wolf Coding

We introduce here another CF scheme based on Slepian–Wolf Coding (CF-SWC) [453]. The SWC deals with the compression of two or more correlated data streams. For example, for the case of two correlated sources X and Y, if X is transmitted at the rate $H(X)$, Y can be compressed to $H(Y|X)$ by using SWC. The destination can recover both X and Y through the joint decoding of X and

$H(Y|X)$. Similarly, for the relay channel, the relay received signals are a different noisy copy of source transmitted signals and they are correlated, therefore they can be compressed by using SWC. CF-SWC can be used independently by the relay directly to compress the binary sequence obtained by making a hard decision on the relay received signals [448]. CF-SWC can also be used together with the quantization-and-forward protocol introduced in Section 4.2.2 to compress the quantized digital signals, thus together forming a truly Wyner–Ziv based CF protocol [443, 447]. In this section, we only focus on the CF-SWC and assume that the inputs to the SWC encoder are already a binary sequence, which could be either the binary sequence obtained directly through hard decision or the quantized digital signals after quantization.

Let $\mathbf{B}_R$ be the input to the SWC encoder. One key concept of CF-SWC is the code binning process, which groups the input sequence $\mathbf{B}_R$ into bins and each bin is identified with a bin index. The compression is to map the input sequence into the bin index and decompression is to use the bin index to identify the target bin and find the corresponding sequence in that bin by using the correlated sequence transmitted from the source. In this section, we introduce one of commonly used binning methods, which uses the cosets of a linear block code to construct bins. The coset is obtained by adding a n-dimensional binary sequence $\mathbf{x}$ in $\mathbf{G}_2^n$ to an (n, k) linear block code, where $\mathbf{G}_2^n$ is a n-dimensional binary space. For example, given an (n, k) block code $C \in \mathbf{G}_2^n$ and an arbitrary vector $\mathbf{x}$ in $\mathbf{G}_2^n$, the subset

$$\mathbf{x} + C = \mathbf{x} + \mathbf{c}, \mathbf{c} \in C \qquad (4.41)$$

forms a coset of C, denoted by $\mathbf{x} + C$. We should note that a coset is usually not a subspace and each n-dimensional nonzero binary sequence $\mathbf{z} \in \mathbf{G}_2^n$ belongs to one and only one coset.

An (n, k) block code $C \in \mathbf{G}_2^n$ consists of 2^k codewords, so we can divide $\mathbf{G}_2^n$ into $2^n/2^k = 2^{n-k}$ nonoverlapping cosets. Therefore, we can use a binary sequence of length $(n - k)$ as a bin index to represent each bin. Let $\mathbf{H}$ denote the parity check matrix of an (n, k) block code C. From the coding theory, we know that for any given $\mathbf{c} \in C$, we have $\mathbf{c}\mathbf{H}^T = 0$. Thus, for any vector $\mathbf{y}$ in the coset $\mathbf{x} + C$, we have:

$$\mathbf{y}\mathbf{H}^T = (\mathbf{x} + \mathbf{c})\mathbf{H}^T = \mathbf{x}\mathbf{H}^T = \mathbf{s}, \ \mathbf{c} \in C, \qquad (4.42)$$

where $\mathbf{s} = \mathbf{x}\mathbf{H}^T$ is called the syndrome of $\mathbf{y}$. It is an $n - k$ dimensional binary sequence.

From Equation (4.42), we can see that all sequences within the same coset $\mathbf{x} + C$ have the same syndrome $\mathbf{s} = \mathbf{x}\mathbf{H}^T$. Therefore, for each bin $\mathbf{x} + C$, which is also a coset of C, we can use its syndrome $\mathbf{s} = \mathbf{x}\mathbf{H}^T$ as its bin index. In this way, each bin is uniquely identified by a unique bin index, which is the syndrome of code sequences in this bin. We can therefore compress an n-dimensional binary sequence $\mathbf{z} \in \mathbf{G}_2^n$ into its corresponding $n - k$ dimension syndrome and we can reach the compression rate of $\frac{n-k}{n}$. We should note that the syndrome based code binning is just one way to implement the Slepian–Wolf compression. There are also other means but we will not discuss these methods here.

The SWC encoder thus compresses the input binary sequence $\mathbf{B}_R$ of length n into the bin index $\mathbf{s}$ of length $n - k$, whereupon $\mathbf{s}$ is modulated and forwarded to the destination. At the destination, by jointly decoding compressed signals transmitted from the relay and its correlated noncompressed signals transmitted from the source, we can finally recover the binary sequence transmitted from the source.

4.2.3 Soft Information Relaying

In the previous sections, we have studied various commonly used relaying protocols. For the AF scheme, each relay not only amplifies the information signals but also the noise. Therefore, one major disadvantage of doing so is the noise amplification at relays. When the channel quality from the source to the relays is good enough, some relays can decode correctly and exactly recover the source information. In this case, DF is superior to AF because successfully decoding and re-encoding can completely avoid noise amplification at the relays. However, if the channel quality from the source to the relay is poor so that decoding errors occur at relays, then the AF scheme will probably be superior to the DF scheme because the operation of decoding and re-encoding will introduce serious error propagation and significantly degrade its performance. We thus present some enhanced variations of regenerative relay protocols to further improve the performance of relayed transmission, that is soft information relaying in this section as well as adaptive relaying and selective decode-and-forward in subsequent sections.

As we know, the main performance degradation in a DF protocol consists in the error propagations during the decoding and re-encoding process when decoding errors occur at the relay. One way of avoiding the error propagation is to calculate and forward the corresponding soft information instead of making a decision based on the transmitted information symbols at the relay. Forwarding soft information at the relays provides additional information to the destination decoder to make decisions, instead of making premature decisions at the relay decoder. We refer to such a protocol as soft information relaying (SIR). Various SIR protocols have been developed recently [454–462] with applications to various systems. In this section, we introduce the basic principle of SIR and some mathematical models. We first introduce the system model of the SIR protocol. We then introduce several types of SIR protocols as well as their representations and modeling approaches. The performances of these SIR protocols are then evaluated through simulations.

The underlying system model resembles those already dealt with in previous sections, that is we consider a simple two-hop relay network consisting of one source, one relay and one destination. Without loss of generality, we consider a convolutional coded system. It can be easily extended to an uncoded system by replacing the code free distance in the coded system by the minimum Euclidean distance in the signal constellation of an uncoded system.

At the source node, the binary information sequence is first encoded by a convolutional encoder into a codeword $\mathbf{C} = (C_1, \ldots, C_k, \ldots, C_L)$, where L is the length of codeword, C_k is the kth coded symbol and $C_k \in \{0, 1, \cdots, M\}$ for M-QAM modulation. C_k is modulated onto x_k, where $\mathbb{E}\{|x_k|^2\} = 1$, and x_k is then broadcast to both the relay and destination. The corresponding received signals at the relay and destination, denoted by $y_{SR,k}$ and $y_{SD,k}$, respectively, can be expressed as:

$$y_{SR,k} = \sqrt{p_{SR}} h_{SR} x_k + n_{SR,k} \tag{4.43a}$$

$$y_{SD,k} = \sqrt{p_{SD}} h_{SD} x_k + n_{SD,k}, \tag{4.43b}$$

where p_{SR}, p_{SD} are the received signal power at the relay and destination, respectively. Furthermore, h_{SR} and h_{SD} are the fading coefficients between the source and relay and between the source and destination, respectively. They are modeled as zero-mean, independent circular symmetric complex Gaussian random variables. We consider a quasi-static fading channel, for which the fading coefficients are constant within one frame and change independently from one frame to another. Finally, $n_{SR,k}$ and $n_{SD,k}$ are zero mean complex Gaussian random variables with two sided power spectral density of $N_0/2$ per dimension.

In the SIR, the relay decodes the received signals from the source and calculates the corresponding soft information estimation of x_k. Let $\hat{x}_{R,k}$ represent the derived soft information. The signal transmitted by the relay can then be represented as:

$$x_{R,k} = \beta \hat{x}_{R,k}, \tag{4.44}$$

where β is an amplification factor which can be calculated from the transmission power constraint at the relay as follows:

$$\beta = \frac{p_R}{\mathbb{E}\{|\hat{x}_{R,k}|^2\}}, \tag{4.45}$$

where p_R is the transmission power limit at the relay. The corresponding received signal at the destination, denoted by $y_{RD,k}$, is given by:

$$y_{RD,k} = G_{RD} h_{RD} x_{R,k} + n_{RD,k} \tag{4.46}$$

where G_{RD} and h_{RD} are the channel gain and fading coefficient between the relay and destination, $n_{RD,k}$ is a zero mean complex Gaussian random variable with two sided power spectral density of

$N_0/2$ per dimension. The destination decoder then combines the signals transmitted from the source and relay and decodes them to recover the transmitted information bits.

There are generally various ways to represent the soft information. In the following sections we introduce some commonly used representations of soft information, that is the SIR based on soft symbol estimation (SIR-SSE) and on log-likelihood ratio (SIR-LLR).

4.2.3.1   SIR Based on Soft Symbol Estimation

We first consider a SIR protocol based on the soft symbol estimation [454, 456, 459–462]. Let $C_k, k = 1, \cdots, L$, represent the kth symbol at the source node. Let x_k be the modulated signal of C_k, denoted by $x_k = \text{Mod}(C_k)$, where $\text{Mod}(\cdot)$ represents the modulation function. Let denote by $P_r(\hat{C}_k = j|\mathbf{Y}_{sr})$ the a posterior probability (APP) that the output at the relay decoder is equal to $j, j = 0, \cdots, M - 1$, where $\mathbf{Y}_{sr}$ is the signal sequence received at the relay. If the MAP decoding is used, the APPs can be calculated as follows:

$$P_r(\hat{C}_k = j|\mathbf{Y}_{sr}) = h \sum_{m,m'=0,\hat{C}_k=j}^{m,m'=M_s-1} \alpha_{k-1}(m')\beta_k(m)\gamma_k(m, m'), \qquad (4.47)$$

where h is a constant such that $\sum_j P_r(\hat{C}_k = j|\mathbf{Y}_{sr}) = 1$, m and m' are a pair of states connected with $\hat{C}_k = j$ in the trellis, M_s is the number of states in the trellis, $\gamma_k(m, m')$ is the metric of the branch connecting the state m and m' in the trellis, $\alpha_i(m')$ and $\beta_i(m)$ are the feed-forward and the feedback recursive variables. The latter can be calculated as:

$$\alpha_i(m') = \sum_m \alpha_{i-1}(m)\gamma_i(m, m'), \qquad (4.48a)$$

$$\beta_{i-1}(m) = \sum_{m'} \beta_i(m')\gamma_i(m, m'). \qquad (4.48b)$$

Then, the soft symbol estimate of x_k at the relay, denoted by $\hat{x}_k$, can be calculated as:

$$\hat{x}_k = \mathbb{E}\{x_k|\mathbf{Y}_{sr}\} = \sum_{x_k} x_k P_r(x_k|\mathbf{Y}_{sr}) = \sum_{j=0}^{M-1} P_r(\hat{C}_k = j|\mathbf{Y}_{sr})\text{Mod}(j). \qquad (4.49)$$

As shown in the following theorem, SSE is actually a minimum mean square error (MMSE) estimate [461, 462].

Theorem 4.2.1 *The soft symbol estimate in Equation (4.49) is an MMSE estimate. Thus, the SIR-SSE is an optimum unconstraint MMSE relay protocol.*

Proof: Let $\hat{x}_k$ be the estimate of x_k. The MSE is given by:

$$MSE = \mathbb{E}\left\{(\hat{x}_k - x_k)^2|\mathbf{Y}_{sr}\right\}. \qquad (4.50)$$

By taking the derivative of above equation with respect to $\hat{x}_k$, we arrive at:

$$\hat{x}_k = E(x_k|\mathbf{Y}_{sr}). \qquad (4.51)$$

This proves that if we do not place any constraint on $\hat{x}_k$, then the SSE in Equation (4.49) is an MMSE estimate. For simplicity, in the following we consider BPSK modulation. We assume that in the BPSK modulation signal set, the binary bit 0 and 1 are mapped into 1 and -1, respectively. The soft symbol estimate of x_k is the given by:

$$\hat{x}_k = P_r(\hat{C}_k = 0|\mathbf{Y}_{sr}) \cdot 1 + P_r(\hat{C}_k = 1|\mathbf{Y}_{sr}) \cdot (-1). \tag{4.52}$$

We can represent $\hat{x}_k$ as $\hat{x}_k = x_k + \hat{n}_k$, where x_k is the exact transmitted source symbol and $\hat{n}_k = \hat{x}_k - x_k$ is an equivalent noise, which can be modeled as a random variable with a zero mean and variance $\sigma_{\hat{n}}^2$. If x_k and $\hat{n}_k$ are independent, then the average power of $\hat{x}_k$ is:

$$\mathbb{E}\left\{|\hat{x}_k|^2\right\} = \mathbb{E}\left\{|x_k|^2 + |\hat{n}_k|^2\right\} = 1 + \sigma_{\hat{n}}^2 > 1. \tag{4.53}$$

However, we note from Equation (4.52) that for BPSK modulation $-1 \le \hat{x}_k \le 1$ and $|\hat{x}_k|^2 \le 1$, which contradicts Equation (4.53). This means that x_k and $\hat{n}_k$ actually are not independent. This has been observed [454, 461], independently. Next, let us introduce two modeling approaches described in these two papers to address the correlation effects. Although these two modeling methods are quite different, as we can see later, they are actually essentially the same.

SSE Modeling Approach #1 Gomadam and Jafar [461], introduced a scalar factor to take into account the correlations between the desired signal and the equivalent noise in the SSE and to represent $\hat{x}_k$ in the following way:

$$\hat{x}_k = \alpha_x(x_k + \grave{n}_k) \tag{4.54}$$

where $\grave{n}_k$ is the equivalent noise in this new model, α_x is the scalar, called the *correlation scalar*, utilized to take into account the correlation between the desired signal x_k and equivalent noise $\grave{n}_k$ in the SSE, such that x_k and $\grave{n}_k$ are uncorrelated. As we will see later, $\alpha_x \le 1$, which indicates that the effective signal power at the relay is reduced from 1 to α_x^2. The reduction of effective signal power is actually due to the erroneous detection at the relay. In the following, we refer to such SSE modeling as the 'SSE-scalar' and the SIR relay protocol using this modeling as 'SIR-SSE-scalar'. Let us now move on to calculate the *correlation scalar* α_x, such that:

$$\mathbb{E}\left\{x_k^* \grave{n}_k\right\} = 0. \tag{4.55}$$

From Equation (4.54), we have:

$$\grave{n}_k = \hat{x}_k/\alpha_x - x_k. \tag{4.56}$$

By substituting Equation (4.56) into Equation (4.55), we get:

$$\mathbb{E}\left\{x_k^* \grave{n}_k\right\} = \mathbb{E}\left\{x_k^*\left(\frac{\hat{x}_k}{\alpha_x} - x_k\right)\right\} = \frac{1}{\alpha_x}\mathbb{E}\left\{x_k^* \hat{x}_k\right\} - \mathbb{E}\left\{|x|^2\right\} = 0. \tag{4.57}$$

By solving the above equation, we obtain:

$$\alpha_x = \frac{\mathbb{E}\left\{x_k^* \hat{x}_k\right\}}{\mathbb{E}\left\{|x|^2\right\}} = \mathbb{E}\left\{x_k^* \hat{x}_k\right\}. \tag{4.58}$$

Now, by using Equations (4.56) and (4.58), we can easily verify that:

$$\mathbb{E}\left\{x_k^* \grave{n}_k\right\} = 0. \tag{4.59}$$

By substituting Equation (4.58) into Equation (4.54), we rewrite $\hat{x}_k$ as:

$$\hat{x}_k = \mathbb{E}\left\{x_k^* \hat{x}_k\right\}(x_k + \grave{n}_k) = \mathbb{E}\left\{x_k^* \mathbb{E}\left\{x_k|\mathbf{Y}_{sr}\right\}\right\}(x_k + \grave{n}_k). \tag{4.60}$$

The signal transmitted from the relay can then be written as:

$$x_{R,k} = \beta \hat{x}_k = \beta \alpha_x(x_k + \grave{n}_k), \tag{4.61}$$

where β is a normalization factor. Here, $x_{R,k}$ satisfies the following transmission power constraint at the relay:

$$\mathbb{E}\left\{|x_{R,k}|^2\right\} = \beta^2 \alpha_x^2 \left(1 + \sigma_{\hat{n}}^2\right) = p_R, \tag{4.62}$$

where p_R is the transmission power limit at the relay and $\sigma_{\hat{n}}^2$ is the variance of $\hat{n}_k$, which can be calculated as follows:

$$\sigma_{\hat{n}}^2 = \sum_{k=1}^{L} (\hat{x}_k/\alpha_x - x_k)^2. \tag{4.63}$$

From Equation (4.62), we can derive:

$$\beta = \sqrt{\frac{p_R}{\alpha_x^2 \left(1 + \sigma_{\hat{n}}^2\right)}}. \tag{4.64}$$

The destination received signal, corresponding to the relay signal $x_{R,k}$, is given by

$$Y_{RD,k} = G_{RD} h_{RD} x_{R,k} + n_{RD,k} = G_{RD} h_{RD} \beta \alpha_x x_k + \ddot{n}_{RD,k}, \tag{4.65}$$

where

$$\ddot{n}_{RD,k} = n_{RD,k} + G_{RD} h_{RD} \beta \alpha_x \hat{n}_k \tag{4.66}$$

is the equivalent noise at the destination with a zero mean and variance of σ_D^2 given by:

$$\sigma_D^2 = N_0 + G_{RD}^2 |h_{RD} \beta|^2 \alpha_x^2 \sigma_{\hat{n}}^2. \tag{4.67}$$

Let us now examine the distribution of $\hat{n}_k$. Since h_{SR} only affects the received SNR at the relay and does not change the distribution of $\hat{n}_k$, we only need to evaluate the distribution of $\hat{n}_k$ for a fixed h_{SR}. For simplicity, we set $h_{SR} = 1$. That is, we evaluate its distribution over an AWGN channel. Figure 4.4 shows the PDF of $\hat{n}_k$ for various E_b/N_0. Here, we use a four-state recursive systematic

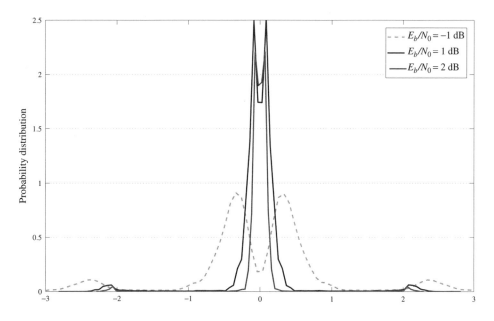

Figure 4.4 The distribution of $\hat{n}_k$, an equivalent noise in the soft symbol estimate at the relay

convolutional code (RSC) with the code rate of $1/2$. The generator matrix of the RSC is $(1, 5/7)$. We note that $\hat{n}_k$ does not follow a Gaussian distribution, but a symmetric bimodal distribution that is a mixture of two normal distributions with the same variance but opposite means. As E_b/N_0 increases, the two peaks get closer and finally converge to a normal distribution at high SNRs. As we will see later, it exhibits a very similar distribution as the equivalent noise in SSE in the second modeling method to be introduced below. Therefore, we will not discuss the distribution of $\hat{n}_k$ here in detail and will leave the detailed discussions to the second modeling approach discussed below.

SSE Modeling Approach #2 Another equivalent model of $\hat{x}_k$ has been introduced in [454] to take into consideration the correlations between the desired signal and the equivalent noise in the SSE. Here, $\hat{x}_k$ is expressed in the following manner:

$$\hat{x}_k = x_k(1 - \hat{n}_k), \tag{4.68}$$

where $\hat{n}_k \geq 0$ is an equivalent noise in the new model with the mean:

$$\mu_{\hat{n}} = \frac{1}{L}\sum_{k=1}^{L} \hat{n}_k = \frac{1}{L}\sum_{k=1}^{L}(1 - \hat{x}_k x_k) = \frac{1}{L}\sum_{k=1}^{L}|\hat{x}_k - x_k| \tag{4.69}$$

and variance:

$$\sigma_{\hat{n}}^2 = \frac{1}{L}\sum_{k=1}^{L}(1 - \hat{x}_k x_k - \mu_{\hat{n}})^2. \tag{4.70}$$

In the following, we refer to this SSE modeling as the SSE-new and the SIR protocol using this model is referred to as SIR-SSE-new.

Under this model, the signal transmitted from the relay can be expressed as:

$$x_{R,k} = \beta \hat{x}_k = \beta x_k(1 - \hat{n}_k), \tag{4.71}$$

where β can be calculated from the following power constraint at the relay:

$$\mathbb{E}\left\{|x_{R,k}|^2\right\} = \beta^2\left(1 - 2\mu_{\hat{n}} + \mathbb{E}\left\{\hat{n}_k^2\right\}\right) = \beta^2\left((1 - \mu_{\hat{n}})^2 + \sigma_{\hat{n}}^2\right) = p_R \tag{4.72}$$

Let $\overline{\omega} = (1 - \mu_{\hat{n}})^2 + \sigma_{\hat{n}}^2$, then from Equation (4.72) we can obtain:

$$\beta = \sqrt{\frac{p_R}{\overline{\omega}}}. \tag{4.73}$$

The corresponding destination received signal can thus be written as:

$$Y_{RD,k} = G_{RD}h_{RD}x_{R,k} + n_{RD,k} = G_{RD}h_{RD}\beta x_k(1 - \mu_{\hat{n}}) + \overline{n}_{RD,k}, \tag{4.74}$$

where

$$\overline{n}_{RD,k} = n_{RD,k} - G_{RD}h_{RD}\beta x_k(\hat{n}_k - \mu_{\hat{n}}) \tag{4.75}$$

is the equivalent noise at the destination with a zero mean and variance of σ_E^2 given by:

$$\sigma_E^2 = N_0 + G_{RD}^2|h_{RD}\beta|^2\sigma_{\hat{n}}^2. \tag{4.76}$$

From Equation (4.74), we can see that in the received signal there is a factor $(1 - \mu_{\hat{n}}) \leq 1$ in front of the transmitted signal x_k, which actually reduces the effective signal power of the received signal. This is again due to the erroneous detection at the relay. Its effect is very similar to the *correlation scalar* α_x in the prior discussed SIR-SSE-scalar modeling approach. Actually, we can prove from the following theorem that α_x is equal to $(1 - \mu_{\hat{n}})$.

Theorem 4.2.2 *The mean value of the equivalent noise in SSE in the 'SSE-new' and the* correlation scalar α_x *in the 'SSE-scalar' have the following relationship:*

$$1 - \mu_{\hat{n}} = \alpha_x. \tag{4.77}$$

Proof: If we compare the two SSE modeling approaches reflected in Equations (4.54) and (4.68), we have:

$$\hat{x}_k = \alpha_x(x_k + \hat{n}_k) = x_k(1 - \hat{n}_k). \tag{4.78}$$

By multiplying x_k on both sides of Equation (4.78), we obtain:

$$\alpha_x + \alpha_x x_k \hat{n}_k = 1 - \hat{n}_k. \tag{4.79}$$

We can further write the above equation as follows

$$\hat{n}_k = 1 - \alpha_x - \alpha_x x_k \hat{n}_k. \tag{4.80}$$

Taking the expectation of both sides of above equation, we have:

$$\mu_{\hat{n}} = 1 - \alpha_x. \tag{4.81}$$

This theorem essentially indicates that 'SSE-new' and 'SSE-scalar' are the same. From one model we can calculate the parameters of the other model. We will also see in the following that the noise terms of these two models have very similar distributions.

Let $\tilde{n}_{R,k} = x_k \hat{n}_k$. Now let us examine the distribution of $\tilde{n}_{R,k}$. Figure 4.5 shows the distribution density of $\tilde{n}_{R,k}$ for various E_b/N_0. The simulation parameters are the same as in Figure 4.4. From the figure, we can see that $\tilde{n}_{R,k}$ can be roughly approximated by a Gaussian distribution.

Similarly, let $\check{n}_{R,k} = x_k(\hat{n}_k - \mu_{\hat{n}})$. Figure 4.6 shows the distribution of $\check{n}_{R,k}$. By comparing this figure with the distribution of the equivalent noise in SSE of the previously introduced SSE-scalar modeling shown in Figure 4.4, we note that they have, surprisingly, a very similar distribution with

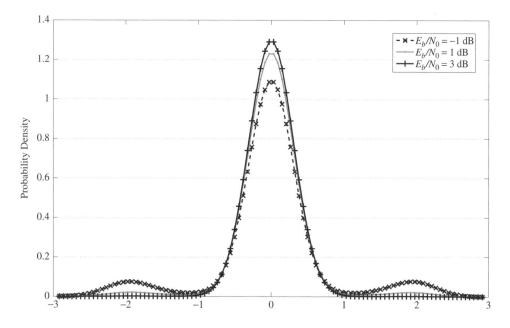

Figure 4.5 The distribution of $\tilde{n}_{R,k} = x_k \hat{n}_k$, an equivalent noise in the soft symbol estimate at the relay

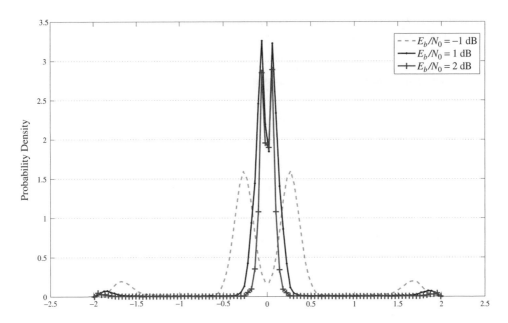

Figure 4.6 The distribution of $\check{n}_{R,k} = x_k(\hat{n}_k - \mu_{\hat{n}})$, an equivalent noise in the soft symbol estimate at the relay

both following a symmetric bimodal distribution. Therefore, in the following, we will mainly focus on the SIR-SSE-new modeling as similar observations and behaviors apply to SIR-SSE-scalar.

Figure 4.7 compares the distribution of $\check{n}_{R,k}$ and $\tilde{n}_{R,k}$, where the curve '$E_b/N_0 = 1$ dB (including mean)' represents the distribution of $\tilde{n}_{R,k}$ at $E_b/N_0 = 1$ dB and '$E_b/N_0 = 1$ dB (excluding mean)' represents the distribution of $\check{n}_{R,k}$ at $E_b/N_0 = 1$ dB.

Figure 4.8 compares the distribution of $\bar{n}_{RD,k}$, $\check{n}_{R,k}$ and $n_{RD,k}$ when the variance of $n_{RD,k}$ is 0.1 times of the variance of $\check{n}_{R,k}$ at $E_b/N_0 = -1$ dB. Similarly, Figure 4.9 compares these distributions when the variance of $n_{RD,k}$ is equal to the variance of $\check{n}_{R,k}$ at $E_b/N_0 = -1$ dB. In these two figures, 'Distribution of total noise at the destination', 'Distribution of noise term in soft symbol estimate' and 'Distribution of destination receiver noise' represents the distribution of $\bar{n}_{RD,k}$, $\check{n}_{R,k}$ and $n_{RD,k}$, respectively.

We note from these figures that, although $\check{n}_{R,k}$ does not follow a Gaussian distribution, $\bar{n}_{RD,k}$ follows approximately a Gaussian distribution even when the destination receiver noise $n_{RD,k}$ is much smaller than $\check{n}_{R,k}$. Since the exact distribution of $\check{n}_{R,k}$ is very difficult to derive, for simplicity, we assume that $\bar{n}_{RD,k}$ follows a Gaussian distribution. As we can see later from the simulation results, although the Gaussian distribution is not the optimal approximation of $\bar{n}_{RD,k}$, it does give very good performance results and outperforms other SSE modeling approaches. Under the assumption of a Gaussian distribution for $\bar{n}_{RD,k}$, the branch metric associated with x_k in a Viterbi decoding can be formulated as:

$$\frac{|y_{SD,k} - \sqrt{p_{SD}}h_{SD}x_k|^2}{N_0} + \frac{|y_{RD,k} - G_{RD}h_{RD}\beta x_k(1 - \mu_{\hat{n}})|^2}{\sigma_E^2}. \qquad (4.82)$$

Similarly, for MAP decoding, the branch metric associated with x_k in $\gamma_k(m, m')$ in Equation (4.48a) is calculated as:

$$\exp\left[-\left(\frac{|y_{SD,k} - \sqrt{p_{SD}}h_{SD}x_k|^2}{N_0} + \frac{|y_{RD,k} - G_{RD}h_{RD}\beta x_k(1 - \mu_{\hat{n}})|^2}{\sigma_E^2}\right)\right]. \qquad (4.83)$$

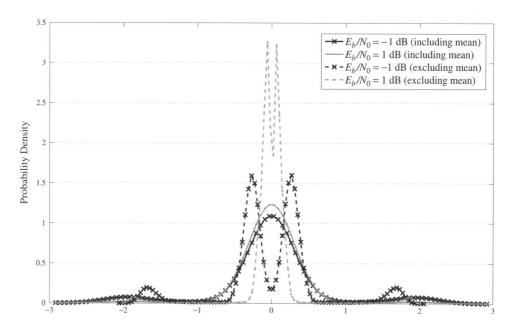

Figure 4.7 The distribution comparison of $\tilde{n}_{R,k}$ and $\check{n}_{R,k}$ in the soft symbol estimate at the relay

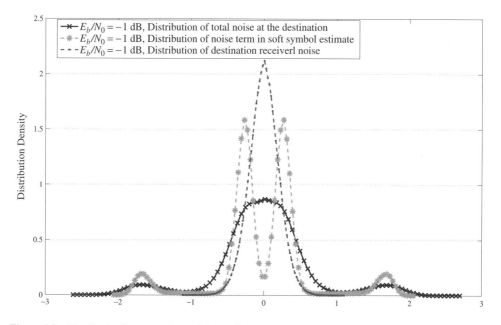

Figure 4.8 The distribution comparison of $\bar{n}_{RD,k}$, $\check{n}_{R,k}$ and $n_{RD,k}$ when the variance of $n_{RD,k}$ is 0.1 times of variance of $\check{n}_{R,k}$ at $E_b/N_0 = -1$ dB

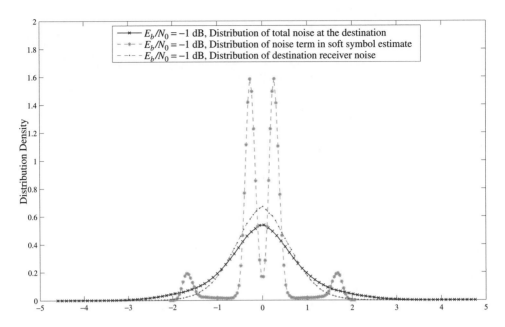

Figure 4.9 The distribution comparison of $\bar{n}_{RD,k}$, $\check{n}_{R,k}$ and $n_{RD,k}$ when the variance of $n_{RD,k}$ is equal to the variance of $\check{n}_{R,k}$ at $E_b/N_0 = -1$ dB

Here we should note that the variance of approximated Gaussian distribution of $\bar{n}_{RD,k}$ is actually not exactly equal to the sum of the variances of two noise terms in $\bar{n}_{RD,k}$ in Equation (4.75). That is, σ_E^2 calculated in Equation (4.76) is not the actual exact variance of approximated Gaussian distribution of $\bar{n}_{RD,k}$. This is mainly due to $\check{n}_{R,k}$ not strictly following a Gaussian distribution. To rectify the variance of the approximated Gaussian distributed $\bar{n}_{RD,k}$, we modify σ_E^2 in Equation (4.76) as follows:

$$\sigma_E^2 = N_0 + G_{RD}^2 |h_{RD}\beta|^2 \sigma_{\hat{n}}^2 * L_v, \tag{4.84}$$

where L_v is a factor used for the variance rectification.

Figures 4.10–4.12 show the BER performance of SIR-SSE versus various L_v values at $\overline{\gamma}_{SR} = 0$ dB, 5 dB and 10 dB, respectively, where 'R-D SNR' represents $\overline{\gamma}_{RD}$. From these figures, we can see that the BER performance decreases quite fast initially as L_v increases until $L_v = 2$. When $2 < L_v < 8$, the BER performance remains almost invariable and become insensitive to the L_v values in this region, which is advantageous for choosing optimal L_v values. Therefore, we can choose L_v in a fairly wide range between $2 < L_v < 8$ with almost no impact on performance. This also indicates that Equation (4.76) actually underestimates the actual variance of $\bar{n}_{RD,k}$ when assuming that it is an approximated Gaussian distribution. To get a satisfactory BER performance, we have to increase its value by at least a factor of two. In the following, we choose $L_v = 4$ as the default optimal L_v value.

Figures 4.13–4.15 compare the BER performance of SIR-SSE for various L_v values at $\overline{\gamma}_{SR} = 0$ dB, 5 dB and 10 dB, respectively. We can observe from these figures that the SIR-SSE with $L_v = 4$ is superior by about 1–2 dB compared with that without variance rectification ($L_v = 1$) and it is much better than the SIR-SSE with $L_v = 0.3$. From these figures, we can also see that the decoder is very sensitive to the variance when $L_v < 2$. Such an inaccurate variance can significantly degrade the system performance. Unfortunately, finding analytically the exact variance of $\bar{n}_{RD,k}$ in the SIR-SSE scheme seems very challenging and is still an open problem. Therefore, we have to mainly rely on simulations to find the optimized rectification factors and get a good estimate of the variance. On other side, fortunately, SIR-SSE becomes insensitive to the L_v values over a wide of range of $2 < L_v < 8$, which gives large freedom and flexibility for selecting the optimized L_v values.

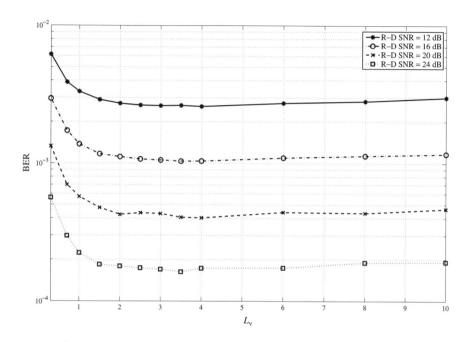

Figure 4.10 BER performance for various rectification factors at $\overline{\gamma}_{SR} = 0$ dB

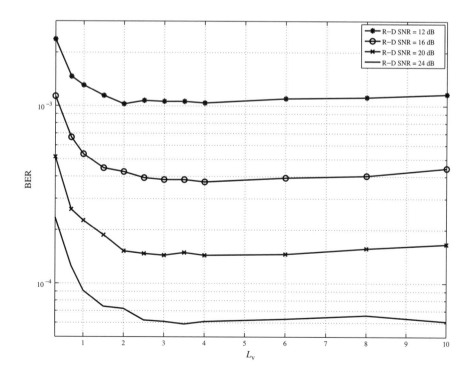

Figure 4.11 BER performance for various rectification factors at $\overline{\gamma}_{SR} = 5$ dB

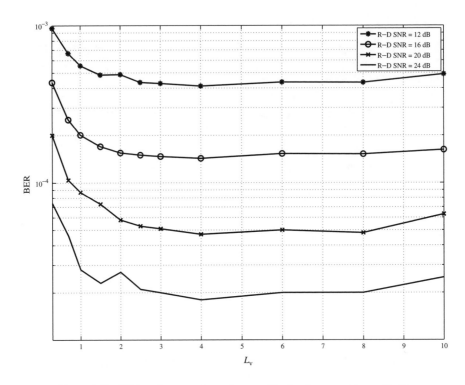

Figure 4.12   BER performance for various rectification factors at $\overline{\gamma}_{SR} = 10$ dB

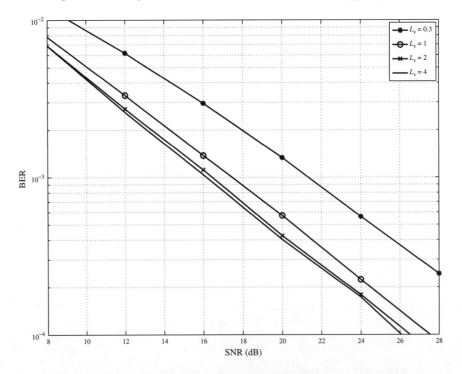

Figure 4.13   BER performance for various rectification factors at $\overline{\gamma}_{SR} = 0$ dB

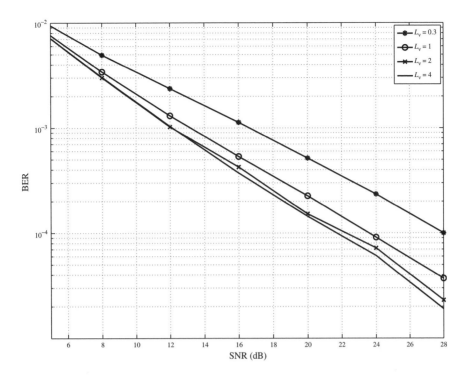

Figure 4.14 BER performance for various rectification factors at $\overline{\gamma}_{SR} = 5$ dB

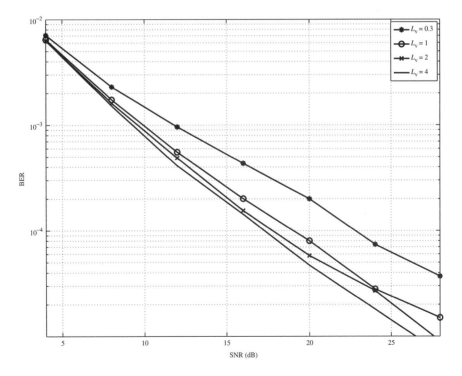

Figure 4.15 BER performance for various rectification factors at $\overline{\gamma}_{SR} = 10$ dB

4.2.3.2 SIR Based on Log-Likelihood Ratio

In this section, we introduce another representation of soft information based on the log-likelihood ratio (SIR-LLR) [455, 457, 458]. As before, we consider a BPSK modulation. Let $L_r(k)$ represent the LLR value of $\hat{C}_k$ at the relay decoder, defined as:

$$L_r(k) = \log\left(\frac{L_r(\hat{C}_k = 0|\mathbf{Y}_{sr})}{L_r(\hat{C}_k = 1|\mathbf{Y}_{sr})}\right). \tag{4.85}$$

Let P_L be the average power of $L_r(k)$ which can be calculated as:

$$P_L = \mathbb{E}\left\{L_r^2(k)\right\} = \frac{1}{L}\sum_{i=1}^{L}(L_r(k))^2. \tag{4.86}$$

Further let μ_L and σ_L^2 be the mean value and variance of $L_r(k)$ (assume $x_k = 1$ is transmitted), which can be calculated as follows:

$$\mu_L = \mathbb{E}\{L_r(k)\} = \frac{1}{L}\sum_{i=1}^{L}(L_r(k) \tag{4.87}$$

and

$$\sigma_L^2 = \mathbb{E}\left\{(L_r(k) - \mu_L)^2\right\} = \frac{1}{L}\sum_{i=1}^{L}(L_r(k) - \mu_L)^2 = P_L - \mu_L^2. \tag{4.88}$$

The statistical distribution of $L_r(k)$ has been widely studied and is well known to follow an approximated Gaussian distribution for very long codes [463]. Let us now assume that the channel code is long enough such that the Gaussian distribution approximately holds. The PDF of $L_r(k)$, denoted by $P_L(x)$, can then be approximated as:

$$P_L(x) = \frac{1}{\sqrt{2\pi\sigma_L^2}}\exp\left(-\frac{(x - \mu_L)^2}{2\sigma_L^2}\right). \tag{4.89}$$

Similarly to the model used in the previous section, $L_r(k)$ can be expressed as [455, 457]:

$$L_r(k) = \mu_L x_k + n_{L,k}, \tag{4.90}$$

where x_k is the exact transmitted symbol, $n_{L,k}$ is an equivalent Gaussian noise with variance σ_L^2. Figure 4.16 shows the distribution of $n_{L,k}$ with the same system parameter as in Figure 4.5. It can be observed that $n_{L,k}$ follows the approximated Gaussian distribution at both low and high SNRs. We also note that the shape of the PDF curve gets thinner as E_b/N_0 increases. This indicates that the variance of $n_{L,k}$, which also represents the average power of equivalent noise, decreases as E_b/N_0 increases.

From Equation (4.90), we have:

$$P_L = E(L_r(k)^2) = \mathbb{E}\left\{(\mu_L x_k + n_{L,k})^2\right\} = \mathbb{E}\left\{(\mu_L x_k)^2\right\} + \mathbb{E}\left\{(n_{L,k})^2\right\} = \mu_L^2 + \sigma_L^2. \tag{4.91}$$

which is consistent with Equation (4.88). Then the corresponding signal transmitted from the relay can be expressed as:

$$x_{R,k} = \beta L_r(k) = \beta\left(\mu_L x_k + n_{L,k}\right), \tag{4.92}$$

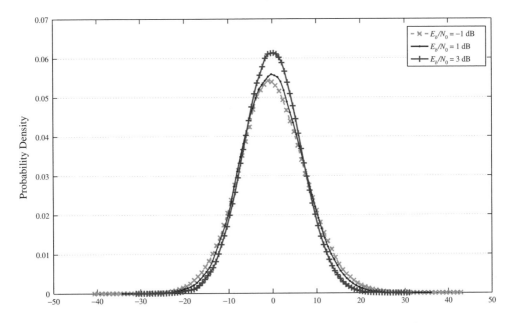

Figure 4.16   The distribution of $n_{L,k}$, an equivalent noise in LLR at the relay

where β is a normalization factor such that the $x_{R,k}$ obeys the power constraint p_R at the relay:

$$\mathbb{E}\left\{|x_{R,k}|^2\right\} = p_R.$$ (4.93)

By substituting Equations (4.91) and (4.92) into Equation (4.93), β can be calculated as:

$$\beta = \sqrt{\frac{p_R}{P_L}}.$$ (4.94)

The corresponding received signal at the destination is thus given by:

$$y_{RD,k} = G_{RD}h_{RD}x_{R,k} + n_{RD,k} = G_{RD}h_{RD}\beta\mu_L x_k + \bar{n}_{RD,k},$$ (4.95)

where $\bar{n}_{RD,k} = n_{RD,k} + G_{RD}h_{RD}\beta n_{L,k}$ is the equivalent noise at the destination with a zero mean and variance of σ_E^2 given by:

$$\sigma_E^2 = N_0 + G_{RD}^2|h_{RD}\beta|^2\sigma_L^2.$$ (4.96)

Since we assumed that $n_{L,k}$ follows a Gaussian distribution, $\bar{n}_{RD,k}$ is a linear combination of two Gaussian random variables and also follows a Gaussian distribution. Similarly to the SIR-SSE approach, the branch metric associated with x_k in the Viterbi decoding procedure is now given by:

$$\frac{|y_{SD,k} - \sqrt{p_{SD}}h_{SD}x_k|^2}{N_0} + \frac{|y_{RD,k} - G_{RD}h_{RD}\beta\mu_L x_k|^2}{\sigma_E^2}.$$ (4.97)

For MAP decoding, the branch metric associated with x_k in $\gamma_k(m, m')$ finally becomes:

$$\exp\left[-\left(\frac{|y_{SD,k} - \sqrt{p_{SD}}h_{SD}x_k|^2}{N_0} + \frac{|y_{RD,k} - G_{RD}h_{RD}\beta\mu_L x_k|^2}{\sigma_E^2}\right)\right].$$ (4.98)

4.2.3.3   Mean Square Errors of Signal Estimation at Relay

In this section, we first compare the MSE of the signal estimation at relay in already discussed relaying schemes. Let $x_{R,k}$ be the signal to be transmitted by the relay represented in the following way:

$$x_{R,k} = \beta_r(x_k + n_r(k)), \tag{4.99}$$

where β_r is the amplification factor, x_k is the signal transmitted by the source and $n_r(k)$ is the equivalent noise in $x_{R,k}$. Then, the MSE is defined as the variance of the equivalent noise term $\beta_r n_r(k)$, that is:

$$MSE = \frac{1}{L}\sum_{k=1}^{L}|\beta_r n_r(k)|^2. \tag{4.100}$$

For simplicity, we consider an AWGN channel and BPSK modulation. We use a four-state recursive systematic convolutional code (RSC) with code rate of $1/2$ as channel code in the link from the source to relay; its generator matrix is $(1, 5/7)$. We assume that the noise variance at the relay is 0.5 per dimension and we only consider the real part of the received signal. We compare the MSE of five different relaying schemes, namely AF, DF, EF, SIR-SSE and SIR-LLR. The difference between the DF and EF protocols is that in DF the relay fully decodes the received signals and makes hard decisions on each symbol, including the information and parity symbols, whilst in EF the relay only demodulates the received signal without actual decoding it. The MSE of these schemes can be calculated as follows:

$$MSE - AF = \frac{0.5}{P_{SR} + 0.5} \tag{4.101a}$$

$$MSE - DF = \frac{1}{L}\sum_{k=1}^{L}|\hat{x}_c(k) - x_k|^2 \tag{4.101b}$$

$$MSE - EF = \frac{1}{L}\sum_{k=1}^{L}|\hat{x}_m(k) - x_k|^2 \tag{4.101c}$$

$$MSE - SIR - SSE = \sigma_{\hat{n}}^2 / \left((1 - \mu_{\hat{n}})^2 + \sigma_{\hat{n}}^2\right) \tag{4.101d}$$

$$MSE - SIR - LLR = \frac{\sigma_L^2}{\mu_L^2 + \sigma_L^2} \tag{4.101e}$$

where in Equation (4.101a), P_{SR} is the received power at the relay and 0.5 is the power of the noise real part; in Equation (4.101b), $\hat{x}_c(k)$ is the symbol estimate of x_k obtained through decoding and hard decision; in Equation (4.101c), $\hat{x}_m(k)$ is the demodulated symbol; $\mu_{\hat{n}}$ and $\sigma_{\hat{n}}$ are the mean and variance of equivalent noise in the soft symbol estimate of x_k; and in Equation (4.101e)), μ_L and σ_L are the mean and variance of LLR values.

Figure 4.17 compares the MSE of relay symbol estimates of various relay schemes. Here the x-axis represents the SNR in the link from the source to the relay. From the figure we can see that SIR-SSE is optimal in terms of minimum MSE. DF has a very similar MSE as the SIR-SSE. AF, EF and SIR-LLR have very large MSEs compared with the SIR-SSE and DF.

4.2.3.4   Simulation Results

In this section, we provide the performance comparison of various SIR schemes. All simulations are performed for BPSK modulation and a frame size of 130 symbols over quasi-static fading channels.

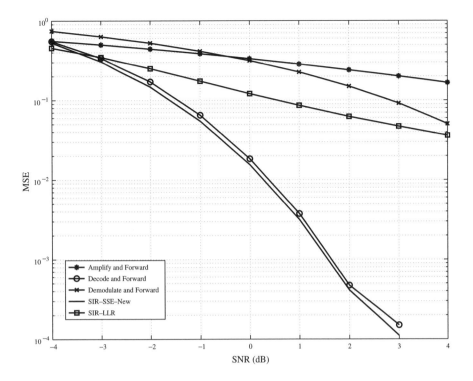

Figure 4.17 Mean square errors of signal estimation at relay for various relaying schemes

For simplicity, we also assume that $\overline{\gamma}_{SD}$ is equal to $\overline{\gamma}_{RD}$, where $\overline{\gamma}_{SD}$ and $\overline{\gamma}_{RD}$ are the average SNRs in the link from the source to the destination and that from the relay to the destination. We use again a four-state RSC code with the code rate of $1/2$; the generator matrix of the RSC is $(1, 5/7)$.

Here, we investigate two SIR-SSE schemes referred to as 'SIR-SSE-new' and 'SIR-SSE-normal'. The latter refers to the SIR-SSE scheme using the following conventional modeling of soft symbol estimate (SSE):

$$\hat{x}_k = x_k + \hat{n}_k, \tag{4.102}$$

where $\hat{x}_k$ is the SSE at the output of relay decoder, x_k is the exact transmitted symbol and $\hat{n}_k$ is modeled as a Gaussian random variable. The former refers to the SIR-SSE scheme introduced in Section 4.2.3.1 using the following modeling approach:

$$\hat{x}_k = x_k(1 - \hat{n}_k). \tag{4.103}$$

Figures 4.18–4.20 compare the BER performance of AF, SIR-SSE-normal, SIR-SSE-new, and SIR-LLR schemes. Here, the x-axis is the average SNR in the link from the relay to the destination $\overline{\gamma}_{RD}$. We evaluate two SIR-SSE-new schemes, one without variance rectification ($L_v = 1$) and one with the rectified variance. For the rectified variance, we choose the rectification factor $L_v = 4$ in the simulations.

From these figures, we can see that the SIR-SSE-new with rectified variance and SIR-LLR have almost the same performance over the entire SNR range and considerably outperform other SIR schemes. For example, at $\overline{\gamma}_{SR} = 0$ dB, they are superior by about 1.5 dB and 2 dB to the SIR-SSE-new without variance rectification and SIR-SSE-normal at a BER of 10^{-3}. All SIR schemes considerably

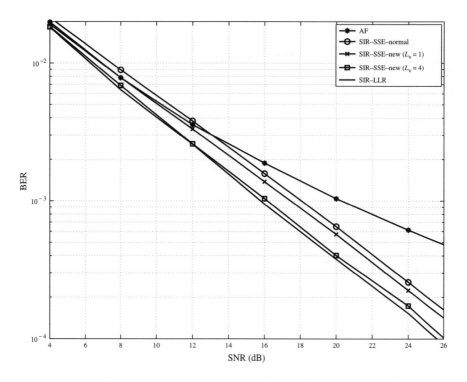

Figure 4.18   BER performance for various rectification factors at $\overline{\gamma}_{SR} = 0$ dB

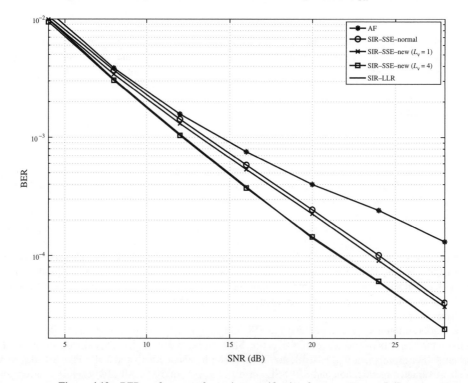

Figure 4.19   BER performance for various rectification factors at $\overline{\gamma}_{SR} = 5$ dB

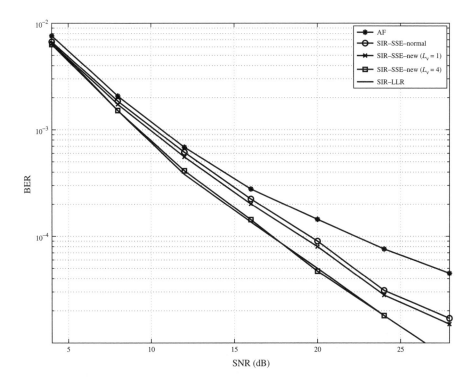

Figure 4.20 BER performance for various rectification factors at $\overline{\gamma}_{SR} = 10$ dB

outperform the AF scheme, especially at high SNRs. For example, SIR-SSE-new ($L_v = 4$) and SIR-SSE-new ($L_v = 1$) provide gains by about 5 dB and 3 dB compared with the AF scheme at a BER of 10^{-4} and $\overline{\gamma}_{SR} = 10$ dB. These gains further increase as $\overline{\gamma}_{RD}$ increases. We can also note from the figures that the SIR-SSE-new schemes are always better than the SIR scheme with conventional modeling of soft symbol estimate (SIR-SSE-normal). This verifies the superiority of soft estimate modeling of SIR-SSE-new introduced in Section 4.2.3.1.

If we compare Figure 4.17 and the above three figures, we can see that although SIR-SSE-new has a much smaller MSE compared with the SIR-LLR, their BER performances are very similar. The reason is that we use a Gaussian distribution to approximate the equivalent noise in the SSE in SIR-SSE-new scheme and it is not a very accurate approximation. The performance of SIR-SSE-new can possibly be further improved if a more accurate approximation on the equivalent noise in the SIR-SSE-new scheme can be found. How to find and formulate such a distribution is very important, but also very challenging and unfortunately this has not been addressed to date.

Soft information is an analog signal. In practical systems, to transmit such analog signals, compression and/or quantization or other modulation techniques should be performed at the relays. The compression and forward protocol based on Wyner–Ziv coding (CF-WZC) introduced in Section 4.2.2 can be used to quantize analog soft information into digital signals and compress the quantized signals before transmitting. Therefore, soft information relaying can be combined with compression-and-forward into a single relay protocol to be used in a practical system. Furthermore, the soft information can also be directly transmitted by using an analog phase modulation scheme [464], called continuous modulation. Besides these approaches, there will be other possible ways to transmit the soft information. Unfortunately, what is the optimal way to transmit the soft information for achieving the optimal performance is still an open question.

4.2.4 Adaptive Relaying

In this section, we introduce another variation of the AF and DF protocols, which we refer to as the adaptive relaying protocol (ARP) [465]. It has advantages of both AF and DF protocols and minimizes their negative effects. In the ARP, each relay adaptively selects the AF or DF protocol based on whether its decoding result is correct or not. All the relays that fail to decode correctly use the AF protocol to amplify the received signals and forward them to the destination. On the other hand, all the relays that can successively decode the received signals, use the DF protocol. The signals received at the destination, forwarded from all relays by using either the AF or DF protocol, are combined into one signal to recover the source information.

In a practical system, in order to determine whether a relay can decode correctly or not, some CRC bits can be appended to each information block (frame). After decoding each frame, the relay can examine the CRC checks to determine if the received signals are decoded correctly or not. This scheme could be complex as each relay has to decode the received signal first no matter whether it can decode correctly or not. As introduced in Section 4.2.2, for a given channel coding scheme, there exists a threshold calculated based on Bhattacharyya code parameter [452]. When the received SNR is greater than this threshold, the received signals can be decoded asymptotically correctly when the block length goes to infinity. This method has been used for checking the decoding results [452]. A similar method can also be used here to determine whether a relay can decode correctly or not. That is, each relay only needs to compare its received SNR with a threshold. Only those relays, the received SNR of which is above this threshold, do the decoding and the rest of the relays are assumed not to be able to decode correctly and will not therefore try to decode.

One important feature of the ARP scheme in a practical application is that the relay can automatically adapt to the channel quality by simply switching between the AF and the DF protocols without requiring the CSI to be fed back from the destination to the relays or the source. This feature is very important in practical relay networks, especially in a large multihop network in which the feedback of CSI for adaptation is very expensive. Another important feature is that the processing at relays and destination for the ARP scheme is the same as for the AF and DF and it does not add much extra complexity to the system.

In this section, we therefore introduce and analyze the ARP scheme as well as evaluate its performance through simulations and compared it to other relaying protocols.

4.2.4.1 Adaptive Relay Protocol

We consider a general two-hop relay network, consisting of one source, n relays and one destination. Figure 4.21 depicts a block diagram of a two-hop relay system with a direct link from the source to the destination. We assume that the source and relays transmit data through orthogonal channels.

The source first broadcasts the information symbol $x(k)$ to both the destination and relays. The corresponding received signals at the relay i and destination, denoted by $y_{SR,k}^i$ and $y_{SD,k}$, respectively, can be expressed as:

$$y_{SR,k}^i = \sqrt{P_{SR}^i} h_{SR}^i x(k) + n_{SR,k}^i \qquad (4.104a)$$

$$y_{SD,k} = \sqrt{P_{SD}} h_{SD} x(k) + n_{SD,k} \qquad (4.104b)$$

where $p_{SR}^i = p_S \cdot (G_{SR}^i)^2$, $p_{SD} = p_S \cdot (G_{SD})^2$ is the received signal power at the relay i and destination, respectively, p_S is the source transmit power, G_{SR}^i and G_{SD} are the channel gains between the source and relay i and that between the source and destination, respectively. Also h_{SR}^i and h_{SD} are the fading coefficients between the source and the relay i and between the source and destination,

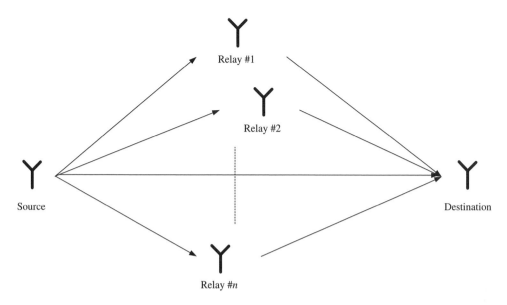

Figure 4.21   A general two-hop relay network

respectively. Furthermore, $n_{SR,k}^i$ and $n_{SD,k}$ are noise vectors and each noise term is a zero mean complex Gaussian random variable with a two-sided power spectral density of $N_0/2$ per dimension.

At each transmission, based on whether relays can make a correct decoding or not, each relay is included in either an AF or a DF relay group, denoted by Ω_{AF} and Ω_{DF}, respectively. These two groups are discussed below.

(i) AF Relay Group. An AF relay group consists of all the relays that could not decode correctly. Upon receiving signals from the source, each relay in the AF relay group simply amplifies the received signals and forwards it to the destination. Let $\tilde{x}_{R,k}^i$, $i \in \Omega_{AF}$, represent the signal transmitted from the relay i at time k, then it can be expressed as:

$$\tilde{x}_{R,k}^i = \mu_i y_{SR,k}^i, i \in \Omega_{AF}, \tag{4.105}$$

where μ_i is an amplification factor such that $\tilde{x}_{R,k}^i$ satisfies the power constraint of p_R at each relay and it can be calculated as:

$$\mu_i \leq \sqrt{\frac{p_R}{|h_{SR}^i|^2 p_{SR}^i + N_0}}. \tag{4.106}$$

The corresponding received signal at the destination, transmitted from the ith relay is given by:

$$y_{RD,k}^i = G_{RD}^i h_{RD}^i \mu_i \left(\sqrt{p_{SR}^i} h_{SR}^i x_k + n_{SR,k}^i \right) + n_{RD,k}^i. \tag{4.107}$$

At destination, all signals forwarded from the AF relay group are then combined with the signals directly transmitted from the source as follows:

$$y_{RD-AF,k} = W_{SD} y_{SD,k} + \sum_{i \in \Omega_{AF}} W_R^i y_{RD,k}^i, \tag{4.108}$$

where $y_{RD-AF,k}$ is the combined signal in the AF group at the destination, and W_{SD} and W_R^i, $i \in \Omega_{AF}$ are the combining coefficients. The optimal values of W_{SD} and W_R^i are given by [100, 466]:

$$W_{SD} = \frac{\sqrt{p_{SD}} h_{SD}^*}{N_0}, \qquad (4.109a)$$

$$W_R^i = \frac{\mu_i G_{RD}^i \sqrt{p_{SR}^i} \left(h_{RD}^i h_{SR}^i\right)^*}{(\mu_i^2 |G_{RD}^i h_{RD}^i|^2 + 1) N_0}, i \in \Omega_{AF}. \qquad (4.109b)$$

The corresponding destination SNR for the combined signals in the AF relay group, denoted by γ_{AF}, can be approximated by:

$$\gamma_{AF} \approx \overline{\gamma}_{SD} |h_{SD}|^2 + \frac{1}{2} \sum_{i \in \Omega_{AF}} H_2^i, \qquad (4.110)$$

where $\overline{\gamma}_{SD} = \frac{p_{SD}}{N_0}$, $\overline{\gamma}_{RD}^i = \frac{p_{RD}^i}{N_0}$, $P_{RD}^i = p_R(G_{RD}^i)^2$, $\overline{\gamma}_{SR}^i = \frac{p_{SR}^i}{N_0}$ and $H_2^i = (\frac{1}{2} \sum_{p=1}^2 \frac{1}{\lambda_{i,p}})^{-1}$ is the harmonic mean of variables $\lambda_{i,p}$, $p = 1, 2$, $\lambda_{i,1} = |h_{SR}^i|^2 \overline{\gamma}_{SR}^i$, and $\lambda_{i,2} = |h_{RD}^i|^2 \overline{\gamma}_{RD}^i$.

(ii) DF Relay Group. The DF relay group consists of all the relays that can decode the received information correctly. Each relay in the DF relay group decodes the received signals from the source, re-encodes and forwards them to the destination with power p_R. The corresponding received signals at the destination are given by:

$$y_{RD,k}^i = G_{RD}^i h_{RD}^i \sqrt{p_R} x(k) + n_{RD,k}^i, i \in \Omega_{DF}. \qquad (4.111)$$

At destination, all signals forwarded from the DF relay group are then combined. Let $y_{RD-DF,k}$ represent the combined signal given by:

$$y_{RD-DF,k} = \sum_{i \in \Omega_{DF}} W_R^i y_{RD,k}^i. \qquad (4.112)$$

Similarly to the calculation in the AF group, the destination SNR of the combined signal in DF is given by:

$$\gamma_{DF} = \sum_{i \in \Omega_{DF}} \overline{\gamma}_{RD}^i |h_{RD}^i|^2. \qquad (4.113)$$

In the ARP scheme, all the signals forwarded from both AF and DF groups are combined into one signal and the overall SNR of the combined signal is thus finally given by:

$$\gamma_{ARP} = \gamma_{AF} + \gamma_{DF} \approx \overline{\gamma}_{SD} |h_{SD}|^2 + \frac{1}{2} \sum_{i \in \Omega_{AF}} H_2^i + \sum_{i \in \Omega_{DF}} \overline{\gamma}_{RD}^i |h_{RD}^i|^2. \qquad (4.114)$$

4.2.4.2 Performance Analysis of ARP

In this section, we analyze the performance of the ARP and compare with AF and perfect DF schemes. For simplicity, we will focus on the average upper bound, which could be very tight for independent fast fading channels, but in general is very loose for quasi-static block fading channels. In order to get a tight upper bound over quasi-static fading channels, we should limit the conditional upper bound before averaging over the fading coefficients, but no closed-form expression can be obtained. Since we are only interested in the relative performance of various relaying protocols, the average upper bound should be sufficient. For simplicity, we assume that $G_{SR}^i = G_{SR}$ for $i = 1, \cdots, n$. Then we have $p_{SR}^i = p_{SR}$ for $i = 1, \cdots, n$. Without loss of generality, we consider a coded system since the analysis for the uncoded system can be derived from it in a straightforward manner. Unlike other relaying protocols, the analysis for ARP is not straightforward. The reason is that the channel from the source to each relay varies with time and thus the number of relay nodes in the AF and DF groups also varies with time. The uncertain number of relay nodes in the AF and DF makes the performance analysis involved.

To solve this problem, we first calculate the PEP for a scenario where the AF relay group consists of q relays numbered from 1 to q and the DF relay group consists of $(n - q)$ relays numbered from $(q + 1)$ to n. As shown in Equation (4.114), the overall SNR of the combined signals at the destination is given by:

$$\gamma_{ARP,(q)} = \gamma_{AF,(q)} + \gamma_{DF,(n-q)}, \qquad (4.115)$$

where $\gamma_{AF,(q)}$ and $\gamma_{DF,(n-q)}$ represent the instantaneous received SNR of the combined signals in the AF and DF relay groups. From Equations (4.110) and (4.113) we have:

$$\gamma_{AF,(q)} \approx \overline{\gamma}_{SD}|h_{SD}|^2 + \frac{1}{2}\sum_{i=1}^{q} H_2^i \qquad (4.116a)$$

$$\gamma_{DF,(n-q)} = \overline{\gamma}_{RD} \sum_{i=q+1}^{n} |h_{RD}^i|^2. \qquad (4.116b)$$

Let $P_{P,sr}^i(d_{sr}^H, \overline{\gamma}_{SR}|h_{SR}^i)$ be the conditional PEP of incorrectly decoding a codeword into another codeword with Hamming distance of d_{sr}^H in the channel from the source to the relay i. It can be expressed by:

$$P_{P,sr}^i\left(d_{sr}^H, \overline{\gamma}_{SR}|h_{SR}^i\right) = Q\left(\sqrt{2d_{sr}^H\overline{\gamma}_{SR}|h_{SR}^i|^2}\right), \qquad (4.117)$$

Further, let $P_{F,sr}^i(\overline{\gamma}_{SR}|h_{SR}^i)$ represent the conditional word error probability in the channel from the source to the ith relay, given by:

$$P_{F,sr}^i\left(\overline{\gamma}_{SR}|h_{SR}^i\right) = \sum_{d_{sr}^H = d_{sr,\min}}^{2L} A\left(d_{sr}^H\right) P_{P,sr}^i\left(d_{sr}^H, \overline{\gamma}_{SR}|h_{SR}^i\right), \qquad (4.118)$$

where $d_{sr,\min}$ is the code minimum Hamming distance, $A(d_{sr}^H)$ is the weight enumerating function with Hamming weight d. The conditional PEP of incorrectly decoding a codeword into another codeword with Hamming distance of d for this scenario, which we denote by $P_{(q)}^{ARP}(d|h_{SD}, \mathbf{h_{SR}}, \mathbf{h_{RD}})$, can thus be calculated as:

$$P_{(q)}^{ARP}\left(d|h_{SD}, \mathbf{h_{SR}}, \mathbf{h_{RD}}\right) \leq \prod_{i=1}^{q} P_{F,sr}^i\left(\overline{\gamma}_{SR}|h_{SR}^i\right)$$

$$\prod_{i=q+1}^{n}\left(1 - P_{F,sr}^i(\overline{\gamma}_{SR}|h_{SR}^i)\right) Q\left(\sqrt{2d\gamma_{AF,(q)} + 2d\gamma_{DF,(n-q)}}\right).$$

Furthermore, the probability that the AF relay group consists of any q relays and the DF relay group consists of the rest $(n - q)$ relays is given by:

$$\sum_{(i_1,\dots,i_n)\in(1,\dots,n)} \prod_{k=1}^{q} P_{F,sr}^{i_k}(\overline{\gamma}_{SR}|h_{SR}^{i_k}) \prod_{k=q+1}^{n}\left(1 - P_{F,sr}^{i_k}\left(\overline{\gamma}_{SR}|h_{SR}^{i_k}\right)\right).$$

Due to the spatially uniform distribution of relays and assumption of $\overline{\gamma}_{SR}^i = \overline{\gamma}_{SR}$ for all $i = 1, \ldots, n$, the average PEP at high SNR, denoted by $P^{ARP}(d)$, can be calculated as:

$$P^{ARP}(d) \leq \sum_{q=0}^{n} \binom{n}{q} \mathbb{E}\left\{ \prod_{i=1}^{q} P_{F,sr}^i (\overline{\gamma}_{SR}|h_{SR}^i) \right. \tag{4.119a}$$

$$\left. \prod_{i=q+1}^{n} \left(1 - P_{F,sr}^i (\overline{\gamma}_{SR}|h_{SR}^i)\right) Q\left(\sqrt{2d\gamma_{AF,(q)} + 2d\gamma_{DF,(n-q)}}\right) \right\}$$

$$\approx \sum_{q=0}^{n} \binom{n}{q} \mathbb{E}\left\{ \prod_{i=1}^{q} P_{F,sr}^i (\overline{\gamma}_{SR}|h_{SR}^i) \right. \tag{4.119b}$$

$$\left. \prod_{i=q+1}^{n} \left(1 - P_{F,sr}^i (\overline{\gamma}_{SR}|h_{SR}^i)\right) Q\left(\sqrt{2d\gamma_{DF,(n-q)}}\right) Q\left(\sqrt{2d\gamma_{AF,(q)}}\right) \right\}$$

$$\leq (d\overline{\gamma}_{SD})^{-1} \sum_{q=0}^{n} \binom{n}{q} (f(d))^q \left(\frac{1 - P_{F,sr}}{d\overline{\gamma}_{RD}}\right)^{n-q} \tag{4.119c}$$

$$= (d\overline{\gamma}_{SD})^{-1} \left(f(d) + \frac{1 - P_{F,sr}}{d\overline{\gamma}_{RD}}\right)^n, \tag{4.119d}$$

with $f(d) = \mathbb{E}\left\{ P_{F,sr}^i \left(\overline{\gamma}_{SR}|h_{SR}^i\right) Q\left(\sqrt{dH_2^i}\right) \right\}$ and $P_{F,sr} = \overline{\gamma}_{SR}^{-1} \sum_{d_{sr}^H = d_{sr,min}}^{2l} A(d_{sr}^H)/(d_{sr}^H)$. The exact closed form expression of $f(d)$ is too complex to be presented here. At high SNRs, however, it can be approximated as follows:

$$f(d) \leq \sum_{d_{sr}^H = d_{sr,min}}^{2L} A\left(d_{sr}^H\right) \overline{\gamma}_{SR}^{-1} \left(\frac{1}{d_{sr}^H + d} + \frac{1}{d_{sr}^H \overline{\gamma}_{RD}(d + d_{sr}^H)}\right). \tag{4.120}$$

By substituting Equation (4.120) into $P^{ARP}(d)$, we have:

$$P^{ARP}(d) \leq (d\overline{\gamma}_{SD})^{-1} \left[\frac{1 + \frac{1}{\overline{\gamma}_{SR}} \sum_{d_{sr}^H = d_{sr,min}}^{2l} A(d_{sr}^H)\left[\frac{d\overline{\gamma}_{RD}}{d+d_{sr}^H} + \frac{d}{d_{sr}^H(d+d_{sr}^H)} - \frac{1}{d_{sr}^H}\right]}{d\overline{\gamma}_{RD}}\right]^n$$

$$= \overline{\gamma}_{SD}^{-1} \overline{\gamma}_{RD}^{-n} d^{-(n+1)} \left[1 + \frac{1}{\overline{\gamma}_{SR}} \sum_{d_{sr}^H = d_{sr,min}}^{2l} A(d_{sr}^H)\left(\frac{d\overline{\gamma}_{RD} - 1}{d + d_{sr}^H}\right)\right]^n$$

$$< \overline{\gamma}_{SD}^{-1} \overline{\gamma}_{RD}^{-n} d^{-(n+1)} \left(1 + \frac{\overline{\gamma}_{RD}}{\overline{\gamma}_{SR}} \phi(d)\right)^n, \tag{4.121}$$

where $\phi(d) = \sum_{d_{sr}^H = d_{sr,min}}^{2l} \frac{A(d_{sr}^H)d}{d_{sr}^H + d}$ now solely depends on the channel code used at the source.

Similarly, for a perfect DF, in which all relay are assumed to be decoding correctly, the average PEP of incorrectly decoding to a codeword with weight d, denoted by $P_{DF}^{Perfect}(d)$, can be calculated as:

$$P_{DF}^{Perfect}(d) \leq (\overline{\gamma}_{SD})^{-1} (\overline{\gamma}_{RD})^{-n} d^{-(n+1)}. \tag{4.122}$$

Equation (4.121) can thus be further expressed as:

$$P^{ARP}(d) \leq \left(\frac{\overline{\gamma}_{RD}}{\overline{\gamma}_{SR}} \phi(d) + 1\right)^n P_{DF}^{Perfect}(d)$$

$$= G_{ARP} P_{DF}^{Perfect}(d), \tag{4.123}$$

where

$$G_{ARP} = \left(\frac{\overline{\gamma}_{RD}}{\overline{\gamma}_{SR}} \phi(d) + 1 \right)^n > 1 \tag{4.124}$$

represents the performance loss of the ARP compared to the perfect DF. It can be seen from Equation (4.124) that as $\overline{\gamma}_{SR} \to \infty$, $G_{ARP} \to 1$, $P^{ARP}(d) \to P_{DF}^{Perfect}(d)$ and the performance of the ARP approaches that of the perfect DF.

4.2.4.3 Performance Evaluations

In this section, we provide simulation results for the ARP and compare it with other relaying protocols for a varying number of relays. All simulations are performed assuming BPSK modulation and a frame size of 130 symbols over quasi-static fading channels. We use again a four-state recursive systematic convolutional code (RSC) with the code rate of 1/2 and a generator matrix of (1, 5/7). For simplicity, we assume that $\overline{\gamma}_{SR,i} = \overline{\gamma}_{SR}$ and $\overline{\gamma}_{RD,i} = \overline{\gamma}_{RD}$ for all $i = 1, \cdots, n$, and $\overline{\gamma}_{RD}$ and $\overline{\gamma}_{SD}$ are the same.

Figures 4.22–4.25 compare the performance of the AF, the ARP and the perfect DF for various numbers of relays, where 'S-R SNR' represents $\overline{\gamma}_{SR}$ and 'R-D SNR' represents $\overline{\gamma}_{RD}$. It can be noted that as the number of relays increases, the ARP significantly outperforms the AF in all SNR regions and performs very closely to the perfect DF as $\overline{\gamma}_{SR}$ increases. This can easily be explained in the following way. For low $\overline{\gamma}_{SR}$ values, the channels from the source to the relays are very noisy and probabilities of decoding errors at each relay are very high, so most of relays cannot correctly decode the received signals. In this case, as the number of relays is small, most of relays are included in the AF relay group with a very high probability and the DF relay group only occasionally includes a few relays. However, even such a limited contribution of coding gain from the DF relay group is significant if the channel from the relay to the destination is poor (corresponding to the low $\overline{\gamma}_{RD}$ values), because in this case the relays in the DF relay group can considerably improve the overall channel quality. As the number of relays increases, the probability that the DF relay group contains at least one relay also increases and the contribution of coding gain from the DF relay group becomes significant. This explains why the ARP can provide a significant coding gain over AF, even at low $\overline{\gamma}_{SR}$ values, as the number of relays increases.

4.2.5 Selective Decode and Forward

In this section, we briefly introduce another variation of the DF protocol, referred to as the selective decode and forward (S-DF) protocol. In the S-DF protocol, those relays that can decode the received signals correctly use DF protocols to forward the signals to the destination and the remaining relays will stay in an idle state. If we compare the S-DF protocol with ARP, we can see that the only difference between these two protocols is that in the ARP, those relays that cannot decode correctly will use the AF protocol, whereas in the S-DF these relays will keep silent and will not forward. Given the analysis for the ARP, we can easily derive the performance for the S-DF.

We consider a system with n relay nodes. For simplicity, we also assume that $\overline{\gamma}_{SR,i} = \overline{\gamma}_{SR}$ and $\overline{\gamma}_{RD,i} = \overline{\gamma}_{RD}$ for all $i = 1, \cdots, n$. Similarly to the ARP scheme, for each transmission, we separate all relays into two groups. All relays that can decode the received signals correctly, are included in a DF group, denoted by Ω_{DF}, and the remaining relays are included in a N-DF group, denoted by Ω_{N-DF}. All relays in the DF group use DF protocol to forward the received signal and those relays in the N-DF group will keep silent. All the received signals forwarded from the DF group will be combined with the signal directly transmitted from the source.

Following the similar analysis for the ARP scheme, we first calculate the PEP for a scenario where the N-DF relay group consists of q relays numbered from 1 to q, and the DF relay group consists of $(n - q)$ relays numbered from $(q + 1)$ to n. The instantaneous received SNR of the combined signals

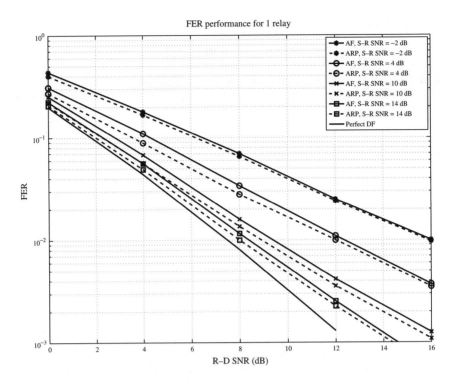

Figure 4.22   FER performance for one relay

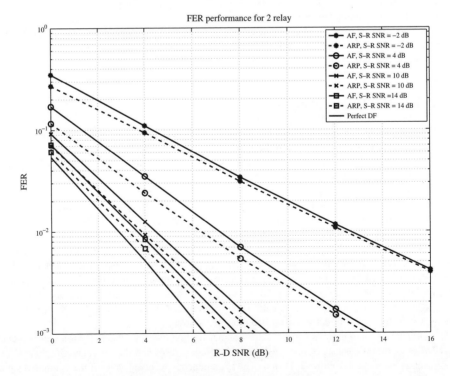

Figure 4.23   FER performance for two relays

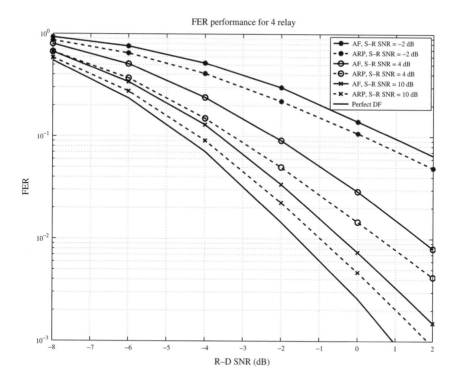

Figure 4.24 FER performance for four relays

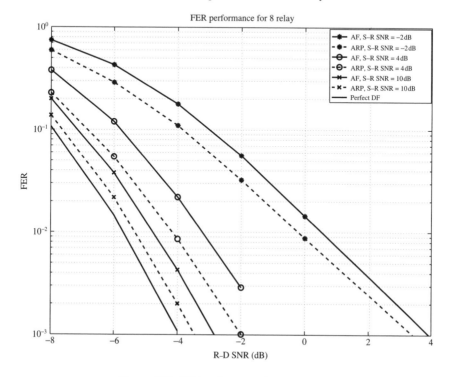

Figure 4.25 FER performance for eight relays

at the destination is given by:

$$\gamma_{S-DF,(n-q)} = \gamma_{sd} + \gamma_{DF,(n-q)}, \tag{4.125}$$

where $\gamma_{sd} = \overline{\gamma}_{SD}|h_{SD}|^2$ and $\gamma_{DF,(n-q)} = \overline{\gamma}_{RD}\sum_{i=q+1}^{n}|h_{RD}^i|^2$. Then the conditional PEP of incorrectly decoding a codeword into another codeword with Hamming distance of d for this scenario, denoted by $P_{(q)}^{S-DF}(d|h_{SD}, \mathbf{h_{SR}}, \mathbf{h_{RD}})$, can be calculated as:

$$P_{(q)}^{S-DF}(d|h_{SD}, \mathbf{h_{SR}}, \mathbf{h_{RD}}) \leq \prod_{i=1}^{q} P_{F,sr}^i\left(\overline{\gamma}_{SR}|h_{SR}^i\right) \tag{4.126}$$

$$\times \prod_{i=q+1}^{n}\left(1 - P_{F,sr}^i\left(\overline{\gamma}_{SR}|h_{sr}^i\right)\right) Q\left(\sqrt{2d\gamma_{sd} + 2d\gamma_{DF,(n-q)}}\right).$$

The average PEP at high SNR, denoted by $P^{S-DF}(d)$, can thus be calculated as:

$$P^{S-DF}(d) \leq \sum_{q=0}^{n}\binom{n}{q}\mathbb{E}\left\{\prod_{i=1}^{q} P_{F,sr}^i(\overline{\gamma}_{SR}|h_{SR}^i)\right. \tag{4.127a}$$

$$\left.\times \prod_{i=q+1}^{n}\left(1 - P_{F,sr}^i\left(\overline{\gamma}_{SR}|h_{SR}^i\right)\right) Q\left(\sqrt{2d\gamma_{sd} + 2d\gamma_{DF,(n-q)}}\right)\right\}$$

$$\leq (d\overline{\gamma}_{SD})^{-1}\sum_{q=0}^{n}\binom{n}{q}\left(P_{F,sr}\right)^q\left(\frac{1 - P_{F,sr}}{d\overline{\gamma}_{RD}}\right)^{n-q} \tag{4.127b}$$

$$= (d\overline{\gamma}_{SD})^{-1}\left(P_{F,sr} + \frac{1 - P_{F,sr}}{d\overline{\gamma}_{RD}}\right)^n \tag{4.127c}$$

$$\approx \overline{\gamma}_{SD}^{-1}\overline{\gamma}_{RD}^{-n}d^{-(n+1)}\left(P_{F,sr}d\overline{\gamma}_{RD} + 1\right)^n \tag{4.127d}$$

$$= G_{S-DF}P_{DF}^{Perfect}(d), \tag{4.127e}$$

where $P_{F,sr} = \overline{\gamma}_{SR}^{-1}\sum_{d_{sr}^H=d_{sr,\min}}^{2l} A(d_{sr}^H)(d_{sr}^H)^{-1}$. Furthermore,

$$G_{S-DF} = \left(P_{F,sr}d\overline{\gamma}_{RD} + 1\right)^n > 1 \tag{4.128}$$

represents the performance loss of the S-DF compared with the perfect DF. As $\overline{\gamma}_{RD} \to \infty$, $P_{F,sr} \to 0$ and the performance of the S-DF approaches the perfect DF. Comparing Equations (4.127) and (4.121), we can easily verify that $P^{S-DF}(d) > P^{ARP}(d)$. That is, the performance of S-DF is always worse than that of the ARP scheme. The performance loss is due to the fact that in the S-DF protocol unsuccessfully decoded relays do not forward the signals. As a result, the SNR of the combined signal at the destination in the S-DF is always lower than that of the ARP, as can been seen from Equations (4.115) and (4.125).

4.3   Distributed Space–Time Coding

A general topology of a regenerative distributed space–time system is depicted in Figure 4.26. This general topology essentially realizes a general regenerative distributed space–time topology with one (or generally several but interfering) information flow(s). Again, each node may have multiple antennas. Compared with the regenerative relaying case discussed in Section 4.2, there is now a formation of virtual antenna arrays (VAA); compared with the transparent case already discussed in

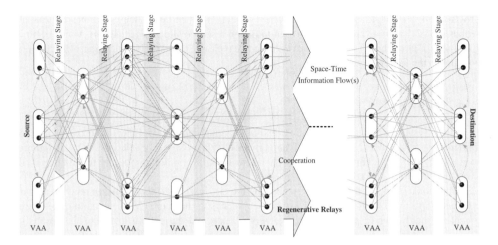

Figure 4.26 General regenerative space–time processing topology with (interfering) information flow(s), realizing a multihop distributed space–time topology

Chapter 3, there are now source and destination VAAs and cooperation between nodes of the same kind is not only possible but also encouraged.

This therefore facilitates the deployment of truly distributed space–time protocols. In Section 4.2, we have discussed various regenerative relaying protocols. Unlike the transparent relaying protocols, in regenerative relaying protocols, the relays try to regenerate the original symbols transmitted by the source and construct new symbols. To provide the system with more coding advantages, the relay nodes can transmit additional incremental redundancy. For the optimal design, the coding at the source and relay should be designed jointly. Such joint coding schemes are referred to as distributed coding.

Designing distributed coding is quite different from the ordinary code design duo for the distributed nature of relay networks. In distributed coding, the overall codeword is constructed in a distributed manner. That is, different parts of the codeword in distributed coding are transmitted by different nodes through different wireless links. This will create an additional freedom, but will also present a great challenge for code construction. Even though we can directly apply the concepts of channel and space–time coding to construct distributed codes in wireless relay networks, some practical issues in implementing these coding schemes have to be taken into account in the distributed code design, such as decoding errors at relays, channel variations in different parts of codeword, and rate and power allocations at the source and relays, etc. The main challenge is to discover the criteria and derive the approach for constructing efficient codes at the source and relays when taking into consideration these factors.

In this section, we present the design and analysis of various distributed coding schemes, which have been successively developed in the past for wireless relay networks. We will mainly concentrate on a distributed space–time block, distributed space–time trellis and distributed turbo coding. In Section 4.4, we will dwell on more advanced and recently emerged topics of distributed network–channel coding and network coding division multiplexing.

4.3.1 Distributed Space–Time Block Coding

Following the approach outlined elsewhere [51, 467], the aim here is to derive fractional resource allocation strategies tailored to distributed multihop networks utilizing estimate-and-forward (EF) protocols. Of prime interest is the derivation of fractional frame duration, power and modulation order for each relaying stage to achieve maximum end-to-end throughput.

To this end, we will first dwell on the system model. We then derive the error rates for spatially distributed STBCs. These are eventually used to obtain resource allocation strategies, which optimize the end-to-end throughput for topologies with complete as well as partial cooperation between nodes belonging to the same relaying stage.

4.3.1.1 System Model

The general system model obeys the same topology as that depicted in Figure 4.26 that is a source communicates with a destination via a given number of relays. Spatially adjacent relays are grouped into relaying Virtual Antenna Arrays (VAAs), where we will briefly describe the functioning of the transmitting, relaying and receiving VAA stages.

The functional blocks of the transceivers forming the distributed-MIMO multistage relaying network are depicted in Figure 4.27. The top of Figure 4.27 relates to the source VAA containing the source; the center panel relates to an arbitrary relaying VAA tier; and the bottom relates to the destination VAA containing the destination. In the figure, each VAA tier is shown to consist of three terminals; it is, however, understood that any reasonable number of terminals can be accommodated. The core blocks are:

- **Source VAA.** Specifically, the information source passes the information to a cooperative transceiver, which relays the data to spatially adjacent relays belonging to the same VAA. In contrast to other protocols dealt with in this book, this communication is assumed to happen over an air interface distinct from the interface used for interstage communication or an air interface not requiring any optimization, and is not considered further. It is also assumed that these cooperative links are error free due to the short communication distances. Each of the terminals in the VAA perform distributed space–time block encoding of the information according to some prior specified codebook. That information is then transmitted from the spatially distributed terminals after having been synchronized. Note that the problem related to synchronization is beyond the scope of this section but is increasingly dealt with in the literature [388, 468–479].
- **Relaying VAA.** Any of the relaying VAA tiers functions as follows. First, each relay within that VAA receives the data, which is optionally decoded before being passed on to the cooperative transceiver. Ideally, every terminal cooperates with every other terminal; however, any degree of cooperation is feasible. If no decoding is performed, then an unprocessed or a sampled version of the received signal is exchanged with the other relays. Note that unprocessed relaying is equivalent to transparent relaying. After cooperation, appropriate decoding is performed. The obtained information is then re-encoded in a distributed manner, synchronized and retransmitted to the subsequent relaying VAA tier.
- **Destination VAA.** As for the destination VAA, the functional blocks are exactly the opposite to the source VAA. All terminals receive the information, possibly decode it, then pass it onto the cooperative transceivers, which relay the data to the target terminal. The data is processed and finally delivered to the information sink.

The functional blocks of the distributed transcoder, that is encoder and decoder, are now elaborated on in more detail. To this end, the encoder and decoder are shown in Figure 4.28 and described as follows:

- **Distributed Encoder.** A channel encoder within a distributed encoder does not normally differ from a nondistributed encoder; however, as has become evident throughout this book, it is generally possible and advisable to design channel codes that reflect the distributed nature of the encoding process. The role of a space–time encoder is to utilize the additional spatial dimension created by the sufficiently spaced antenna elements to increase the system performance. The functionality of distributed space–time codes (STCs) differs from a traditional deployment because only a fraction

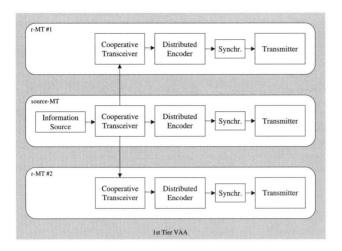

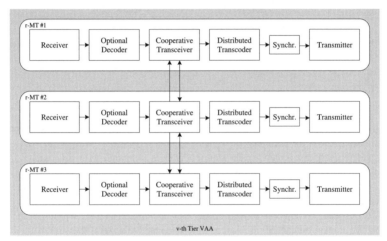

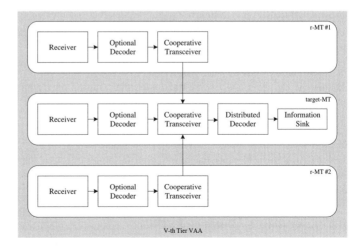

Figure 4.27 Functional blocks of the source VAA (top), the vth relaying VAA (center) and the target VAA (bottom)

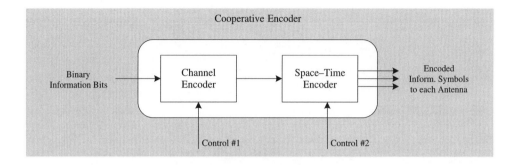

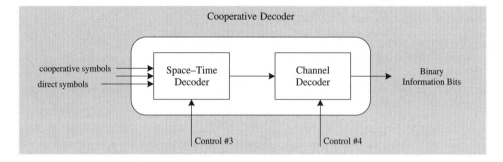

Figure 4.28   Distributed encoder and decoder

of the entire space–time codeword is transmitted from any of the spatially distributed terminals. The transmission across all terminals then yields the complete space–time codeword. Therefore, a control signal to each distributed space–time encoder is essential, as it tells each of them which fraction of the entire space–time codeword to pass onto the transmitting antenna(s). This is indicated as Control #2 in Figure 4.28. This control information is assumed to be available to the space–time encoder, and is hence not discussed further.

- **Distributed Decoder.** The cooperative decoder can be realized as the inversion of all processes at the cooperative transmitter. Here, the space–time decoder is fed with the signals directly received from the available antenna(s), as well as the information received via the cooperative links from adjacent terminals. Again, a control signal is needed that specifies the type of information fed into the space–time decoder, to allow for optimum decoding. For example, the control signal could inform the decoder that the relayed signals are a one bit representation of the sampled soft information available at the respective cooperative relaying terminals. After the space–time decoding process, the information is passed on to the channel decoder, which performs the inverse process to the channel encoder. In a cooperative transcoder, the resulting binary information output may then be fed into the cooperative encoder, to get relayed to the next VAA tier.

In the following analysis, a relaying access scheme based on TDMA is assumed. Therefore, the entire bandwidth W is utilized by all relaying links, whereas only a fraction of the total frame duration T is used by each stage to relay the information to the consecutive stage. We now first characterize the error rates of a distributed STBC, before using these expressions to derive near-optimal resource allocation strategies.

4.3.1.2   Error Rates for Distributed STBCs

The space–time encoder, as depicted in Figure 4.28, is realized by means of a traditional STBC, which has not been further adapted to the distributed topology. This is clearly sub-optimum but yields interesting insights into the behavior of the said systems. With the error expression of modified distributed STBCs at hand, the same analysis can be repeated to obtain suitable resource allocation strategies.

A typical space-time encoded MIMO system with t transmit and r receive antennas is shown in Figure 4.29. Here, $b \cdot s$ information bits are fed into the modulator, which Gray-maps $b = \log_2 M$ consecutive bits onto an M-PSK or M-QAM signal constellation, thereby producing s symbols, that is $x_1, x_2, \dots, x_s$. To remind the reader, these are subsequently space–time encoded with an orthogonal space–time coding matrix $\mathcal{G}$ of size $d \times t$, where d is the number of symbol durations required to transmit the space–time code word, and t is the number of transmit elements. At each time instant $k = 1, \dots, d$, the space–time encoded symbol $c_{k,i} \in \mathcal{G}$ is transmitted from the ith transmit element, where $i = 1, \dots, t$. The reduction in transmission rate is thus $R = s/d$. The space–time code generator matrix $\mathcal{G}$ therefore maps the symbols $x_1, x_2, \dots, x_s$ onto a transmitted space–time matrix $\mathbf{X}$ of dimensions $t \times d$, that is:

$$x_1, x_2, \dots, x_s \overset{\mathcal{G}}{\longmapsto} \mathbf{X}, \tag{4.129}$$

which is transmitted over a flat fading $r \times t$ space–time channel. The latter can be cast into a matrix $\mathbf{H}$ given in Chapter 2, which allows the received signal to be written in matrix form [480]:

$$\mathbf{Y} = \mathbf{HX} + \mathbf{N}, \tag{4.130}$$

with $\mathbf{N}$ being the $r \times d$ receive noise matrix. The covariance matrix of the noise obeys $\mathrm{E}\left\{\mathbf{NN}^H\right\} = d \cdot N \cdot \mathbf{I}_{r \times r}$, where N is the total noise power per sample in space and time. Under the condition of perfect CSI at the receiver, the problem of detecting $\mathbf{X}$ given $\mathbf{Y}$ has been shown [121, 480] to be equal to minimizing the ML decision metric $\|\mathbf{Y} - \mathbf{HX}\|$ over all possible symbols $x_1, x_2, \dots, x_s$. The complexity therefore increases linearly with the total number of antennas $r \times t$ and exponentially with the modulation order M and the codeword duration d. However, it has been proved [480] that the orthogonality of the space–time code generator matrix $\mathcal{G}$ allows the ML problem to be broken down into s parallel ML decision metrics for each of the originally sent symbols $x_{l \in (1,s)}$. The optimum decision metrics can be found in Larsson and Stoica [480, page 102]. The complexity therefore increases exponentially only with the modulation order M, which constitutes a great simplification of the detection process.

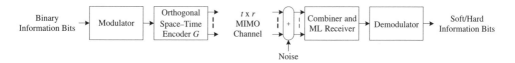

Figure 4.29   A space–time block encoded MIMO system (O-MIMO)

Using Larsson and Stoica's Theorem 7.3 [480], it can been shown that the instantaneous SNR ρ per symbol $x_{l \in (1,s)}$ at detection is given as:

$$\rho = \frac{1}{R} \frac{\lambda}{t} \frac{S}{N} = \frac{\lambda}{t} \frac{E_s}{N_0} = \log_2(M) \frac{\lambda}{t} \frac{E_b}{N_0}, \tag{4.131}$$

where $\lambda \triangleq \|\mathbf{H}\|^2$, S is the average transmitted signal power, E_s is the average transmitted symbol energy, E_b is the average transmitted bit energy, N is the average receiver power, and N_0 is the average receiver noise power density. From above, the MGF can be derived as:

$$\phi_\rho(s) \equiv \phi_{\frac{1}{R} \frac{\lambda}{t} \frac{S}{N}}(s). \tag{4.132}$$

The closed solution to the symbol error rate for generic M-PSK and M-QAM schemes is based Hasna and Alouini [271], and Shin and Lee [481]. From Hasna and Alouini [271], one can obtain the average

SER for coherent M-PSK to be:

$$P_s(e) = \frac{1}{\pi} \int_0^{\pi \frac{M-1}{M}} \phi_\rho \left(\frac{g_{PSK}}{\sin^2 \theta} \right) d\theta,$$ (4.133)

where $g_{PSK} \triangleq \sin^2(\pi/M)$. Considering Equation (4.132), this can be rewritten as:

$$P_s(e) = \frac{1}{\pi} \int_0^{\pi \frac{M-1}{M}} \phi_{\frac{1}{R} \frac{\lambda}{t} \frac{S}{N}} \left(\frac{g_{PSK}}{\sin^2 \theta} \right) d\theta.$$ (4.134)

Similarly, the SER for coherent M-QAM is given as [271]:

$$P_s(e) = \frac{4q}{\pi} \int_0^{\pi/2} \phi_{\frac{1}{R} \frac{\lambda}{t} \frac{S}{N}} \left(\frac{g_{QAM}}{\sin^2 \theta} \right) d\theta - \frac{4q^2}{\pi} \int_0^{\pi/4} \phi_{\frac{1}{R} \frac{\lambda}{t} \frac{S}{N}} \left(\frac{g_{QAM}}{\sin^2 \theta} \right) d\theta,$$ (4.135)

where $g_{QAM} \triangleq 3/2/(M-1)$ and $q \triangleq 1 - 1/\sqrt{M}$. The SERs are now obtained in closed form for various orthogonal STBC MIMO communication scenarios.

(i) Rayleigh Fading–Equal Subchannel Gains. For equal subchannel gains $\gamma_1 = \ldots = \gamma_u \triangleq \gamma$, the MGF of the instantaneously experienced SNR can be expressed as:

$$\phi_{\frac{1}{R} \frac{\lambda}{t} \frac{S}{N}}(s) = \frac{1}{\left(1 - \frac{1}{R} \frac{\gamma}{t} \frac{S}{N} \cdot s \right)^u}$$ (4.136)

where $u \triangleq t \cdot r$. SER of M-PSK the can be expressed [467, 481] in closed form as:

$$P_s(e) = \phi_{\frac{1}{R} \frac{\lambda}{t} \frac{S}{N}}(-g_{PSK}) \left[\frac{1}{2\sqrt{\pi}} \frac{\Gamma(u+1/2)}{\Gamma(u+1)} \, {}_2F_1 \left(u, 1/2; u+1; \left(1 + \frac{g_{PSK}}{R} \frac{\gamma}{t} \frac{S}{N} \right)^{-1} \right) \right.$$

$$\left. + \frac{\sqrt{1-g_{PSK}}}{\pi} F_1 \left(1/2, u, 1/2-u; 3/2; \frac{1-g_{PSK}}{1 + \frac{g_{PSK}}{R} \frac{\gamma}{t} \frac{S}{N}}, 1-g_{PSK} \right) \right]$$ (4.137)

where ${}_2F_1(\cdot, \cdot; \cdot; \cdot)$ is the generalized hypergeometric function with two parameters of type 1 and one parameter of type 2 [268] (Section 9.14.1); it is sometimes referred to as the Gauss hypergeometric function [268] (Section 9.14.2). The function $F_1(\cdot, \cdot, \cdot; \cdot; \cdot, \cdot)$ is a hypergeometric function of two variables [268] (Section 9.180.1); it is sometimes referred to as the Appell hypergeometric function. To simplify subsequent analysis, Equation (4.137) is denoted as $P_{PSK}(u, t, R, \gamma, S/N, M)$. Similarly, the SER of M-QAM is shown to be [467, 481]:

$$P_s(e) = \phi_{\frac{1}{R} \frac{\lambda}{t} \frac{S}{N}}(-g_{QAM}) \frac{2q}{\sqrt{\pi}} \frac{\Gamma(u+1/2)}{\Gamma(u+1)} \, {}_2F_1 \left(u, 1/2; u+1; \left(1 + \frac{g_{QAM}}{R} \frac{\gamma}{t} \frac{S}{N} \right)^{-1} \right)$$

$$- \phi_{\frac{1}{R} \frac{\lambda}{t} \frac{S}{N}}(-2g_{QAM}) \frac{2q^2}{\pi(2u+1)} F_1 \left(1, u, 1; u+3/2; \frac{1 + \frac{g_{QAM}}{R} \frac{\gamma}{t} \frac{S}{N}}{1 + 2\frac{g_{QAM}}{R} \frac{\gamma}{t} \frac{S}{N}}, 1/2 \right)$$ (4.138)

To simplify later analysis, Equation (4.138) is denoted as $P_{QAM}(u, t, R, \gamma, S/N, M)$.

(ii) Rayleigh Fading–Unequal Subchannel Gains. The MGF has been shown to be [467]:

$$\phi_{\frac{1}{R} \frac{\lambda}{t} \frac{S}{N}}(s) = \sum_{i=1}^{u} K_i \cdot \phi_{\frac{1}{R} \frac{\lambda_i}{t} \frac{S}{N}}(s)$$ (4.139)

with constants K_i

$$K_i = \prod_{i'=1,i'\neq i}^{u} \frac{\gamma_i}{\gamma_i - \gamma_{i'}} \tag{4.140}$$

where γ_i is the average channel gain of the ith path. This allows us to derive the closed form SER for distributed STBCs, where all the channel gains differ. With the simplified notation, the respective error rates can be expressed as [467]:

$$P_s(e) = \sum_{i=1}^{u} K_i \cdot P_{\text{PSK/QAM}}(1, t, R, \gamma_i, S/N, M). \tag{4.141}$$

(iii) Rayleigh Fading–Generic Subchannel Gains. In this case there are $g \leq u$ subchannel gains with distinct SNR, which are referred to as $\hat{\gamma}_{i\in(1,g)}$ with each of them being repeated $v_{i\in(1,g)}$ times. In this case, the MGF was shown to be:

$$\phi_{\frac{1}{R}\frac{\lambda}{t}\frac{S}{N}}(s) = \sum_{i=1}^{g}\sum_{j=1}^{v_g} K_{i,j} \cdot \phi_{\frac{1}{R}\frac{\hat{\lambda}_i}{t}\frac{S}{N}}^{j}(s) \tag{4.142}$$

where the coefficients $K_{i,j}$ are given as [467]:

$$K_{i,j} = \frac{1}{(v_i - j)!\left(-\hat{\gamma}_i\right)^{v_i-j}} \frac{\partial^{v_i-j}}{\partial s^{v_i-j}} \left[\prod_{\substack{i'=1,\\i'\neq i}}^{g} \frac{1}{\left(1-s\hat{\gamma}_{i'}\right)^{v_{i'}}}\right]_{s=1/\hat{\gamma}_i} . \tag{4.143}$$

The respective error rates can now be expressed as:

$$P_s(e) = \sum_{i=1}^{g}\sum_{j=1}^{v_g} K_{i,j} \cdot P_{\text{PSK/QAM}}(j, t, R, \hat{\gamma}_i, S/N, M). \tag{4.144}$$

(iv) Nakagami Fading–Equal Subchannel Gains. Given a Nakagami fading factor m, the respective error rates can be similarly derived as above, yielding:

$$P_s(e) = P_{\text{PSK/QAM}}(mu, mt, R, \gamma, S/N, M). \tag{4.145}$$

(v) Nakagami Fading–Unequal Subchannel Gains. The respective error rates for a Nakagami fading channel with different subchannel gains $\gamma_{i\in(1,u)}$ and different fading factors $m_{i\in(1,u)}$ can be derived as above to yield:

$$P_s(e) = \sum_{i=1}^{u}\sum_{j=1}^{f_i} K_{i,j} \cdot P_{\text{PSK/QAM}}(j, jt, R, \gamma_i, S/N, M) \tag{4.146}$$

where the coefficients $K_{i,j}$ are obtained by performing partial fractions on the MGF of the Nakagami fading channel [467].

(vi) Nakagami Fading–Generic Subchannel Gains. The respective SERs for the case of generic subchannel coefficients are obtained similarly as for the Rayleigh channel with unequal subchannel gains, and are therefore omitted here.

4.3.1.3 Maximum Throughput for End-to-End Transmission

The error rates obtained in the previous section are utilized here to derive fractional resource allocation rules, assuming that a decision on the correctness of the received signal is carried out at the destination and that the relays use the estimate-and-forward (EF) or demodulate-and-forward protocol. This should not be confused with transparent relaying, where the information is simply amplified and forwarded. It is also in contrast to the DF protocols and derivatives, where a decision on the correctness of the received signal is done at each stage.

If all relays per stage cooperate and cooperation takes place at a sufficiently high SNR, then the signal samples from the previous stage are the same for all relays. Therefore, if an error occurs in the signal from the previous stage, then that error is the same in all relays belonging to the same stage.

Such a scenario greatly simplifies to the analysis, since the errors in consecutive stages become independent. This is in contrast to a generic relaying process with partial cooperation (clustering), where one relay may have a more reliable estimate than another relay in the same relaying VAA, leading to error dependencies between the stages.

Subsequently, the problem of maximizing the end-to-end throughput is shown to be equivalent to the problem of minimizing the end-to-end BER. The fractional resource allocation rules are then derived for the cases of full and partial cooperation.

It is assumed here that the source MS (source) transmits B bits per frame to the target MS (destination) via K relaying stages. The normalized end-to-end throughput can be expressed as [467]:

$$\Theta = \min_{v \in (1,K)} \left\{ \alpha'_v R_v \log_2(M_v) \right\} \cdot \left(1 - P_{f,\text{e2e}}(e) \right), \tag{4.147}$$

where α'_v, R_v and M_v are the fractional frame duration, STBC rate and modulation index of the vth stage respectively, and $P_{f,\text{e2e}}(e)$ is the end-to-end FER. Equation (4.147) has to be understood as follows. If there were no losses between a directly communicating source and destination, then all of the B bits would reach the receiver; the throughput normalized by the total number of sent bits hence amounts to 1. The use of a modulation scheme with index M and a STBC with rate R during a fractional frame duration α' to accomplish such a link results in a throughput, normalized by the utilized time and frequency, as $1 \cdot \alpha' \cdot R \cdot \log_2(M)$ [bits/s/Hz]. It is then diminished by the loss caused by the end-to-end FER $P_{f,\text{e2e}}(e)$. For a communication system with K relaying stages, the weakest link in the chain determines the throughput, hence $\min_{v \in (1,K)} \left\{ \alpha'_v R_v \log_2(M_v) \right\}$. It is thus the aim to derive optimum resource allocation strategies, which maximize the end-to-end throughput.

To this end, note that $P_{f,\text{e2e}}(e)$ is a function of $M_{v \in (1,K)}$ and $R_{v \in (1,K)}$ (hence also $\min_{v \in (1,K)}$ $\left\{ \alpha'_v R_v \log_2(M_v) \right\}$) and the fractional transmission power allocated to each stage. Optimizing Equation (4.147) w.r.t. these parameters is very complex, which is the reason why the optimization process is performed in three stages:

- **First Stage.** The modulation indices $M_{v \in (1,K)}$ are fixed first and the limiting case where SNR $\to \infty$ is considered. This reduces Equation (4.147) to:

$$\Theta = \min_{v \in (1,K)} \left\{ \alpha'_v R_v \log_2(M_v) \right\}, \tag{4.148}$$

where the fractional frame durations α'_v need to be chosen such as to maximize Θ under constraint $\sum_{v=1}^{K} \alpha'_v = 1$. This is clearly achieved by equating all $\alpha'_v R_v \log_2(M_v)$, which results in:

$$\alpha'_v = \frac{\prod_{w=1, w \neq v}^{K} R_w \cdot \log_2(M_w)}{\sum_{k=1}^{K} \prod_{w=1, w \neq k}^{K} R_w \cdot \log_2(M_w)}. \tag{4.149}$$

- **Second Stage.** It is shown now that under the current assumptions, the throughput is maximized by minimizing the end-to-end BER. To this end, the normalized throughput is expressed in terms of the end-to-end BER $P_{b,\text{e2e}}(e)$ which, assuming independent bit errors, gives:

$$\Theta = \min_{v \in (1,K)} \left\{ \alpha'_v R_v \log_2(M_v) \right\} \cdot \left(1 - P_{b,\text{e2e}}(e) \right)^B. \tag{4.150}$$

Here, B is the number of bits per frame. Defining a nondecreasing (in Θ) metric

$$\Theta' \triangleq \exp\left[\frac{\log\left(\Theta / \min_{v \in (1,K)} \left\{ \alpha'_v R_v \log_2(M_v) \right\} \right)}{B} \right] - 1, \tag{4.151}$$

it can easily be shown that:

$$\Theta' \approx -P_{b,\text{e2e}}(e). \tag{4.152}$$

Therefore, to maximize the end-to-end throughput, one has to minimize the end-to-end bit error rate by optimally assigning fractional transmission power to each relaying stage. The BER at each stage is related with the occurring SER via [278]:

$$P_b(e) \approx \frac{P_s(e)}{\log_2(M)} \tag{4.153}$$

at low BERs or SERs, where one symbol error causes one bit error.

- **Third Stage.** The optimum modulation order $M_{v \in (1,K)}$ has to be determined depending on the previously derived fractional resource allocations. This is easily done by permuting all possible modulation orders at each stage such as to maximize the end-to-end throughput. Since the number of modulation orders will be limited, such optimization is feasible without consuming too much computational power.

Subsequently, the second step is performed assuming either total or partial (clustered) cooperation in each stage. The near-optimum fractional power allocation rules are first derived, then assessed in terms of their precision; finally, the maximum achievable throughput will be illustrated by means of a few examples.

4.3.1.4 Full Cooperation at Each Stage

Under the assumption of full cooperation, each of the K relaying stages experiences independent BERs $P_{b,v \in (1,K)}(e)$ caused by independent SERs $P_{s,v \in (1,K)}(e)$. A bit from the source is received correctly at the destination only when at all stages the bit has been transmitted correctly.[1] The end-to-end BER can therefore be expressed as:

$$P_{b,\text{e2e}}(e) = 1 - \prod_{v=1}^{K} \left(1 - P_{b,v}(e) \right), \tag{4.154}$$

which, at low BERs at every stage, can be approximated as:

$$P_{b,\text{e2e}}(e) \approx \sum_{v=1}^{K} P_{b,v}(e) \tag{4.155}$$

$$\approx \sum_{v=1}^{K} \frac{P_{s,v}(e)}{\log_2(M_v)}, \tag{4.156}$$

where M_v is the modulation order at the vth stage. Further analysis concentrates on the case of Rayleigh fading with equal channel gains per relaying stage; other cases are a straightforward extension to

---

[1] The cases where two or more wrong bits may result again in a correct bit is neglected here.

the exposed analysis. Assuming that each stage is allocated a fractional power β'_v, the above-given dependency can be expressed as:

$$P_{b,\text{e2e}}(e) \approx \sum_{v=1}^{K} \frac{P_{s,v}\left(u_v, t_v, R_v, \gamma_v, \beta'_v \cdot S/N, M_v\right)}{\log_2\left(M_v\right)} \tag{4.157}$$

where the SERs $P_{s,v}(\cdot) = P_{\text{PSK/QAM}}(\cdot)$ are given by Equations (4.137) and (4.138), respectively. Furthermore, $u_v \triangleq t_v \cdot r_v$, t_v and r_v are the number of transmit and receive antennas in the vth stage, R_v is the rate of the STBC, γ_v is the average attenuation experienced, S is the total power given to deliver the information from source to sink, and N is the noise power.

The optimization process has only to be performed w.r.t. the fractional power allocation $\beta'_{v \in (1,K)}$. Even so, the optimization process is very intricate. To simplify analysis further, an upper bound to the derived SERs for M-PSK and M-QAM is invoked. To this end, the integrant of Equation (4.134) is upper-bounded by its largest argument, which occurs at $\theta = \pi/2$. The SER for M-PSK in the vth relaying stage is hence upper-bounded as:

$$P_{s,v}(e) \leq \frac{\frac{M_v - 1}{M_v}}{\left(1 + \beta'_v \frac{g_{\text{PSK},v}}{R_v} \frac{\gamma_v}{t_v} \frac{S}{N}\right)^{u_v}}, \tag{4.158}$$

where $g_{\text{PSK},v} = \sin^2(\pi/M_v)$. The end-to-end BER for an M-PSK modulation scheme can finally be upper-bounded as:

$$P_{b,\text{e2e}}(e) \leq \sum_{v=1}^{K} \frac{M_v - 1}{M_v \log_2(M_v)} \left(1 + \beta'_v \frac{g_{\text{PSK},v}}{R_v} \frac{\gamma_v}{t_v} \frac{S}{N}\right)^{-u_v}. \tag{4.159}$$

Following a similar approach, the upper bound for the end-to-end BER of an M-QAM modulation can be derived as:

$$P_{b,\text{e2e}}(e) \leq \sum_{v=1}^{K} \frac{2q_v}{\log_2(M_v)} \left[\left(1 + \beta'_v \frac{g_{\text{QAM},v}}{R_v} \frac{\gamma_v}{t_v} \frac{S}{N}\right)^{-u_v}\right.$$
$$\left. + \frac{q_v}{2} \left(1 + 2\beta'_v \frac{g_{\text{QAM},v}}{R_v} \frac{\gamma_v}{t_v} \frac{S}{N}\right)^{-u_v}\right] \tag{4.160}$$

$$\lesssim \sum_{v=1}^{K} \frac{2q_v}{\log_2(M_v)} \left(1 + \beta'_v \frac{g_{\text{QAM},v}}{R_v} \frac{\gamma_v}{t_v} \frac{S}{N}\right)^{-u_v}, \tag{4.161}$$

where $g_{\text{QAM},v} = 3/2/(M_v - 1)$ and $q_v = 1 - 1/\sqrt{M_v}$. Note that in Equation (4.161) the second summand appearing in Equation (4.160) was neglected due to $q_v/2 < 1$ and $(1 + 2x)^{-u_v}$ being much less than $(1 + x)^{-u_v}$ for x sufficiently large and $u_v \geq 1$. Either modulation scheme results in an upper bound unified below as:

$$P_{b,\text{e2e}}(e) \leq \sum_{v=1}^{K} A_v \left(1 + B_v \beta'_v\right)^{-u_v} \tag{4.162}$$

The constants A_v and B_v are obtained by comparing Equation (4.162) with Equations (4.159) or (4.161) to arrive at:

$$A_v = \begin{cases} \dfrac{M_v - 1}{M_v \log_2(M_v)} & \text{for M-PSK} \\ \dfrac{2q_v}{\log_2(M_v)} & \text{for M-QAM} \end{cases} \tag{4.163}$$

and

$$
B_v = \begin{cases} \dfrac{g_{\text{PSK},v}}{R_v} \dfrac{\gamma_v}{t_v} \dfrac{S}{N} & \text{for M-PSK} \\[3mm] \dfrac{g_{\text{QAM},v}}{R_v} \dfrac{\gamma_v}{t_v} \dfrac{S}{N} & \text{for M-QAM} \end{cases} \tag{4.164}
$$

It has been shown [467] that the fractional power allocations $\beta'_{v \in (1,K)}$ have to obey:

$$
\beta'_v \approx \left[\sum_{w=1}^{K} \alpha'_w \left(\frac{u_v^{-1} A_v^{-1} B_v^{u_v}}{u_w^{-1} A_w^{-1} B_w^{u_w}} \right)^{\frac{1}{u_{\max}+1}} \right]^{-1}, \tag{4.165}
$$

where $u_{\max} = \max(u_1, \ldots, u_K)$. The performance of the developed algorithm is assessed by means of Figures 4.30–4.33 for M-QAM schemes only. Note that if reference is made to the nonoptimized scenario, then only the fractional transmission power is meant not to be optimized since the frame duration is easily related to the modulation order.

Explicitly, Figure 4.30 depicts the end-to-end BER versus the SNR in the first link (in dB) for various two-stage communication scenarios deploying the developed fractional power allocation strategy (Equation 4.165), which is also compared with a numerically obtained optimum and a nonoptimum allocation.

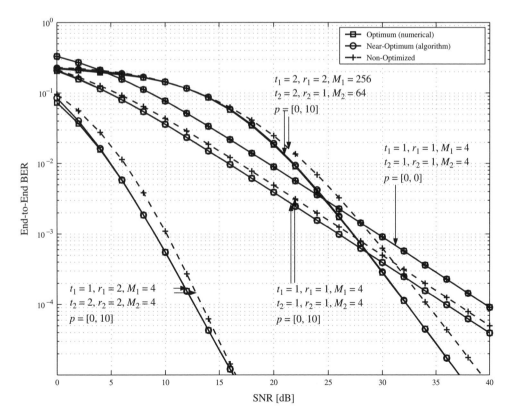

Figure 4.30 Comparison between optimum and near-optimum, as well as nonoptimized end-to-end BER for various configurations of a two-stage relaying network

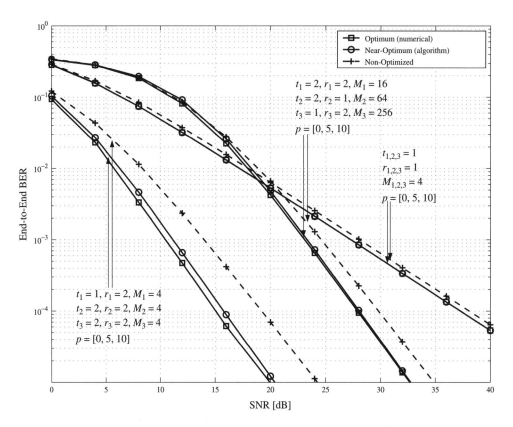

Figure 4.31 Comparison between optimum and near-optimum, as well as nonoptimized end-to-end BER for various configurations of a three-stage relaying network

The first scenario, where $t_{1,2} = r_{1,2} = 1$, $M_{1,2} = 4$ (QPSK) and $p = [0, 0]$ dB, is entirely symmetrical, which leads to the same performance for all three allocation strategies. The second scenario is the same as the first, with the only difference being that the channel in the second stage is now 10-times stronger than in the first stage, that is $p = [0, 10]$ dB. The resulting nonsymmetric scenario reveals a performance difference between the optimized (solid lines) and nonoptimized (dashed line) power allocation.

It can be observed that the optimum and developed allocation strategies yield the same performance for any of the depicted configurations. Furthermore, the gain of an optimized system over a nonoptimized system is highest for very asymmetric cases; here, for $t_1 = 2, r_1 = 2, t_2 = 2, r_2 = 1, M_1 = 256$, $M_2 = 64$ and $p = [0, 10]$ dB. At a target end-to-end BER of 10^{-5}, about 1 dB in power can be saved.

Figure 4.31 is similar in nature to Figure 4.30, with the only difference being that a three-stage network is scrutinized. Similar observations can be made for these scenarios, where gains of almost 4 dB can be observed. This corroborates the importance of the derived allocation strategy.

The throughput of a two-stage system is illustrated by means of Figure 4.32, which utilizes the fractional resource allocation strategies in Equations (4.149) and (4.165). The system deployed has the number of bits fixed to $B = 100$; furthermore, for all configurations $M_{1,2} = 4$ (QPSK) and $p = [0, 10]$ dB. It can be seen that in the region of low SNR, the developed allocation strategy performs worse than the optimum one. This is obvious, as the fractional frame durations have been derived assuming the SNR $\rightarrow \infty$.

For most of the transitional region from zero throughput to maximum throughput, however, the derived allocations yield near optimum throughput. In contrast, no optimization exhibits drastic losses

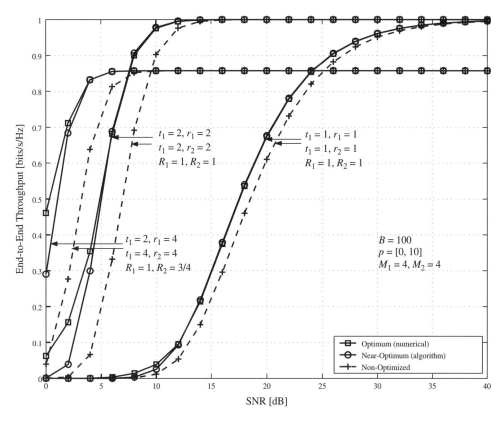

Figure 4.32 Comparison between optimum and near-optimum, as well as nonoptimized end-to-end throughput for various configurations of a two-stage relaying network

in the transitional region. For example, given the scenario with $t_{1,2} = r_{1,2} = 2$ operating at an SNR in the first link of 6 dB, around 0.4 bits/s/Hz are lost which mounts to approximately 40%.

Note also that the cases of full-rate STBC in each stage yield the same maximum throughput, whereas the case with the 3/4-rate STBC has a lower maximum throughput, notwithstanding the fact that it has the strongest link with $(2 \times 4)/(4 \times 4)$. This is due to the limiting spectral efficiency of the STBC with a rate less than unity. Clearly, the strength of the link determines the rate with which the system approaches the limiting throughput for the SNR $\rightarrow \infty$.

The precision of the fractional allocation algorithm for fixed modulation indexes allows a final numerical optimization to be performed in each relaying stage over all possible modulation indexes. The low complexity of Equations (4.149) and (4.165) guarantees that such optimization comes at little additional computational power.

Such numerical optimization was performed for a two-stage network with $p = [0, 10]$ dB and $t_{1,2} = r_{1,2} = 2$. Each stage could choose a modulation index belonging to the set $M_{1,2} = (2, 4, 16, 64, 256)$; this leads to 25 possible combinations, which are calculated in a fraction of a second. The performance gains in terms of increased throughput are clear from Figure 4.33, where the near-optimum adaptive modulation per stage is compared with various fixed combinations. At any SNR, the developed algorithm clearly outperforms any of the fixed configurations. For example, if the system were to operate at an SNR in the first link of 20 dB, then the best but fixed modulation index can only reach 2 bits/s/Hz; in this case either 16-QAM or 64-QAM in both stages. The optimum selection is 64-QAM with an optimized fractional power allocation, which yields a performance benefit of 30%.

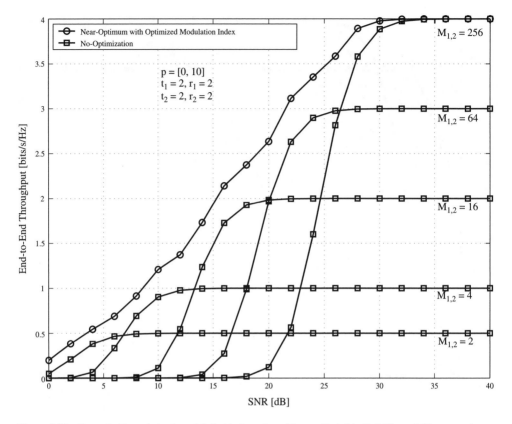

Figure 4.33 Numerically optimized modulation index where $M_{1,2} = (2, 4, 16, 64, 256)$ to yield near-optimum end-to-end throughput, compared with nonoptimized systems

4.3.1.5 Partial Cooperation at Each Stage

Partial cooperation at each relaying stage results in parallel MIMO channels, all possibly of different strengths. An example of such a clustering process is depicted in Figure 4.34 where none of the involved relays is cooperating with any other. Here, the first stage spans two independent SISO channels with average attenuation $\gamma_{1,1}$ and $\gamma_{1,2}$, respectively. Each of these channels causes independent BERs, denoted as $P_{1,1}$ and $P_{1,2}$, respectively. Similarly, the second stage spans two independent MISO channels, where the first MISO channel consists of channels with average attenuations $\gamma_{2,1}$ and $\gamma_{2,3}$, and the second MISO channel consists of channels with average attenuations $\gamma_{2,2}$ and $\gamma_{2,4}$. Furthermore, assuming an error free input into the second VAA relaying tier, the BERs at the output of the MISO channels are $P_{2,1}$ and $P_{2,2}$. Finally, the third stage spans a single MISO channel with a BER $P_{3,1}$.

Note that the relays belonging to the same stage need to communicate at the same rate; furthermore, they obviously need to know which part of the space–time block code to transmit. Although already previously stated, it is assumed that synchronization among all terminals is perfect.

To obtain the exact end-to-end BER is not trivial, as an error in the first stage may propagate to the destination; however, it may also be corrected at the next stage. Referring to Figure 4.34, for example, it is assumed that the same information bit is erroneously received over the link denoted as (1, 1) and correctly for (1, 2). Then, the STBC formed by (2,1) and (2,3) has as its input one erroneous and one correct information bit. Assuming that $\gamma_{2,3} \gg \gamma_{2,1}$, then the error does not propagate further since it

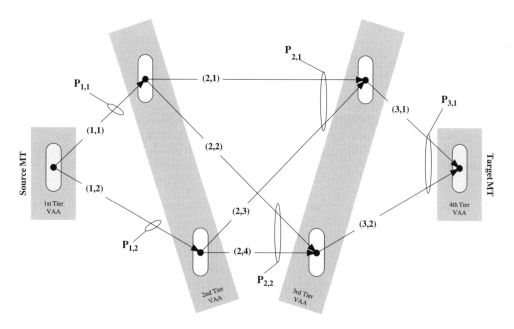

Figure 4.34   Three-stage distributed STBC MIMO communication system without cooperation

will be outweighed by the correct bit. Alternatively, if $\gamma_{2,3} \ll \gamma_{2,1}$, then there is a large likelihood that the error propagates. This creates dependencies between the error events at each stage in dependency of the modulation scheme used, the prevailing channel statistics, and the average channel attenuations, as well as the STBC chosen. These fairly complex interdependencies call for suitable simplifications, which are exposed and justified below.

Generally, it is desirable to develop an approximation that decouples the error events at the respective stages. To this end, it is assumed that the system operates at low error rates that cause only one error event at a time in the entire network. Let us assume that an error occurs in link (1,1); however, (1,2) is error free. Then the probability that the error propagates further is related to the strengths of channels (2,1) and (2,3). Corroborated by the elegant asymptotic analysis of Ribeiro et al. and Wang and Giannakis [275, 383], the probability that such error propagates is proportional to the strength of the STBC branch it departs from, here (2,1) for one of two MISO channels, and (2,2) for the other one. Therefore, the probability that an error that occurred in link (1,1) with probability $P_{1,1}$ propagates through the O-MISO channel spanned by (2,1) and (2,3) is approximated as $P_{1,1} \cdot \gamma_{2,1}/(\gamma_{2,1} + \gamma_{2,3})$, where the strength of the erroneous channel (2,1) is normalized by the total strength of both subchannels. To capture the probability that such an error propagates until the destination, all possible paths in the network have to be found and the original probability of error weighed with the ratios between the respective path gains.

Taking the above-said into account and assuming that at high SNRs, only one such error will occur at any link, the end-to-end BER for the network depicted in Figure 4.34 can be expressed as:

$$P_{b,\text{e2e}}(e) \approx \left[P_{1,1}(e) \left(\frac{\gamma_{2,1}}{\gamma_{2,1} + \gamma_{2,3}} \frac{\gamma_{3,1}}{\gamma_{3,1} + \gamma_{3,2}} + \frac{\gamma_{2,2}}{\gamma_{2,2} + \gamma_{2,4}} \frac{\gamma_{3,2}}{\gamma_{3,1} + \gamma_{3,2}} \right) \right.$$

$$\left. + P_{1,2}(e) \left(\frac{\gamma_{2,4}}{\gamma_{2,2} + \gamma_{2,4}} \frac{\gamma_{3,2}}{\gamma_{3,1} + \gamma_{3,2}} + \frac{\gamma_{2,3}}{\gamma_{2,1} + \gamma_{2,3}} \frac{\gamma_{3,1}}{\gamma_{3,1} + \gamma_{3,2}} \right) \right]$$

$$+ \left[P_{2,1}(e) \left(\frac{\gamma_{3,1}}{\gamma_{3,1} + \gamma_{3,2}} \right) + P_{2,2}(e) \left(\frac{\gamma_{3,2}}{\gamma_{3,1} + \gamma_{3,2}} \right) \right] + \left[P_{3,1}(e) \right]. \qquad (4.166)$$

This can be simplified to:

$$P_{b,e2e}(e) \approx \left[\xi_{1,1}P_{1,1}(e) + \xi_{1,2}P_{1,2}(e)\right]$$
$$+ \left[\xi_{2,1}P_{2,1}(e) + \xi_{2,2}P_{2,2}(e)\right]$$
$$+ \left[\xi_{3,1}P_{3,1}(e)\right], \tag{4.167}$$

where $\xi_{v,i}$ is the probability that an error occurring in link (v,i) will propagate to the destination. This result is easily generalized for networks of any size and any form of partial cooperation. To this end, remember that there are $Q_{v\in(1,K)}$ cooperative clusters at the vth stage, each of which will yield an error probability of $P_{v\in(1,K),i\in(1,Q_v)}$. The end-to-end BER is hence finally approximated as:

$$P_{b,e2e}(e) \approx \sum_{v=1}^{K}\sum_{i=1}^{Q_v}\xi_{v,i}P_{v,i}(e), \tag{4.168}$$

where the probabilities $\xi_{v,i}$ are easily found from the specific network topology. The BERs $P_{v,i}(e)$ can be found from Equation (4.153) and any of the previously derived SERs with an appropriate number of transmit and receive antennas per cluster, as well as prevailing channel conditions. The applicability of the derived end-to-end BER is assessed by means of Figures 4.35 and 4.36.

Explicitly, Figure 4.35 compares the numerically obtained and derived end-to-end BER to the SNR in the first link for a two-stage network, as depicted in Figure 4.34, without the second stage. For all simulations, QPSK has been used. The graphs are labeled on the respectively utilized channel gains.

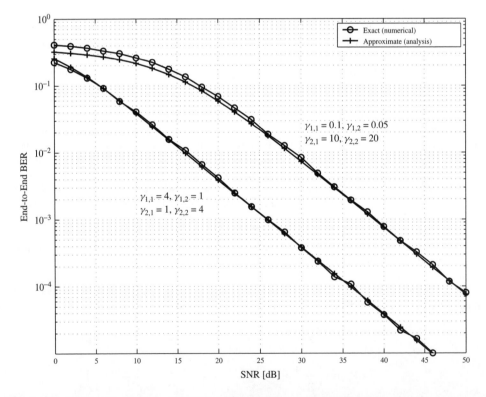

Figure 4.35 Numerically obtained and derived end-to-end BER versus the SNR in the first link for a two-stage network without cooperation

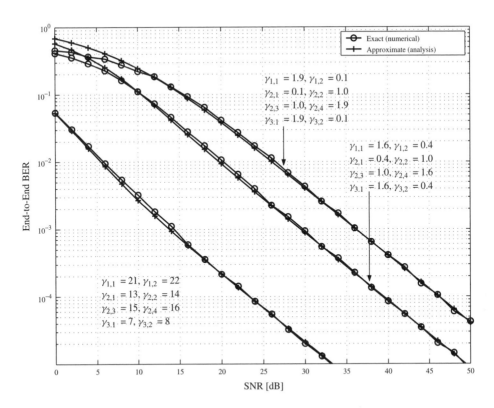

Figure 4.36 Numerically obtained and derived end-to-end BER versus the SNR in the first link for a three-stage network without cooperation

It can be observed that the derived BER differs from the exact one for low SNRs; however, for an increasing SNR, both curves converge.

Figure 4.36 compares the numerically obtained and derived end-to-end BER versus to the SNR in the first link for a three-stage network as depicted in Figure 4.34. The curves are again labeled on the channel gains. From Figure 4.36 it is clear that the derived end-to-end BER holds with high precision for a variety of different scenarios.

The derived end-to-end BERs in form of Equation (4.168) allow one to assign optimum fractional powers $\beta'_{v\in(1,K)}$ such that, together with the fractional frame durations $\alpha'_{v\in(1,K)}$, near-optimum end-to-end throughput is achieved. The fractional frame durations are clearly independent of the channel statistics or the degree of cooperation in the high SNR mode; therefore, Equation (4.149) holds for $\alpha'_{v\in(1,K)}$. The fractional power allocations are derived as follows.

Without loss of generality, let us assume that all links obey Rayleigh fading and have a different channel gain. The error rates are then governed by Equation (4.141), where u has to be replaced by the number of subchannels created in each of the Q_v clusters. The fractional power allocations have been derived as [467]:

$$
\beta'_v \approx \left[\sum_{w=1}^{K} \alpha'_w \sqrt{\frac{\sum_{i=1}^{Q_v} \sum_{j \in i} \xi_{v,i}^{-1} K_{v,i,j}^{-1} A_v^{-1} B_{v,i,j}}{\sum_{i=1}^{Q_w} \sum_{j \in i} \xi_{w,i}^{-1} K_{w,i,j}^{-1} A_w^{-1} B_{w,i,j}}} \right]^{-1},
\tag{4.169}
$$

where the notation $j \in i$ represents the jth subchannel belonging to the ith cluster. The partial expansion coefficients $K_{v,i,j}$ in the vth stage for the ith cluster can be written as:

$$K_{v,i,j} = \prod_{j' \in i, j' \neq j} \frac{\gamma_{v,j}}{\gamma_{v,j} - \gamma_{v,j'}} \qquad (4.170)$$

which has $u_{v,i}$ multiplicative terms. The constant A_v is given by Equation (4.163), whereas

$$B_{v,i,j} = \begin{cases} \dfrac{g_{PSK,v}}{R_v} \dfrac{\gamma_{v,j \in i}}{t_v} \dfrac{S}{N} & \text{for M-PSK} \\[2ex] \dfrac{g_{QAM,v}}{R_v} \dfrac{\gamma_{v,j \in i}}{t_v} \dfrac{S}{N} & \text{for M-QAM} \end{cases} \qquad (4.171)$$

This case is not further illustrated. Note that no waterfilling has been deployed prior to the above optimization; however, numerically obtained simulation results indicate that waterfilling for the transmit power at each stage does not yield notable performance gains, which is the reason why it has been omitted for the above analysis.

4.3.2 Distributed Space–Time Trellis Coding

In wireless relay networks, source and relays communicate cooperatively with a common destination. The cooperation between the source and relays actually forms a virtual antenna array, therefore intuitively the conventional space–time coding schemes can be extended to the relay networks in order to explore the cooperative diversity and coding gain. Various distributed space–time coding schemes have been developed in the past a few years, such as distributed space time block coding discussed in the previous section, distributed space time trellis coding [405, 482–485], distributed space time frequency coding [486, 487], etc.

In the below analysis, we assume that in distributed space–time coding each codeword is split into two parts, one transmitted by the source and the other by the relay nodes. In addition to the coding advantage, distributed space–time coding is based on incremental redundancy and thus allows a more flexible distribution of channel symbols between source and relay nodes, compared to repetition algorithms based on amplify-and-forward relaying protocols.

We will subsequently focus on the design of distributed STTC (DSTTC). We will discuss two DSTTC schemes based on two regenerative relaying protocols [482]. One is based on the decode-and-forward (DF) relaying protocol and assumes error free decoding at the relay. The other is based on the estimate-and-forward (EF) protocol, sometimes also referred to as detect-and-forward or demodulate-and-forward relaying protocol, and takes into account the erroneous detections at the relay. The design criteria are presented based on the pairwise error probability (PEP) analysis. The DSTTC codes with various numbers of states constructed on the basis of the design criteria are evaluated by means of simulations.

4.3.2.1 Generator Polynomial Description

In this section, we first briefly introduce the generator matrix description of STTCs [123], which is needed to understand subsequent material. We consider a system with n_S transmit antennas and M-PSK modulation. An STTC encoder consists of an $m = \log_2 M$-branch feedforward shift register with a total memory order of v. Let $\mathbf{b} = (\mathbf{b}_1, \cdots, \mathbf{b}_L)$ be the input sequence fed into the encoder, where $\mathbf{b}_t = (b_t^1, \cdots, b_t^m)$ is the m binary input streams fed into the m branches of the encoder at time t. The memory order of the kth branch, v_k, is given by:

$$v_k = \left\lfloor \frac{v + k - 1}{\log_2 M} \right\rfloor, \qquad (4.172)$$

where $\lfloor x \rfloor$ denotes the maximum integer not larger than x. The m binary input streams of input bits are simultaneously passed through their respective shift register branches and multiplied by the coefficient vectors:

$$\mathbf{g}^1 = \left[\left(g_{0,1}^1, g_{0,2}^1, \ldots, g_{0,n_S}^1 \right), \ldots, \left(g_{v_1,1}^1, g_{v_1,2}^1, \ldots, g_{v_1,n_S}^1 \right) \right] \tag{4.173a}$$

$$\vdots$$

$$\mathbf{g}^m = \left[\left(g_{0,1}^m, g_{0,2}^m, \ldots, g_{0,n_S}^m \right), \ldots, \left(g_{v_m,1}^m, g_{v_m,2}^m, \ldots, g_{v_m,n_S}^m \right) \right], \tag{4.173b}$$

where $g_{j,i}^k$, $k = 1, \ldots, m$, $j = 0, \ldots, v_k$, $i = 1, \ldots, n_S$, is an element of the MPSK constellation set, and $g_{j,i}^k \in \{0, 1, \ldots, M - 1\}$. $\mathbf{g}^k$, $k = 1, \cdots, m$, are called the STTC generator sequences, or STTC generator polynomials. The encoder structure of STTC using the above generator sequences is shown in Figure 4.37. The encoder output in stream i at time t, denoted by $c_i(t)$, can be computed as:

$$c_i(t) = \sum_{k=1}^{m} \sum_{j=0}^{v_k} g_{j,i}^k b_{t-j}^k \quad \text{mod } M, \quad i \in \{1, \ldots, n_S\}, \tag{4.174}$$

where $c_i(t) \in \{0, 1, \cdots, M\}$ is the encoder output. We henceforth denote the modulated complex symbol of $c_i(t)$ by $x_i(t)$. The encoder outputs $x_i(t)$, $i = 1, \ldots, n_S$, $t = 1, \ldots, L$, are arranged in a

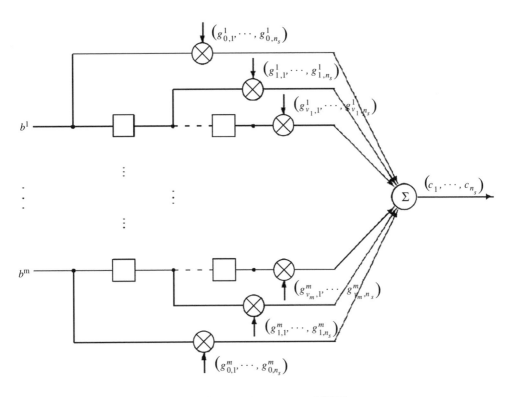

Figure 4.37   Encoder structure of STTC

code matrix, denoted by $\mathbf{X}$, as follows:

$$\mathbf{X} = \begin{bmatrix} x_1(1) & \cdots & x_1(L) \\ \vdots & \ddots & \\ x_{n_S}(1) & & x_{n_S}(L) \end{bmatrix}, \tag{4.175}$$

where L is the frame length. Here, $\mathbf{X}$ is the space time codeword of $\mathbf{b}$, generated by the generator polynomial $[\mathbf{g}^1, \cdots, \mathbf{g}^m]$.

4.3.2.2 DSTTC with Decode-and-Forward Relaying

We will now concentrate on the perfect DF protocol, where we will deal with the encoder structure, various error probability analyses, derivation of design criteria and resulting code design, and finally some performance comparisons.

(i) Encoder Structure. We consider a two-hop relay network, consisting of one source, one relay and one destination. They are equipped with n_S, n_R and n_D antennas, respectively. Figure 4.38 depicts the block diagram of a DSTTC system. Let $\mathbf{b}$ be the source transmitted information sequence. We consider M-PSK modulations at each node and a half-duplex transmission. The overall transmission is divided into two phases. $\mathbf{b}$ is first encoded into two space time trellis codewords:

$$\mathbf{X}_1 = [\mathbf{x}_1, \cdots, \mathbf{x}_{k_1}, \cdots, \mathbf{x}_L] \tag{4.176a}$$

$$\mathbf{X}_2 = \left[\mathbf{x}_{L+1}, \cdots, \mathbf{x}_{k_2}, \cdots, \mathbf{x}_{2L}\right], \tag{4.176b}$$

where $\mathbf{x}_{k_1} = \left[x_{k_1}^1, \cdots, x_{k_1}^{n_S}\right]^T$, $\mathbf{x}_{k_2} = \left[x_{k_2}^1, \cdots, x_{k_2}^{n_S}\right]^T$, and x_k^i denotes the signal transmitted from antenna i of the source node. They are generated by the STC generator polynomials $\mathbf{G}_{s_1} = (\mathbf{g}_{s_1}^1, \cdots, \mathbf{g}_{s_1}^m)$ and $\mathbf{G}_{s_2} = (\mathbf{g}_{s_2}^1, \cdots, \mathbf{g}_{s_2}^m)$, respectively, where $m = \log_2 M$ and

$$\mathbf{g}_{s_i}^j = \left[\left(g_{s_i,(0,1)}^j, g_{s_i,(0,2)}^j, \cdots, g_{s_i,(0,n_S)}^j\right), \cdots, \left(g_{s_i,(v_j,1)}^j, g_{s_i,(v_j,2)}^j, \cdots, g_{s_i,(v_j,n_S)}^j\right)\right] \tag{4.177}$$

for $i = 1, 2$. During the first half frame of the transmission, the source first broadcasts the STC codeword $\mathbf{X}_1$ from n_S antennas to both the relay and destination.

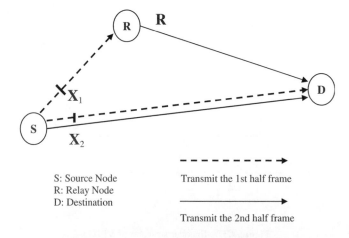

S: Source Node
R: Relay Node
D: Destination

Transmit the 1st half frame

Transmit the 2nd half frame

Figure 4.38 Block diagram of a DSTTC system

The received signals at n_R receive antennas of the relay node and n_D receive antennas of the destination node, at time k_1, denoted by $\mathbf{Y}_{R,k_1} = \left[y_{R,k_1}^1, \cdots, y_{R,k_1}^{n_R} \right]^T$ and $\mathbf{Y}_{D,k_1} = \left[y_{D,k_1}^1, \cdots, y_{D,k_1}^{n_D} \right]^T$, can be expressed as:

$$\mathbf{Y}_{R,k_1} = \mathbf{H}_{SR,k_1}\mathbf{x}_{k_1} + \mathbf{N}_{R,k_1}, k_1 = 1, \cdots, L \tag{4.178a}$$

$$\mathbf{Y}_{D,k_1} = \mathbf{H}_{SD,k_1}\mathbf{x}_{k_1} + \mathbf{N}_{D,k_1}, k_1 = 1, \cdots, L, \tag{4.178b}$$

where $\mathbf{N}_{R,k_1} = \left[n_{R,k_1}^j, j = 1, \cdots, n_R \right]$ and $\mathbf{N}_{D,k_1} = \left[n_{D,k_1}^j, j = 1, \cdots, n_D \right]$ are noise vectors, and their elements are independent samples of the zero mean complex Gaussian random variable, with the one sided power spectral density of N_0. $\mathbf{H}_{SD,k_1} = \left[h_{SD,k_1}^{j,i} \right]$ and $\mathbf{H}_{SR,k_1} = \left[h_{SR,k_1}^{j,i} \right]$ are the MIMO channel matrices from n_S source antennas to n_D destination antennas and that from n_S source antennas to n_R relay antennas in the first half-frame, $h_{SD,k_1}^{j,i}$ and $h_{SR,k_1}^{j,i}$ are the corresponding fading coefficients from the source antenna i to the destination antenna j and that from the source antenna i to the relay antenna j.

Upon receiving signals from the source, the relay decodes the received signals and re-encodes. We assume that the link quality between the source and relay node is good enough for the relay node to decode the message without error. After decoding, the relay node re-encodes the previously decoded message using separate generator polynomials $\mathbf{g}_r = (\mathbf{g}_r^1, \cdots, \mathbf{g}_r^m)$, where:

$$\mathbf{g}_r^j = \left[\left(g_{r,(0,1)}^j, g_{r,(0,2)}^j, \cdots, g_{r,(0,n_R)}^j \right), \cdots, \left(g_{r,(v_j,1)}^j, g_{r,(v_j,2)}^j, \cdots, g_{r,(v_j,n_R)}^j \right) \right], \tag{4.179}$$

and generates a new codeword given as:

$$\mathbf{R} = \left[\mathbf{r}_{L+1}, \cdots, \mathbf{r}_{k_2}, \cdots, \mathbf{r}_{2L} \right], \tag{4.180}$$

where $\mathbf{r}_{k_2} = \left[r_{k_2}^1, \cdots, r_{k_2}^{n_R} \right]^T$.

During the second half-frame, $\mathbf{X}_2$ and $\mathbf{R}$ are transmitted by the source and the relay nodes at the same time. The corresponding received signals at n_D receive antennas of the destination node, at time k_2, denoted by $\mathbf{Y}_{D,k_2} = \left[y_{D,k_2}^1, \cdots, y_{D,k_2}^{n_D} \right]^T$, can be expressed as:

$$\mathbf{Y}_{D,k_2} = \mathbf{H}_{SD,k_2}\mathbf{x}_{k_2} + \mathbf{H}_{RD,k_2}\mathbf{r}_{k_2} + \mathbf{N}_{D,k_2}, k_2 = L + 1, \cdots, 2L, \tag{4.181}$$

where $\mathbf{N}_{D,k_2} = \left[n_{D,k_2}^j, j = 1, \cdots, n_D \right]$ is a noise vector. $\mathbf{H}_{SD,k_2} = [h_{SD,k_2}^{j,i}]$ and $\mathbf{H}_{RD,k_2} = [h_{RD,k_2}^{j,i}]$ are the MIMO channel matrices from the source to the destination and that from the relay to the destination in the second half-frame, $h_{SD,k_2}^{j,i}$ and $h_{SR,k_2}^{j,i}$ are the corresponding fading coefficients from the source antenna i to the destination antenna j and that from the relay antenna i to the destination antenna j.

In general, the code sequences generated by the source node at the first and second half-frame and that by the relay node can generally be different in order to achieve better performance. Therefore, the entire coded sequences generated in the first and second transmission result in an $(n_S + n_R) \times 2L$ space–time code matrix:

$$\mathbf{X} = \begin{bmatrix} \mathbf{X}_1 & \mathbf{X}_2 \\ \mathbf{0} & \mathbf{R} \end{bmatrix} = \begin{bmatrix} x_1^1 & \cdots & x_L^1 & x_{L+1}^1 & \cdots & x_{2L}^1 \\ \vdots & \ddots & \vdots & \vdots & \ddots & \vdots \\ x_1^{n_S} & \cdots & x_L^{n_S} & x_{L+1}^{n_S} & \cdots & x_{2L}^{n_S} \\ 0 & \cdots & 0 & r_{L+1}^1 & \cdots & r_{2L}^1 \\ \vdots & \ddots & \vdots & \vdots & \ddots & \vdots \\ 0 & \cdots & 0 & r_{L+1}^{n_R} & \cdots & r_{2L}^{n_R} \end{bmatrix}, \tag{4.182}$$

which has length $2L$ and an equivalent number of transmit antennas of $n_S + n_R$. However, the coded symbols are transmitted from different node antennas and/or at different times.

We assume that the decoder at the destination node uses a maximum likelihood (ML) algorithm with perfect channel state information to estimate the transmitted information sequence. At the destination receiver, the decision metric is computed on the basis of the squared Euclidean distance between the hypothesized received sequence and the actual received sequence as:

$$
\Delta(\mathbf{Y}, \mathbf{X}) = \sum_{k_1=1}^{L} \left| \mathbf{Y}_{D,k_1} - \mathbf{H}_{SD,k_1} \mathbf{x}_{k_1} \right|^2
$$

$$
+ \sum_{k_2=L+1}^{2L} \left| \mathbf{Y}_{D,k_2} - \mathbf{H}_{SD,k_2} \mathbf{x}_{k_2} - \mathbf{H}_{RD,k_2} \mathbf{r}_{k_2} \right|^2. \tag{4.183}
$$

The decoder selects a codeword $\hat{\mathbf{X}}$ with the minimum decision metric as the decoded sequence. To formulate the appropriate code design criteria for the proposed scheme, we analyze the pairwise error probability (PEP) of the cooperative space–time coded system in slow and quasi-slow channels.

(ii) Error Probability of DSTTC with Perfect DF on Slow Fading Channels. In slow fading channels, the channel fading coefficients remain the same over the entire $2L$ time slots of one frame, and change independently from one frame to another. Therefore, we have:

$$
h_{SD,k_1}^{j,i} = h_{SD,k_2}^{j,i} = h_{SD}^{j,i} \tag{4.184}
$$

and

$$
h_{RD,k_2}^{j,i} = h_{RD}^{j,i}. \tag{4.185}
$$

Here, we define a modified channel matrix:

$$
\mathbf{H} = \left[\mathbf{h}^1, \cdots, \mathbf{h}^{n_D} \right]^T \tag{4.186}
$$

with each row vector, denoted by $\mathbf{h}^j$, given as:

$$
\mathbf{h}^j = \left[h_{SD}^{j,1}, \cdots, h_{SD}^{j,n_S}, h_{RD}^{j,1}, \cdots, h_{RD}^{j,n_R} \right]. \tag{4.187}
$$

The PEP, denoted by $P(\mathbf{X}, \hat{\mathbf{X}})$, is the probability that the decoder at the destination selects as it estimates an erroneous sequence $\hat{\mathbf{X}}$ when the transmitted sequence was in fact $\mathbf{X}$. In ML decoding, this occurs if

$$
\Delta(\mathbf{Y}, \mathbf{X}) \geq \Delta(\mathbf{Y}, \hat{\mathbf{X}}). \tag{4.188}
$$

This inequality is equivalent to

$$
\sum_{j=1}^{n_D} \sum_{t_1=1}^{L} 2Re \left\{ \left(n_{t_1}^j \right)^* \sum_{i_1=1}^{n_S} h_{SD}^{j,i_1} \left(x_{t_1}^{i_1} - \hat{x}_{t_1}^{i_1} \right) \right\}
$$

$$
+ \sum_{j=1}^{n_D} \sum_{t_2=L+1}^{2L} 2Re \left\{ \left(n_{t_2}^j \right)^* \left(\sum_{i_2=1}^{n_S} h_{SD}^{j,i_2} \left(x_{t_2}^{i_2} - \hat{x}_{t_2}^{i_2} \right) + \sum_{k=1}^{n_R} h_{RD}^{j,k} \left(r_{t_2}^k - \hat{r}_{t_2}^k \right) \right) \right\}
$$

$$
\geq \sum_{j=1}^{n_D} \sum_{t_1=1}^{L} \left| \sum_{i_1=1}^{n_S} h_{SD}^{j,i_1} \left(x_{t_1}^{i_1} - \hat{x}_{t_1}^{i_1} \right) \right|^2
$$

$$
+ \sum_{j=1}^{n_D} \sum_{t_2=L+1}^{2L} \left| \sum_{i_2=1}^{n_S} h_{SD}^{j,i_2} \left(x_{t_2}^{i_2} - \hat{x}_{t_2}^{i_2} \right) + \sum_{k=1}^{n_R} h_{RD}^{j,k} \left(r_{t_2}^k - \hat{r}_{t_2}^k \right) \right|^2, \tag{4.189}
$$

where $Re\{\cdot\}$ and $(\cdot)^*$ denote the real part and the complex conjugate of a complex number, respectively. The term on the left-hand side of Equation (4.189) is a zero mean Gaussian random variable and that on the right-hand side is a modified Euclidean distance between the two codeword matrices $\mathbf{X}$ and $\hat{\mathbf{X}}$, given by:

$$d_h^2(\mathbf{X}, \hat{\mathbf{X}}) = \sum_{j=1}^{n_D} \sum_{t_1=1}^{L} \left| \sum_{i_1=1}^{n_S} h_{SD}^{j,i_1} \left(x_{t_1}^{i_1} - \hat{x}_{t_1}^{i_1} \right) \right|^2$$

$$+ \sum_{j=1}^{n_D} \sum_{t_2=L+1}^{2L} \left| \sum_{i_2=1}^{n_S} h_{SD}^{j,i_2} \left(x_{t_2}^{i_2} - \hat{x}_{t_2}^{i_2} \right) + \sum_{k=1}^{n_R} h_{RD}^{j,k} \left(r_{t_2}^k - \hat{r}_{t_2}^k \right) \right|^2 . \tag{4.190}$$

The PEP conditioned on channel matrix $\mathbf{H}$ is thus given by

$$P(\mathbf{X}, \hat{\mathbf{X}}|\mathbf{H}) = Q\left(\sqrt{\frac{E_s}{2N_0} d_h^2(\mathbf{X}, \hat{\mathbf{X}})} \right) \leq \frac{1}{2} \exp\left(-\frac{E_s}{4N_0} d_h^2(\mathbf{X}, \hat{\mathbf{X}}) \right), \tag{4.191}$$

where $Q(\cdot)$ is the Q-function, and E_s is the energy per symbol at the transmit antenna at the source or the relay node. Let us now define an $(n_S + n_R) \times 2L$ codeword difference matrix $\mathbf{B}(\mathbf{X}, \hat{\mathbf{X}})$, expressed as:

$$\mathbf{B}(\mathbf{X}, \hat{\mathbf{X}}) = \mathbf{X} - \hat{\mathbf{X}}$$

$$= \begin{bmatrix} x_1^1 - \hat{x}_1^1 & \cdots & x_L^1 - \hat{x}_L^1 & x_{L+1}^1 - \hat{x}_{L+1}^1 & \cdots & x_{2L}^1 - \hat{x}_{2L}^1 \\ \vdots & \ddots & \vdots & \vdots & \ddots & \vdots \\ x_1^{n_S} - \hat{x}_1^{n_S} & \cdots & x_L^{n_S} - \hat{x}_L^{n_S} & x_{L+1}^{n_S} - \hat{x}_{L+1}^{n_S} & \cdots & x_{2L}^{n_S} - \hat{x}_{2L}^{n_S} \\ 0 & \cdots & 0 & r_{L+1}^1 - \hat{r}_{L+1}^1 & \cdots & r_{2L}^1 - \hat{r}_{2L}^1 \\ \vdots & \ddots & \vdots & \vdots & \ddots & \vdots \\ 0 & \cdots & 0 & r_{L+1}^{n_R} - \hat{r}_{L+1}^{n_R} & \cdots & r_{2L}^{n_R} - \hat{r}_{2L}^{n_R} \end{bmatrix}, \tag{4.192}$$

and an $(n_S + n_R) \times (n_S + n_R)$ matrix $\mathbf{A}(\mathbf{X}, \hat{\mathbf{X}}) = \mathbf{B}(\mathbf{X}, \hat{\mathbf{X}})\mathbf{B}^H(\mathbf{X}, \hat{\mathbf{X}})$. It can be verified that $\mathbf{A}(\mathbf{X}, \hat{\mathbf{X}})$ is a nonnegative definite Hermitian. That is, $\mathbf{A}(\mathbf{X}, \hat{\mathbf{X}}) = \mathbf{A}^H(\mathbf{X}, \hat{\mathbf{X}})$ and the eigenvalues of $\mathbf{A}(\mathbf{X}, \hat{\mathbf{X}})$, denoted by $\lambda_i > 0, i = 1, \cdots, r_s$ are positive real numbers, where r_s is the rank of $\mathbf{A}(\mathbf{X}, \hat{\mathbf{X}})$. Then, there exists a unitary matrix $\mathbf{V} = [\mathbf{v}_1, \mathbf{v}_2, \cdots, \mathbf{v}_{(n_S+n_R)}]$, such that:

$$\mathbf{V}\mathbf{A}(\mathbf{X}, \hat{\mathbf{X}})\mathbf{V}^H = \Lambda \tag{4.193}$$

where Λ is a diagonal matrix with $\lambda_i > 0, i = 1, \cdots, r_s$ as its diagonal elements and $\mathbf{v}_i$ is the eigenvector corresponding to the eigenvalue of λ_i. Then the modified squared Euclidean distance in Equation (4.190) can be expressed in a matrix form as:

$$d_h^2(\mathbf{X}, \hat{\mathbf{X}}) = \sum_{j=1}^{n_D} \mathbf{h}^j \mathbf{A}(\mathbf{X}, \hat{\mathbf{X}}) \mathbf{h}^{jH} = \sum_{j=1}^{n_D} \sum_{l=1}^{r_s} \lambda_l |\beta_{j,l}|^2, \tag{4.194}$$

where $\beta_{j,l} = \mathbf{h}^j \cdot \mathbf{v}^l$. Substituting Equation (4.194) into Equation (4.191), we get:

$$P(\mathbf{X}, \hat{\mathbf{X}}|\mathbf{H}) \leq \frac{1}{2} \exp\left(-\frac{E_s}{4N_0} \sum_{j=1}^{n_D} \sum_{l=1}^{r_s} \lambda_l |\beta_{j,l}|^2 \right) \tag{4.195}$$

Generally, $|\beta_{j,l}|$ obeys a Rician distribution and following the approach outlined by Tarokh *et al.* [122], we average the conditional PEP in Equation (4.195) with respect to the Rician random variables $|\beta_{j,l}|$ to arrive at:

$$P(\mathbf{X}, \hat{\mathbf{X}}) \le \left(\prod_{l=1}^{r_s} \left(1 + \frac{E_s}{4N_0} \lambda_l \right) \right)^{-n_D} \le \left(\prod_{l=1}^{r_s} \frac{1}{\lambda_l} \right)^{n_D} \left(\frac{E_s}{4N_0} \right)^{-r_s n_D}. \tag{4.196}$$

The error performance of a communication system is usually determined by its diversity order and coding gains. The diversity order determines the slope of an error rate curve plotted as a function of SNR, while the coding gain determines the horizontal shift of the uncoded system error rate curve to the coded error rate curve obtained for the same diversity order. From Equation (4.196), we can see that the DSTTC achieves a diversity advantage of $r_s n_D$, and a coding advantage [122] of $(\lambda_1 \lambda_2 \cdots \lambda_{r_s})^{1/r_s}$. It is clear that the maximum diversity order that this DSTTC can achieve is $n_D(n_S + n_R)$.

Let us dwell on the achievability of this maximum diversity order $n_D(n_S + n_R)$. From Equations (4.182) and (4.192), we can see that matrix $\mathbf{B}$ is a block low triangular matrix. Therefore, for slow fading channels, as long as in matrix $\mathbf{B}$, $\mathbf{X}_1 - \hat{\mathbf{X}}_1$ has the full rank n_S and that matrix $\mathbf{R} - \hat{\mathbf{R}}$ has the full rank n_R, then matrix $\mathbf{A}$ can achieve the maximum diversity order of $(n_S + n_R)$, and the system can achieve the full diversity order of $n_D(n_S + n_R)$. Therefore, to design a full diversity distributed STTC, we can design the codes $\mathbf{X}_1$ and $\mathbf{R}$ at the source and relay separately so that $\mathbf{X}_1$ and $\mathbf{R}$ are each a full rank code. However, such a design approach can only guarantee the full diversity order, and it can not guarantee the optimal coding gain. In order to design the optimal code with the maximum coding gain, we have to design the code $\mathbf{X}_1$, $\mathbf{X}_2$ and $\mathbf{R}$ jointly.

(iii) Error Probability of DSTTC with Perfect DF on Quasi-slow Fading Channels. In quasi-slow fading channels, the channel fading coefficients remain the same over half-frame, but change from the first half to the second half of one frame. Therefore, we have $h_{SD,k_1}^{j,i} = h_{SD1}^{j,i}$, $h_{SD,k_2}^{j,i} = h_{SD2}^{j,i}$, and $h_{RD,k_2}^{j,k} = h_{RD}^{j,k}$. Hence, Equation (4.190) can be written as:

$$d_h^2(\mathbf{X}, \hat{\mathbf{X}}) = \sum_{j=1}^{n_D} \sum_{k_1=1}^{L} \left| \sum_{i_1=1}^{n_S} h_{SD1}^{j,i_1} \left(x_{k_1}^{i_1} - \hat{x}_{k_1}^{i_1} \right) \right|^2$$

$$+ \sum_{j=1}^{n_D} \sum_{k_2=L+1}^{2L} \left| \sum_{i_2=1}^{n_S} h_{SD2}^{j,i_2} \left(x_{k_2}^{i_2} - \hat{x}_{k_2}^{i_2} \right) + \sum_{k=1}^{n_R} h_{RD}^{j,k} \left(r_{k_2}^k - \hat{r}_{k_2}^k \right) \right|^2. \tag{4.197}$$

We define an $n_S \times L$ codeword difference matrix:

$$\mathbf{Q}(\mathbf{X}, \hat{\mathbf{X}}) = \mathbf{X}_1 - \hat{\mathbf{X}}_1 \tag{4.198}$$

and an $n_S \times n_S$ matrix:

$$\mathbf{D}(\mathbf{X}, \hat{\mathbf{X}}) = \mathbf{Q}(\mathbf{X}, \hat{\mathbf{X}}) \mathbf{Q}^H(\mathbf{X}, \hat{\mathbf{X}}). \tag{4.199}$$

Similarly, we define an $(n_S + n_R) \times L$ codeword difference matrix:

$$\mathbf{C}(\mathbf{X}, \hat{\mathbf{X}}) = \left[\mathbf{X}_2 - \hat{\mathbf{X}}_2, \mathbf{R} - \hat{\mathbf{R}} \right]^T \tag{4.200}$$

and an $(n_S + n_R) \times (n_S + n_R)$ matrix:

$$\mathbf{G}(\mathbf{X}, \hat{\mathbf{X}}) = \mathbf{C}(\mathbf{X}, \hat{\mathbf{X}}) \mathbf{C}^H(\mathbf{X}, \hat{\mathbf{X}}). \tag{4.201}$$

Then, Equation (4.198) can be further written as:

$$d_h^2(\mathbf{X}, \hat{\mathbf{X}}) = \sum_{j=1}^{n_D} \mathbf{h}_1^j \mathbf{D}(\mathbf{X}, \hat{\mathbf{X}}) \mathbf{h}_1^{j\,H} + \sum_{j=1}^{n_D} \mathbf{h}_2^j \mathbf{G}(\mathbf{X}, \hat{\mathbf{X}}) \mathbf{h}_2^{j\,H}, \tag{4.202}$$

where $\mathbf{h}_1^j = \left(h_{SD1}^{j,1}, \cdots, h_{SD1}^{j,n_S} \right)$ and $\mathbf{h}_2^j = \left[h_{SD2}^{j,1}, \cdots, h_{SD2}^{j,n_S}, h_{RD}^{j,1}, \cdots, h_{RD}^{j,n_R} \right]$.

Let r_{q_1} and r_{q_2} denote the ranks of matrices $\mathbf{D}(\mathbf{X}, \hat{\mathbf{X}})$ and $\mathbf{G}(\mathbf{X}, \hat{\mathbf{X}})$, respectively. Furthermore, $\beta_m > 0$, $m = 1, \cdots, r_{q_1}$ and $\lambda_n > 0$, $n = 1, \cdots, r_{q_2}$ are the positive real eigenvalues of $\mathbf{D}(\mathbf{X}, \hat{\mathbf{X}})$ and $\mathbf{G}(\mathbf{X}, \hat{\mathbf{X}})$, respectively. Using a similar method to that in the slow fading channels, we can obtain the average PEP as:

$$P(\mathbf{X}, \hat{\mathbf{X}}) \leq 2 \left(\prod_{m=1}^{r_{q_1}} \frac{1}{\left(1 + \frac{E_s}{4N_0}\beta_m\right)} \prod_{n=1}^{r_{q_2}} \frac{1}{\left(1 + \frac{E_s}{4N_0}\lambda_n\right)} \right)^{n_D}. \tag{4.203}$$

At high SNRs, the above upper bound can be simplified to:

$$P(\mathbf{X}, \hat{\mathbf{X}}) \leq 2 \left(\prod_{m=1}^{r_{q_1}} \frac{1}{\beta_m} \right)^{n_D} \left(\prod_{n=1}^{r_{q_2}} \frac{1}{\lambda_n} \right)^{n_D} \left(\frac{E_s}{4N_0} \right)^{-n_D(r_{q_1}+r_{q_2})} \tag{4.204}$$

It is clear from Equation (4.204) that a diversity order of $n_D(r_{q_1} + r_{q_2})$ and a coding advantage given by $\left(\prod_{m=1}^{r_{q_1}} \beta_m \prod_{n=1}^{r_{q_2}} \lambda_n \right)^{\frac{1}{r_{q_1}+r_{q_2}}}$ can be achieved. Note that the maximal values of r_{q_1} and r_{q_2} are n_S and $n_S + n_R$, respectively. It indicates that the maximal transmit diversity order of the DSTTC in the first and second half-frame is n_S and $n_S + n_R$. Therefore, the maximal cooperation diversity to be achieved is $n_D(2n_S + n_R)$. For this case, as long as $\mathbf{X}_1 - \hat{\mathbf{X}}_1$ has rank n_S and that $\mathbf{C}(\mathbf{X}, \hat{\mathbf{X}})$ has the full rank of $n_S + n_R$, the code can achieve the full diversity order of $n_D(2n_S + n_R)$. Therefore, to design a full diversity code, we can design the codes $\mathbf{X}_1$ and $(\mathbf{X}_2, \mathbf{R})$ separately to make sure that each is a full rank code. However, $\mathbf{X}_2$ and $\mathbf{R}$ must be designed jointly in order to achieve its optimal error performance in terms of the coding and diversity gains.

(iv) Design Criteria for DSTTC with Perfect Decode and Forward. Based on Equations (4.196) and (4.204), the design criteria for DSSTC based on perfect decode-and-forward in slow and quasi-slow Rayleigh fading channels are summarized below [482].

- **Design criteria for DSTTC with perfect DF on slow fading channels:**
  I. Maximize the minimum rank of matrix $\mathbf{A}(\mathbf{X}, \hat{\mathbf{X}})$ over all pairs of distinct codewords.
  II. Maximize the minimum product $\prod_{i=l}^{r_s} \lambda_i$ of matrix $\mathbf{A}(\mathbf{X}, \hat{\mathbf{X}})$ along the pairs of distinct codewords with the minimum rank. If a full rank of $n_S + n_R$ is achievable, this is equivalent to the maximization of the minimum determinant of matrix $\mathbf{A}(\mathbf{X}, \hat{\mathbf{X}})$ along all the pairs of distinct codewords.

- **Design criteria for DSTTC with perfect DF on quasi-slow fading channels:**
  I. Maximize the minimum ranks of matrix $\mathbf{D}(\mathbf{X}, \hat{\mathbf{X}})$ and $\mathbf{G}(\mathbf{X}, \hat{\mathbf{X}})$ over all pairs of distinct codewords.
  II. Maximize the minimum product $\prod_{m=1}^{r_{q_1}} \beta_m \prod_{n=1}^{r_{q_2}} \lambda_n$ along the pairs of distinct codewords with the minimum rank. If a full rank of $2n_S + n_R$ is achievable, this is equivalent to the maximization of the minimum product of the determinants of matrix $\mathbf{D}(\mathbf{X}, \hat{\mathbf{X}})$ and $\mathbf{G}(\mathbf{X}, \hat{\mathbf{X}})$ along all the pairs of distinct codewords.

Let us exemplify the above by considering a special case where $n_S = n_R = 1$, Equation (4.204) can be simplified to:

$$P(\mathbf{X}, \hat{\mathbf{X}}) \leq \frac{2}{d_E^2} \prod_{n=1}^{r_{q2}} \frac{1}{\lambda_n} \left(\frac{E_s}{4N_0} \right)^{-(1+r_{q2})} \tag{4.205}$$

where $d_E^2 = \sum_{t=1}^{L} |x_t^i - \hat{x}_t^i|^2$ is the squared Euclidean distance of the coded sequence transmitted in the first half-frame from the source node. The design criteria can then be simplified to:

I. Maximize the minimum rank of matrix $\mathbf{G}(\mathbf{X}, \hat{\mathbf{X}})$ over all pairs of distinct codewords, and the maximum rank value is 2.
II. Maximize the minimum product $d_E^2 \prod_{n=1}^{r_{q2}} \lambda_n$ along the pairs of distinct codewords with the minimum rank. The product $d_E^2 \prod_{n=1}^{r_{q2}} \lambda_n$ is referred to as the *product coding gain* (PCG). If $\mathbf{C}(\mathbf{X}, \hat{\mathbf{X}})$ can achieve a full rank of 2, PCG is the product between the minimum squared Euclidean distance d_E^2 of $\mathbf{X}_1$ and the minimum determinant of $\mathbf{C}(\mathbf{X}, \hat{\mathbf{X}})$.

Based on the above criteria, one can find the new DSTTC codes. Figure 4.39 gives the code trellis structures for an eight-state DSTTC code on slow fading channels. Each branch is labeled by a/bcd where a is the input symbol at the source, b, c, d are the encoder output symbols transmitted by the source at the first and second half-frame and by the relay, respectively. We can easily check that this code has the full rank of 2 and determinant of 32. The comparisons of various DSTTC codes with determinant will be given in the next section and we will see that the code with larger determinant exhibits a better performance.

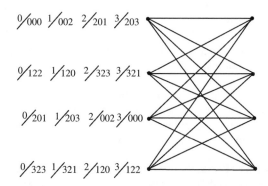

Figure 4.39 Trellis structure for an eight-state DSTTC code

(v) Code Design. For simplicity, we consider a system with a single antenna at both source and relay nodes. Following the notation introduced in Equation (4.177), let $\mathbf{G}_{s_1} = (\mathbf{g}_{s_1}^1, \cdots, \mathbf{g}_{s_1}^m)$ and $\mathbf{G}_{s_2} = (\mathbf{g}_{s_2}^1, \cdots, \mathbf{g}_{s_2}^m)$ denote the generator polynomials used to generate the coded sequence $\mathbf{X}_1$ and $\mathbf{X}_2$, which are transmitted by the source node during the first and second half-frames. Let $\mathbf{g}_r = (\mathbf{g}_r^1, \cdots, \mathbf{g}_r^m)$ be the generator polynomial used by the relay during the second half-frame. Then for the single antenna system, we have:

$$\mathbf{g}_{s_i}^j = \left[g_{s_i,(0,1)}^j, \cdots, g_{s_i,(v_j,1)}^j \right], i = 1, 2 \tag{4.206a}$$

$$\mathbf{g}_r^j = \left[g_{r,(0,1)}^j, \cdots, g_{r,(v_j,1)}^j \right] \tag{4.206b}$$

Let $\mathbf{g}^j = [\mathbf{g}_0^j, \cdots, \mathbf{g}_{v_j}^j], j = 1, \cdots, m$, where:

$$\mathbf{g}_i^j = \left[g_{i,1}^j, g_{i,2}^j \right] = [g_{s_1,(i,1)}^j, g_{s_2,(i,1)}^j], i = 0, \cdots, v_j \tag{4.207}$$

Figure 4.40 shows the encoder structure for QPSK modulation. For QPSK modulation, $m = 2$. $\mathbf{G}_{s_1} = (\mathbf{g}_{s_1}^1, \mathbf{g}_{s_1}^2) = ([g_{0,1}^1, \cdots, g_{v_1,1}^1], [g_{0,1}^2, \cdots, g_{v_2,1}^2])$ are used to generate the coded sequence $\mathbf{X}_1$, which is transmitted by the source node during the first half-frame. And $\mathbf{G}_{s_2} = (\mathbf{g}_{s_2}^1, \mathbf{g}_{s_2}^2) = ([g_{0,2}^1, \cdots, g_{v_1,2}^1], [g_{0,2}^2, \cdots, g_{v_2,2}^2])$ are used to generate the coded sequence $\mathbf{X}_2$, which is transmitted by the source node during the second half-frame. The encoder of the relay node takes a similar structure to that in Figure 4.40. However, the coefficient vectors $\mathbf{g}_p^1$ and $\mathbf{g}_q^2$ are replaced by coefficients $g_{r,(p,1)}^1$ and $g_{r,(q,1)}^2$, respectively.

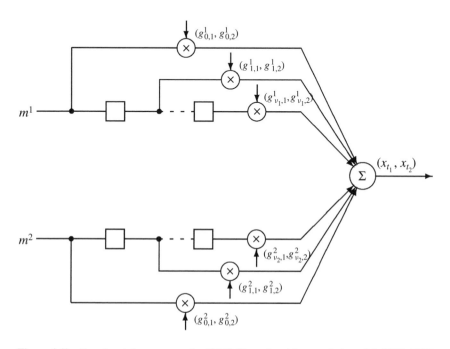

Figure 4.40   Encoder at the source node, QPSK (Reproduced by permission of © IEEE 2009)

Based on the criterion developed for slow and quasi-slow fading channels, new QPSK DSTTC for systems with a single antenna at both source and relay nodes are obtained through an exhaustive computer search. We present the encoder coefficients and parameters of the new codes with memory orders 2 to 4. Table 4.1 presents the generator polynomials, corresponding rank and determinant for slow fading channels. Note from the table that all codes have the rank of 2. The determinant of the code increases as the memory order increases. The optimal code is the one with largest determinant.

Similarly, Table 4.2 presents the encoder coefficients together with code parameters for quasi-slow fading channels. For each memory order, three different codes with the same rank and determinant but different d_E^2 (equivalently, different PCG value) are presented. The purpose is to show the impact of PCG on the code error performance. Note that the code with the largest PCG is the optimal one in Table 4.2.

(vi) Performance of DSTTC with Perfect DF. The code performance is evaluated by means of simulations. In the simulations, each frame consists of 130 symbols. A ML Viterbi decoder with perfect CSI is employed at the receiver. The frame error rate (FER) curves are plotted against the SNR per receive antenna.

Table 4.1 QPSK codes for DSTTC based on perfect DF with $n_S = n_R = 1$ over slow fading channels; v_s is the memory order of the source [482]. (Reproduced by permission of © IEEE 2009)

v_s	Generator (source)	Generator (relay)	Rank	Determinant
2	$\mathbf{g}^1 = [(0\ 0), (1\ 2)]$ $\mathbf{g}^2 = [(2\ 0), (2\ 0)]$	$\mathbf{g}_r^1 = [2, 2]$ $\mathbf{g}_r^2 = [1, 1]$	2	32
3	$\mathbf{g}^1 = [(2\ 2), (2\ 0)]$ $\mathbf{g}^2 = [(0\ 0), (1\ 2), (0\ 2)]$	$\mathbf{g}_r^1 = [1, 2]$ $\mathbf{g}_r^2 = [2, 0, 2]$	2	60
4	$\mathbf{g}^1 = [(2\ 0), (2\ 0), (2\ 2)]$ $\mathbf{g}^2 = [(0\ 2), (1\ 0), (1\ 2)]$	$\mathbf{g}_r^1 = [1, 0, 2]$ $\mathbf{g}_r^2 = [2, 2, 1]$	2	80

Table 4.2 QPSK codes for DSTTC based on perfect DF with $n_S = n_R = 1$ over quasi-slow fading channels; v_s is the memory order of the source [482]. (Reproduced by permission of © IEEE 2009)

v_s	Generator (source)	Generator (relay)	Rank	Determinant	d_E^2	PCG
3	$\mathbf{g}^1 = [(1\ 0), (3\ 2)]$ $\mathbf{g}^2 = [(1\ 2), (0\ 3), (0\ 0)]$	$\mathbf{r}^1 = [2, 3]$ $\mathbf{r}^2 = [2, 0, 1]$	2	16	2	32
3	$\mathbf{g}^1 = [(2\ 2), (0\ 0)]$ $\mathbf{g}^2 = [(1\ 0), (1\ 3), (1\ 2)]$	$\mathbf{r}^1 = [1, 2]$ $\mathbf{r}^2 = [2, 3, 0]$	2	16	6	96
3	$\mathbf{g}^1 = [(1\ 2), (1\ 0)]$ $\mathbf{g}^2 = [(2\ 0), (1\ 1), (2\ 2)]$	$\mathbf{r}^1 = [1, 2]$ $\mathbf{r}^2 = [3, 1, 0]$	2	16	10	160
4	$\mathbf{g}^1 = [(0\ 2), (1\ 2), (0\ 2)]$ $\mathbf{g}^2 = [(3\ 0), (0\ 1), (0\ 2)]$	$\mathbf{r}^1 = [2, 3, 0]$ $\mathbf{r}^2 = [2, 0, 2]$	2	32	2	64
4	$\mathbf{g}^1 = [(0\ 0), (3\ 1), (0\ 2)]$ $\mathbf{g}^2 = [(3\ 2), (1\ 2), (3\ 2)]$	$\mathbf{r}^1 = [2, 0, 2]$ $\mathbf{r}^2 = [2, 3, 0]$	2	32	6	192
4	$\mathbf{g}^1 = [(0\ 2), (2\ 0), (2\ 2)]$ $\mathbf{g}^2 = [(2\ 2), (2\ 1), (1\ 0)]$	$\mathbf{r}^1 = [0, 3, 2]$ $\mathbf{r}^2 = [2, 2, 2]$	2	32	10	320

Figure 4.41 shows the performance of the new QPSK DSTTCs for various numbers of states from Table 4.1 designed for perfect DF protocol in slow Rayleigh fading channels. The figure shows that all the codes can achieve the same diversity of two, which is reflected in the same slope of the FER performance curves. From this figure, we can also see that as the number of states increases, the error performance improves accordingly. This is consistent with the previous analysis.

Figure 4.42 compares the QPSK DSTTCs of Table 4.2 designed for perfect DF protocol in quasi-slow Rayleigh fading channels. From the figure, we can see that all the codes achieve the same diversity of three. We can also see that for the codes with the same rank and determinant value, larger squared Euclidean distance (and, equivalently, larger PCG value) result in a better error performance. This clearly demonstrates the impact of PCG on the error behavior of DSTTCs in quasi-slow Rayleigh fading channels.

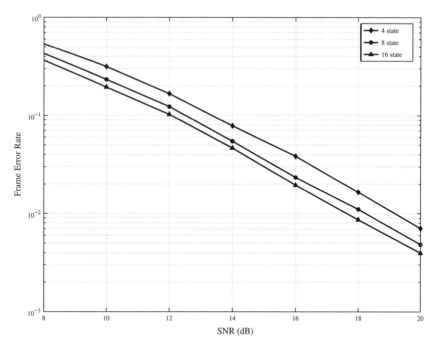

Figure 4.41 The performance comparison of the new QPSK DSTTCs with perfect DF slow Rayleigh fading channels [482]. (Reproduced by permission of © IEEE 2009)

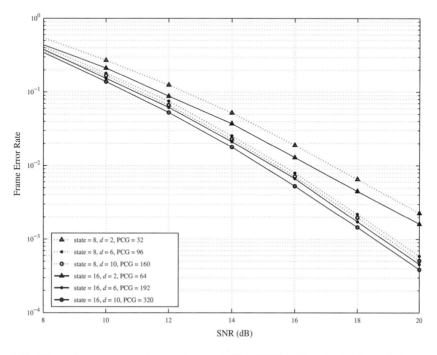

Figure 4.42 The performance comparison of the new QPSK DSTTCs with perfect DF in quasi-slow Rayleigh fading channels [482]. (Reproduced by permission of © IEEE 2009)

4.3.2.3 DSTTC with Estimate-and-Forward Relaying

In perfect DF relaying protocol, we assumed that the relays can always decode correctly and this would happen when the link from the source to the relay is much better than the link from the relay to the destination. In general relay networks, however, the relay may be located in any position w.r.t. the source and destination. Therefore, when decoding errors occur at the relay, the process of decoding and re-encoding at the relay in DF relaying protocols may lead to error propagation and seriously degrade the system performance. To avoid the error propagation inherited in the DF protocol, other relaying protocols should be considered.

Similarly to Section 4.3.1, we therefore study in this section the estimate-and-forward relaying (EF) protocol and design the DSTTC for EF for practical systems when taking into account the detection errors at the relay. The difference between the DF and EF is that in the EF the relay only demodulates the received signal from the source and remodulates it before transmission to the destination. The relay nodes no longer decode the demodulated signal nore re-encode it. This protocol is employed when relay nodes cannot fully decode the information from the source node. This happens very often in practical systems when the link from the source to the relay is not very strong. In this section, we study the design and performance of a DSTTC system with EF. The encoder structure and its error probability is presented by taking into account the detection errors at the relay. Based on the analysis, the design criteria for DSTTC with EF is introduced. New sets of codes obtained on the basis of the design criteria are presented and their performance is evaluated.

(i) Encoder Structure. The encoding processing for a DSTTC with EF protocol can be described as follows. For simplicity, we only consider $n_S = n_R = n_D = 1$. During the first half-frame $t_1 \in [1, L]$, the source broadcasts the encoded symbol x_{t_1} to both the destination and the relay. The corresponding received signals at the relay and destination can be expressed as:

$$y_{R,t_1} = h_{SR,t_1} x_{t_1} + n_{R,t_1} \tag{4.208a}$$

$$y_{D,t_1} = h_{SD,t_1} x_{t_1} + n_{D,t_1} \tag{4.208b}$$

where n_{R,t_1} and n_{D,t_1} are the noise components at the relay and destination antenna at time t_1, respectively. Upon receiving signals from the source, the relay node demodulates the received symbol and remodulates into $\tilde{x}_{t_1}$. During the second half frame $t_2 \in [L + 1, 2L]$, the source node transmits x_{t_2} and relay transmits $\tilde{x}_{t_1}$ to the destination node simultaneously. The destination received signal is given by:

$$y_{D,t_2} = h_{SD,t_2} x_{t_2} + h_{RD,t_2} \tilde{x}_{t_1} + n_{t_2} \tag{4.209}$$

where n_{t_2} is the destination noise at time t_2. The entire coded sequences received in the first and second half-frame can form a DSTTC matrix:

$$\mathbf{X} = \begin{bmatrix} x_1 & \cdots & x_L & x_{L+1} & \cdots & x_{2L} \\ 0 & \cdots & 0 & \tilde{x}_1 & \cdots & \tilde{x}_L \end{bmatrix}, \tag{4.210}$$

where the remodulated signal $\tilde{x}_{t_1}$, $t_1 = 1, \cdots, L$, is typically different from the source signal x_{t_1} due to the errors introduced in the source-relay link. It is clear that this code has an equivalent number of transmit antennas of 2.

(ii) Equivalent Channel Model of Source–Relay–Destination Link. Before analyzing the error performance of DSTTC with EF when taking into account the detection errors at relay, we first introduce the concept of equivalent channel model [442], which was used in Section 4.2.1 to approximate the two hop source–relay–destination link by an equivalent single-hop link when considering detection errors at the relay in the EF protocol. As we will see later, such a model can facilitate the performance analysis of DSTTC with EF. We first define the SNRs for each link:

$$\gamma_{SR,t_1} = \left|h_{SR,t_1}\right|^2 \bar{\gamma} \tag{4.211a}$$

$$\gamma_{SD,t_1} = \left|h_{SD,t_1}\right|^2 \bar{\gamma} \tag{4.211b}$$

$$\gamma_{SD,t_2} = \left|h_{SD,t_2}\right|^2 \bar{\gamma} \tag{4.211c}$$

$$\gamma_{RD,t_2} = \left|h_{RD,t_2}\right|^2 \bar{\gamma}, \tag{4.211d}$$

where $\bar{\gamma}$ is the average SNR, h_{SR,t_1}, h_{SD,t_1}, h_{SD,t_2} and h_{RD,t_2} are the channel fading coefficients at time t_1 and t_2.

The error performance of this whole system depends on the error performance of each link. Let us now focus on the source–relay–destination link. We consider a system employing BPSK modulation. At the destination, errors occur either when the source–relay transmission is received correctly and relay–destination transmission is received in error, or the source–relay transmission is received in error and the relay–destination transmission is received correctly. Therefore the conditional end-to-end PEP of this two-hop source–relay–destination link can be calculated as:

$$P\left(\gamma_{SR,t_1}, \gamma_{RD,t_2}\right) = (1 - P(\gamma_{SR,t_1}))P(\gamma_{RD,t_2}) + (1 - P(\gamma_{RD,t_2}))P(\gamma_{SR,t_1}), \tag{4.212}$$

where $P(\gamma_{SR,t_1})$ and $P(\gamma_{RD,t_2})$ are the error probabilities in the links from the source to the relay and that from the relay to the destination. An equivalent one-hop AWGN link has been introduced [442] to replace the two-hop source–relay–destination non-Gaussian and nonlinear link. It can be expressed as:

$$\gamma_{eq} = \frac{1}{2}\left\{Q^{-1}\left[P\left(\gamma_{SR,t_1}, \gamma_{RD,t_2}\right)\right]\right\}^2. \tag{4.213}$$

The output SNR of this equivalent link at time t_2 can be further simplified as $\gamma_{eq,t_2} \leq \gamma_{\min}$, where $\gamma_{\min} = \min\left\{\gamma_{RD,t_2}, \gamma_{SR,t_1}\right\}$. By using this equivalent link, we analyze the performance of the single relay system not only based on the relay–destination link, but also taking the source–relay link performance into consideration.

Similarly, we assume that the channel fading coefficients undergo a Rayleigh distribution. If we assume the source–relay link is perfect, the relay can perfectly demodulate the received signal from the source. In such a scenario, the received signal at the destination is $\mathbf{Y} = \begin{bmatrix} y_{D,t_1} & y_{D,t_2} \end{bmatrix}^T$, and the transmitted codeword matrix is:

$$\mathbf{X} = \begin{bmatrix} x_1 \cdots x_L & x_{L+1} \cdots x_{2L} \\ 0 \cdots 0 & x_1 \cdots x_L \end{bmatrix}. \tag{4.214}$$

Then the decision metric at the destination is computed as

$$\Psi(\mathbf{Y}, \mathbf{X}) = \sum_{t_1=1}^{L}\left|y_{D,t_1} - h_{SD,t_1}x_{t_1}\right|^2 + \sum_{t_2=L+1}^{2L}\left|y_{D,t_2} - h_{SD,t_2}x_{t_2} - h_{RD,t_2}x_{t_2-L}\right|^2. \tag{4.215}$$

However, if the source–relay channel is not perfect and detection errors occur at the relay node, considering the equivalent link model of Wang *et al.* [442], the decision metric in Equation (4.215) should be modified to:

$$\Psi(\mathbf{Y}, \mathbf{X}) = \sum_{t_1=1}^{L}\left|y_{D,t_1} - h_{SD,t_1}x_{t_1}\right|^2$$

$$+ \sum_{t_2=L+1}^{2L}\left|y_{D,t_2} - h_{SD,t_2}x_{t_2} - \sqrt{\frac{\gamma_{eq,t_2}}{\gamma_{RD,t_2}}}h_{RD,t_2}x_{t_2-L}\right|^2, \tag{4.216}$$

where $\sqrt{\frac{\gamma_{eq,t_2}}{\gamma_{RD,t_2}}}$ is the extra factor after taking into account the detection errors at the relay.

Let us define the equivalent channel matrix $\mathbf{H} = [\mathbf{h}_1, \mathbf{h}_2, \cdots, \mathbf{h}_L, \mathbf{h}_{L+1}, \cdots, \mathbf{h}_{2L}]^T$, with each row vector defined as

$$\mathbf{h}_t = \begin{cases} \begin{bmatrix} h_{SD,t} & 0 \end{bmatrix}, & t = 1, \cdots, L \\ \begin{bmatrix} h_{SD,t} & \sqrt{\frac{\gamma_{eq,t}}{\gamma_{RD,t}}} h_{RD,t} \end{bmatrix}, & t = L+1, \cdots, 2L. \end{cases} \tag{4.217}$$

We know that the amplitudes of the fading coefficient $h_{SD,t}$ follow a Rayleigh distribution. The distribution of the fading coefficient $\sqrt{\frac{\gamma_{eq,t}}{\gamma_{RD,t}}} h_{RD,t}$ is determined by the distribution of $\sqrt{\gamma_{eq,t}}$, which is further affected by the distribution of $\min\{|h_{RD,t}|, |h_{SR,t}|\}$. Using the knowledge of order statistics [319], we find that the PDF of $Y = \sqrt{\frac{\gamma_{eq,t}}{\gamma_{RD,t}}} |h_{RD,t}|$ can be expressed as $p_Y(y) = 4ye^{-2y^2}$.

If we compare the PDF of direct link $h_{SD,t}$ and equivalent source–relay–destination link $\sqrt{\frac{\gamma_{eq,t}}{\gamma_{RD,t}}} h_{RD,t}$, we can see that compared with the direct link, the average power gain for the source–relay–destination link is reduced by half. This will eventually affect the code design as the codeword matrix will be affected by this power loss. At the second half of the frame, the relay transmits the codeword sequence $x_t, t = 1, \cdots, L$, through the channel, where the channel coefficients are independent complex Gaussian random variables with zero mean and variance $1/4$ per dimension. This operation at the relay is equivalent to transmitting a modified codeword sequence $x_t' = \frac{1}{\sqrt{2}} x_t, t = 1, \cdots, L$ through an equivalent channel $h_{eq,t}', t = L+1, \cdots, 2L$, where the equivalent channel fading coefficients $h_{eq,t}'$ are independent Gaussian random variables with zero mean and variance $1/2$ per dimension. In other words, $h_{eq,t}', t = L+1, \cdots, 2L$, follow the same distribution as $h_{SD,t}$. Figure 4.43 shows the block diagram of DSTTC based on EF using the equivalent one-hop link. As we can see from the figure, by considering the power loss for the equivalent one-hop link, the transmitted codeword matrix in Equation (4.214) can be represented as:

$$\mathbf{X} = \begin{bmatrix} x_1 \cdots x_L & x_{L+1} \cdots x_{2L} \\ 0 \cdots 0 & \frac{1}{\sqrt{2}} x_1 \cdots \frac{1}{\sqrt{2}} x_L \end{bmatrix}. \tag{4.218}$$

If we modify the channel matrix $\mathbf{H}$ to $\mathbf{H}'$, with each row vector modified to $\mathbf{h}_t'$, expressed as:

$$\mathbf{h}_t' = \begin{cases} \begin{bmatrix} h_{SD,t} & 0 \end{bmatrix}, & t = 1, \cdots, L \\ \begin{bmatrix} h_{SD,t} & h_{eq,t}' \end{bmatrix}, & t = L+1, \cdots, 2L, \end{cases} \tag{4.219}$$

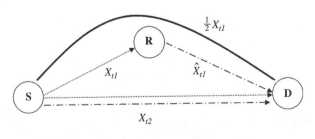

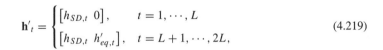 Transmission in the first half-frame

Transmission in the second half-frame

Equivalent one-hop link

Figure 4.43 The block diagram of DSTTC based on EF using the equivalent one-hop link

we can find the error probability conditioned on $\mathbf{H}'$ as:

$$P(\mathbf{X}, \hat{\mathbf{X}}|\mathbf{H}') = Q\left(\sqrt{\frac{E_s}{2N_0}d_{h'}^2(\mathbf{X}, \hat{\mathbf{X}})}\right),\tag{4.220}$$

where $d_{h'}^2(\mathbf{X}, \hat{\mathbf{X}})$ is a modified squared Euclidean distance between the two codeword matrices $\mathbf{X}$ and $\hat{\mathbf{X}}$, given by $d_{h'}^2(\mathbf{X}, \hat{\mathbf{X}}) = \parallel \mathbf{H}'\left(\mathbf{X} - \hat{\mathbf{X}}\right)\parallel^2$.

By using this equivalent model of source–relay–destination path, the relay system can be regarded as transmitting a distributed space–time coded symbol from the source and the relay, with the codeword matrix defined in Equation (4.218), through the virtual channel model to the destination, that is, $\mathbf{h}'_t = \begin{bmatrix} h_{SD,t} & 0 \end{bmatrix}$ at time $t = 1, \cdots, L$, and $\mathbf{h}'_t = \begin{bmatrix} h_{SD,t} & h'_{eq,t} \end{bmatrix}$ at time $t = L + 1, \cdots, 2L$.

(iii) Error Probability of DSTTC with EF on Slow Fading Channels. For the slow fading scenario, we have $h_{SD,t_1} = h_{SD,t_2} = h_{SD}$ and $h'_{eq,t_2} = h'_{eq}$. We define $\mathbf{B}(\mathbf{X}, \hat{\mathbf{X}}) = \mathbf{X} - \hat{\mathbf{X}}$ and $\mathbf{A}(\mathbf{X}, \hat{\mathbf{X}}) = \mathbf{B}(\mathbf{X}, \hat{\mathbf{X}})\mathbf{B}^H(\mathbf{X}, \hat{\mathbf{X}})$, where the codeword matrix in Equation (4.218) has a dimension of $2 \times 2L$. Let $\lambda_t > 0$, $t = 1, \cdots, r_s$, be the positive real eigenvalues of $\mathbf{A}(\mathbf{X}, \hat{\mathbf{X}})$ and r_s denotes the rank of matrix $\mathbf{A}(\mathbf{X}, \hat{\mathbf{X}})$. Following a similar approach to that in the previous sections, at high SNRs, we have the PEP of the DSTTC with EF as:

$$P(\mathbf{X}, \hat{\mathbf{X}}) \leq \left(\prod_{t=1}^{r_s} \lambda_t\right)^{-1}\left(\frac{E_s}{4N_o}\right)^{-r_s}.\tag{4.221}$$

It is clear from Equation (4.221) that a diversity order of r_s is achieved with $\left(\prod_{t=1}^{r_s}\lambda_t\right)^{\frac{1}{r_s}}$ as the coding advantage.

(iv) Error Probability of DSTTC with EF on Quasi-slow Fading Channels. We define:

$$\mathbf{G}(\mathbf{X}, \hat{\mathbf{X}}) = \mathbf{C}(\mathbf{X}, \hat{\mathbf{X}})\mathbf{C}^H(\mathbf{X}, \hat{\mathbf{X}}),\tag{4.222}$$

where

$$\mathbf{C}(\mathbf{X}, \hat{\mathbf{X}}) = \begin{bmatrix} x_{L+1} - \hat{x}_{L+1} & \cdots & x_{2L} - \hat{x}_{2L} \\ \frac{1}{\sqrt{2}}(x_1 - \hat{x}_1) & \cdots & \frac{1}{\sqrt{2}}(x_L - \hat{x}_L) \end{bmatrix}.\tag{4.223}$$

Using a similar method to that in the previous section, we can get the asymptotic average PEP as:

$$P(\mathbf{X}, \hat{\mathbf{X}}) \leq \frac{1}{d_E^2}\prod_{n=1}^{r_{q2}}\frac{1}{\lambda_n}\left(\frac{E_s}{4N_0}\right)^{-(1+r_{q2})},\tag{4.224}$$

in which $d_E^2 = \sum_{t_1=1}^{L}|x_{t_1} - \hat{x}_{t_1}|^2$ is the squared Euclidean distance of the coded sequence transmitted in the first half of the frame from the source node, r_{q2} denotes the rank of matrix $\mathbf{G}(\mathbf{X}, \hat{\mathbf{X}})$, and $\lambda_n > 0, n = 1, \cdots, r_{q2}$, are the positive real eigenvalues of $\mathbf{G}(\mathbf{X}, \hat{\mathbf{X}})$. In Equation (4.224), $1 + r_{q2}$ is the diversity advantage, while the product $\left(d_E^2\prod_{n=1}^{r_{q2}}\lambda_n\right)^{\frac{1}{1+r_{q2}}}$ is the coding advantage. Therefore, a diversity order of 3 is achievable for $n_S = n_R = n_D = 1$.

(v) Design Criteria for DSTTC with Detection and Forward Relaying Protocol. Based on Equations (4.221) and (4.224), the following design criteria for DSTTC based on EF in slow and quasi-slow Rayleigh fading channels are formulated below.

- **Design criteria for DSTTC with EF on slow fading channels:**
  - I. Maximize the minimum rank of matrix $\mathbf{A}(\mathbf{X}, \hat{\mathbf{X}})$ over all pairs of distinct codewords.
  - II. Maximize the minimum product $\prod_{l=l}^{r_s} \lambda_l$ of matrix $\mathbf{A}(\mathbf{X}, \hat{\mathbf{X}})$ along the pairs of distinct codewords with the minimum rank.
- **Design criteria for DSTTC with EF on quasi-slow fading channels:**
  - I. Maximize the minimum rank of matrix $\mathbf{G}(\mathbf{X}, \hat{\mathbf{X}})$ over all pairs of distinct codewords.
  - II. Maximize the minimum PCG $d_E^2 \prod_{n=1}^{r_{q2}} \lambda_n$ along the pairs of distinct codewords with the minimum rank. If a full rank of 2 is achievable, this is equivalent to the maximization of the minimum product of minimum squared Euclidean distance d_E^2 and the determinant of matrix $\mathbf{G}(\mathbf{X}, \hat{\mathbf{X}})$ along all the pairs of distinct codewords.

(vi) Code Design. The encoder structure at the source is the same as the one in Figure 4.40. Based on the design criteria for DSTTC with EF over slow fading channels, new QPSK DSTTCs for systems with a single antenna at both source and relay nodes are obtained through an exhaustive computer search. The encoder coefficients and code parameters, including the rank and determinant, are listed in Table 4.3. Note that for a given memory order, these QPSK DSTTCs have the same rank as the Tarokh/Seshadri/Calderbank (TSC) codes [122], but a much larger minimum determinant, which in turn results in a larger coding gain.

Table 4.3 QPSK codes for DSTTC based on EF with $n_S = n_R = 1$ over slow fading channels; v_s is the memory order of the source [482]. (Reproduced by permission of © IEEE 2009)

v_s	Generator sequence (source)	Rank	Determinant
2	$\mathbf{g}^1 = [(3\ 3), (3\ 1)]$ $\mathbf{g}^2 = [(2\ 2), (2\ 2)]$	2	12
3	$\mathbf{g}^1 = [(2\ 0), (2\ 0)]$ $\mathbf{g}^2 = [(3\ 2), (3\ 0), (1\ 2)]$	2	24
4	$\mathbf{g}^1 = [(3\ 3), (2\ 0), (2\ 0)]$ $\mathbf{g}^2 = [(2\ 0), (0\ 2), (3\ 2)]$	2	34

Similarly, new DSTTCs with EF constructed for quasi-slow fading channels are listed in Table 4.4. We also include the determinant, the squared Euclidean distance (d_E^2) and the PCG of each code in the table. For codes with the same rank, a larger determinant and a larger Euclidean distance can result in a larger PCG and therefore better performance.

Table 4.4 QPSK codes for DSTTC based on EF with $n_S = n_R = 1$ over quasi-slow fading channels; v_s is the memory order of the source [482]. (Reproduced by permission of © IEEE 2009)

v_s	Generator sequence (source)	Rank	Determinant	d_E^2	PCG
3	$\mathbf{g}^1 = [(2\ 0), (1\ 2)]$ $\mathbf{g}^2 = [(2\ 2), (2\ 1), (2\ 0)]$	2	8	12	96
4	$\mathbf{g}^1 = [(2\ 0), (3\ 1), (0\ 2)]$ $\mathbf{g}^2 = [(2\ 2), (2\ 1), (2\ 0)]$	2	16	12	192

(vii) Performance of DSTTC with EF. The performance of DSTTCs with EF is evaluated through simulations. In the simulations, each frame consists of 130 symbols. The frame error rate (FER) curves are plotted against the signal-to-noise ratio per receive antenna.

Figure 4.44 compares the FER performances of the QPSK DSTTCs with EF for various states from Table 4.3 in slow Rayleigh fading channels. In the simulation, we consider three scenarios, in which the relay node is located either equidistant from both the source and the destination node, or it is closer to the destination node, or it is closer to the source node. The corresponding output SNRs $(\gamma_{SR}, \gamma_{RD}, \gamma_{SD})$ are $(\bar{\gamma}, \bar{\gamma}, \bar{\gamma}), (\bar{\gamma}, \bar{\gamma} + 20\,dB, \bar{\gamma})$, and $(\bar{\gamma} + 20\,dB, \bar{\gamma}, \bar{\gamma})$, respectively. It can be seen from the figure that the first scenario has the worst performance, because both S-R and R-D links have a low SNR at the same time. The third scenario $(\bar{\gamma} + 20\,dB, \bar{\gamma}, \bar{\gamma})$ outperforms the second scenario $(\bar{\gamma}, \bar{\gamma} + 20\,dB, \bar{\gamma})$, because the S-R link has a higher SNR, therefore less error propagation to the destination. From the figure, it is clear that, the three codes can achieve the same diversity order.

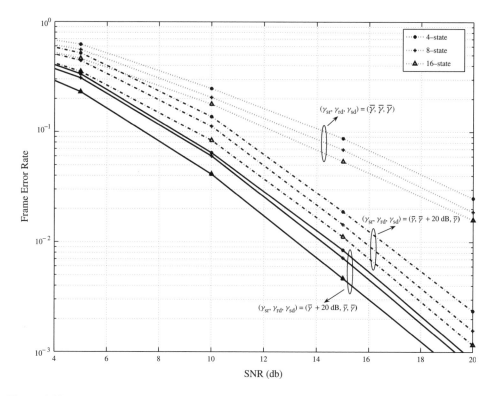

Figure 4.44 The performance comparison of the QPSK DSTTCs based on EF in slow Rayleigh fading channels [482]. (Reproduced by permission of © IEEE 2009)

Figure 4.45 compares the error performance of the two new four-state codes from Table 4.1 and Table 4.3 in slow Rayleigh fading channels when EF protocol is employed. We should note that the code from Table 4.1 is designed for the perfect DF protocol, but the code from Table 4.3 is specifically designed for the EF protocol. For the code designed for the perfect DF protocol, we only utilize the generator sequence at the source. It is assumed that the relay node is closer to the source node than to the destination node, such that $(\gamma_{SR}, \gamma_{RD}, \gamma_{SD}) = (\bar{\gamma} + 30\,dB, \bar{\gamma}, \bar{\gamma})$. It can be seen from Figure 4.45 that the code from Table 4.3 designed for the EF is superior by 3.6 dB to the one from Table 4.1

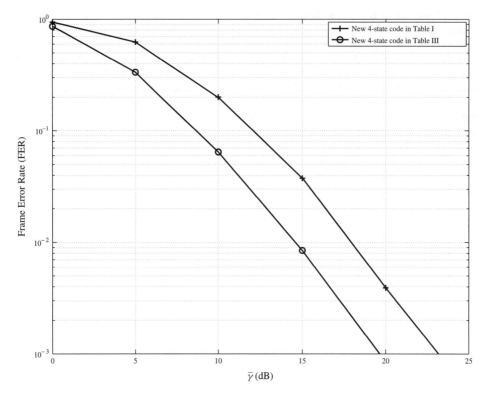

Figure 4.45 The performance comparison of the new four-state DSTTCs based perfect DF and EF in slow Rayleigh fading channels [482], EF protocol is employed in the simulation. $(\gamma_{SR}, \gamma_{RD}, \gamma_{SD}) = (\overline{\gamma} + 30 \text{ dB}, \overline{\gamma}, \overline{\gamma})$. (Reproduced by permission of © IEEE 2009)

designed for the perfect DF, at a FER of 10^{-3}. This clearly indicates that the code designed for this specific protocol significantly outperforms other codes.

4.3.3 Distributed Turbo Coding

Distributed STBCs and distributed STTCs can only provide limited coding gain due to their coding structure. To improve the system performance further, some capacity approaching distributed coding schemes, such as distributed LDPC coding [488, 489] and distributed turbo coding (DTC) schemes have been developed recently. Basically, these codes borrow the concept from the conventional LDPC and turbo coding and apply them to distributed networks. One challenge in designing a practical capacity approaching distributed coding schemes is how to deal with the decoding errors and avoid error propagation.

In this section, we consider the design and performance of DTCs. DTC was originally proposed by Zhao and Valenti [490] and has been further developed [454, 491–494]. In this section, we will introduce two efficient DTC schemes and analyze their performance.

4.3.3.1 Turbo Encoder Structure

The concept of turbo codes and iterative decoding was originally proposed by Berrou *et al.* in 1993 [495]. These can achieve very low bit error rates with signal-to-noise ratios that are only slightly above the minimum SNR for a given channel and code rate as established by Shannon's original capacity theorems. In this sense, turbo codes are said to be near capacity achieving codes.

Figure 4.46 depicts the block diagram of a turbo encoder. It is formed by parallel concatenation of two recursive systematic convolutional (RSC) encoders separated by a random interleaver, which is why it is also called a parallel concatenated convolutional code (PCCC). In the encoder, the same information sequence is encoded twice in parallel but in different orders. The first encoder directly operates on the input sequence and the second encoder operates on the interleaved input sequence. The outputs of these two encoders together form a turbo codeword. Similar to the encoder structure, the turbo decoder consists of two concatenated decoders of the component codes separated by the same interleaver. The component decoders are based on a maximum *a posteriori* (MAP) probability algorithm. The iterative process is performed between two component decoders to exchange information. Two essential elements in turbo coding are the RSC constituent encoder and random interleaver. The generator matrix of a RSC code can be represented as:

$$\mathbf{G}(D) = \left[1, \frac{\mathbf{g}_1(D)}{\mathbf{g}_0(D)}\right],\qquad(4.225)$$

where $\mathbf{g}_0(D)$ and $\mathbf{g}_1(D)$ are a feedback and feedforward polynomial, respectively. We can see that in an (n, k) RSC code the first k output sequences are exact replicas of the k input information sequences and the rest $n - k$ bits are the parity check bits generated by the recursive encoder. Figure 4.47 shows an example of a rate 1/3 turbo code encoder, based on a (2,1,3) RSC code. The generator matrix of the RSC component code is $\left[1, \frac{1+D+D^2}{1+D^2}\right]$.

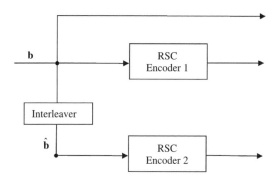

Figure 4.46   Turbo encoder

4.3.3.2   Distributed Turbo Coding with Perfect DF

We will now treat the case of perfect DF. We apply the turbo coding principle to construct the DTCs for wireless relay networks. For simplicity, we first introduce the DTC design when assuming the perfect decodings at the relay and later will discuss the design of practical DTC schemes when taking into account the imperfect decodings at the relay. We still consider a two-hop relay system, consisting of one source, one relay and one destination. We will first develop the encoder and decoder structures and then gauge their performance.

(i) Encoder Structure of Perfect DTC.   Figure 4.48 shows a block diagram of a distributed turbo coded system. Let $\mathbf{B} = (b_1, \ldots, b_k, \ldots, b_L)$ denote the information binary sequence transmitted by the source node, where b_k is the kth transmitted bit and L is the frame length. The binary information sequence $\mathbf{B}$ is first encoded by a channel encoder. For simplicity, we consider a recursive systematic

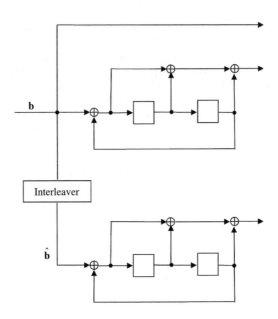

Figure 4.47 Turbo encoder with [1, 7/5] RSC code

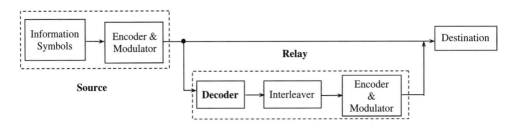

Figure 4.48 Block diagram of a perfect DTC system

convolutional code (RSCC) with a code rate of 1/2. To this end, let $\mathbf{C}$ represent the corresponding codeword, given by:

$$\mathbf{C} = (\mathbf{C}_1, \ldots, \mathbf{C}_k, \ldots, \mathbf{C}_L) \tag{4.226}$$

where $\mathbf{C}_k = (b_k, c_k)$ is the codeword of b_k, and $c_k \in \{0, 1\}$ is the parity bit of b_k. $\mathbf{C}$ is then mapped into a modulated signal stream $\mathbf{X}$. For simplicity, we consider BPSK modulation. The modulated codeword, denoted by $\mathbf{X}$, is given by:

$$\mathbf{X} = (\mathbf{X}_1, \ldots, \mathbf{X}_k, \ldots, \mathbf{X}_L) \tag{4.227}$$

where $\mathbf{X}_k = \left(x_k^I, x_k^P\right)$, $x_k^I, x_k^P \in \{-1, 1\}$, are the modulated information and parity signals transmitted by the source at time $2k - 1$ and $2k$, respectively.

We consider a half duplex transmission, that is, the source and relay transmit data in separate time slots. The source first broadcasts the information to both the destination and relay. The received signals at the relay and destination, at time $2k - 1$ and $2k$, denoted by $\mathbf{Y}_{SR,k}$ and $\mathbf{Y}_{SD,k}$, respectively, can be expressed as:

$$\mathbf{Y}_{SR,k} = \sqrt{p_{SR}} h_{SR} \mathbf{X}_k + \mathbf{N}_{SR,k} \tag{4.228a}$$

$$\mathbf{Y}_{SD,k} = \sqrt{p_{SD}} h_{SD} \mathbf{X}_k + \mathbf{N}_{SD,k}, \tag{4.228b}$$

where $\mathbf{Y}_{SR,k} = (y_{SR,k}^I, y_{SR,k}^P)$, $\mathbf{Y}_{SD,k} = (y_{SD,k}^I, y_{SD,k}^P)$, $p_{SR} = p_s \cdot (G_{SR})^2$, $p_{SD} = p_s \cdot (G_{SD})^2$ is the received signal power at the relay and destination, respectively, p_s is the source transmit power, and $G_{SR} = G_0 (d_{SR}/d_0)^{-\kappa/2}$, $G_{SD} = G_0 (d_{SD}/d_0)^{-\kappa/2}$ are the channel gains between the source and relay and between the source and destination, respectively; d_{SR} and d_{SD} are the distances between the source and relay, and between the source and destination, respectively; d_0 is a reference distance, G_0 is a constant depending on the carrier wavelength and κ is a path loss factor with values typically in the range $2 \leq \kappa \leq 6$. Also, h_{SR} and h_{SD} are the fading coefficients between the source and relay and between the source and destination, respectively. They are modeled as zero-mean, independent circular symmetric complex Gaussian random variables. In this section, we consider a quasi-static fading channel, for which the fading coefficients are constant within one frame and change independently from one frame to another. Furthermore, $\mathbf{N}_{SR,k} = (n_{SR,k}^I, n_{SR,k}^P)$, $\mathbf{N}_{SD,k} = (n_{SD,k}^I, n_{SD,k}^P)$ and they are zero mean complex Gaussian random variables with two sided power spectral density of $N_0/2$ per dimension. We also henceforth assume that all noise processes have the same variances, without loss of generality. Different noise variances can be taken into account by appropriately adjusting the channel gain.

In the perfect DTC, the relay decodes the received signals from the source and interleaves the decoded information bits. Let $\hat{\mathbf{B}}$ be the interleaved bit sequence of $\mathbf{B}$. $\hat{\mathbf{B}}$ is then re-encoded by the same encoder as at the source node and modulated into:

$$\mathbf{X}_R = (\mathbf{X}_{R,1}, \ldots, \mathbf{X}_{R,k}, \ldots, \mathbf{X}_{R,L}), \tag{4.229}$$

where $\mathbf{X}_{R,k} = (x_{R,k}^I, x_{R,k}^P)$. The parity symbol $\tilde{x}_{R,k} = \sqrt{p_R} x_{R,k}^P$ is then transmitted from the relay at time k and it satisfies the following transmit power constraint,

$$E(|\tilde{x}_{R,k}|^2) = p_R \tag{4.230}$$

where p_R is the transmission power limit at the relay. The corresponding received signal at the destination, denoted by $y_{RD,k}$, can be written as:

$$y_{RD,k} = G_{RD} h_{RD} \tilde{x}_{R,k} + n_{RD,k} \tag{4.231}$$

where G_{RD} and h_{RD} are the channel gain and fading coefficient between the relay and destination, $n_{rd,k}$ is a zero mean complex Gaussian random variable with two sided power spectral density of $N_0/2$ per dimension.

The overall received signals at the destination consist of two parts. One is the codeword transmitted from the source, denoted by $\mathbf{Y}_{SD} = (\mathbf{Y}_{SD}^I, \mathbf{Y}_{SD}^P)$, where the received information sequence is denoted by $\mathbf{Y}_{SD}^I = (y_{SD,1}^I, \cdots, y_{SD,L}^I)$ and the the received parity sequence by $\mathbf{Y}_{SD}^P = (y_{SD,1}^P, \cdots, y_{SD,L}^P)$. The other is the codeword of interleaved information bits transmitted from the relay, denoted by $\mathbf{Y}_{RD} = (y_{RD,1}, \cdots, y_{RD,L})$. These two parts of the signal sequences form a standard distributed turbo code.

(ii) Decoder of Perfect DTC. The decoder of the considered perfect DTC system is shown in Figure 4.49. It is very similar to the conventional turbo decoder. It consists of two concatenated component decoders. One operates on the received signals transmitted from the source $\mathbf{Y}_{SD}$ and one operates on the signals from the relay $\mathbf{Y}_{RD}$. The component decoder is based on a BCJR MAP decoding algorithm [496]. Here, $\mathbf{Y}_{SD}^I$ and $\mathbf{Y}_{SD}^P$ form the input to the first MAP decoder. It computes the *a posteriori* probability (APP) of $b_k, k = 1, \ldots, L$, denoted by $P_{SD}(b_k = w|\mathbf{Y}_{SD})$, $w \in \{0, 1\}$ as

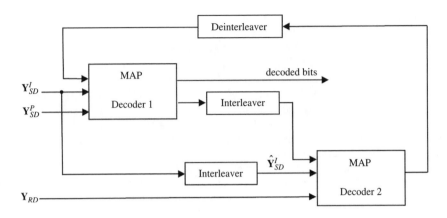

Figure 4.49 Iterative decoder of a perfect DTC system

follows:

$$P_{SD}(b_k = w | \mathbf{Y}_{SD}) = h \sum_{m,m'=0, b(k)=w}^{m,m'=M_s-1} \alpha_{k-1}(m')\beta_k(m)\gamma_k(m, m') \tag{4.232}$$

where h is a constant such that $\sum_w P_{SD}(b_k = w | \mathbf{Y}_{SD}) = 1$, and

$$\gamma_k(m, m') = \exp\left(-\frac{\left|y_{SD,k}^I - \sqrt{P_{SD}}h_{SD}x_k^I\right|^2 + \left|y_{SD,k}^P - \sqrt{P_{SD}}h_{SD}x_k^P\right|^2}{N_0}\right) \tag{4.233}$$

where m and m' are a pair of states connected with $b_k = w$ in the trellis, M_s is the number of states in the trellis, $\alpha_i(m')$ and $\beta_i(m)$ are the feedforward and the feedback recursive variables. These can be calculated as:

$$\alpha_i(m') = \sum_m \alpha_{i-1}(m)\gamma_i(m, m'), \tag{4.234a}$$

$$\beta_{i-1}(m) = \sum_{m'} \beta_i(m')\gamma_i(m, m'). \tag{4.234b}$$

We assume that the encoder clears the register at the end of the encoding operation for each codeword. Then the boundary values for $\alpha_i(m')$ and $\beta_i(m)$ are $\alpha_0(0) = 1$ and $\alpha_0(m) = 0$ for $m \neq 0$ and $\beta_l(0) = 1$ and $\beta_l(m) = 0$ for $m \neq 0$.

Similarly, $\mathbf{Y}_{RD}$, together with the deinterleaved information sequence of $\mathbf{Y}_{SD}$, denoted by $\hat{\mathbf{Y}}_{SD}^I$, form the input to the second MAP decoder. It computes the APPs for $\hat{b}_k, k = 1, \ldots, L$, which are the interleaved information bits of $\mathbf{B}$. These APPs are computed as follows:

$$P_{RD}(\hat{b}_k = w | \mathbf{Y}_{RD}, \hat{\mathbf{Y}}_{SD}^I) = h \sum_{m,m'=0, b(k)=w}^{m,m'=M_s-1} \alpha_{k-1}(m')\beta_k(m)\gamma_k(m, m') \tag{4.235}$$

where

$$\gamma_k(m, m') = \exp\left(-\frac{\left|\hat{y}_{SD,k}^I - \sqrt{P_{SD}}h_{SD}x_{R,k}^I\right|^2 + \left|y_{RD,k}^P - G_{RD}h_{RD}\sqrt{P_R}x_{R,k}^P\right|^2}{N_0}\right) \tag{4.236}$$

A turbo iterative decoding algorithm is then performed between these two decoders as shown in Figure 4.49. The first MAP decoder calculates the APPs and the extrinsic probabilities of the information bits and the second computes those of the interleaved information bits, respectively. The extrinsic information of the first decoder is then interleaved and used to produce an improved estimate of the *a priori* probabilities for the second decoder in the next iteration. Similarly, the extrinsic information of the second decoder is deinterleaved and passed to the first decoder to update its *a priori* probabilities. After a few iterations, the decision is made based on the APPs of the first decoder.

(iii) Performance Analysis. In this section, we analyze the performance of perfect DTC. The calculation of the traditional union bound requires knowledge of the number of codewords with various Hamming weights, which requires an exhaustive search of the code trellis. Due to the high complexity of this search, we consider an average upper bound [497].

As stated above, a DTC codeword consists of the codeword of information bits transmitted from the source and the parities of the interleaved information bits sent from the relay. These two signal sequences have different SNRs at the destination. Let γ_{sd} and γ_{rd} denote the corresponding instantaneous SNRs of these two signal sequences. They are given by:

$$\gamma_{sd} = \overline{\gamma}_{SD}|h_{SD}|^2, \tag{4.237a}$$

$$\gamma_{rd} = \overline{\gamma}_{RD}|h_{RD}|^2, \tag{4.237b}$$

where $\overline{\gamma}_{SD} = \frac{p_{SD}}{N_0}$, $\overline{\gamma}_{RD} = \frac{p_{RD}}{N_0}$ and $p_{RD} = p_R G_{RD}^2$ is the received power at the destination for the signals transmitted from the relay. Let $\varphi_{SD} = |h_{SD}|^2$ and $\varphi_{RD} = |h_{RD}|^2$. The PDF of φ_{SD} and φ_{RD}, denoted by $p(\varphi_{SD})$ and $p(\varphi_{RD})$, are given by $p(\varphi_{SD}) = e^{-\varphi_{SD}}$ and $p(\varphi_{RD}) = e^{-\varphi_{RD}}$, respectively. Let us now calculate the average error probability upper bound of the perfect DTC scheme. Assume that an all-zero codeword is transmitted, then the PEP that the decoder decides in favor of another erroneous codeword with Hamming weight d, for linear codes, is given by [491, 498]:

$$P(d|h_{SD}, h_{RD}) = Q(\sqrt{2d_1\gamma_{sd} + 2d_2\gamma_{rd}}) < \frac{1}{2}\exp\left(-(d_1\gamma_{sd} + d_2\gamma_{rd})\right)$$

$$= \frac{1}{2}\exp\left(-(d_1\overline{\gamma}_{SD}\varphi_{SD} + d_2\overline{\gamma}_{RD}\varphi_{RD})\right) \tag{4.238}$$

where d_1 and d_2 are the Hamming weights of the erroneous codewords with Hamming weight d, transmitted from the source and the relay, respectively, such that $d = d_1 + d_2$.

The average sequence error probability of decoding an erroneous code sequence with weight d, denoted by $P^{perfect}(d)$, can be derived by averaging Equation (4.238) over the fading coefficients φ_{SD} and φ_{RD}. It can be calculated as:

$$P^{perfect}(d) \leq \frac{1}{2}(d_1\overline{\gamma}_{SD})^{-1}(d_2\overline{\gamma}_{RD})^{-1} \tag{4.239}$$

Let $P_b^{Perfect}$ be the average upper bound on the BER of perfect DTC. It can be expressed as:

$$P_b^{Perfect} \leq \sum_{d=d_{min}}^{3L} \sum_{i=1}^{L} \frac{i}{L}\binom{L}{i} p(d|i) P^{Perfect}(d) \tag{4.240}$$

$$\simeq \frac{1}{2}(\overline{\gamma}_{SD})^{-1}(\overline{\gamma}_{RD})^{-1} \sum_{d=d_{min}}^{3L} \sum_{i=1}^{L} \frac{i}{L}\binom{L}{i} p(d|i)(d_1 d_2)^{-1}$$

where $\binom{L}{i}$ is the number of words with Hamming weight i and $p(d|i)$ is the probability that an input word with Hamming weight i produces a codeword with Hamming weight d. We see from Equation (4.240) that the perfect DTC can achieve a diversity order of 2. We should point out that the union bound is tight for independent fast fading channels only [498, 499]. In order to get a tight upper bound over quasi-static fading channels, we should limit the conditional upper bound in Equation (4.238) before averaging over the fading coefficients [498], but no closed-form expression can be obtained. Here we only use the perfect DTC as a performance low bound and a reference to compare other nonideal DTC schemes, which will be introduced in the next two sections. Since we are only interested in the relative performance loss of other nonideal DTC schemes compared to perfect DTC, the average union bound is sufficient for evaluating this performance loss.

4.3.3.3 Distributed Turbo Coding with Soft Information Relaying

In the perfect DTC scheme, we assumed that the relay can always decode correctly. In practical systems, when the source–relay link suffers from deep fading, decoding errors may occur at the relay. In this case, if the relay re-encodes these incorrect bits, then error propagation will take place and lead to even worse performance. This error propagation can be quite severe in practical systems and the assumption of perfect decoding may not be realistic. Although the source–relay link reliability can be improved by using automatic repeat request (ARQ) in the link from the source to the relay, the use of ARQ will considerably reduce the system throughput. Therefore, how to deal with imperfect decoding at the relay in the distributed turbo coding has been a challenging problem.

In Section 4.2.3, we introduced a soft information relaying (SIR) protocol to mitigate the error propagation. In this section, we apply the SIR protocol to distributed turbo coding and introduce a distributed turbo coding with soft information relaying (DTC-SIR). However, the extension is not so straightforward, as will be explained below. We know from Section 4.2.3 that in the SIR the decoder at the relay can calculate the soft outputs for the information symbols. However, in distributed turbo coding, the soft outputs for parity symbols, corresponding to the interleaved information sequence, cannot be obtained by interleaving soft outputs of the parity symbols. A soft encoding method has been proposed [454] to calculate this estimate by using the APP of the information symbol. The scheme is referred to as the distributed turbo coding scheme with soft information relaying (DTC-SIR). The basic principle behind the DTC-SIR is a probability inference method. The relay first calculates the APP of the information bits. It then uses these APPs to infer the APPs of parity symbols for the interleaved information sequence. The probability inference method follows the code trellis to calculate the probability of each parity symbol of the interleaved information sequence at time k based on the probability of all symbols prior to time k. After obtaining the probabilities of the parity symbols of the interleaved information, the corresponding parity symbol soft estimates can then be calculated as the mean value of their modulated signals.

In this section, we present the encoder structure of the DTC-SIR scheme and analyze its performance. The performance of the DTC-SIR is also evaluated through simulations.

(i) Encoder Structure of DTC-SIR. The block diagram of the DTC-SIR system is shown in Figure 4.50. In the DTC-SIR, the processing at the relay can be divided into two steps. The decoder first uses a MAP decoding to calculate the APPs of the information symbols. At the second step,

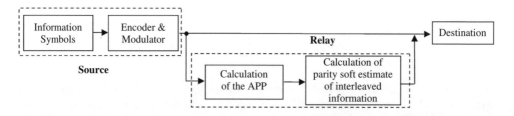

Figure 4.50   Encoder structure of DTC-SIR

the relay utilizes the derived APPs to calculate the parity symbol soft estimates for the interleaved information symbols. These two steps are now dealt with in greater detail.

- **Calculation of the APPs for the information symbols.** After receiving signals from the source, the relay first uses $\mathbf{Y}_{SR}$ to calculate the APPs of the information symbols. Let $P(b_k = w|\mathbf{Y}_{SR})$, $w \in \{0, 1\}$ be the APP of b_k, $k = 1, \ldots, l$ and they can be calculated as follows:

$$P(b_k = w|\mathbf{Y}_{SR}) = h \sum_{\substack{m,m'=0,b(k)=w}}^{m,m'=M_s-1} \alpha_{k-1}(m')\beta_k(m)\gamma_k(m, m') \qquad (4.241)$$

where

$$\gamma_k(m, m') = \exp\left(-\frac{|y_{SR,k}^I - \sqrt{p_{SR}}h_{SR}x_k^I|^2 + |y_{SR,k}^P - \sqrt{p_{SR}}h_{SR}x_k^P|^2}{N_0}\right). \qquad (4.242)$$

- **Calculation of parity symbol soft estimates for the interleaved source information.** In the next step, the relay uses the APPs of the information symbols, $P(b_k = w|\mathbf{Y}_{SR})$, $k = 1, \ldots, l$, to calculate the parity symbol soft estimates for the interleaved information symbols. Let

$$\hat{\mathbf{B}} = (\hat{b}_1, \ldots, \hat{b}_k, \ldots, \hat{b}_L) = (b_{p_1}, \ldots, b_{p_k}, \ldots, b_{p_L}) \qquad (4.243)$$

represent the interleaved version of the binary information stream $\mathbf{B}$, where $\hat{b}_k$, $k = 1, \ldots, l$, is the kth symbol of $\hat{\mathbf{B}}$, and $\hat{b}_k = b_{p_k}$ represents the interleaving operation in which the p_kth bit in $\mathbf{B}$ becomes the kth symbol in $\hat{\mathbf{B}}$. Let $\hat{\mathbf{C}}$ be the vector of parity symbols of $\hat{\mathbf{B}}$, denoted by:

$$\hat{\mathbf{C}} = (\hat{c}_1, \ldots, \hat{c}_k, \ldots, \hat{c}_L) \qquad (4.244)$$

where $\hat{c}_k$ is the parity symbol of $\hat{b}_k$. Let

$$\mathbf{P}_\mathbf{B} = \{P(b_k = w|\mathbf{Y}_{SR}), w = 0, 1, k = 1, \ldots, L\} \qquad (4.245)$$

represent the set of the APPs of the information symbols and let

$$\mathbf{P}_{\hat{\mathbf{B}}} = \{P(\hat{b}_k = w|\mathbf{Y}_{SR}), w = 0, 1, k = 1, \ldots, L\} \qquad (4.246)$$

where $P(\hat{b}_k = w|\mathbf{Y}_{SR}) = P(b_{p_k} = w|\mathbf{Y}_{SR})$ is the APP of the kth interleaved information symbol, which is also the APP of the p_kth un-interleaved information symbol.

Now let us calculate the APP of $\hat{c}_k$ given $\mathbf{P}_\mathbf{B}$, or equivalently, given $\mathbf{P}_{\hat{\mathbf{B}}}$, denoted by:

$$P(\hat{c}_k = \hat{w}_c|\mathbf{Y}_{SR}, \mathbf{P}_{\hat{\mathbf{B}}}), w_c \in \{0, 1\} \qquad (4.247)$$

We calculate this probability in the following recursive format:

$$
\begin{aligned}
&P(\hat{c}_k = \hat{w}_c|\mathbf{Y}_{SR}, \mathbf{P}_{\hat{\mathbf{B}}}) \\
&= \sum_{m \in U(\hat{c}_k = \hat{w}_c)} P(\hat{b}_k = \hat{w}_b|\mathbf{Y}_{SR}, \mathbf{P}_{\hat{\mathbf{B}}}, g(k-1) = m)P(g(k-1) = m|\mathbf{Y}_{SR}, \mathbf{P}_{\hat{\mathbf{B}}}) \quad (4.248) \\
&= \sum_{m \in U(\hat{c}_k = \hat{w}_c)} P(\hat{b}_k = \hat{w}_b|\mathbf{Y}_{SR})P(g(k-1) = m|\mathbf{Y}_{SR}, \mathbf{P}_{\hat{\mathbf{B}}})
\end{aligned}
$$

$$
\begin{aligned}
&P(g(k) = m|\mathbf{Y}_{SR}, \mathbf{P}_{\hat{\mathbf{B}}}) \\
&= \sum_{m'} P(g(k) = m|g(k-1) = m', \mathbf{Y}_{SR}, \mathbf{P}_{\hat{\mathbf{B}}})P(g(k-1) = m'|\mathbf{Y}_{SR}, \mathbf{P}_{\hat{\mathbf{B}}}) \\
&= \sum_{m'} P(b(m, m')|\mathbf{Y}_{SR})P(g(k-1) = m'|\mathbf{Y}_{SR}, \mathbf{P}_{\hat{\mathbf{B}}}) \qquad (4.249)
\end{aligned}
$$

where $U(\hat{c}_k = \hat{w}_c)$ is the set of branches, for which the output parity symbol is equal to $\hat{w}_c$, $\hat{w}_b$ is the output information symbol corresponding to the parity symbol $\hat{w}_c$ and the trellis state $g(k-1) = m$. $P(\hat{b}_k = \hat{w}_b|\mathbf{Y}_{SR}, \mathbf{P}_{\hat{\mathbf{B}}}, g(k-1) = m)$ represents the APP of information symbol $\hat{w}_b$ at time k, which is equivalent to $P(\hat{b}_k = \hat{w}_b|\mathbf{Y}_{SR})$. $P(g(k-1) = m|\mathbf{Y}_{SR}, \mathbf{P}_{\hat{\mathbf{B}}})$ is the probability of the state m at time $k-1$, $b(m, m')$ represents the input information symbol resulting in the transition from state m' at time $(k-1)$ to m at time k and $P(b(m, m')|\mathbf{Y}_{SR})$ is the APP of information symbol $b(m, m')$, at time k. Finally, we assume that the encoder clears the register at the end of the encoding operation for each codeword, resulting in the initial state probabilities

$$P(g(0) = 0) = 1 \text{ and } P(g(0) = m, m \neq 0) = 0. \tag{4.250}$$

- **Example.** To better understand the principle of soft encoding processing in Equations (4.248) and (4.249), let us consider the following example where we consider the trellis shown in Figure 4.51. In this figure, $m_i, i = 0, 1, 2, 3$ denote the trellis state, b/bp in each branch of the trellis represents that state when the input bit is b, the encoder output information bit and parity bit are b and p, respectively. Figure 4.52 shows the corresponding trellis when calculating $P(\hat{c}_k = 0|\mathbf{Y}_{SR}, \mathbf{P}_{\hat{\mathbf{B}}})$ in Equation (4.248), where the summation in Equation (4.248) is over all branches, whose output parity bit is equal to 0 ($p = 0$). In the example shown in Figure 4.52, $P(\hat{c}_k = 0|\mathbf{Y}_{SR}, \mathbf{P}_{\hat{\mathbf{B}}})$ can be further written as:

$$P(\hat{c}_k = 0|\mathbf{Y}_{SR}, \mathbf{P}_{\hat{\mathbf{B}}}) = P(\hat{b}_k = 0|\mathbf{Y}_{SR})P(g(k-1) = m_0|\mathbf{Y}_{SR}, \mathbf{P}_{\hat{\mathbf{B}}})$$
$$+ P(\hat{b}_k = 0|\mathbf{Y}_{SR})P(g(k-1) = m_1|\mathbf{Y}_{SR}, \mathbf{P}_{\hat{\mathbf{B}}})$$
$$+ P(\hat{b}_k = 1|\mathbf{Y}_{SR})P(g(k-1) = m_2|\mathbf{Y}_{SR}, \mathbf{P}_{\hat{\mathbf{B}}})$$
$$+ P(\hat{b}_k = 1|\mathbf{Y}_{SR})P(g(k-1) = m_3|\mathbf{Y}_{SR}, \mathbf{P}_{\hat{\mathbf{B}}}) \tag{4.251}$$

where $P(\hat{b}_k = w|\mathbf{Y}_{SR})$ is given in Equation (4.246) and is obtained in the first step of calculation. And $P(g(k) = m|\mathbf{Y}_{SR}, \mathbf{P}_{\hat{\mathbf{B}}})$ can be calculated recursively using Equation (4.249). Figure 4.53 gives an example of calculating $P(g(k) = m_2|\mathbf{Y}_{SR}, \mathbf{P}_{\hat{\mathbf{B}}})$, where the summation in Equation (4.249) is over all branches, whose next state is m_2. In this example, $P(g(k) = m_2|\mathbf{Y}_{SR}, \mathbf{P}_{\hat{\mathbf{B}}})$ can be further written as:

$$P(g(k) = m|\mathbf{Y}_{SR}, \mathbf{P}_{\hat{\mathbf{B}}}) = P(\hat{b}_k = 1)P(g(k-1) = m_0|\mathbf{Y}_{SR}, \mathbf{P}_{\hat{\mathbf{B}}})$$
$$+ P(\hat{b}_k = 0)P(g(k-1) = m_1|\mathbf{Y}_{SR}, \mathbf{P}_{\hat{\mathbf{B}}}). \tag{4.252}$$

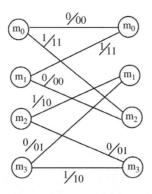

Figure 4.51   Trellis diagram of example component code

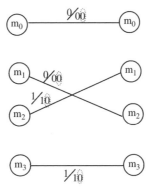

Figure 4.52 Trellis diagram when calculating $P(\hat{c}_k = 0|\mathbf{Y}_{SR}, \mathbf{P}_{\hat{\mathbf{B}}})$ in Equation (4.248)

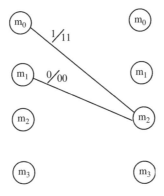

Figure 4.53 Trellis diagram when calculating $P(g(k) = m_2|\mathbf{Y}_{SR}, \mathbf{P}_{\hat{\mathbf{B}}})$ in Equation (4.249)

Therefore, by using Equations (4.248) and (4.249), we can recursively calculate the APPs of $\hat{\mathbf{C}}$ by using the APPs of $\mathbf{B}$. In Equations (4.248)–(4.250), we only consider a forward recursive calculation of the APP of $\hat{\mathbf{C}}$. A more exact calculation method can be derived by using both forward and backward recursive operations, in which more calculations are needed.

We assume that in the considered BPSK modulation signal set, the binary bit 0 and 1 are mapped into 1 and -1, respectively. The parity symbol soft estimates of $\hat{c}_k$, $k = 1, \ldots, l$, denoted by $\hat{x}^P_{R,k}$, can then be calculated as follows:

$$\hat{x}^P_{R,k} = P(\hat{c}_k = 0|\mathbf{Y}_{SR}, \mathbf{P}_{\hat{\mathbf{B}}}) \cdot 1 + P(\hat{c}_k = 1|\mathbf{Y}_{SR}, \mathbf{P}_{\hat{\mathbf{B}}}) \cdot (-1). \tag{4.253}$$

Following the modeling of soft symbol estimate (SSE) in Section 4.2.3, we represent $\hat{x}^P_{R,k}$ in the following manner:

$$\hat{x}^P_{R,k} = x^P_{R,k}(1 - \hat{n}_k) \tag{4.254}$$

where $\hat{n}_k \geq 0$ is an equivalent noise, with the mean:

$$\mu_{\hat{n}} = \frac{1}{l} \sum_{k=1}^{l} \hat{n}_k = \frac{1}{l} \sum_{k=1}^{l} (1 - \hat{x}^P_{R,k} x^P_{R,k}) = \frac{1}{l} \sum_{k=1}^{l} |\hat{x}^P_{R,k} - x^P_{R,k}| \tag{4.255}$$

and variance:

$$\sigma_{\hat{n}}^2 = \frac{1}{l} \sum_{k=1}^{l} \left(1 - \hat{x}_{R,k}^P x_{R,k}^P - \mu_{\hat{n}}\right)^2 . \tag{4.256}$$

The signals transmitted from the relay can then be written as:

$$\tilde{x}_{R,k} = \beta \hat{x}_{R,k}^P = \beta x_{R,k}^P (1 - \hat{n}_k) \tag{4.257}$$

where β is a normalization factor calculated from the transmission power constraint at the relay:

$$E(|\hat{x}_{R,k}|^2) = \beta^2 \left(1 - 2\mu_{\hat{n}} + E(\hat{n}_k^2)\right) = \beta^2 \left((1 - \mu_{\hat{n}})^2 + \sigma_{\hat{n}}^2\right) \le p_R \tag{4.258}$$

where p_R is the transmission power limit at the relay. Let $\overline{\omega} = (1 - \mu_{\hat{n}})^2 + \sigma_{\hat{n}}^2$, then from Equation (4.258), we can obtain:

$$\beta \le \sqrt{\frac{p_R}{\overline{\omega}}}. \tag{4.259}$$

The destination received signal, corresponding to the relay signal at time k, can be written as:

$$Y_{RD,k} = G_{RD} h_{RD} \tilde{x}_{R,k} + n_{RD,k} = G_{RD} h_{RD} \beta x_{R,k}^P (1 - \mu_{\hat{n}}) + \overline{n}_{RD,k} \tag{4.260}$$

where $\overline{n}_{RD,k} = n_{RD,k} - G_{RD} h_{RD} \beta x_{R,k}^P (\hat{n}_k - \mu_{\hat{n}})$ is the equivalent noise at the destination, with a zero mean and variance of σ_E^2, given by:

$$\sigma_E^2 = N_0 + G_{rd}^2 |h_{RD}\beta|^2 \sigma_{\hat{n}}^2 . \tag{4.261}$$

The overall distributed codeword consists of the coded information symbols transmitted from the source and the parity symbols of the interleaved information sequence sent from the relay. These two signal sequences are decoded by two separate decoders at the destination. Similar to the perfect DTC, a turbo iterative decoding algorithm is performed between these two decoders.

For the optimal decoder, we have to find the distribution of overall destination noise $\overline{n}_{RD,k}$. The distributions of $\overline{n}_{RD,k}$ in SIR-SSE and DTC-SIR could be much different. This is because in the SIR the relay decoder directly calculates the SSE of the information symbols, while in DTC-SIR the relay has to calculate the SSE for parity symbols corresponding to the interleaved information sequence and this SSE is heavily dependent on the constituent code used at the source encoder. If the constituent code is a nonrecursive convolutional code, the SSE for parity symbols of the interleaved information will have the similar distribution as in SIR-SSE. However, if the constituent code is a recursive systematic convolutional code (RSCC), the distribution of SSE could become different and it is very difficult to evaluate the distribution of $\overline{n}_{RD,k}$ in this case. However, for turbo codes, the constituent code has to be the RSCC and this make the modeling of $\overline{n}_{RD,k}$ and quest for the optimal decoding metrics in DTC-SIR very complicated and challenging.

To make the decoder tractable and simple, we still assume that $\overline{n}_{RD,k}$ follows an approximated Gaussian distribution, the branch metric $\gamma_k(m, m')$ in the MAP decoder, associated with $y_{RD,k}$, is somewhat different from Equation (4.236) and should be modified as:

$$\gamma_k(m, m') = \exp\left(-\frac{||y_{RD,k} - G_{RD} h_{RD} \beta x_{R,k}^P (1 - \mu_{\hat{n}})|^2}{\sigma_E^2}\right). \tag{4.262}$$

Similar to the SIR-SSE, we may have to rectify σ_E^2, the variance of $\overline{n}_{RD,k}$, when assuming that $\overline{n}_{RD,k}$ is Gaussian distributed. The optimal σ_E^2 can be found through simulations and this requires quite a considerable amount of work. For simplicity, in the following simulation results, we do not take into account the rectification of σ_E^2, thus the system performance could possibly be further improved if the accurate estimate of σ_E^2 is used at the decoder. We should also note that the decoding metrics in

Equation (4.262) is not optimal as the actual distribution of $\bar{n}_{RD,k}$ may not be the Gaussian distribution. Therefore, the DTC-SIR can be further improved by accurately modeling the distribution of $\bar{n}_{RD,k}$ and designing the decoders accordingly. It is still an open issue to find such an optimal decoder for the DTC-SIR.

(iii) Performance Evaluation. In this section, we provide simulation result comparisons for various DTC schemes. All simulations are performed for BPSK modulation and a frame size of 130 symbols over quasi-static fading channels. For simplicity, we also assume that $\bar{\gamma}_{SD}$ is equal to $\bar{\gamma}_{RD}$. We use a four-state recursive systematic convolutional code (RSC) with the code rate of $1/2$ as the turbo component code. The generator matrix of the RSC is $(1, 5/7)$ with d_{free} of 5. We carry out the performance comparisons under the following two scenarios.

- **Comparison for various $\bar{\gamma}_{SR}$.** In this case, we investigate the performance of the DTC schemes when signal power from the source to the relay is fixed, while the power from the source to the destination and from the relay to the destination are varied.

  For the conventional DTC with no ARQ scheme, the relay re-encodes the decoded codeword and passes it to the destination regardless of whether the relay can make a correct decoding or not. Therefore, when decoding errors occur at the relay, the relay will make an erroneous symbol decision, and re-encoding the erroneous decoded symbols will cause error propagation. We refer to such a scheme as the DTC with no ARQ. In the simulations, we use a log-map decoding algorithm at the destination. Unlike the DTC-SIR, in which the equivalent noise variance, corresponding to the destination noise variance and the error variance in soft estimates, can be calculated as shown in Equations (4.255) and (4.256), for DTC with no ARQ, the destination decoder has no knowledge of erroneous codewords introduced by the relay. Thus the decoder only gets the destination noise variance from the destination in the decoding. This will further degrade its performance.

  As discussed before, the performance of DTC can be improved further by using an ARQ scheme in the link from the source to the relay. Such a scheme performs very close to perfect DTC. We refer to this scheme as DTC with ARQ. In the following simulation results, for DTC with ARQ, we assume that the maximum number of retransmissions (MNR) is three and a transmission is said to be a failure if it is still unsuccessful after the maximum number of retransmissions. In this case the relay will discard the frame.

  Figures 4.54–4.56 compare the BER performance of DTCs with and without ARQ, and the proposed DTC-SIR, as a function of $\bar{\gamma}_{RD}$ and with $\bar{\gamma}_{SR}$ as a parameter. It can be seen from these figures that due to error propagation, DTC with no ARQ performs badly and achieves an error floor for high $\bar{\gamma}_{RD}$, especially at lower $\bar{\gamma}_{SR}$. In contrast, DTC-SIR can effectively mitigate the effect of error propagation. Its performance approaches DTC with ARQ as $\bar{\gamma}_{SR}$ increases.

- **Comparison for various SNR gaps between $\bar{\gamma}_{SR}$ and $\bar{\gamma}_{RD}$.** Here we compare the performance of DTC for various SNR gaps, which is defined as the difference between $\bar{\gamma}_{SR}$ and $\bar{\gamma}_{RD}$ in dBs. We assume that the signal energy decays exponentially with distance between two nodes as shown in Section 2. Obviously, SNR_{Gap} is determined by the ratio of power transmitted from the source and the relay, as well as the relative distance of d_{SR} and d_{RD}. More specifically,

$$SNR_{Gap} = 10\log\left(\frac{p_s(G_{SR})^2}{p_R(G_{RD})^2}\right) = 10\log\left(\frac{p_s}{p_r}\left(\frac{d_{SR}}{d_{RD}}\right)^{-\kappa}\right) \qquad (4.263)$$

where p_s is the source transmission power, p_{SR}/p_R and $(d_{SR}/d_{RD})^{-\kappa}$ are referred to as the power amplification factor and the geometrical distribution factor, denoted by ρ_P^{sr} and ρ_d^{sr}, respectively. ρ_P^{sr} denotes the ratio of the source and relay transmit power while ρ_d^{sr} represents the ratio of d_{SR} and d_{RD}, which depends on the network geometrical distribution. We evaluate here the performance of DTC-SIR for various SNR gaps to investigate the effect of the power amplification factor and the network geometrical distribution on system performance and throughput.

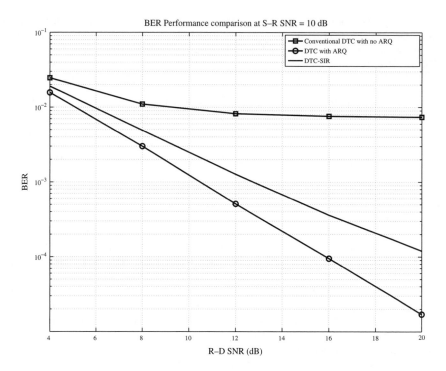

Figure 4.54   BER performance comparison at $\overline{\gamma}_{SR} = 10$ dB

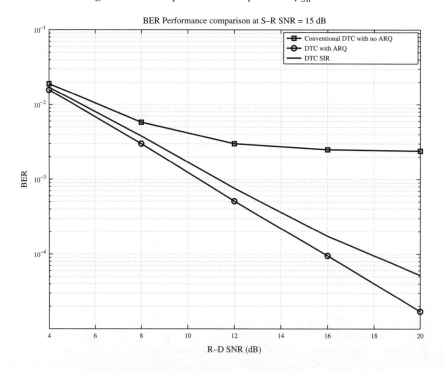

Figure 4.55   BER performance comparison at $\overline{\gamma}_{SR} = 15$ dB

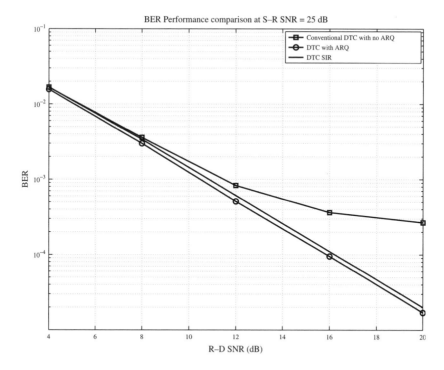

Figure 4.56   BER performance comparison at $\overline{\gamma}_{SR} = 25$ dB

Figures 4.57 and 4.58 show the BER performance comparison for various SNR gaps. It can be observed that DTC-SIR significantly outperforms DTC with no ARQ and it approaches the performance of DTC with ARQ as SNR_{Gap} increases. The performance of the DTC-SIR is worse than that of DTC with ARQ by about 1.5 dB and 0.5 dB for SNR_{Gap} of 0 dB and 4 dB, at a BER of 10^{-4}, respectively. Therefore if the SNR_{Gap} between $\overline{\gamma}_{SR}$ and $\overline{\gamma}_{RD}$ is large enough, then the performance of DTC-SIR approaches the DTC with ARQ.

As shown in Equation (4.263), the SNR_{Gap} is determined by the power amplification factor ρ_P^{sr} and the geometrical distribution factor ρ_d^{sr}. In order to increase the SNR_{Gap} enabling DTC-SIR to approach the perfect DTC, we should increase ρ_P^{sr} and ρ_d^{sr}. This can be achieved by placing the relay closer to the source than to the destination and/or making the transmit power from the source larger than that from the relay.

4.3.3.4 Generalized Distributed Turbo Coding

In the previous section, we presented a practical distributed coding scheme designed for the system when taking into account decoding errors at the relay. This scheme is mainly designed for a relay network with a single relay node. It is very important to develop a DTC scheme for a general relay network with any number of relays. In this section we thus present a generalized DTC scheme (GDTC) for a general two hop relay network with an arbitrary number of relays. In GDTC, at each transmission, based on whether relays can make correct decoding or not, all relays are divided into two groups, which we refer to as the DF relay group and the AF relay group. This is similar to the approach already taken in Section 4.2.4. All relays, which can correctly decode the signals transmitted from the source, are included in the DF relay group and the remaining relays that fail

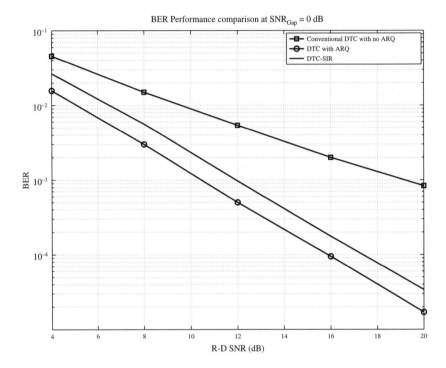

Figure 4.57   BER performance of the proposed scheme at $SNR_{Gap} = 0$ dB

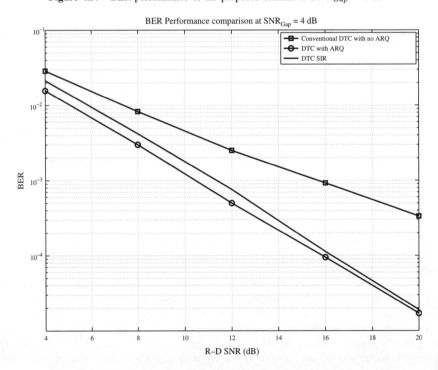

Figure 4.58   BER performance of the proposed scheme at $SNR_{Gap} = 4$ dB

to make a correct decoding are included in the AF relay group. Each relay in the AF relay group amplifies the received signals from the source and forwards it to the destination, while each relay in the DF relay group decodes the received signals, interleaves, re-encodes the interleaved symbols and forwards them to the destination. At the destination, all signals forwarded from the relays in the AF relay group are combined into one signal and that from the DF relay group are combined into another signal. After combination, the overall codeword consists of combined coded information symbols transmitted from the relays in the AF relay group and combined coded interleaved information symbols transmitted from the relays in the DF relay group. These two signals thus form a generalized DTC codeword.

Similarly to the ARP, in a practical system, in order to determine whether a relay can decode correctly or not, some CRC bits can be appended to each information block (frame). After decoding each frame, the relay can examine the CRC checks to determine whether the received signals are decoded correctly or not. To reduce the complexity, each relay only needs to compare its received SNR with a threshold calculated based on code Bhattacharyya parameter to determine whether a relay can decode correctly or not. That is, only those relays whose received SNR is above the threshold do the decoding, and the rest of the relays are assumed not to be able to decode correctly and thus will not perform decoding.

In this section, we introduce the encoder structure of the GDTC scheme and present its performance analysis. The performance of the GDTC is also evaluated through simulations.

(i) Encoder Structure of Generalized DTC. We consider a general two-hop relay network, consisting of one source, n relays and one destination and assume that there is a direct link from the source to the destination. We follow the same notations as in the previous sections. We also consider a recursive systematic convolutional code (RSCC) with a code rate of $1/2$ and BPSK modulation. We also assume that all the relays have the same transmission power of p_R. Figure 4.59 shows the block diagram of the GDTC scheme.

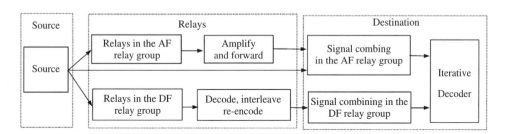

Figure 4.59   Block diagram of the GDTC scheme

Let $\mathbf{B} = (b_1, \ldots, b_k, \ldots, b_L)$ be the transmitted source binary stream, where L is the information block length. $\mathbf{B}$ is first encoded by a channel encoder and modulated into $\mathbf{X} = (\mathbf{X}_1, \ldots, \mathbf{X}_k, \ldots, \mathbf{X}_L)$, where $\mathbf{X}_k = \left(x_k^I, x_k^P \right)$, $x_k^I, x_k^P \in \{-1, 1\}$ are the modulated information and parity signals, which are transmitted by the source at time $2k - 1$ and $2k$, respectively. $\mathbf{X}_k$ is then broadcast to both the destination and relays. The corresponding received signals at the relay i and destination are given by:

$$\mathbf{Y}_{SR,k}^i = \sqrt{p_{SR}^i} h_{SR}^i \mathbf{X}_k + \mathbf{N}_{SR,k}^i \tag{4.264a}$$

$$\mathbf{Y}_{SD,k} = \sqrt{p_{SD}} h_{SD} \mathbf{X}_k + \mathbf{N}_{SD,k} \tag{4.264b}$$

where $p_{SR}^i = p_s \cdot (G_{SR}^i)^2$, $p_{SD} = p_s \cdot (G_{SD})^2$ are the received signal power at the relay i and destination, respectively, p_s is the source transmit power, G_{SR}^i and G_{SD} are the channel gains between the source and relay i and that between the source and destination, respectively. Also h_{SR}^i and h_{SD} are the fading coefficients between the source and the relay i and between the source and destination, respectively. Furthermore, $\mathbf{N}_{SR,k}^i$ and $\mathbf{N}_{SD,k}$ are noise vectors and each noise term is a zero mean complex Gaussian random with two-sided power spectral density of $N_0/2$ per dimension.

Similarly to ARP scheme introduced in Section 4.2.4, in GDTC, in each transmission, based on whether relays can make correct decoding or not, each relay is included in either an AF or a DF relay group, denoted by Ω_{AF} and Ω_{DF}, respectively.

Each relay in the AF group will simply forward a scaled version of the received signals to the destination. At the destination, all signals forwarded from the AF relay group are then combined with the signals directly transmitted from the source as follows:

$$\mathbf{Y}_{RD-AF,k} = W_{SD}\mathbf{Y}_{SD,k} + \sum_{i \in \Omega_{AF}} W_R^i \mathbf{Y}_{RD,k}^i \tag{4.265}$$

where $\mathbf{Y}_{RD-AF,k}$ is the combined signal in the AF group at the destination, and W_{SD} and $W_R^i, i \in \Omega_{AF}$ are the combining coefficients. The corresponding destination SNR for the combined signals in the AF relay group, denoted by γ_{AF}, can be approximated by:

$$\gamma_{AF} = \overline{\gamma}_{SD}|h_{SD}|^2 + \frac{1}{2}\sum_{i \in \Omega_{AF}} H_2^i \tag{4.266}$$

where $\overline{\gamma}_{SD} = \frac{p_{SD}}{N_0}$, $\overline{\gamma}_{RD}^i = \frac{P_{RD}^i}{N_0}$, $P_{RD}^i = p_R (G_{RD}^i)^2$, $\overline{\gamma}_{SR}^i = \frac{P_{SR}^i}{N_0}$, $H_2^i = (\frac{1}{2}\sum_{p=1}^2 \frac{1}{\lambda_{i,p}})^{-1}$ is the harmonic mean of variables $\lambda_{i,p}$, $p = 1, 2$, $\lambda_{i,1} = |h_{SR}^i|^2 \overline{\gamma}_{SR}^i$, and $\lambda_{i,2} = |h_{RD}^i|^2 \overline{\gamma}_{RD}^i$.

Each relay in the DF relay group decodes the received signals and recovers the information bits $\mathbf{B}$. $\mathbf{B}$ is then interleaved into $\hat{\mathbf{B}} = (\hat{b}_1, \ldots, \hat{b}_k, \ldots, \hat{b}_L)$, encoded and modulated into $\mathbf{X}_R$, where:

$$\mathbf{X}_R = (\mathbf{X}_{R,1}, \ldots, \mathbf{X}_{R,k}, \ldots, \mathbf{X}_{R,L}) \tag{4.267}$$

where $\mathbf{X}_{R,k} = (x_{R,k}^I, x_{R,k}^P)$ is the modulated codeword of $\hat{b}_k$, $x_{R,k}^I$ and $x_{R,k}^P$ are the corresponding modulated information and parity symbol. The relay i in the DF relay group then forwards $\mathbf{X}_{R,k}$ with power p_R^i, to the destination. The corresponding received signals at the destination are given by:

$$\mathbf{Y}_{RD,k}^i = G_{RD}^i h_{RD}^i \sqrt{p_R^i} \mathbf{X}_{R,k} + \mathbf{N}_{RD,k}^i, i \in \Omega_{DF}. \tag{4.268}$$

At the destination, all signals forwarded from the DF relay group are then combined. Let $\mathbf{Y}_{RD-DF,k}$ represent the combined signal, and it is given by

$$\mathbf{Y}_{RD-DF,k} = \sum_{i \in \Omega_{DF}} W_R^i \mathbf{Y}_{RD,k}^i. \tag{4.269}$$

The destination SNR of the combined signal in DF is thus given by:

$$\gamma_{DF} = \sum_{i \in \Omega_{DF}} \overline{\gamma}_{RD}^i |h_{RD}^i|^2. \tag{4.270}$$

(ii) Iterative Decoding of GDTC. From Equations (4.265) and (4.269), we can observe that an overall codeword for a generalized distributed turbo coding, consists of the combined coded information symbols transmitted from the AF relay group, given in Equation (4.265), and the combined coded symbols of the interleaved information sequence sent from the DF relay group, shown in Equation (4.269). These two signals at the destination are denoted by $\mathbf{Y}_{RD-AF}$ and $\mathbf{Y}_{RD-DF}$, respectively. Let $\mathbf{Y}^I_{RD-AF}$, $\mathbf{Y}^I_{RD-DF}$, $\mathbf{Y}^P_{RD-AF}$, $\mathbf{Y}^P_{RD-DF}$ represent the information and parity signal sequence of $\mathbf{Y}_{RD-AF}$ and $\mathbf{Y}_{RD-DF}$, respectively. Since $\mathbf{Y}^I_{RD-AF}$ and $\mathbf{Y}^I_{RD-DF}$ carry the same information, they should be properly combined before decoding.

Let $\mathbf{I}(\mathbf{a})$ and $\mathbf{I}^{-1}(\mathbf{a})$ be the interleaving and de-interleaving version of a sequence $\mathbf{a}$. Let $\mathbf{Y}^I_{AF}$ represent the combined information signal sequence and $\mathbf{Y}^I_{DF}$ denote its interleaved one. Then we have:

$$\mathbf{Y}^I_{AF} = \mathbf{Y}^I_{RD-AF} + \mathbf{I}^{-1}(\mathbf{Y}^I_{RD-DF}) \qquad (4.271a)$$

$$\mathbf{Y}^I_{DF} = \mathbf{I}(\mathbf{Y}^I_{RD-AF}) + \mathbf{Y}^I_{RD-DF} \qquad (4.271b)$$

Further, let $\mathbf{Y}_{AF} = \left(\mathbf{Y}^I_{AF}, \mathbf{Y}^P_{RD-AF}\right)$ and $\mathbf{Y}_{DF} = \left(\mathbf{Y}^I_{DF}, \mathbf{Y}^P_{RD-DF}\right)$. Then $\mathbf{Y}_{RD-AF}$ and $\mathbf{Y}_{RD-DF}$ are fed into two convolutional decoders. A turbo iterative decoding algorithm is performed between these two decoders. Similarly to the conventional iterative turbo decoding, two MAP decoders associated with input symbols $\mathbf{Y}_{AF}$ and $\mathbf{Y}_{DF}$ calculate the *a posteriori* probability (APP) of the transmitted information symbols and the interleaved information symbols and respective extrinsic information, respectively. The extrinsic information of one decoder is used to update the *a priori* probability of the other decoder in the next iteration. After several iterations, the decision is made based on the APPs of the first decoder. Figure 4.60 shows the block diagram of a GDTC decoder.

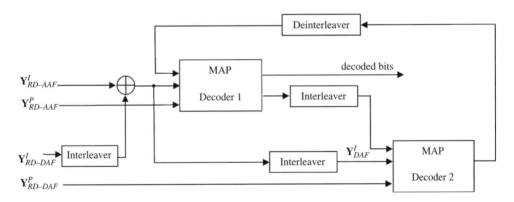

Figure 4.60 Block diagram of the GDTC iterative decoder

There is one possible scenario where for some transmission blocks there are no relays that can make correct decoding. This could happen at the low signal to noise region. In this case, there are no relays in the DF relay group and all relays are in the AF relay group, so we only need to decode $\mathbf{Y}_{RD-AF}$, from which we get the information symbol estimates.

(iii) Performance Analysis. In this section, we analyze the performance of the GDTC. Without loss of generality, we assume that $G^i_{SR} = G_{SR}$ for $i = 1, \cdots, n$. Then we have $p^i_{SR} = p_{SR}$ for $i = 1, \cdots, n$. Similarly to the analysis for the ARP in Section 4.2.4, we first calculate the PEP for a scenario where the AF relay group consists of q relays numbered from 1 to q and the DF relay group consists of $(n - q)$ relays numbered from $(q + 1)$ to n. Following the same analysis as in Equations

(4.117)–(4.119), the conditional PEP for GDTC at high SNR for this scenario, is given by:

$$P_{(q)}^{GDTC}(d|h_{SD}, \mathbf{h_{SR}}, \mathbf{h_{RD}}) \leq \prod_{i=1}^{q} P_{F,sr}^{i}(\overline{\gamma}_{SR}|h_{SR}^{i})$$

$$\prod_{i=q+1}^{n} \left(1 - P_{F,sr}^{i}(\overline{\gamma}_{SR}|h_{sr}^{i})\right) Q\left(\sqrt{2d_1 \gamma_{AF,(q)} + 2d_2 \gamma_{DF,(n-q)}}\right)$$

where d_1 and d_2 are the Hamming weights of the erroneous codewords with Hamming weight d, transmitted from the AF and DF group, respectively, such that $d = d_1 + d_2$, $\gamma_{AF,(q)}$ and $\gamma_{DF,(n-q)}$ represent the instantaneous received SNR of the combined signals in the AF and DF relay groups, given by:

$$\gamma_{AF,(q)} \approx \overline{\gamma}_{SD}|h_{SD}|^2 + \frac{1}{2}\sum_{i=1}^{q} H_2^i \tag{4.272a}$$

$$\gamma_{DF,(n-q)} = \overline{\gamma}_{RD}\sum_{i=q+1}^{n} |h_{RD}^i|^2 \tag{4.272b}$$

and $P_{F,sr}^{i}(\overline{\gamma}_{SR}|h_{SR}^{i})$ represent the conditional word error probability in the channel from the source to the ith relay, given in Equation (4.118).

Following the analysis as in Section 4.2.4, the average PEP at high SNR, denoted by $P^{GDTC}(d)$, can be calculated as:

$$P^{GDTC}(d) \leq (d_1\overline{\gamma}_{SD})^{-1}(\overline{\gamma}_{RD}d_2)^{-n}\left(1 + \frac{d_2\overline{\gamma}_{RD}}{d_1\overline{\gamma}_{SR}}\right)^n \tag{4.273}$$

Let P_b^{GDTC} be the BER upper bound for the GDTC. At the high SNR, it can be approximated as:

$$P_b^{GDTC} \leq \sum_{d=d_{\min}}^{4L} \overline{A}(d) P^{GDTC}(d) \tag{4.274}$$

where $\overline{A}(d) = \sum_{j=1}^{l} \frac{j}{l}\binom{L}{j} p(d|j)$ and $p(d|i)$ is the probability that an input word with Hamming weight i produces a codeword with Hamming weight d. From Equation (4.273), we can observe that a diversity order of $(n + 1)$ can be achieved for the GDTC scheme in a relay network with n relays, when $\overline{\gamma}_{SR}^{-1}\overline{\gamma}_{RD} \to 0$. Similarly, for a perfect DTC, in which all relay are assumed to decode correctly, the average PEP of incorrectly decoding to a codeword with weight d, denoted by $P_{DTC}^{Perfect}(d)$, can be calculated as:

$$P_{DTC}^{Perfect}(d) \leq (d_1\overline{\gamma}_{SD})^{-1}(d_2\overline{\gamma}_{RD})^{-n} \tag{4.275}$$

The average BER upper bound of perfect DTC, at high SNR, denoted by $P_b^{Perfect}$, can be finally approximated as:

$$P_b^{Perfect} \leq \sum_{d=d_{\min}}^{4l} \overline{A}(d) P_{DTC}^{Perfect}(d). \tag{4.276}$$

Equation (4.274) can be further expressed as:

$$P_b^{GDTC} \leq \sum_{d=d_{\min}}^{4l} \overline{A}(d) P_{DTC}^{Perfect}(d)\left(1 + \frac{d_2\overline{\gamma}_{RD}}{d_1\overline{\gamma}_{SR}}\right)^n. \tag{4.277}$$

It can be noted from Equation (4.277) that as $\overline{\gamma}_{SR}^{-1}\overline{\gamma}_{RD} \to 0$, $P_b^{GDTC} \to P_b^{Perfect}$ and the performance of the GDTC approaches the perfect DTC. This is the same as for the DTC-SIR scheme.

(iv) Performance Evaluation. In this section, we provide simulation results for the GDTC scheme. All simulations are performed for a BPSK modulation and a frame size of 130 symbols over quasi-static fading channels. We use a four-state recursive systematic convolutional code (RSC) with the code rate of 1/2 and the generator matrix of $(1, 5/7)$. We assume that $\gamma_{sr,i} = \gamma_{sr}$ and $\gamma_{rd,i} = \gamma_{rd}$ for all $i = 1, \cdots, n$, and γ_{rd} and γ_{sd} are the same.

Figures 4.61–4.64 compare the performance of the GDTC, distributed coding with AF (DC-AF) and the perfect DTC for various numbers of relays. In these figures, 'S-R SNR' represents $\overline{\gamma}_{SR}$ and 'R-D SNR' represents $\overline{\gamma}_{RD}$. It can be seen that as the number of relays increases, the GDTC significantly outperforms the DC-AF in all SNR regions, and performs very close to the perfect DTC as $\overline{\gamma}_{SR}$ increases. This conclusion is consistent with the analysis.

It can be seen from the above results that for low values of $\overline{\gamma}_{SR}$ and high values of $\overline{\gamma}_{RD}$, the GDTC and the DC-AF exhibit a similar performance in this case. However, as n increases, such as $n = 4, 8$, the GDTC can achieve considerable performance gain compared to the DC-AF. This can be easily explained in the following way. For low $\overline{\gamma}_{SR}$ values, the channel from the source to the relays is very noisy and probabilities of decoding errors at each relay are very high, so most of relays cannot correctly decode the received signals. In this case, as the number of relays is small, most of them are included in the AF relay group at a very high likelihood and the DF relay group only occasionally includes few relays. However even such a limited contribution of coding gain from the DF relay group is significant if the channel from the relay to the destination is poor (corresponding to the low $\overline{\gamma}_{RD}$ values), because in this case the relays in the DF relay group can considerably improve the overall channel quality. As the number of relays increases, the probability that the DF relay group contains at least one relay also increases and the contribution of coding gain from the DF relay group becomes significant. This explain the reason why the GDTC can provide a significant coding gain over DC-AF, even at low γ_{sr} values, as the number of relays increases.

Above simulation results generally confirm that GDTC can provide significant coding gains at all SNR regions and perform very closely to the perfect DTC as $\overline{\gamma}_{SR}$ increases.

4.4 Distributed Network Coding

This section treats distributed network coding, an advanced regenerative distributed space–time processing approach. We first deal with distributed network coding for a single source–destination pair and then with network coding division multiplexing for multiple source–destination pairs.

4.4.1 Distributed Network–Channel Coding

The distributed coding schemes discussed above are mainly developed for a two-hop unicast relay network, in which information bits are sent from one source to one destination through single or multiple relays. The extension of these distributed coding schemes to general multihop relay networks seems very difficult and unfeasible as these schemes do not scale well when the number of hops and network size increases. These distributed coding schemes are based on the conventional space–time coding or turbo coding paradigms, which can only be used with fixed and predefined code trellises and are only limited to a fairly small number of component codes. This clearly limits the application of these coding schemes only to rather small and static network topologies. In wireless networks, however, the wireless links between any two nodes are impaired by fading and noise, and the physical layer topology will change due to the failure of some wireless links. Although the generalized DTC scheme introduced in the previous section can provide real-time adaptation to the changing network topology to a certain degree, the structure inherited in turbo coding determines that it can only be used in two-hop networks.

In this section, we extend the distributed coding design from a simple two-hop relay system to the general network context with time-varying network topology. Network coding has recently been

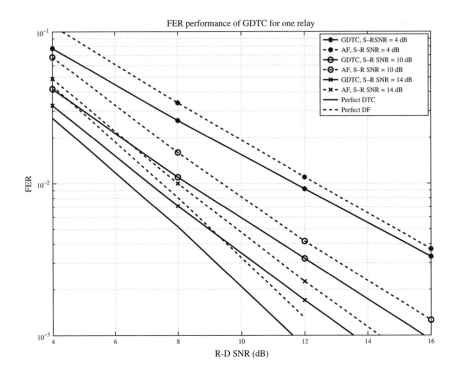

Figure 4.61   FER performance for one relay

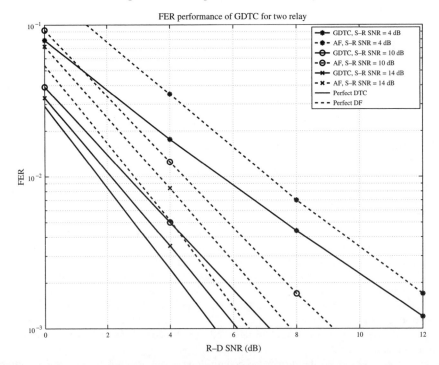

Figure 4.62   FER performance for two relays

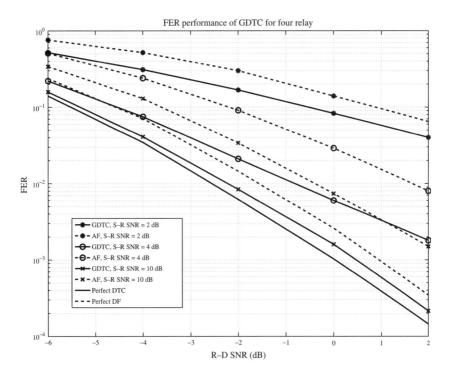

Figure 4.63 FER performance for four relays

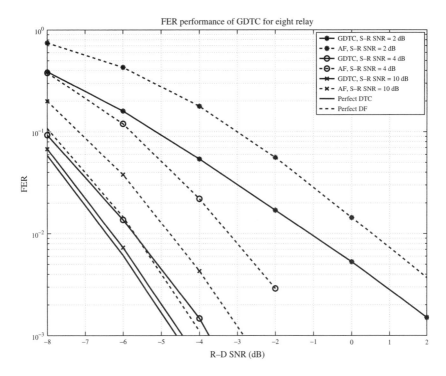

Figure 4.64 FER performance for eight relays

shown to be an effective technique to increase the network's spectral efficiency by using simple coding and routing process through the network [500]. Network coding was originally proposed for lossy networks and enables the achieving of optimal throughput in such networks. Recently, it has been shown that the properly designed network coding can also provide strong error correcting capabilities for packets flowing through lossy networks [501]. The broadcast nature of wireless channels actually facilitates the application of network coding. Some research works have been carried out to combine the concept of network coding and channel coding for wireless networks [502–505]. We refer to these schemes as distributed network–channel coding (DNCC) schemes.

In this section, we introduce the distributed network and channel coding scheme designed for general wireless relay networks. The coding schemes are based on the general graph codes, such as low density generator matrix (LDGM) codes, low density parity check (LDPC) codes, etc. We will start with a brief review of graph codes and their decoding. The principle of graph code based distributed network and channel coding scheme is then introduced. The performance of these schemes is also evaluated through computer simulations.

4.4.1.1 Introduction to LDPC Codes

Before introducing the graph-code-based DNCC scheme, in this section we first give a brief introduction of LDPC codes and their message passing decoding algorithm.

LDPC codes were originally introduced by Gallager in 1963 [506] and shown to be a good class of binary linear block codes. These codes are specified by a sparse parity check matrix with a small number of 1's randomly distributed throughout the matrix. The limited computational power of computers at that time restricted computer simulations to codes with small block lengths, and their near-capacity achieving performance was not recognized until their rediscovery [507, 508].

An LDPC code is defined by its parity check matrix. For a regular LDPC code, its parity check matrix has exactly γ 1's in each column and exactly ρ 1's in each row, where $\gamma < \rho$ are both small compared to n. For an irregular LDPC code, the parity check matrix is still sparse, but not all rows and columns contain the same number of 1's.

A Gallager (n, γ, ρ) LDPC code is like a regular LDPC code of length n defined by a parity check matrix $\mathbf{H}$ with a fixed column weight γ and a fixed row weight ρ. The number of columns in $\mathbf{H}$ is equal to the block length n and the number of rows is equal to the number of parity check symbols $J = n - k \leq \frac{n\gamma}{\rho}$, where k is the message length and the symbol $\leq$ is used as the rows of $\mathbf{H}$ may not be independent. The code rate is given as $R \geq 1 - \gamma/\rho$. 'Sparse' means that γ and ρ are small with respect to n. An example of the Gallager LDPC code is shown in Figure 4.65. It shows an (20,3,4) LDPC code parity check matrix. It is divided horizontally into $\gamma = 3$ equal submatrices, each containing a single 1 in each column. The first submatrix looks like a 'flattened' identity matrix, where each '1' in the identity matrix is replaced by ρ '1's in one row and where the number of columns is multiplied accordingly. The subsequent submatrices are random column permutations of the first one.

The LDPC code can also be represented by a bipartite graph with n variable nodes on one side of graph and $n - k$ check nodes on other side when assuming the rows of H are independent. n variable nodes, represented by circles on the graph, denote the n coordinates of the codeword. $n - k$ check nodes on the graph, represented by squares on the graph, denote the $n - k$ check equations. From the parity check matrix, we can straightforwardly get the corresponding bipartite graph. If the entry (i, j) in the parity check matrix is 1, then the jth variable node is connected to the ith check node in the graph. An example of such a code is the (7,4) Hamming code. Its parity check matrix is shown in Figure 4.66, whereas Figure 4.67 shows the bipartite graph of this code.

The decoding algorithms of a LDPC code are based on its bipartite graph. They are called message passing decoding algorithms because in each iteration, messages are passed from variable nodes to check nodes and also from check nodes back to variable nodes. The sparsity of the code graph allows for the efficient decoding of LDPC codes. Sum–product decoding algorithm is an important subclass

$$\begin{bmatrix}
1 & 1 & 1 & 1 & 0 & 0 & 0 & 0 & 0 & 0 & 0 & 0 & 0 & 0 & 0 & 0 & 0 & 0 & 0 & 0 \\
0 & 0 & 0 & 0 & 1 & 1 & 1 & 1 & 0 & 0 & 0 & 0 & 0 & 0 & 0 & 0 & 0 & 0 & 0 & 0 \\
0 & 0 & 0 & 0 & 0 & 0 & 0 & 0 & 1 & 1 & 1 & 1 & 0 & 0 & 0 & 0 & 0 & 0 & 0 & 0 \\
0 & 0 & 0 & 0 & 0 & 0 & 0 & 0 & 0 & 0 & 0 & 0 & 1 & 1 & 1 & 1 & 0 & 0 & 0 & 0 \\
0 & 0 & 0 & 0 & 0 & 0 & 0 & 0 & 0 & 0 & 0 & 0 & 0 & 0 & 0 & 0 & 1 & 1 & 1 & 1 \\
1 & 0 & 0 & 0 & 1 & 0 & 0 & 0 & 1 & 0 & 0 & 0 & 1 & 0 & 0 & 0 & 0 & 0 & 0 & 0 \\
0 & 1 & 0 & 0 & 0 & 1 & 0 & 0 & 0 & 1 & 0 & 0 & 0 & 0 & 0 & 0 & 1 & 0 & 0 & 0 \\
0 & 0 & 1 & 0 & 0 & 0 & 1 & 0 & 0 & 0 & 0 & 0 & 0 & 1 & 0 & 0 & 0 & 1 & 0 & 0 \\
0 & 0 & 0 & 1 & 0 & 0 & 0 & 0 & 0 & 0 & 1 & 0 & 0 & 0 & 1 & 0 & 0 & 0 & 1 & 0 \\
0 & 0 & 0 & 0 & 0 & 0 & 0 & 1 & 0 & 0 & 0 & 1 & 0 & 0 & 0 & 1 & 0 & 0 & 0 & 1 \\
1 & 0 & 0 & 0 & 0 & 1 & 0 & 0 & 0 & 0 & 0 & 1 & 0 & 0 & 0 & 0 & 0 & 1 & 0 & 0 \\
0 & 1 & 0 & 0 & 0 & 0 & 1 & 0 & 0 & 0 & 1 & 0 & 0 & 0 & 0 & 1 & 0 & 0 & 0 & 0 \\
0 & 0 & 1 & 0 & 0 & 0 & 0 & 1 & 0 & 0 & 0 & 0 & 1 & 0 & 0 & 0 & 0 & 0 & 1 & 0 \\
0 & 0 & 0 & 1 & 0 & 0 & 0 & 0 & 1 & 0 & 0 & 0 & 1 & 0 & 0 & 1 & 0 & 0 & 0 & 0 \\
0 & 0 & 0 & 0 & 1 & 0 & 0 & 0 & 0 & 1 & 0 & 0 & 0 & 0 & 1 & 0 & 0 & 0 & 0 & 1
\end{bmatrix}$$

Figure 4.65   Example: parity check matrix for the Gallager (20,3,4) code

$$\begin{bmatrix}
1 & 0 & 0 & 1 & 0 & 1 & 1 \\
0 & 1 & 0 & 1 & 1 & 1 & 0 \\
0 & 0 & 1 & 0 & 1 & 1 & 1
\end{bmatrix}$$

Figure 4.66   Example: parity check matrix of the (7,4) Hamming code

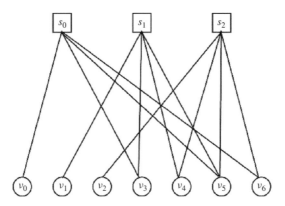

Figure 4.67   Example bipartite graph of the above (7,4) Hamming code

of message passing algorithm used for decoding LDPC codes. It consists of the initialization, iterative updating operation and decision evaluation. The algorithm can be summarized as follows.

(i) Initialization. Let $\mathbf{x} = (x_1, \ldots, x_n)$ denote the transmitted modulated codeword, where $x_i \in \{-1, 1\}$. Let us denote by $\mathbf{r} = (r_1, \ldots, r_n)$ the corresponding received signals corrupted by the noise and channel fading. The initial information about each binary data symbol of the codeword derived from the received sequence is expressed as a likelihood ratio:

$$\Delta_i = \frac{P(r_i | x_i = 1)}{P(r_i | x_i = 0)}. \tag{4.278}$$

We then create an matrix $\mathbf{J}$ by assigning Δ_i to all nonzero elements in ith column of the parity check matrix $\mathbf{H}$. As a result, in the same column of $\mathbf{J}$, any nonzero entries Δ_{ji} have the same value Δ_i.

(ii) Iterative Process. The iterative process consists of a horizontal step and a vertical step. In the horizontal step, the decoder updates the likelihood ratio passed from check node j and variable node i as follows:

$$\lambda_{j,i} = \prod_{j' \in N(s_j) \setminus x_i} \frac{1 - \Delta_{j',i}}{1 + \Delta_{j',i}} \tag{4.279}$$

where $N(s_j)$ is a set of variable nodes participating in check j, that is the columns in the $\mathbf{H}$ which have 1's in row j, and $N(s_j) \setminus x_i$ is the set of variable nodes participating in check j excluding variable node i. In the vertical step, the decoder updates the probability ratio passed from the variable node i to check node j as follows:

$$\Delta_{j,i} = \frac{P(r_i|x_i = 1)}{P(r_i|x_i = 0)} \prod_{i' \in M(x_i) \setminus s_j} \frac{1 - \lambda_{j,i'}}{1 + \lambda_{j,i'}}. \tag{4.280}$$

For each variable node i, the '*pseudo-posteriori*' probabilities are calculated as:

$$\Delta_i = \frac{P(r_i|x_i = 1)}{P(r_i|x_i = 0)} \prod_{j \in M(x_i)} \frac{1 - \lambda_{j,i}}{1 + \lambda_{j,i}}, \tag{4.281}$$

where $M(x_i)$ is a set of parity checks connected to a variable node x_i, that is the rows in $\mathbf{H}$ that have 1's in column i and $M(x_i) \setminus s_j$ is the set of checks connected to variable node x_i excluding check s_j.

(iii) Decision Evaluation. Based on the estimated Δ_i, the decoder first creates a tentative decoding as follows:

$$\hat{x}_i = \begin{cases} -1 & \text{if } \Delta_i \leq 1 \\ 1 & \text{if } \Delta_i > 1 \end{cases} \tag{4.282}$$

and

$$\hat{v}_i = \begin{cases} 1 & \text{if } \hat{x}_i = 1 \\ 0 & \text{if } \hat{x}_i = -1 \end{cases} \tag{4.283}$$

If $\hat{v}\mathbf{H}^T = 0$, then the decoding operation stops and $\hat{v}$ is considered a valid codeword. Otherwise, the decoder repeats the above iterative decoding process and a failure is pronounced if a maximum number of iterations occurs without a valid decoding decision.

This procedure is exemplified in Figure 4.68 which shows the message passing process in sum-product decoding for the Hamming code shown in Figure 4.67.

4.4.1.2 Adaptive Network Coded Cooperation

It has been shown that to achieve the optimum performance in wireless networks, a joint design of network and channel coding is required. In this section, we introduce the basic principles of distributed network and channel coding schemes based on the graph codes introduced in the previous section. The graph code based DNCC scheme was initially proposed by Bao and Li [504] and is referred to as the adaptive network coded cooperation (ANCC). The basic idea behind the ANCC is to match the code-on-graph with network-on-graph.

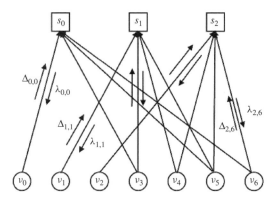

Figure 4.68 The sum product decoding process for the Hamming code shown in Figure 4.67

In the ANCC, nodes communicate with each other through wireless links. We assume that there are M source nodes, denoted by $S_1, \ldots, S_M$, and L relay nodes, denoted by $r_1, \ldots, r_L$, as shown in Figure 4.69. All nodes transmit on orthogonal channels. Initially, each source node broadcasts its symbols to other source nodes, all relays and also destination. Upon receiving signals from other nodes, each relay tries to decode each received signal, assuming that the relay r_i can decode correctly signals from a group of source nodes, denoted by $D(i) \subset \{S_1, \ldots, S_M\}$, which we called the decoding set of relay r_i. Each relay then randomly selects symbols from a subset $T(i) \subset D(i)$ of nodes in its decoding set and performs module-2 binary summation in the field of GF(2) to generate a single parity check bit. We refer to this subset $T(i)$ as network coding subset of relay r_i. Based on the encoding operations at relays, a graph code, such as LDPC or low density generate matrix (LDGM) network code is finally formed at the destination.

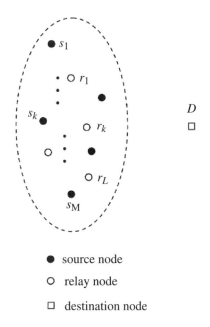

● source node

○ relay node

□ destination node

Figure 4.69 A general multiple access relay network

To explain in more detail the basic principle of ANCC, let us consider a small network with four source and five relay nodes. We assume that at each time slot, only one node (either source or relay node) can transmit. Figure 4.70 shows the network graph that describes the instantaneous topology of this network at a certain time, where the filled nodes denote the source nodes, unfilled nodes denote the relay nodes, and each dashed line represents a virtual connection between a relay node and a source node in its decoding sets. That is, the signals transmitted over this link can be decoded correctly by this relay at this instant. For example, as we can see from the figure, the decoding set of r_1 includes the source nodes S_1 and S_3, that is $D(1) = \{S_1, S_3\}$. Similarly, we have $D(2) = \{S_1, S_2, S_4\}$, $D(3) = \{S_2, S_3\}$, $D(4) = \{S_1, S_4\}$ and $D(5) = \{S_2, S_3, S_4\}$.

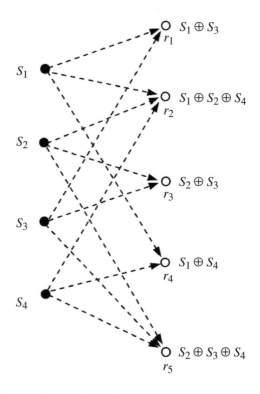

Figure 4.70   The example network graph to describe an instantaneous network topology

The decoding set of each relay and these virtual connections will vary with time because the network topology will change at different time instants. Let us take the network graph shown in Figure 4.70 as an example to explain the principle of ANCC. Let $b_{s,i}(k)$ denote the kth bit sent by the source S_i. We assume that the source nodes and relay nodes transmit in the order of their subscripts. That is, in the first four time slots, four source nodes $S_1, \cdots, S_4$ sequentially transmit to both relays and destination. After receiving the signals from all source nodes, each relay then decodes the received signals. Source nodes that are decoded correctly by relay r_i, are included in its decoding set $D(i)$. Based on the representation of the network graph shown in Figure 4.70, only the source nodes, which are virtually connected to a relay node, can be successfully decoded by this relay.

Relay i then randomly selects symbols from a network coding subset $T(i) \subset D(i)$ of nodes in its decoding set, performs binary summation of these symbols in GF(2) and forwards it to the destination. To understand clearly how the network graph can be matched to the code graph, let us first consider the case $T(i) = D(i)$. Let $b_{r,j}(k)$ be the kth bit sent by the relay r_j. Then, based on the network graph in Figure 4.70, the relay r_1 can correctly recover the bits $b_{s,1}(k)$ and $b_{s,3}(k)$, so the bit transmitted by the relay r_1 is given by $b_{r,1}(k) = b_{s,1}(k) \oplus b_{s,3}(k)$. Similarly, we have $b_{r,2}(k) = b_{s,1}(k) \oplus b_{s,2}(k) \oplus b_{s,4}(k)$, $b_{r,3}(k) = b_{s,2}(k) \oplus b_{s,3}(k)$, $b_{r,4}(k) = b_{s,1}(k) \oplus b_{s,4}(k)$, and $b_{r,5}(k) = b_{s,2}(k) \oplus b_{s,3}(k) \oplus b_{s,4}(k)$.

In the next five time slots, $b_{r,1}(k), \cdots, b_{r,5}(k)$ are sequentially transmitted by five relay nodes r_1, $\cdots, r_5$. The overall codeword received at the destination consists of four information symbols transmitted by the four source nodes in the first four time slots and five parity symbols sent by the five relays in the subsequent five time slots. The information and parity symbols together form an LDGM codeword. Let C represent the LDGM code constructed at destination. Let $\mathbf{G}$ denote the generator matrix of code C and $\mathbf{D}(k) = (d_1(k), \cdots, d_9(k)) = (b_{s,1}(k), b_{s,2}(k), b_{s,3}(k), b_{s,4}(k), b_{r,1}(k), b_{r,2}(k), b_{r,3}(k), b_{r,4}(k), b_{r,5}(k))$ is the codeword in C corresponding to the source symbols $\mathbf{B}_s(k) = \big(b_{s,1}(k), b_{s,2}(k), b_{s,3}(k), b_{s,4}(k)\big)$. It can be further written as:

$$\mathbf{D}(k) = \mathbf{B}_s(k)\mathbf{G} \qquad (4.284)$$

where:

$$\mathbf{G} = \begin{bmatrix} \mathbf{I}_4 & \mathbf{P} \end{bmatrix} = \left[\begin{array}{cccc|ccccc} 1 & 0 & 0 & 0 & 1 & 1 & 0 & 1 & 0 \\ 0 & 1 & 0 & 0 & 0 & 1 & 1 & 0 & 1 \\ 0 & 0 & 1 & 0 & 1 & 0 & 1 & 0 & 1 \\ 0 & 0 & 0 & 1 & 0 & 1 & 0 & 1 & 1 \end{array}\right]. \qquad (4.285)$$

Since it is a systematic code, its parity check matrix is given by:

$$\mathbf{H} = \begin{bmatrix} \mathbf{P}^T & \mathbf{I}_5 \end{bmatrix} = \left[\begin{array}{cccc|ccccc} 1 & 0 & 1 & 0 & 1 & 0 & 0 & 0 & 0 \\ 1 & 1 & 0 & 1 & 0 & 1 & 0 & 0 & 0 \\ 0 & 1 & 1 & 0 & 0 & 0 & 1 & 0 & 0 \\ 1 & 0 & 0 & 1 & 0 & 0 & 0 & 1 & 0 \\ 0 & 1 & 1 & 1 & 0 & 0 & 0 & 0 & 1 \end{array}\right]. \qquad (4.286)$$

Given the parity check matrix in Equation (4.286), we can draw its bipartite graph. Figure 4.71 shows the corresponding bipartite code graph of the resulted LDGM code, which is matched to the instantaneous network graph shown in Figure 4.70. After deriving the code graph, the conventional message passing iterative decoding algorithms can then be used to recover the source information symbols.

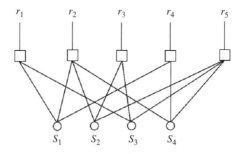

Figure 4.71   Bipartite code graph matching the network graph in Figure 4.70

However, as the number of source nodes increases, if each relay performs the check sum on all successfully decoded bits, the variable and check node degrees of the resulted distributed code will be very large and this will significantly increase the decoding complexity. Also as the node degrees increase, the probability of creating short cycles also increases, which will considerably degrade the system performance. To reduce the decoding complexity and enable efficient decoding, we have to reduce the node degrees. To do so, relay r_i only needs random pick-ups symbols from its network coding subset $T(i) \subset D(i)$ of nodes in its decoding set to perform a check sum and this will 'thin' the code graph [504]. For example, in Figure 4.70, the relays r_2 and r_5 may drop the source symbols sent by the source S_1 and S_2 in its check sum calculations. The network coding subsets for these relays are given by $T(1) = \{S_1, S_3\}$, $T(2) = \{S_2, S_4\}$, $T(3) = \{S_2, S_3\}$, $T(4) = \{S_1, S_4\}$ and $T(5) = \{S_3, S_4\}$. The resulted thinned code graph is shown in Figure 4.72.

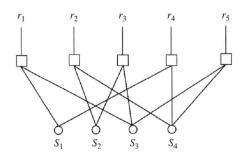

Figure 4.72   Thinned bipartite code graph

In general relay networks with M source and L relay nodes, the parity check matrix of ANCC based on LDGM code can be represented in the following format:

$$\mathbf{H} = \begin{bmatrix} \mathbf{P}^T & \mathbf{I}_L \end{bmatrix} \tag{4.287}$$

where $\mathbf{P}^T$ is a $L \times M$ sparse matrix and $\mathbf{I}_L$ is a $L \times L$ identity matrix. For each transmission, a distributed LDGM code will be formed at the destination, which is now a time varying code. That is, for different transmissions, different LDGM codes with different parity check matrices result. If the relay r_i puts a constraint on its network coding subset $T(i)$ such that the degree of the network coded symbol is not larger than a specific degree D, then all the resulted LDGM will form a degree-D LDGM ensemble.

Let us now consider a more general multihop network, where each relay not only helps to forward the source symbols, but also helps other relays that transmit prior to it, to forward their signals. Figure 4.73 shows an example, where the decoding set of relay r_i consists of not only the source nodes, but also early transmitted relay node r_j, $j < i$. For example, the networking set of relay r_4 includes both the source nodes S_1, S_4 and the relays r_1 and r_3. Similarly to the LDGM based ANCC, we can also write the corresponding parity check equations of the resulting graph code, which is given by:

$$\mathbf{H} = \begin{bmatrix} \mathbf{P}^T & \mathbf{P}_{tri} \end{bmatrix} = \begin{bmatrix} 1 & 0 & 1 & 0 & 1 & 0 & 0 & 0 & 0 \\ 0 & 1 & 0 & 1 & 1 & 1 & 0 & 0 & 0 \\ 0 & 1 & 1 & 0 & 0 & 1 & 1 & 0 & 0 \\ 1 & 0 & 0 & 1 & 1 & 0 & 1 & 1 & 0 \\ 0 & 0 & 1 & 1 & 0 & 1 & 0 & 1 & 1 \end{bmatrix}. \tag{4.288}$$

where $\mathbf{P}_{tri}$ is a lower triangular matrix. We note that compared with the LDGM code, the right part of the parity check equation is a lower triangular matrix with all 1s in the main diagonal. Such a code is referred to as a lower-triangular LDPC code [504]. The corresponding bipartite tanner graph is shown in Figure 4.74. Similarly, other general LDPC codes can also be constructed in other network topologies.

In practical systems, in order for the destination to decode the LDGM or LDPC code, the code graph has to be known at the destination. To help the destination determine the code graph, each source

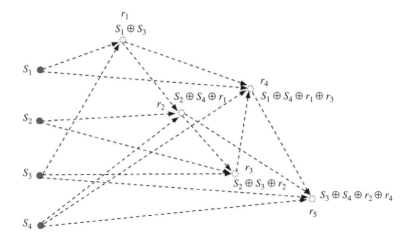

Figure 4.73 An example multihop network graph to describe an instantaneous network topology

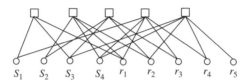

Figure 4.74 Bipartite code graph matching to the network graph in Figure 4.73

node has to transmit some header information ahead of the data symbols. For the relay network with M source nodes, each source has to transmit M header symbols, followed by K data symbols. Let $\mathbf{X}_i$ be the symbol vector transmitted by source S_i, given by:

$$\mathbf{b}_{s,i} = \left(b_{s,i}(1), \ldots, b_{s,i}(M+K)\right) = \left[\ \mathbf{b}_{s,i}^h\ \ \mathbf{b}_{s,i}^d\ \right] \tag{4.289}$$

where $\mathbf{b}_{s,i}^h = \left(b_{s,i}^h(1), \ldots, b_{s,i}^h(M)\right)$ and $\mathbf{b}_{s,i}^d = \left(b_{s,i}^d(1), \ldots, b_{s,i}^d(K)\right)$ represent the header and data portion of $\mathbf{b}_{s,i}$, respectively. Then the signal vectors transmitted by M source nodes can be written in matrix form as follows:

$$\mathbf{B}_s = \left[\ \mathbf{B}_s^h\ \ \mathbf{B}_s^d\ \right] \tag{4.290}$$

where $\mathbf{B}_s^h = \left((\mathbf{b}_{s,1}^h)^T, \ldots, (\mathbf{b}_{s,M}^h)^T\right)^T$ and $\mathbf{B}_s^d = \left((\mathbf{b}_{s,1}^d)^T, \ldots, (\mathbf{b}_{s,K}^d)^T\right)^T$.

Let $\mathbf{D}$ represent the LDGM codeword obtained at the destination corresponding to the source symbols $\mathbf{B}_s$. Let $\mathbf{G}$ denote the generator matrix of LDGM constructed at the destination. Then we have:

$$\mathbf{D} = \left[\ \mathbf{D}^h\ \ \mathbf{D}^d\ \right] = \mathbf{B}_s\mathbf{G} = \left[\ \mathbf{B}_s^h\mathbf{G}\ \ \mathbf{B}_s^d\mathbf{G}\ \right] \tag{4.291}$$

where $\mathbf{D}^h = \mathbf{B}_s^h\mathbf{G}$ and $\mathbf{D}^d = \mathbf{B}_s^d\mathbf{G}$ represent the header and data parts of received codewords. From the above equation, we can see that if we set $\mathbf{B}_s^h = \mathbf{I}_M$, then we have $\mathbf{D}^h = \mathbf{G}$. The destination can easily obtain the generator matrix of LDGM code and thus the corresponding code graph. Obviously, the transmission of header information will occupy some system resources. However, if $K \gg M$, that is, if the block length is much larger than the header length, the percentage of resource occupied by the headers can be neglected. We should also note that in generating each distributed LDGM or LDPC code, there are no coordinations between relay nodes, so each relay randomly selects source symbols to generate the parity check. It could happen that some source symbols are not included in any parity checks. Then the columns corresponding to these source symbols in the parity check matrix **H** will be all-zero vectors. That is, those source symbols that are not involved in parity checks are not protected by any parity symbols; therefore, the whole system performance will be seriously affected by these symbols. Furthermore, these distributed graph codes are constructed in a decentralized way; as a result, significant number of short cycles may exist. This will also affect the system performance. To solve these problems, certain rules have to be applied to each relay to make sure that each source node has participated in at least one check equation and short cycles can be eliminated. How to design such a centralized or distributed mechanism is significant in practical systems, but is still an open problem.

4.4.1.3 Simulation Results

In this section, we present some performance evaluation for the ANCC. We consider a wireless sensor network scenario in which 1000 source nodes communicate with a single destination node through 1000 relays. Each packet sent from each source node is appended with a CRC check. After receiving packets from the source nodes, each relay node decodes the received packets and checks its CRC. If the CRC check is correct, the packet will be kept at the relay nodes, otherwise it will be discarded. Each relay then performs network coding on its network coding subset and forwards it to the destination. The signals sent from the source and relay nodes form a distributed LDGM codeword. We assume that each source node has participated in at least one check equation and there are no all-zero columns in the resulting parity check equations. Short cycles, however, are not eliminated and thus they may exist in the LDGM code graph. We also assume that all the nodes in the network have equal transmission SNR.

Figures 4.75 and 4.76 show the BER performance of ANCC based on LDGM codes with various degrees over AWGN and Rayleigh fading channels. The final BER performance is obtained by averaging the results over 100000 realizations of the LDGM based network construction. The code rate of the constructed systematic LDGM code is 1/2. For comparison, we also plot the BER performance for repetition code. From the figure, we can see that the performance of ANCC significantly outperforms the repetition-based distributed coding schemes. The performance improvement comes from the coding gain inherited in ANCC.

It is also well known that the degree distribution for LDPC-like codes has a significant impact on the system performance. The influence of the degree distribution can also be seen from these two figures. We note that there is a tradeoff between the error floor and the waterfall performance. The codes with low degrees have better waterfall performance but high error floors compared to the high-degree codes. It is thus important in practice to select the proper degrees in order to achieve a good waterfall performance with desired error floor.

4.4.2 Network Coding Division Multiplexing

In the previous section, we introduced a distributed network and channel coding (DNCC) scheme based on graph codes for wireless relay networks. We have seen that the DNCC can bring system significant coding gains. However, the DNCC is designed for the scenario where a single group of source nodes communicate with a single destination node, which we refer to as the multi-access relay channel (MARC). In many wireless networks, multiple groups of source nodes need to communicate simultaneously with multiple respective destinations through a common relay network, which we call a multi-access relay interference channel (MARIC). For example, smart metering networks are an example of MARIC, where all water meters communicate with a water supplier data measurement center, while all power meters communicate with an energy supplier center. However, all these sensors

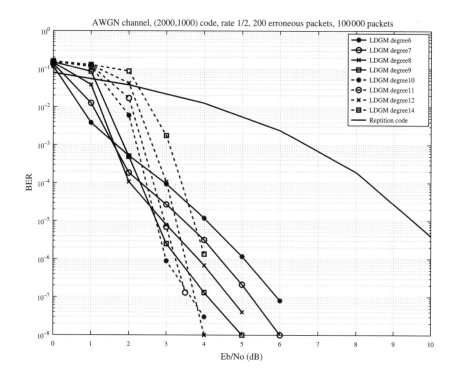

Figure 4.75   Performance of ANCC over AWGN channels

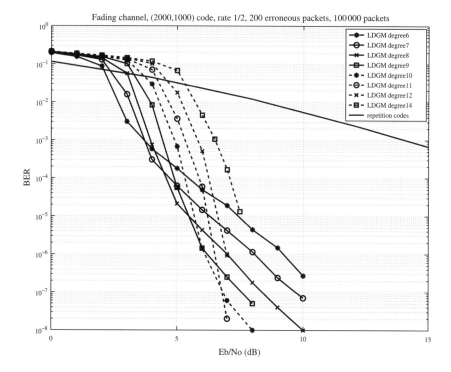

Figure 4.76   Performance of ANCC over Rayleigh fading channels

share the same wireless relay network and thus multiple groups of source nodes interfere with each other at each destination.

In this section, we introduce a network coding-based multiple access division multiplexing scheme, which we henceforth refer to as the distributed network–channel coding (DNCC) scheme for MARIC, to explore both network and channel coding gains as well as enabling multiple source groups to communicate with multiple destination nodes independently [509]. In the DNCC-MARIC, each relay performs a linear network coding and a graph code is formed at each destination. A code-nulling algorithm is then proposed to eliminate the intergroup interference at each destination. This enables multiple groups of source nodes to communicate with multiple destination nodes simultaneously without any interference between the groups. Results show that the DNCC in MARIC achieves the same level of performance but at a higher network throughput than in MARC.

4.4.2.1 System Model

For simplicity, we consider the MARIC system shown in Figure 4.77, which consists of two source groups A and B, denoted by Ω_A and Ω_B, respectively. All the source nodes in group A communicate with a common destination node d_A and all the source nodes in group B communicate with the destination d_B, through the same relay network. We aim to design a distributed coding scheme, so that the nodes in each group can communicate with their common destination without interference from the other group of source nodes.

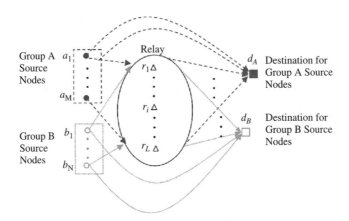

Figure 4.77   Multi-access relay interference channel

We assume that there are M source nodes in group A, denoted by $a_1, \cdots, a_M$, N source nodes in group B, denoted by $b_1, \cdots, b_N$, and L relay nodes, denoted by $r_1, \cdots, r_L$. We assume that all nodes transmit on orthogonal channels. Let A_i and B_j represent the bits sent by the source nodes a_i and b_j, respectively. Let $\mathbf{A} = (A_1, \ldots, A_M)$ and $\mathbf{B} = (B_1, \ldots, B_N)$ be the bit sequence sent from M source nodes in group A and N source nodes in group B.

Initially, each source node broadcasts its information to all relays and destinations. Similarly to the DNCC, upon receiving signals from other nodes, each relay randomly selects a small number of successfully detected bits and performs module-2 binary summation to generate a single parity check bit. Based on the encoding operations at relays, a graph code is formed at each destination, which is called the DNCC in Section 4.4.1.

Each graph code can be described by a bipartite graph. The bipartite code graphs formed at destinations d_A and d_B consist of $M + N$ variable nodes, corresponding to all the $M + N$ source nodes, and

L check nodes corresponding to the L parity check equations, used to calculate the parity checks at L relay nodes. The drawings on the left-hand side of Figures 4.78 and 4.79 show an example bipartite graph at destinations d_A and d_B for a MARIC with $M = N = 3$ and $L = 4$. Let C_A and C_B represent the graph code constructed at destinations d_A and d_B. Let $\mathbf{G}_A$ and $\mathbf{G}_B$ denote the generator matrix of code C_A and C_B. We represent $\mathbf{G}_A$ and $\mathbf{G}_B$ in the form of block matrices according to the dimensions of matrices $\mathbf{A}$ and $\mathbf{B}$ as follows:

$$\mathbf{G}_A = \begin{pmatrix} \mathbf{G}_{A,1} \\ \mathbf{G}_{A,2} \end{pmatrix}, \mathbf{G}_B = \begin{pmatrix} \mathbf{G}_{B,1} \\ \mathbf{G}_{B,2} \end{pmatrix} \tag{4.292}$$

where $N_{row}(\mathbf{G}_{A,1}) = N_{row}(\mathbf{G}_{B,1}) = N_{col}(\mathbf{A}) = M$, and $N_{row}(\mathbf{G}_{A,2}) = N_{row}(\mathbf{G}_{B,2}) = N_{col}(\mathbf{B}) = N$, $N_{col}(\mathbf{X})$ and $N_{row}(\mathbf{X})$ represent the number of columns and rows of matrix $\mathbf{X}$, respectively. Let $\mathbf{D}_A = (D_A^1, \ldots, D_A^{M+N+L})$ and $\mathbf{D}_B = (D_B^1, \ldots, D_B^{M+N+L})$ be a codeword in C_A and C_B, respectively, then we have:

$$\mathbf{D}_A = (\mathbf{A} \ \mathbf{B}) \, \mathbf{G}_A = (\mathbf{A} \ \mathbf{B}) \begin{pmatrix} \mathbf{G}_{A,1} \\ \mathbf{G}_{A,2} \end{pmatrix} \tag{4.293a}$$

$$\mathbf{D}_B = (\mathbf{A} \ \mathbf{B}) \, \mathbf{G}_B = (\mathbf{A} \ \mathbf{B}) \begin{pmatrix} \mathbf{G}_{B,1} \\ \mathbf{G}_{B,2} \end{pmatrix} \tag{4.293b}$$

From Equation (4.293a), we can see that the received codewords $\mathbf{D}_A$ and $\mathbf{D}_B$ at destinations d_A and d_B are linear combinations of both $\mathbf{A}$ and $\mathbf{B}$. This means that signals from group A and that from

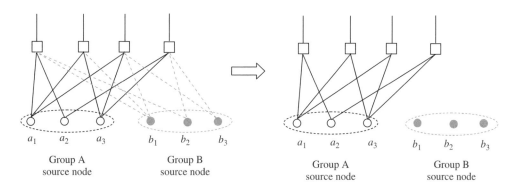

Figure 4.78 Bipartite graph before and after code-nulling process at the destination d_A

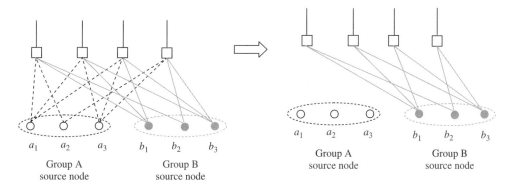

Figure 4.79 Bipartite graph before and after code-nulling process at the destination d_B

group B interfere with each other at each destination. However, the destination d_A should receive only messages from the source nodes in group A while d_B should receive only messages sent from the source nodes in group B. Therefore, we should develop a coding scheme to eliminate the interference coming from the other source groups and at the same time explore the network and channel coding gains.

4.4.2.2 Network Coding Division Multiplexing

In this section, we introduce a network coding-based multiple access division multiplexing to enable simultaneous transmission of message vectors $\mathbf{A}$ and $\mathbf{B}$ through a MARIC without interference at each respective destination.

By using the above block matrix representation in Equations (4.293a) and (4.293b), we further divide each block code into two subcodes. That is, we divide the code C_A generated by $\mathbf{G}_A$ into two subcodes $C_{A,1}$ and $C_{A,2}$, generated by $\mathbf{G}_{A,1}$ and $\mathbf{G}_{A,2}$, and the code C_B generated by $\mathbf{G}_B$ into two subcodes $C_{B,1}$ and $C_{B,2}$, generated by $\mathbf{G}_{B,1}$ and $\mathbf{G}_{B,2}$, respectively. Let $\mathbf{H}_{A,2}$ and $\mathbf{H}_{B,1}$ denote the parity check matrix of $\mathbf{G}_{A,2}$ and $\mathbf{G}_{B,1}$ Then we have:

$$\mathbf{G}_{A,2}\mathbf{H}_{A,2}^T = \mathbf{0}_{N\times(M+L)} \tag{4.294a}$$

$$\mathbf{G}_{B,1}\mathbf{H}_{B,1}^T = \mathbf{0}_{M\times(N+L)} \tag{4.294b}$$

From the above two equations, we have:

$$\tilde{\mathbf{D}}_A = \mathbf{D}_A\mathbf{H}_{A,2}^T = (\mathbf{A}\ \mathbf{B})\begin{pmatrix} \mathbf{G}_{A,1}\mathbf{H}_{A,2}^T \\ \mathbf{G}_{A,2}\mathbf{H}_{A,2}^T \end{pmatrix} = (\mathbf{A}\ \mathbf{B})\begin{pmatrix} \mathbf{G}_{A,1}\mathbf{H}_{A,2}^T \\ \mathbf{0} \end{pmatrix} = \mathbf{A}\tilde{\mathbf{G}}_{A,1} \tag{4.295a}$$

$$\tilde{\mathbf{D}}_B = \mathbf{D}_B\mathbf{H}_{B,1}^T = (\mathbf{A}\ \mathbf{B})\begin{pmatrix} \mathbf{G}_{B,1}\mathbf{H}_{B,1}^T \\ \mathbf{G}_{B,2}\mathbf{H}_{B,1}^T \end{pmatrix} = (\mathbf{A}\ \mathbf{B})\begin{pmatrix} \mathbf{0} \\ \mathbf{G}_{B,2}\mathbf{H}_{B,1}^T \end{pmatrix} = \mathbf{B}\tilde{\mathbf{G}}_{B,2} \tag{4.295b}$$

where

$$\tilde{\mathbf{G}}_{A,1} = \mathbf{G}_{A,1}\mathbf{H}_{A,2}^T \tag{4.296a}$$

$$\tilde{\mathbf{G}}_{B,2} = \mathbf{G}_{B,2}\mathbf{H}_{B,1}^T \tag{4.296b}$$

From Equations (4.295a) and (4.295b), we can see that after multiplying the received codeword $\mathbf{D}_A$ by $\mathbf{H}_{A,2}^T$, we get a new codeword $\tilde{\mathbf{D}}_A$ at the destination d_A with an equivalent generator matrix of $\tilde{\mathbf{G}}_{A,1}$. $\tilde{\mathbf{D}}_A$ is a codeword of $\mathbf{A}$ and is free from the interference coming from $\mathbf{B}$. Similarly, after multiplying the received codeword $\mathbf{D}_B$ by $\mathbf{H}_{B,1}^T$, we get a new codeword with an equivalent generator matrix of $\tilde{\mathbf{G}}_{B,2}$ at the destination d_B, which is free of the interference coming from $\mathbf{A}$.

Through the above processing, the whole network is decomposed into two independent networks with transport matrices of $\tilde{\mathbf{G}}_{A,1}$ and $\tilde{\mathbf{G}}_{B,2}$, respectively, and the intergroup interference is completely removed. We call this process code-nulling. It converts a MARIC into two independent multi-access relay channels (MARCs). Each of them consists of a group of source nodes communicating with a single destination, and it enables two groups of nodes to communicate with two destination nodes simultaneously without any interference between the two groups. In other words, through the above process, we convert two linear block codes, an $(M + N + L, M + N)$ code denoted by C_A and C_B into two independent linear block codes, an $(M + L, M)$ code $\tilde{C}_A$ and a $(N + L, N)$ code $\tilde{C}_B$, generated by $\tilde{\mathbf{G}}_{A,1}$ and $\tilde{\mathbf{G}}_{B,2}$, respectively. This code nulling process is illustrated in Figure 4.80.

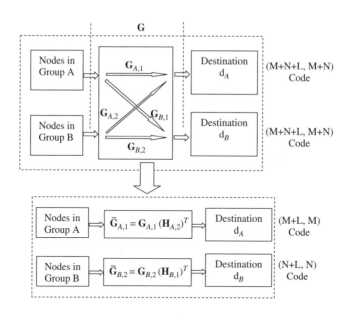

Figure 4.80 The decomposition of MARIC into two MARCs through code nulling

For simplicity, we assume further that C_A and C_B are systematic graph codes, such as low density generator matrix (LDGM) codes. Then we have:

$$\mathbf{G}_A = \begin{pmatrix} \mathbf{G}_{A,1} \\ \mathbf{G}_{A,2} \end{pmatrix} = \begin{pmatrix} \mathbf{P}_{A,1} & \mathbf{I}_M & \mathbf{0}_{M \times N} \\ \mathbf{P}_{A,2} & \mathbf{0}_{N \times M} & \mathbf{I}_N \end{pmatrix} \tag{4.297a}$$

$$\mathbf{G}_B = \begin{pmatrix} \mathbf{G}_{B,1} \\ \mathbf{G}_{B,2} \end{pmatrix} = \begin{pmatrix} \mathbf{P}_{B,1} & \mathbf{I}_M & \mathbf{0}_{M \times N} \\ \mathbf{P}_{B,2} & \mathbf{0}_{N \times M} & \mathbf{I}_N \end{pmatrix} \tag{4.297b}$$

and

$$\mathbf{H}_{A,2} = \begin{pmatrix} \mathbf{I}_{M+L} & \mathbf{P}_{A,2}^T \\ & \mathbf{0}_{M \times N} \end{pmatrix} = \begin{pmatrix} \mathbf{I}_L & \mathbf{0}_{L \times M} & \mathbf{P}_{A,2}^T \\ \mathbf{0}_{M \times L} & \mathbf{I}_M & \mathbf{0}_{M \times N} \end{pmatrix} \tag{4.298a}$$

$$\mathbf{H}_{B,1} = \begin{pmatrix} \mathbf{I}_L & \mathbf{P}_{B,1}^T & \mathbf{0}_{L \times N} \\ \mathbf{0}_{N \times L} & \mathbf{I}_{N \times M} & \mathbf{I}_N \end{pmatrix} \tag{4.298b}$$

By substituting Equations (4.298a) and (4.298b) into (4.296a) and (4.296b), we have:

$$\tilde{\mathbf{G}}_{A,1} = \mathbf{G}_{A,1} \mathbf{H}_{A,2}^T = \begin{pmatrix} \mathbf{P}_{A,1} & \mathbf{I}_M & \mathbf{0}_{M \times N} \end{pmatrix} \begin{pmatrix} \mathbf{I}_L & \mathbf{0}_{L \times M} \\ \mathbf{0}_{M \times L} & \mathbf{I}_M \\ \mathbf{P}_{A,2} & \mathbf{0}_{N \times M} \end{pmatrix} \tag{4.299a}$$

$$= \begin{pmatrix} \mathbf{P}_{A,1} & \mathbf{I}_M \end{pmatrix}$$

$$\tilde{\mathbf{G}}_{B,2} = \mathbf{G}_{B,2} \mathbf{H}_{B,1}^T = \begin{pmatrix} \mathbf{P}_{B,2} & \mathbf{0}_{N \times M} & \mathbf{I}_N \end{pmatrix} \begin{pmatrix} \mathbf{I}_L & \mathbf{0}_{L \times N} \\ \mathbf{P}_{B,1} & \mathbf{0}_{M \times N} \\ \mathbf{0}_{N \times L} & \mathbf{I}_N \end{pmatrix}$$

$$= \begin{pmatrix} \mathbf{P}_{B,2} & \mathbf{I}_N \end{pmatrix} \tag{4.299b}$$

From the above two equations, we can see that the generator matrix of the equivalent code, after the code-nulling for group A at destination d_A, denoted by $\tilde{\mathbf{G}}_{A,1}$, is essentially the same as its original generator matrix before the code nulling, $\mathbf{G}_{A,1}$, except for some zero matrices. The same holds for group B nodes. Therefore, with respect to each group of source nodes, the bipartite graph code formed at the corresponding destination is essentially the same before and after the code-nulling. This process is shown in Figures 4.78 and 4.79, which show the bipartite graphs before and after code-nulling process at the destinations d_A and d_B. We call this property the transformation invariance. Therefore, regardless of the code generated matrix for each group of source nodes, the two groups will not interfere with each other. This can significantly relax the system design as we can construct the distributed graph code for each group of source nodes independently of the other group of source nodes.

In the above process, we assume that the hard decision is performed on $\mathbf{D}_A$ and $\mathbf{D}_B$ at the destinations. In order to use a belief propagation (BP) algorithm to decode the new code $\tilde{C}_A$, the destination d_A has to calculate the log-likelihood ratios (LLRs) of $\tilde{\mathbf{D}}_A$. However only the LLRs of $\mathbf{D}_A$ are known at d_A. As shown in Equation 4.295a, $\tilde{\mathbf{D}}_A$ is actually the binary linear transformation of $\mathbf{D}_A$, and its LLRs can be calculated by using the following theorem.

Theorem 4.4.1 *Let $L(b_{k_j})$ represent the log-likelihood ratio (LLR) of information bit b_{k_j}. Then the LLR of binary linear summation of $b_{k_1}, \ldots, b_{k_m}$ can be calculated as [510]:*

$$L\left(b_{k_1} \oplus b_{k_2} \oplus \cdots \oplus b_{k_m}\right) = \ln\left(\frac{1 + \prod_{q=1}^{m}\tanh(L(b_{k_q})/2)}{1 - \prod_{q=1}^{m}\tanh(L(b_{k_q})/2)}\right) \tag{4.300}$$

where $\tanh(x/2) = \frac{e^x-1}{e^x+1}$.

4.4.2.3 Simulation Results

In this section, we finally evaluate the performance of the DNCC over multiple access relay interference channels (MARIC). We assume that the graph code formed at each destination is a systematic LDGM code. We assume that the link between any two nodes is either an AWGN channel or a Rayleigh fading channel and all nodes have an equal transmitted E_b/N_0, where E_b is the information symbol energy. We assume that there are two source groups communicating with two destination nodes, and that there are 500 sensor nodes in each source group and 1000 relays. Figure 4.81 compares the BER performance of MARIC and MARC. The degree of the relay node is 4. We can see that the BER performance of MARIC using the proposed DNCC has roughly the same level of performance as MARC for an AWGN channel. This implies that the DNCC can almost eliminate the intergroup interference completely. For a Rayleigh fading channel, the performance of MARIC is a little worse than the MARC channel. The performance loss is due to the process of calculating the LLR of $\tilde{\mathbf{D}}_A$ by using the LLR of $\mathbf{D}_A$. Also as the degree of relay node increases, the performance gap between the MARIC using DNCC and the MARC will also increase due to the processing of a large number of LLR values in $\tilde{\mathbf{D}}_A$. However, it increases the network throughput by 50% compared with MARC in this example. Figure 4.81 also compares the BER performance of DNCC and a repetition coding in which each source's symbols are relayed once in each transmission. It has hence been demonstrated that DNCC provides significant coding gain compared with the repetition coding.

4.5 Concluding Remarks

In this chapter we have introduced some important regenerative physical layer protocols facilitating the deployment of regenerative cooperative relaying systems. It requires a relay to receive a signal, process it and then retransmit it. Regenerative processing essentially implies that the relay is changing the waveform, information representation and/or information contents in the digital domain.

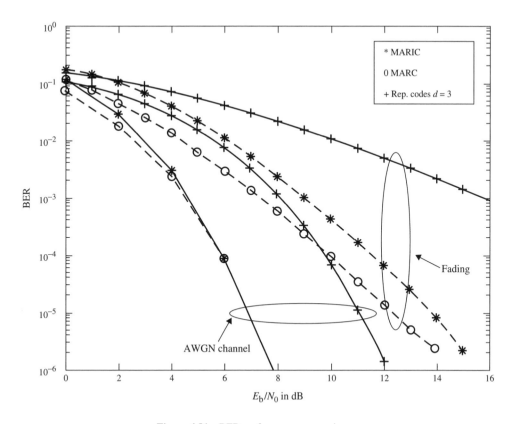

Figure 4.81 BER performance comparisons

We first dealt with regenerative relaying systems where we revised and introduced novel relaying protocols. We then moved on to distributed space–time architectures where the relays retransmit their information in a controlled manner so that known space–time processing mechanisms can be used in an adapted manner. We finally dealt with more advanced topics pertaining to cooperative network coding. Observations pertaining to these parts are summarized below:

- **Regenerative Relaying Protocols.** We have started with the DF protocol where we have shown that the system can be reduced to an equivalent one-hop communication system. We then moved on to the CF protocol, where we devised protocols based on the Wyner–Ziv and Slepian–Wolf Coding. We then dealt with a protocol referred to as soft information relaying, which essentially bridges AF and DF protocols to achieve optimum performance. We then moved on to adaptive relaying protocols where relays adapt to channel conditions so as to choose the optimum relaying technique, mostly operating as AF or DF protocols. Finally, we dwelt briefly on selective DF protocols, where relays only participate in the relaying process if they manage to decode the received packet successfully. Whilst the adaptive relaying protocol exhibits superior performance, the selective protocol is of lower implementational complexity.
- **Distributed Space–Time Coding.** We commenced with distributed STBCs, for which we derived error rates over general fading channels in closed form. We then used these codes in conjunction with the EF protocol over arbitrary relaying topologies. For the said system, we then embarked upon the important, and often neglected, topic of resource allocation, that is how to assign power optimally, frame durations, modulation index, etc., such that the end-to-end throughput is maximized. We then

moved on to distributed STTCs, for which we thoroughly analyzed the code design and performance assuming DF and EF relaying protocols. We have shown that a tailored design can greatly improve the performance of such systems. Finally, we dealt with distributed TCs, which are optimum in the Shannon sense. We have introduced TC designs for different relaying protocols, such as the DF protocol and the soft information relaying protocol. We then generalized the design leading to a superior class of distributed TCs.

- **Distributed Network Coding.** We commenced by dealing with the case of a single source wishing to communicate with a single destination over a given relaying network. For the said system, we introduced some novel approaches referred to as distributed network–channel coding. We have detailed the underlying theory and also extended the approach to an adaptive deployment. We then generalized these insights to topologies where multiple sources communicate with multiple destinations over the same relaying network. The thus designed protocol is referred to as network coding division multiplexing. The performance and superiority of said protocol has also been demonstrated although, due to its novelty, numerous issues remain open.

Most of the analysis carried out in this chapter was without consideration to clock and general timing asynchronisms, the impact of wideband and shadow fading, etc. These are important issues arising in today's modern communication systems and generally impact the behavior of regenerative relaying architectures. We will discuss in Chapter 6 how techniques exposed in this chapter can be used in conjunction with the aforementioned impairments.

Whilst regenerative relaying protocols have been explored in great detail, quite a few problems, mainly related to distributed space–time coding and optimization, remain open at the time of writing of this book. For instance, most existing distributed coding schemes are developed on the basis of the conventional channel coding schemes. Furthermore, most distributed coding schemes rely on some ideal assumptions, such as error free decoding at relays, etc. Some initial works for modeling detection errors in the demodulation and forward have been done and used to construct, for instance, distributed STTCs. It has been shown that detection errors do have some effect on constructing practical distributed coding. However, an accurate mathematical representation of modeling these decoding errors is still absent. As a result, the accurate design criteria for distributed coding with DF protocols when decoding errors occur at relays has not been formulated to date. Note that we will again highlight some more open issues in Chapter 6.

5

Hardware Issues

5.1 Introductory Note

5.1.1 Chapter Contents

In the end, it is the hardware that facilitates as well as limits any implementation of cooperative relaying schemes. These limitations in hardware favor some implementations but also render a low-complexity implementation of some of the recently proposed cooperative protocols infeasible. As of 2009, however, there is very little material available on hardware issues explicitly pertaining to cooperative systems. We will hence treat in this chapter the following issues:

- analog hardware transceiver components and designs;
- digital hardware transceiver components and designs;
- complexity and cost comparisons of these architectures;
- complexity and power consumption of 3G UMTS voice/HSDPA relay;
- complexity and power consumption of LTE/WiMAX relay;
- available hardware demonstrators.

Note that, from a hardware point of view, a preferred way of distinguishing between architectures is to divide them into purely analog and digital architectures. Fortunately, there is a simple mapping of hardware architectures to protocol families of Table 1.2 and multiple access methods of Figure 1.17: All regenerative relaying protocols must be realized by a digital hardware architecture, and all transparent relaying protocols can be realized by either a digital or a purely analog hardware architecture. Furthermore, TDR must be realized by a digital hardware architecture and both FDR or DFR can be realized by either a digital or a purely analog hardware architecture. The choice of the link duplexing method has no direct impact onto the choice of hardware, although it is likely that a system designed with TDD will use TDR and hence requires a digital hardware architecture.

5.1.2 Choice of Notation

We will use $\mathcal{O}$ to denote the number of total arithmetic operations, $\mathcal{L}$ for the load/store operations, and $\mathcal{M}$ for the memory requirements in bytes. The notation of the remaining symbols is very intuitive and has been summarized at the beginning of this book.

Cooperative Communications   Mischa Dohler and Yonghui Li
© 2010 John Wiley & Sons, Ltd

5.2   Analog Hardware Transceivers

5.2.1   Important Hardware Components

As per Figure 5.1, the most important radio components for realizing a purely analog hardware transceiver are briefly discussed below:

- **Receive Antenna.** The receive antenna converts the impinging electromagnetic wave into a current that feeds into the RF chain. The antenna will exhibit certain radiation characteristics, that is it is able to receive waves better from one direction than from another. For instance, the typically used dipole antenna has an isotropic radiation pattern in azimuth but not in elevation. It is also important to note that the efficiency of the antenna in converting electromagnetic wave energy into current is proportional to $(l/\lambda)^2$, where λ is the wavelength and l is the effective length of the antenna. This essentially means that large wavelengths, that is low frequencies, require large antennas for the system to be sufficiently efficient.
- **Receive Bandpass Filter (BPF).** The received signal in form of a current is perturbed by noise. This noise is usually composed of (i) radiation noise from natural and manmade sources, which arrive together with the useful electromagnetic signal; and (ii) thermal noise generated in the antenna and connecting cables. To reduce the impact of these wideband noises on the subsequent low power amplifier, mainly avoiding it's getting saturated, a receive bandpass filter is used. This filter blocks all spectral components but those where the useful signal is placed. Naturally, this useful signal is still perturbed by noise, albeit of much lower power w.r.t. the noise power before the filter. However, the filter operation also diminishes the power of the useful signal.
- **Low Noise Amplifier (LNA).** The role of the LNA is to amplify the received signal from the received power levels (around nA to μA) to power levels at which the RF chain can operate without being perturbed by thermal noise (around mA). Since the initial power levels are very feeble, the amplifier is referred to as low noise amplifier. It has to be of very good quality and is usually one of the expensive components in the RF chain.
- **Synthesizer.** The role of the (sometimes programmable) synthesizer is to produce a relative reference frequency that allows the received signal to be frequency-translated onto a different frequency band. The synthesizer is usually composed of a phased locked loop (PLL) and a simple or voltage

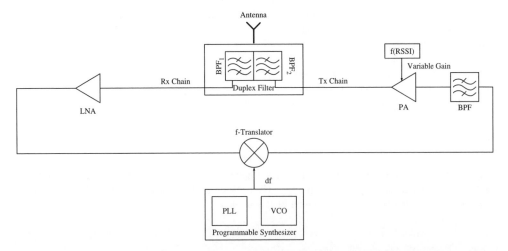

Figure 5.1   Analog relaying hardware architecture realizing, for example, the transparent amplify and forward (AF) relaying protocol

controlled oscillator (VCO). Both PLL and VCO coherently lock onto the received carrier and produce a prior-programmed frequency offset. The latter is usually multiplied by the incoming carrier frequency, thereby producing a translation of the frequency band. Oscillators exhibit two forms of imperfections: (i) a frequency drift due to change in temperature and other factors; and (ii) phase noise due to random variations induced by thermal fluctuations. Whilst the former can often be compensated for, the latter remains a deteriorating performance factor when cheap oscillators with large phase noises are being used.

- **Bandpass Filter.** The frequency shift by means of the programmable synthesizer usually not only translates the signal of interest but also creates mirror spectral components of the desired signal at lower and higher frequencies. These clearly need to be eliminated, which is why a simple bandpass filter is used after the programmable synthesizer.

- **Power Amplifier (PA).** The role of the power amplifier is to amplify the signal power from circuit levels to transmit power levels. Short range systems typically use around 0 dBm output power and long range systems from 20–40 dBm. If the signal to be amplified is wideband in nature, the PA needs to be of sufficiently good quality for no nonlinearities and hence distortions are created at the output. This makes the PA an expensive component of the RF chain if used for wideband systems. Note that the PA can realize a fixed or variable amplification. In the latter case, the amplification factor can, for example, depend on the average or instantaneous receive or even transmit channel fading power.

- **Transmit Bandpass Filter.** Finally, since the PA does not have fully linear amplification characteristics it usually produces spurious emissions that need to be filtered by a suitable transmit BPF centered to the new transmit frequency. The signal is then transmitted using the same antennas as are already used at reception. Note that if the receive and transmit bandpass filters share the same antenna, they are often referred to as duplex filters.

- **Power Supply.** Power supply is a crucial issue in any wireless device. Whilst base stations have access to mains, mobile stations and nodes will need to be powered autonomously. This is usually achieved by means of batteries that provide an almost constant output voltage over their lifetime. The lifetime of a battery depends on its initial energy budget and the amount of energy consumed by the hardware. For instance, having a node continuously powered on drains an AA battery of 3000 mAh in about 4 days. On the other extreme, even if a battery is not used at all, current leakages inherent to each battery limits the battery lifetime to 10–15 years, depending on the operating temperature.

We apply now these hardware components to purely analog relaying architectures.

5.2.2 Analog Relaying Architectures

The typical realization of a purely analog relaying architecture is shown in Figure 5.1. Generally, the transmitter side could use a superheterodyne, image-reject and low-IF hardware architecture and the receiver side could use a superheterodyne, direct-up or two-step-up hardware architecture. The important design points for such an architecture are listed below:

- **Variable Frequency Shift.** Unless only a single relay is used and the bands are well determined prior to roll-out, the frequency shift between the two receive and transmit duplex bands ought ideally to be variable; a programmable synthesizer is hence indispensable and is more easily realized by means of a digital architecture. A variable shift allows the system dynamically to configure the duplexing bands used by all sources, destinations and relays, so that interference is minimized. It is important to note that the bands that lend themselves naturally to the relaying process, such as the ISM bands, are becoming seriously congested with average interference temperatures often preventing viable communications.

- **Variable Power Gain.** As shown in Chapter 3 of this book, full diversity gains in the transparent relaying regime are only attainable if a variable amplification of the re-transmitted signal is used.

This means that a variable power amplifier is required where the amplification factor typically depends either on average or instantaneous channel fading conditions. Both are fairly easy to implement using an analog hardware architecture in that the received power is simply averaged over a long or short observation window in the analog domain.

- **Excellent Duplex Filters.** The severe electromagnetic coupling between transmitter and receiver RF chains is mitigated by means of the duplex filter, that is the receive and transmit BPF, as well as a careful design of mechanical and electrical shielding. The quality, and hence cost, of this filter depends heavily on the system requirements. The smaller the spectral separation between receive and transmit bands, the more sophisticated and expensive the duplex filter needs to be. A good tradeoff can be achieved by using surface (SAW) and bulk acoustic wave (BAW) filters. For instance, let us assume that a Bluetooth-type radio is used for the cooperative communications link, which requires a receiver sensitivity of -70 dBm for the signal to be detected correctly. Let us assume that it retransmits the signal at 0 dBm. This means that the duplex filter needs to guarantee a blockage of at least 70 dB (and typically more to avoid other detrimental effects). Another example is a GSM out-of-band repeater with a slight frequency shift between both bands, which typically requires the receive and retransmit antennas to be spatially separated by about 3–6 m because today's filter isolation and shielding methods simply do not suffice.

- **Difficulty of In-Band Relaying.** As already alluded to in Section 1.6.2, a practical realization of in-band relaying is very difficult as of today. The main problem is that the transmitted signal will couple into the receiver chain, causing a saturation of the LNA and hence rendering a meaningful amplification of the received signal unfeasible [511]. To realize in-band relaying, we need to guarantee that the relay's output-to-input antenna port isolation I is larger than its input-to-output gain G. To avoid oscillations and other detrimental effects, it is commonly recommended that this isolation margin is above 15 dB, that is $I > G + 15$ dB [85]. To ensure high port isolations, today's in-band repeaters not only spatially separate the receive and transmit antennas (10–20 m for typical in-band GSM repeaters), but also ensure that their directive beams point in orthogonal directions. Another method is to utilize orthogonal polarizations for reception and transmission. Either of these two methods, however, work only with a strong line of sight (LOS) component in static environments with some prior planning. Novel techniques are nevertheless emerging now, which mainly manage to reduce the spacing between the receive and transmit antennas to about $\lambda/2$ without sacrificing the port isolation and thereby facilitate implementation into small relays. One promising idea uses a hybrid ring of electrical length adjusted to the working frequency [512], which utilizes similar working principles as the bridge or duplex coil in a full-duplex telephone. Another approach uses echo or self-interference cancelation techniques that subtract the transmitted high-power signal from the received low-power signal and thereby provide port isolations of up to 30 dB [81, 513, 514]; however, the cancelation techniques are currently only available in digital domains with a purely analog design still pending.

- **No Storage of Signal.** A vital point of a purely analog relaying hardware architecture is that the received analog signal cannot be stored. That is, the analog signal is received, possibly frequency translated and retransmitted. No TDR protocols, introduced in Section 1.6.2, can therefore be used in the context of such architectures. This contradicts a large amount of literature available on AF protocols that inherently assumes a time-slotted reception and retransmission at the relay using analog hardware only. Note that the large majority of these contributions can be easily adopted to reality by simply assuming that the retransmission happens in a different frequency band instead of a different time slot. Alternatively, one has to use a digital architectures which is clearly more complex and expensive when compared with a purely analog realization.

The above-discussed, purely analog, hardware architecture would typically realize the AF transparent relaying protocol. Slight modifications to such a hardware architecture are needed to facilitate the implementation of the linear-process and forward (LF) transparent relaying protocol. This is exemplified in Figure 5.2. Typically, one would perform any linear operation of choice right before amplification, the latter also being a linear operation, strictly speaking. If this operation, for example,

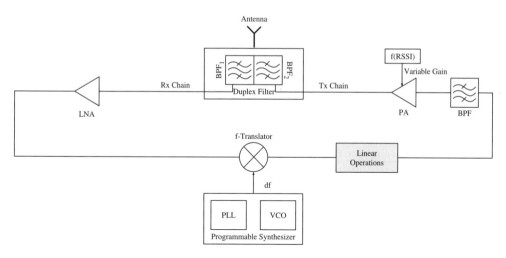

Figure 5.2 Analog relaying hardware architecture realizing, for example, the transparent linear-process and forward (LF) relaying protocol

involves a phase rotation, then a VCO can be used that produces the required phase shift. This is useful if beamforming is deployed or a conjugate complex of a signal needs to be produced, etc.

Finally, as shown in Figure 5.3, the nonlinear-process and forward (nLF) transparent relaying protocol can be constructed in a similar fashion. Since nonlinear operations are usually derived and performed on the baseband representation of a signal without carrier, the signal needs to be down-converted to baseband prior to applying any nonlinear operation. Such down and up conversions can be performed in several steps via a chain of intermediate frequencies (IF).

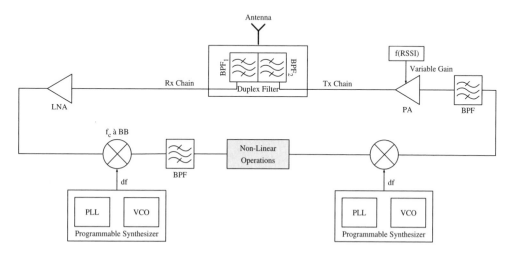

Figure 5.3 Analog relaying hardware architecture realizing, for example, the transparent nonlinear-process and forward (nLF) relaying protocol

One of the shortcomings of purely analog hardware architectures is that the analog signal cannot be stored, thus prohibiting the use of time division relaying (TDR), that is time-slotted relay protocols.

To circumvent this, one would need to use digital architectures as discussed below. It basically involves the signal being sampled and stored in digital form at baseband.

5.3 Digital Hardware Transceivers

5.3.1 Important Hardware Components

As shown in Figure 5.4, the most important radio components for realizing a digital hardware relaying architecture in addition to the ones already discussed for the analog architecture are:

- **Intermediate Frequency Stage (IF).** The received signal is typically downconverted to an intermediate frequency prior to regaining the two orthogonal signal phases. Typical IFs are in the range of a several hundred MHz.
- **I and Q Brancher.** To recuperate the two orthogonal signal dimensions, that is the in (I) and quadrature (Q) phases, an I and Q brancher needs to be deployed, which essentially consists of a sinusoidal oscillator being multiplied with one branch and a cosinusoidal one with the other branch. The resultant signal streams are orthogonal to each other and can hence be processed separately.
- **ADC and DAC Converter.** The analog-to-digital conversion (ADC) is needed to convert the received signal to its digital representation. The precision of representation depends on the number of bits used for digitization, where a higher number yields a lower digitization noise but also a higher cost. Prior to transmission, the digital signal is transformed into an analog signal by means of the digital-to-analog converter (DAC).
- **Digital Signal Processing.** The actual signal operations are carried out in the digital domain, which requires elements such as matched filters, microprocessors, memory, clocks, etc. An important design point is the digital architecture to use, that is FPGA, DSP, ASIC, etc. The field programmable gate array (FPGA) is a very generic programmable device, which in recent years has gained significantly in terms of speed and flexibility w.r.t. competing solutions. A digital signal processor (DSP) is capable of performing very challenging signal processing tasks at reasonable speed and power consumption. Finally, application-specific integrated circuits (ASICs) are integrated circuits customized for a very specific purpose, thereby offering high processing speeds at reasonable power

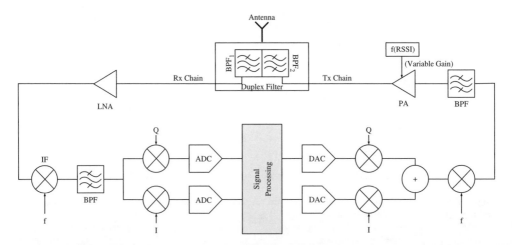

Figure 5.4 Digital relaying hardware architecture realizing, for example, the regenerative decode and forward (DF) relaying protocol

consumption but with little flexibility. Typical transceiver hardware architectures choose to use an FPGA for fairly heavy processing done at sampling level, such as matched filtering, multipath delay estimation, descrambler and despreader, integrator, and sometimes even channel estimation, etc., and one or several DSPs for operations at symbol level, such as detection, rake combining, Fourier transforms, etc.

Another point of importance is whether to use fixed point or floating point architectures. Fixed point architectures usually represent each number with a fixed number of bits; for instance, having 16 bits available and using signed integers, the range of numbers that can be addressed ranges from $-32\,768$ to $32\,767$. Floating point architectures typically use a minimum of 32 bits to store each value, that is $4\,294\,967\,296$ levels can be addressed. The key notion of the floating point notation is that the represented numbers are not uniformly spaced. Typically, the values are unequally spaced between the two extremes such that the gap between any two numbers is about ten-million times smaller than the value of the numbers. This leads to large gaps between large numbers and small gaps between small numbers. Fixed point architectures are generally slightly cheaper, whereas floating point ones have an inherently higher precision, cover a larger dynamic range and also enjoy shorter development times. Whether fixed or floating point, the digital signal processing part is usually the most expensive building block of a regenerative relaying architecture.

We now apply these hardware components to digital relaying hardware architectures.

5.3.2 Digital Relaying Architectures

A typical realization of a digital hardware architecture is shown in Figure 5.4. Generally, the transmitter side could use a superheterodyne, image-reject, zero-IF and low-IF hardware architecture and the receiver side could use a superheterodyne, direct-up or two-step-up hardware architecture. The important design points to be noted are listed below:

- **Synchronization.** To facilitate a coherent reception of the I and Q branches, the receiver needs to be synchronized to the incoming data stream. This requires a stringent design of synchronization algorithms and, if not done properly, leads to performance degradation. Transparent architectures do not suffer from this problem.
- **Large Memory.** To yield a sufficient representation of the information, the analog signal needs to be sampled at least at Nyquist rate and ideally even be oversampled to get a more precise representation. This leads to huge data quantities, which increase linearly with the slot duration and the oversampling factor. To store all this digitized data, the relay would need to have a very large memory.
- **Fast Data Buses.** The data needs to be transferred between sampling device and memory via data buses. Due to today's high data rates, coupled with a likely oversampling of the signal, this requires the use of fast data buses.
- **Powerful Baseband.** The baseband and signal processing components for regenerative architectures need to be fairly powerful. This means that clocking frequencies, data buses, memory, processors, microcontroller, etc., need to be powerful enough to handle the sampled, oversampled and processed data flows. Depending upon which relaying protocol and access methods are being used, these requirements might be more or less stringent. For instance, fixed point architectures might be sufficient for the EF relay protocol but not for the DF protocol.
- **Quantization Noise.** Any form of quantization leads to irreversible signal distortion often referred to as quantization noise. The signal to the quantization noise power ratio of a sine wave using an ideal quantizer is $\text{SNR} = b \times 6.02 + 1.76\,\text{dB}$, where b is the effective number of quantization bits and the $1.76\,\text{dB}$ offset is due to the peak-to-average value of the sine wave; for modulated carriers, this offset changes with modulation type [515]. Clearly, the more bits b are used to represent the digitized samples, the smaller the quantization noise but also the more stringent the requirements on

memory and data buses. Sufficiently complex digital architectures, however, use ADCs, DACs and DSPs with sufficient digitization bits and the thus created quantization noise is usually negligible when compared to the additive thermal noise.

- **In-Band Relaying.** A digital architecture allows for in-band relaying since the received signal can be stored and retransmitted in a later time slot. However, a full-duplex relay, which is capable of receiving and transmitting at the same time and the same band, is still difficult to design as has already been discussed in relation to a purely analog hardware platform. The first steps taken [81, 513, 514] require further exploration so as to facilitate a deployment with one antenna only.

- **Storage of Signal.** Since the signal is digitized, stored and possibly processed, protocols relying on both TDR and FDR can be used. However, since processing delay of digital architectures is in the order of a transmission slot, one would typically use TDR-type protocols.

- **Reuse of Filters.** If TDR and TDD are used then the incoming and relayed data streams do not interfere. This allows the same filters, mixers and synthesizers to be used for the receiving and transmitting chains, and hence do not necessitate duplexers and other shielding techniques.

The requirements for digital architectures are therefore generally much more stringent, which is reflected in complexity and cost as discussed below.

5.4   Architectural Comparisons

We will now qualitatively compare both analog and digital hardware architectures in terms of use of duplex, relay and multiple access protocols as well as in terms of complexity and cost. A quantitative evaluation of power consumption, processing complexity, data bus and memory requirements for a UMTS and LTE/WiMAX-type relay will then be performed in the following sections.

5.4.1   Duplex, Relay and Access Protocols

The restriction of a purely analog architecture in not being able to store the received signal is the major constraining factor in analog hardware architectures. However, neither the analog nor the digital hardware architectures are restricted in the use of the link duplexing method. This is because the link duplex method refers to the way in which a transmitter and receiver communicate and the relay has hence no immediate impact onto the choice of duplexing method. This is corroborated by means of Figures 1.19–1.22, where no storage of the relayed signal is required to facilitate the use of either TDD or FDD.

As for the choice of relaying duplex method, that is TDR, FDR or DFR, the hardware architecture has a clear impact. As discussed above, the analog architecture cannot support TDR nor can it support any form of regenerative relaying protocol. The digital architecture can support all three; however, processing delays at the relay in the case of regenerative relaying protocols are likely to impose TDR.

Finally, as for the choice of access protocols, neither analog nor digital hardware architecture can support multiple access protocols other than TDMA and FDMA in the context of transparent relaying protocols. For instance, to use OFDMA, the cooperative relay would need to perform FFT and IFFT operations, which would require a change of waveform and hence be against the definition of a transparent relaying protocol. The digital hardware architecture can support any other access protocol in the context of regenerative relaying protocols. However, simple relaying protocols such as EF and CF require very low computational complexity; this means that the digital hardware one would use for these protocols does not need to be powerful, which somehow limits the use of complex multiple access schemes, such as CDMA, OFDMA and MC-CDMA. The analog hardware architecture, on the other hand, can only support simple multiple access protocols, such as TDMA or FDMA.

These insights have been summarized for various relaying protocols in Table 5.1 and Table 5.2 for the analog and digital relay hardware architectures, respectively.

Table 5.1 Typical applicability of duplex, relay and access protocols to some transparent (left) and regenerative (right) relaying protocols using a purely analog hardware architecture ('$\checkmark$' = applicable, '$\times$' = not applicable)

ANALOG	AF	LF	nLF	EF	CF	DF	PF	GF
TDD	$\checkmark$	$\checkmark$	$\checkmark$	$\times$	$\times$	$\times$	$\times$	$\times$
FDD	$\checkmark$	$\checkmark$	$\checkmark$	$\times$	$\times$	$\times$	$\times$	$\times$
DFD	$\checkmark$	$\checkmark$	$\checkmark$	$\times$	$\times$	$\times$	$\times$	$\times$
TDR	$\times$	$\times$	$\times$	$\times$	$\times$	$\times$	$\times$	$\times$
FDR	$\checkmark$	$\checkmark$	$\checkmark$	$\times$	$\times$	$\times$	$\times$	$\times$
DFR	$\checkmark$	$\checkmark$	$\checkmark$	$\times$	$\times$	$\times$	$\times$	$\times$
TDMA	$\checkmark$	$\checkmark$	$\checkmark$	$\times$	$\times$	$\times$	$\times$	$\times$
FDMA	$\checkmark$	$\checkmark$	$\checkmark$	$\times$	$\times$	$\times$	$\times$	$\times$
CDMA	$\times$	$\times$	$\times$	$\times$	$\times$	$\times$	$\times$	$\times$
OFDMA	$\times$	$\times$	$\times$	$\times$	$\times$	$\times$	$\times$	$\times$
MC-CDMA	$\times$	$\times$	$\times$	$\times$	$\times$	$\times$	$\times$	$\times$

Table 5.2 Typical applicability of duplex, relay and access protocols to some transparent (left) and regenerative (right) relaying protocols using a digital hardware architecture ('$\checkmark$' = applicable, '$(\checkmark)$' = normally not used, '$\times$' = not applicable)

DIGITAL	AF	LF	nLF	EF	CF	DF	PF	GF
TDD	$\checkmark$	$\checkmark$	$\checkmark$	$\checkmark$	$\checkmark$	$\checkmark$	$\checkmark$	$\checkmark$
FDD	$\checkmark$	$\checkmark$	$\checkmark$	$\checkmark$	$\checkmark$	$\checkmark$	$\checkmark$	$\checkmark$
DFD	$\checkmark$	$\checkmark$	$\checkmark$	$\checkmark$	$\checkmark$	$\checkmark$	$\checkmark$	$\checkmark$
TDR	$\checkmark$	$\checkmark$	$\checkmark$	$\checkmark$	$\checkmark$	$\checkmark$	$\checkmark$	$\checkmark$
FDR	$\checkmark$	$\checkmark$	$\checkmark$	$(\checkmark)$	$(\checkmark)$	$(\checkmark)$	$(\checkmark)$	$(\checkmark)$
DFR	$\checkmark$	$\checkmark$	$\checkmark$	$(\checkmark)$	$(\checkmark)$	$(\checkmark)$	$(\checkmark)$	$(\checkmark)$
TDMA	$\checkmark$	$\checkmark$	$\checkmark$	$\checkmark$	$\checkmark$	$\checkmark$	$\checkmark$	$\checkmark$
FDMA	$\checkmark$	$\checkmark$	$\checkmark$	$\checkmark$	$\checkmark$	$\checkmark$	$\checkmark$	$\checkmark$
CDMA	$\times$	$\times$	$\times$	$(\checkmark)$	$(\checkmark)$	$\checkmark$	$\checkmark$	$\checkmark$
OFDMA	$\times$	$\times$	$\times$	$(\checkmark)$	$(\checkmark)$	$\checkmark$	$\checkmark$	$\checkmark$
MC-CDMA	$\times$	$\times$	$\times$	$(\checkmark)$	$(\checkmark)$	$\checkmark$	$\checkmark$	$\checkmark$

5.4.2 Transceiver Complexity

As per Table 5.3, a purely analog hardware architecture cannot realize any regenerative relaying protocol. Furthermore, memory and signal processing requirements are absent in the case of transparent relaying protocols. Clock frequency is moderate and the requirements for the amplifier are also not very stringent, mainly because one would use the purely analog architecture in the context of fairly narrowband systems. An important design parameter, however, is the duplex filter design, since incoming signal stream has to be well separated from the outgoing stream.

Table 5.3 Qualitative complexity of important hardware blocks for some transparent (left) and regenerative (right) relaying protocols using a purely analog hardware architecture (' $+++$ ' = high, ' $++$ ' = medium, ' $+$ ' = low, ' $-$ ' = none, ' $\times$ ' = not applicable)

ANALOG	AF	LF	nLF	EF	CF	DF	PF	GF
Filter design	$+++$	$+++$	$+++$	$\times$	$\times$	$\times$	$\times$	$\times$
Clock frequency	$+$	$+$	$+$	$\times$	$\times$	$\times$	$\times$	$\times$
Power amplifier	$+$	$+$	$+$	$\times$	$\times$	$\times$	$\times$	$\times$
Signal processing	$-$	$-$	$-$	$\times$	$\times$	$\times$	$\times$	$\times$
Memory	$-$	$-$	$-$	$\times$	$\times$	$\times$	$\times$	$\times$

Table 5.4 qualitatively describes the transceiver complexity in the case of a digital hardware architecture. It can be seen that due to the digital implementation, allowing the use of a slotted operation and hence not requiring to relay the signal immediately on a different frequency, the duplex filter design is greatly relaxed. Having said this, if either transparent or regenerative relaying protocols use FDR, the complexity of the filter immediately increases. Regenerative relaying protocols generally require higher clock rates and more complex signal processing capabilities. Transparent relaying protocols, on the other hand, require a large memory because of the storage of (over-)sampled symbols. Finally, the complexity and cost of a low-noise power amplifier essentially depends on the signal's bandwidth. Since we can assume that less complex narrowband systems are likely to use transparent as well as the low complexity regenerative relaying protocols, the requirement on the power amplifier are fairly loose. In contrast, the high complexity regenerative protocols are likely to operate on wideband signals and hence require a fairly elaborate power amplifier.

Table 5.4 Qualitative complexity of important hardware blocks for some transparent (left) and regenerative (right) relaying protocols using a digital hardware architecture (' $+++$ ' = high, ' $++$ ' = medium, ' $+$ ' = low)

DIGITAL	AF	LF	nLF	EF	CF	DF	PF	GF
Filter design	$+$	$+$	$+$	$+$	$+$	$+$	$+$	$+$
Clock frequency	$+$	$+$	$+$	$+++$	$+++$	$+++$	$+++$	$+++$
Power amplifier	$+$	$+$	$+$	$+$	$+$	$+++$	$+++$	$+++$
Signal processing	$+$	$+$	$+$	$++$	$++$	$+++$	$+++$	$+++$
Memory	$+++$	$+++$	$+++$	$+$	$+$	$++$	$++$	$++$

5.4.3 Cost Estimates

We will now quantify the cost of typical transparent and regenerative relaying architectures. The values given below are in Euros and hold for a production volume of 1000 items as of the third quarter of 2009.

As per Table 5.5, the most expensive elements of a purely analog relaying architecture are the duplex filter and programmable synthesizer. The total estimated cost for 1000 production items produced in 2009 mounts to around 17 Euros. As per Table 5.6, the most expensive elements of a digital relaying

Table 5.5 Approximate cost for the production of
1000 analog relays at 900 MHz as of Q3 2009

List of items	Approximate cost in euros
Bandpass filter	1
Duplex filter	3
Low noise amplifier	1
Programmable synthesizer	4
Variable gain amplifier	3
Printed circuit board	3
Other items	2
Total	17

Table 5.6 Approximate cost for the production of
1000 digital relays at 900 MHz as of Q3 2009

List of items	Approximate cost in euros
Transparent architecture*	13
Programmable synthesizer	2×5
I and Q brancher	2×4
ADCs/DACs	2×4
Signal processing	7
Total	46

*minus the cost of the synthesizer

hardware are the digital building blocks. The total estimated cost for 1000 production items produced in 2009 mounts to around 46 Euros. Comparing both architectures reveals that the digital hardware architecture is about three-times more expensive than the purely analog one. Note however that the use of other hardware peripherals diminishes this cost difference. Furthermore, the following has an important impact on the cost trends:

- **Production Volume.** Every decade in produced items diminishes the cost by approximately 10%. Therefore, if 10 000 digital relays are being produced, the cost diminishes from 46 to 42 Euros.
- **Change of Frequency.** Changing the operational frequency has an impact on the choice of oscillator and hence cost. For instance, going from 900 MHz to 2 GHz increases the price by approximately 5%, whereas going from 900 MHz to 5 GHz increases the price by approximately 20%.
- **Temporal Devaluation.** The impact of the global economy on production costs makes an evaluation of the temporal devaluation of the hardware components difficult. However, according to general consensus, the introductory price of new components is high and generally does not change over product generations. Over time, as new product generations are introduced, the price of the old product generation drops exponentially even though production costs do not change. This price roughly halves every 2 years.
- **Estimation Error.** All of above estimates usually have a 10% error margin.

These trends have been depicted in Figure 5.5. Clearly, the purely analog relaying hardware is significantly cheaper and a change in frequency has no major impact on costs.

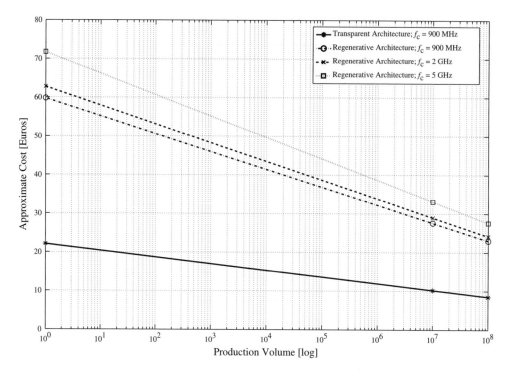

Figure 5.5 Cost trends for various analog and digital hardware architectures

5.5   Complexity of 3G UMTS Voice/HSDPA Relay

The Universal Mobile Telecommunications System (UMTS) is one of the third generation (3G) mobile telecommunications technologies. The most popular form of the available UMTS variants uses W-CDMA in FDD mode. It has been standardized by the 3GPP in response to the ITU IMT-2000 requirements for a 3G cellular radio system. The aim of this section is to quantify the prime factors influencing the complexity and power consumption of a relay based on UMTS specs, where we focus on the basic UMTS voice service as well as on its data-centric evolution, referred to as high-speed downlink packet access (HSDPA):

- **Processing Requirements.** The number of arithmetic operations, usually counted in million instructions per second (Mips), influences the clock rate of the underlying digital signal processors. This, in turn, has a direct impact on the cost and also on the power consumption of the relay.
- **Memory Requirements.** The actual amount of memory needed, usually given in bytes, is maybe not a prime design driver today since memory has become fairly cheap; however, the read and write operations consume power and hence impact the transceiver design.
- **Power Consumption.** If the relay is not powered by mains, the power consumed, usually given in mW, is a prime design driver. It determines battery recharging cycles, battery costs, etc.

A quantification of complexity of realistic transceivers is generally a very difficult task since it depends on many implementation-dependent factors. Nonetheless, the below calculations may serve as a baseline that can be adapted to different architectures. The time horizon of the below study is around 3 years from 2008, when hardware developments will have sufficiently evolved and the UMTS network will have been rolled out and reached maturity. This would allow for more sophisticated transceiver methods to be implemented at the network and terminal side, even though only simple repeaters are part of today's UMTS standard [85, 86]. Indeed, as discussed [(516, pp. 318–322)], relays (repeaters) based on current UMTS specs yield surprising capacity losses. This is mainly because the relay is not power controlled, thus requiring an increased power control margin, and also because the current radio resource management (RRM) algorithms in UMTS are not designed to rely on relays. However, it has been pointed out [516] that they have a beneficial impact onto the coverage of UMTS systems, such as in tunnels, street canyons, offices, etc.

This therefore justifies a discussion of UMTS-based relays and we thus proceed with the outline of the system assumptions, followed by the arithmetic complexity analysis of the baseband, the power consumption analysis by the RF and baseband, and some voice and HSDPA case studies.

5.5.1 System Assumptions

We briefly justify and expose the choice of scenarios, access method and physical layer. This in turn impacts the design of the digital modem.

5.5.1.1 Choice of Scenarios

We will compare the complexity of supportive and cooperative relays using the two AF and DF relaying protocols implemented in purely analog and also digital hardware architectures, leading to four relaying approaches:

- **Supportive AF with Analog Hardware (s-AF$_a$).** The simple amplify and forward protocol does not require any complexity nor memory estimations; however, an integral part of the study will be the power consumption of the RF chains.
- **Supportive AF with Digital Hardware (s-AF$_d$).** Here, we will quantify the large memory requirements of the AF protocol implemented in the digital domain, along with its power consumption.
- **Supportive DF with Digital Hardware (s-DF$_d$).** The core of the study, however, will relate to the more complex decode and forward protocol. It will comprise complexity, memory and power consumption, taking into account important elements of the UMTS PHY layer protocol stack.
- **Cooperative DF with Digital Hardware (c-DF$_d$).** Finally, we will also investigate the complexity of a relay that is itself terminal and hence has its own traffic to transmit. It is important here – in contrast to above cases – that such a relay is battery powered for which voltage decreases nonlinearly over time.

We will assume that the purely analog relay consists of the hardware elements discussed in Section 5.2.1 and the digital relay in addition of the elements discussed in Section 5.3.1.

5.5.1.2 Choice of Access Method

The 3G UMTS system under consideration is based on the UMTS terrestrial radio access (UTRA) FDD mode where both up and downlink have two separate bands and access is guaranteed by direct sequence CDMA (DS-CDMA). The reason why the TDD mode is not considered here is because it does not enjoy the same popularity as the FDD mode and its future, except for niche market applications (safety, emergency, etc.) or the time division synchronous CDMA (TD-SCDMA) variant developed in China, is doubtful [517].

Due to the assumed FDD mode, both up- and downlink have two separate bands, within each of which a cellular operator has typically 1–3 carriers of 5 MHz width. As already discussed in Section 1.6.2, this means that the relay duplexing procedure needs to obey strict rules that neatly fit into the original FDD system design. An example is the CDMA realization in Figure 1.21; alternatively, the relay may retransmit using a second downlink channel dedicated to relay activities.

Whilst also related to the link layer, transmission powers have a profound impact on the performance of access protocols. The transmission power of a base station (BS) is limited to 43 dBm (20 W per carrier) in the downlink, which a relay is unlikely to use being so close to other terminals because interference would be just too high. Therefore, although operating in the downlink band, the relay would rather resort to the maximum transmission powers applicable to the mobile station (MS) in the uplink. Currently, only class 2 (27 dBm maximum transmission power), class 3 (24 dBm) and class 4 (21 dBm) MS have been accredited, and the latter will be used for this study.

5.5.1.3 Choice of Link Layer

Data in UMTS is carried by means of physical channels. Physical channels are characterized by a specific carrier frequency, scrambling code, channelization code (optional) and time start-and-stop (giving a duration). Time durations are measured in integer multiples of chips with each chip lasting approximately 0.26 μs. Suitable multiples of chips also used in the specification are (i) the slot (length of a slot corresponds to 2560 chips); (ii) the subframe for HSDPA data transmissions (length of a subframe corresponds to three slots, equal to 7680 chips); (iii) the radio frame (processing duration which consists of 15 slots, equal to 38400 chips or 10 ms); and (iv) the time transmission interval (TTI) for voice and data transmissions (duration is specific to different realizations but confined to the set of 10, 20, 40 and 80 ms).

Dedicated voice and/or data transfer is facilitated by the downlink dedicated physical channel (DPCH). Data of the DPCH arrives to the coding/multiplexing unit in form of transport block sets once every TTI, that is every 10, 20, 40 or 80 ms. The following steps are of importance when dealing with the link layer frame structure: cyclic redundancy check (CRC), channel coding, interleaving and modulation; more details can be found in the literature [516]. The dedicated channels are transmitted using a user-specific variable spreading sequence (thereby trading reliability against data rate) and quadrature phase shift keying (QPSK) modulation.

The transfer of larger chunks of data by means of HSDPA is facilitated by the high speed physical downlink shared channel (HS-PDSCH). A HS-PDSCH corresponds to one channelization code of fixed spreading factor $SF = 16$ from the set of channelization codes reserved for high-speed transmission. Multicode transmission is allowed, which translates to a user being assigned multiple channelization codes in parallel. The modulation can be either QPSK or 16QAM.

Finally, a channel central to the proper functioning of a UMTS system is the common pilot channel (CPICH). Changing the transmission power allocated to the CPICH can be used to control the cell-size. Furthermore, it can be used to improve the downlink channel estimation. It is transmitted at a fixed rate of 30 kbps using $SF = 256$ and carries a pre-defined bit sequence.

5.5.1.4 Digital Modem Design

Based on the above link layer specs, a typical realization of a digital relay is shown in Figure 5.6. On the receive side, this includes analog-to-digital conversion, matched root-raised cosine (RRC) filtering, channel acquisition (full and partial), channel estimation, various detection algorithms, channel turbo and trellis decoder, de-interleaving, etc. Detection is performed using a Rake receiver with maximum ratio combining (MRC), as shown in Figure 5.6, and a more sophisticated multipath interference canceler (MPIC) with iterative channel decoding/encoding [518].

For complexity calculations, it is important to understand which of these modem blocks happen at which sampling rates. The most complex processes are clearly those performed at sample and chip

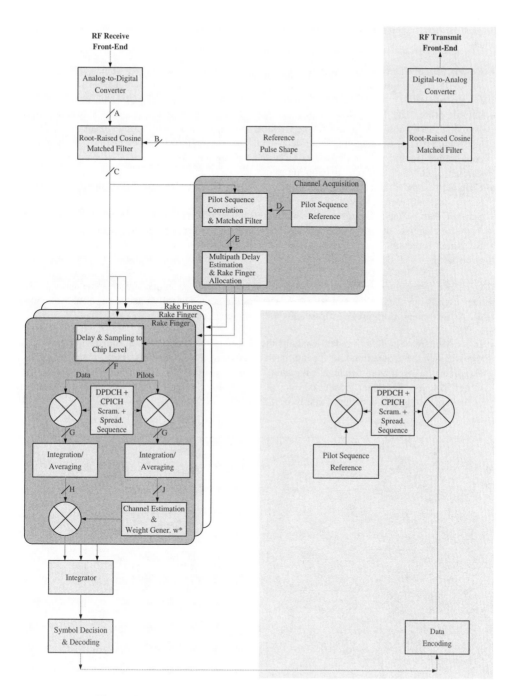

Figure 5.6 UMTS digital relay with Rake receiver and MRC detector

level; it is therefore an ultimate design goal to minimize the number of processes performed at these levels:

- **Analog Signal.** The received and transmitted signal going through the RF chain (antenna in and output, IF stages, low noise amplifiers, analog filters, input to AD and output from DA converter) are analog signals.
- **Sample Rate.** The RRC filter (output samples of AD converter are fed into a matched filter; sampling rate is above chip rate for better time resolution at acquisition unit), acquisition (correlates on CPICH to estimate multipath delays; channel estimation should be avoided here), and the hold/sample (input to Rake finger has to be delayed by appropriate number of samples and then sampled at chip rate) operate at sampling rate.
- **Chip Rate.** Spreading and scrambling operations, and the integrator (output of the despreader has to be integrated/averaged over one symbol).
- **Symbol Rate.** Channel estimation, combining/summation operation of received signal streams from, for example, different Rake fingers, and some interleavers operate at symbol rate.
- **Bit Rate.** Typically, channel coding, rate matching, interleaver and CRC operations are operated at bit level.

Given these signal rates, a modem is generally structured into an inner and an outer modem. The former usually operates below symbol rates, whereas the latter typically operates at or above symbol rates. As illustrated in Figure 5.7, this yields the following design:

- **Inner Modem.** The inner modem contains the root-raised cosine (RRC) matched filter and on the receiver side the acquisition unit for multipath component (MPC) delay estimation and Rake finger assignment, as well as the channel estimator.
- **Outer Modem.** The outer modem consists of the (de-)interleaver, (de-)multiplexing, rate matcher, channel encoding and decoding (Viterbi or Turbo) and CRC.

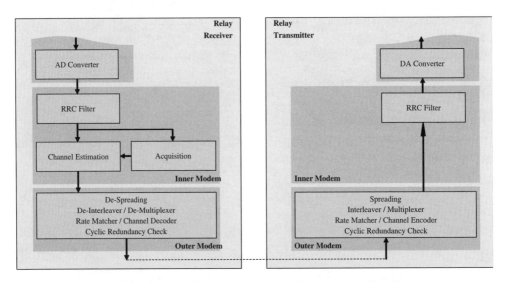

Figure 5.7 Typical digital modem design blocks for UMTS-based relay

As for the digital hardware architecture, we generally assume that the arithmetic logical unit (ALU) of the signal processing board can do four instructions per cycle, which is referred to as 4 ALU. Furthermore, the data is represented by 16 bits per complex dimension, and additions and multiplications have more or less the same associated complexity.[1] Moreover, we assume a single digital architecture for the sake of analytical tractability, thus excluding the use of specialized circuitry as can be found in mobile terminals.

5.5.2 Algorithmic Complexity

Based the literature [518–520], we will include only the most complexity-hungry building blocks. We will focus on the relay receiving side since – as per Figures 5.6 and 5.7 – the transmitting side is a subset of the receive algorithms and hence easy to derive.

5.5.2.1 RRC Matched Filter

The impulse response of the matched chip impulse root-raised cosine (RRC) filter is:

$$RC_0(t) = \frac{\sin\left(\pi \frac{t}{T_c}(t - \beta)\right) + 4\beta \frac{t}{T_c} \cos\left(\pi \frac{t}{T_c}(t + \beta)\right)}{\pi \frac{t}{T_c}\left(1 - \left(4\beta \frac{t}{T_c}\right)^2\right)}, \tag{5.1}$$

where T_c is the chip duration and the roll-off factor is β. The RRC matched filter is traditionally implemented in the digital domain with a finite number of taps, and operates at sample level.

The computational complexity of the RRC matched filter depends upon the following parameters: DS the channel delay spread (in chips), by which impinging MPCs are delayed; OSF the oversampling factor, which allows the channel acquisition unit to process the samples that are closer to the real path delay; N_{chips} the number of chips per processing cycle, for example, $N_{chips} = 2560$ chips per slot; this amounts to $N_{chips} \times OSF$ samples; N_{data} the number of data symbols; SF_{data} the spreading factor for the data symbols; L_{RRC} the length, in chips, of the pulse shaping filter impulse response that is used to match the incoming signal stream.

As for the number of arithmetic operations, the received signal has a length of $OSF \times (N_{chips} + L_{RRC} + DS - 1)$ samples. The pulse shape length is $OSF \times L_{RRC} - 1$ samples. The output of the cross-correlation has a length of $OSF \times (N_{chips} + 2 \times L_{RRC} + DS - 1)$ samples, but we need to compute only the last $OSF \times (N_{chips} + L_{RRC} + DS - 1)$ samples. By considering the fact that the pulse shape is a real valued signal, the computing of one cross-correlation sample needs $2 \times (OSF \times L_{RRC} + 1)$ real multiplications and $2 \times (OSF \times L_{RRC} + 1)$ real additions. Fewer operations are needed for the last $OSF \times L_{RRC}$ samples owing to the border effect (presence of zeros values). In total, $OSF \times (OSF \times L_{RCC} + 1) \times (2 \times (N_{data} \times SF_{data} + DS - 1) + L_{RRC})$ real multiplications and the same number of real additions are needed.

As for the load/store operations, we need to load the received signal by means of $OSF \times (N_{data} \times SF_{data} + L_{RRC} + DS - 1)$ load operations, and also the pulse shape filter impulse response by means of $(OSF \times L_{RRC} + 1)$ load operations. Storing the cross-correlation sequence needs $OSF \times (N_{data} \times SF_{data} + L_{RRC} + DS - 1)$ store operations.

As for the memory requirements to compute the cross-correlation sequences, the required memory includes $4 \times OSF \times (N_{data} \times SF_{data} + L_{RRC} + DS - 1)$ bytes for the received signal, $8 \times (OSF \times L_{RRC} + 1)$ bytes for computing the cross-correlation sample, and $4 \times OSF \times (N_{data} \times SF_{data} + L_{RRC} + DS - 1)$ bytes to store the resulting cross-correlation sequences, assuming a 2×16 bits representation of complex numbers. For the MPIC detector to be discussed below, it is important to note that we have to store the resulting cross-correlation sequences all the time, as they are needed to generate the inputs in and after the second stage as explained later.

---

[1] Note that if one real multiplication is more complex than three real additions, then a complex multiplication can be realized as $(a + jb)(c + jd) = [(ac - bd)] + j[(a + b)(c + d) - (ac - bd)]$ and accordingly changes subsequent complexity analysis.

In summary, the number of total arithmetic operations $\mathcal{O}$ assuming equal complexity of multiplication and addition, as well as the load/store operations $\mathcal{L}$, and memory requirements $\mathcal{M}$ in bytes per slot for the RRC matched filter are given as:

$$\mathcal{O} = 2 \times OSF \times (OSF \times L_{RCC} + 1)$$
$$\times (2 \times (N_{data} \times SF_{data} + DS - 1) + L_{RRC}), \tag{5.2}$$

$$\mathcal{L} = 2 \times OSF \times (N_{data} \times SF_{data} + L_{RRC} + DS - 1)$$
$$+ OSF \times L_{RRC} + 1, \tag{5.3}$$

$$\mathcal{M} = 8 \times OSF \times (N_{data} \times SF_{data} + L_{RRC} + DS - 1)$$
$$+ 8 \times (OSF \times L_{RRC} + 1). \tag{5.4}$$

5.5.2.2 Channel Acquisition

One of the most crucial entities in a UMTS receiver is the unit that identifies the delays of the impinging MPCs and assigns the Rake fingers accordingly; this entity is henceforth referred to as the searcher. In the downlink, the searcher traditionally obtains delays from the CPICH, which is continuously transmitted at higher power level and in parallel with the data streams. For the complexity analysis it is important to know how many pilot symbols are actually used as well as the sample rate, length of the cross-correlation window, and other factors. Also, there are numerous techniques to estimate the channel delays and hence influence the Rake receiver performance, where performance is usually traded against complexity. We have chosen to analyze an iterative-type channel acquisition process without any further signal processing techniques in the form of, for example, weighted multislot averaging, etc.

During power up, loss of session or even handover, the searcher clearly has to perform a complete update on channel delays and Rake assignments, henceforth referred to as full channel acquisition. On the other hand, once the modem is operational, an incremental update with shorter correlation windows and hence lower complexity suffices, and is henceforth referred to as partial channel acquisition. The frequency at which either method is performed will depend on the communication scenario and eventually on the manufacturer.

The computational complexity of full and partial channel acquisition depends upon the following parameters: DS is the delay spread (in chips) is the length of the window in which valid paths are searched; OSF is the oversampling factor, which allows the Rake receiver to process the samples that are closer to the real path delay; SF is the spreading factor of the code used to spread the burst; $N_{bit,pilot}$ is the number of pilot bits used to perform acquisition; $N_{chip,pilot}$ is the number of pilot chips, which is linked to $N_{bit,pilot}$ and SF by $N_{chip,pilot} = N_{bit,pilot} \times SF/2$; L_{RRC} is the length, in chips, of the pulse shaping filter impulse response that is used to regenerate locally the ideal pilot signal; L is the number of identified MPCs; and, finally, R_{acq} is the ratio between partial (preferred) versus full (high complexity) acquisition.

As for the *full channel acquisition*, given the received signal that is supposed to contain the CPICH and the codes that were used to spread and scramble the pilot symbols, the channel estimator must compute the number of valid paths, the delays of the valid paths with respect to the beginning of the slot, and the complex attenuation of the valid paths for MPC cancelation and estimation purposes. This is traditionally performed by means of several iterations of computing the cross-correlation between the received signal and a local copy of the ideal received signal as if no multipath propagation had occurred, selection of the delay that yields the highest magnitude correlation peak, estimation of the identified path's complex attenuation, validation of the delay and complex attenuation and, finally, suppression of a weighted local copy of the ideal pilot signal to allow identification of the next strongest path with the same process. This algorithm is considered of to be of average complexity compared with other available acquisition techniques.

First, the pilots need to be generated. The pilot symbols, unique to each cell/sector, have to be spread and scrambled first by the appropriate codes. Only simple degenerate multiplications occur here, which amounts to $N_{bit,pilot} \times SF$ arithmetic operations and $5/2 \times N_{bit,pilot} \times SF$ load/store operations. Then, the resulting pilot chips have to be pulse shaped by the transmitting RRC filter and by the receiving RRC filter, to ensure that the dynamics of the weighted local replica of the ideal pilot signal to be suppressed is correct. The transmit filter is a polyphase filter, since it oversamples the chips according to OSF. On the other hand, the receive filter is a full RRC filter, the complexity of which grows with OSF^2. This leads to $2 \times N_{chip,pilot} \times OSF \times (L_{RRC}(OSF + 1) + 3/2) + N_{chip,pilot}$ arithmetic operations and $L_{RRC} \times (OSF + 1) + N_{chip,pilot}(2 + OSF \times (6 + L_{RCC} \times (OSF + 1)))$ load/store operations. Thereupon, to reduce the overall complexity of the channel acquisition process, the cross-correlation between the local training sequence and the received signal is performed between two sequences that are undersampled, being thus sampled at the chip rate and not at the sample rate, which would strongly inflate the number of operations to be performed. No precision loss regarding the delay estimation occurs, since the correlation process is performed at each sample, and not at each chip. Furthermore, the complex attenuation of the detected path is now the exact value of the correlation point, which eliminates the need of properly estimating this attenuation. This computation of the cross-correlation requires $DS \times OSF \times (4 \times N_{chip,pilot} + 3) \times L$ arithmetic operations and $DS \times OSF(2 \times N_{chip,pilot} + 1) \times L$ load/store operations. Then, to ensure that a detected correlation peak is effectively a valid path, the channel estimator ought to perform the actual demodulation of the pilot symbols according to the identified path. The received samples are descrambled then despread, and the resulting symbols are rotated according to the measured complex attenuation. The resulting symbols are then compared to the known transmitted pilot bits to estimate the likelihood of the validity of the path currently being detected. This requires $2 \times N_{bit,pilot} \times (SF + 2) \times L$ arithmetic operations and $N_{bit,pilot} \times SF \times L$ load/store operations. Finally, if the path is deemed valid, it has to be suppressed from the received signal in order to allow the same process to be used to estimate the second strongest path, etc. The locally generated pilot sequence, at the sample rate, is weighted by the estimated complex attenuation of the path, and then suppressed from the received signal in order to blank the correlation peak yielded by the presence of a valid path at this delay. This amounts to $5 \times N_{chip,pilot} \times OSF \times (L - 1)$ arithmetic operations and $(3 \times N_{chip,pilot} \times OSF + 1) \times (L - 1)$ load/store operations.

As for the memory requirements, the only temporary storage that has to be allocated to the full channel estimation routine must be large enough to hold the local copy of the portion of the received signal that contains the pilot symbols (a local copy has to be stored since some data is suppressed from it during the iterations of the algorithm), that is $4 \times (N_{chip,pilot} + DS) \times OSF$ bytes assuming a 2×16 bit representation of complex numbers. Further memory is needed for the spread and scrambled pilot chip sequence that is needed to compute the cross-correlation, that is $4 \times N_{chip,pilot}$ bytes, and the double-filtered pilot sequence that is weighted and suppressed from the received signal, that is $4 \times N_{chip,pilot} \times OSF$.

In summary, the number of total arithmetic operations $\mathcal{O}$ assuming equal complexity of multiplication and addition, as well as the load/store operations $\mathcal{L}$, and memory requirements $\mathcal{M}$ in bytes per slot for the full channel acquisition are given as:

$$\mathcal{O} = N_{bit,pilot} \times SF + 2 \times N_{chip,pilot} \times OSF(L_{RRC}(OSF + 1) + 3/2) + N_{chip,pilot}$$

$$+ DS \times OSF(4 \times N_{chip,pilot} + 3) \times L + 2 \times N_{bit,pilot} \times (SF + 2) \times L$$

$$+ 5 \times N_{chip,pilot} \times OSF \times (L - 1), \tag{5.5}$$

$$\mathcal{L} = 5/2 \times N_{bit,pilot} \times SF + L_{RRC} \times (OSF + 1)$$

$$+ N_{chip,pilot}(2 + OSF(6 + L_{RCC} \times (OSF + 1)))$$

$$+ DS \times OSF(2 \times N_{chip,pilot} + 1) \times L + N_{bit,pilot} \times SF \times L$$

$$+ (3 \times N_{chip,pilot} \times OSF + 1) \times (L - 1), \tag{5.6}$$

$$\mathcal{M} = 4 \times (N_{chip,pilot} + DS) \times OSF + 4 \times N_{chip,pilot} \times (OSF + 1). \tag{5.7}$$

Concerning *partial channel acquisition*, it is sometimes preferable to reduce the overall complexity to perform only a simple form of MPC estimation, known as tracking or partial channel acquisition. Given the channel model that was estimated for the previous slot, the tracking module will only adjust the delays and complex attenuations for the current slot without having to deal with a cross-correlation on the whole search window length. For each previously identified path, the tracking module adjusts the delay and the complex attenuation by computing the cross-correlation over a window that is about one chip long.

As for the complexity analysis, it can be assumed that the pilots and RRC pulse shape are generated and stored from the full acquisition process. The cross-correlation computation is now performed over one chip only and hence leads to a reduced arithmetic complexity, that is $L \times OSF \times (4 \times N_{chip,pilot} + 5)$ arithmetic operations and $2 \times L \times (N_{chip,pilot} \times OSF + 2)$ load/store operations. It is further assumed that path validation is not performed during partial acquisition and the weighting and suppression process is hence not performed either. The total memory requirement is thus the same as for the full acquisition.

In summary, the number of total arithmetic operations $\mathcal{O}$ assuming equal complexity of multiplication and addition, as well as the load/store operations $\mathcal{L}$, and memory requirements $\mathcal{M}$ in bytes per slot for the partial channel acquisition are given as:

$$\mathcal{O} = L \times OSF \times (4 \times N_{chip,pilot} + 5), \tag{5.8}$$

$$\mathcal{L} = 2 \times L \times (N_{chip,pilot} \times OSF + 2), \tag{5.9}$$

$$\mathcal{M} = 4 \times (N_{chip,pilot} + DS) \times OSF + 4 \times N_{chip,pilot} \times (OSF + 1). \tag{5.10}$$

Finally, the ratio R_{acq} between the two modes of channel acquisition strongly depends on the operational environment, and eventually on the manufacturer. Given a defined ratio, the overall complexity can be estimated as $R_{acq} \times$ full acquisition $+ (1 - R_{acq}) \times$ partial acquisition.

5.5.2.3 Channel Estimation

Channel estimation has to be performed per resolved multipath; therefore, the estimation process itself has to be done at the lowest possible rate. There are options for estimating the channel from (i) the CPICH channel (extra operations at chip level due to despreading operation, but much higher reliability); and/or (ii) from the pilot symbols embedded in the slot (no extra chip level operations but less reliability). The natural choice is to use the CPICH channel since it is transmitted with higher power than are the data channels and hence audible with a higher reliability. There are situations, however, when channel estimation from the slot pilots becomes an attractive option, namely when the terminal is at the cell edge; this is because the data channels are power controlled, whereas the CPICH is not.

The computational complexity of the channel estimator depends upon the following parameters: SF_{pilot} is the spreading factor of the CPICH, that is $SF_{pilot} = 256$; N_{pilot} is the number of pilot symbols in CPICH, that is $N_{pilot} = 10$ symbols, and L is the number of identified MPCs. The task of channel estimation by means of the CPICH channel requires the despreading of the CPICH channel, pilot sequence multiplication, integration/averaging and a scale factor multiplication. Specifically to carry out an estimate of the complex channel coefficient by means of the CPICH, the despread data sequence is multiplied by the complex conjugate of the pilot symbol sequence. The samples of the resulting chip sequence are summed and the obtained result is scaled by a multiplication factor.

As for the arithmetic complexity, since the codes (including channelization and scrambling codes) are taken from a QPSK modulation alphabet, one complex multiplication needs only two real additions. For the despreading operation of the CPICH, we hence need $2 \times N_{pilot} \times SF_{pilot} \times L$ real additions. For the pilot sequence multiplication, we need $2 \times N_{pilot} \times L$ real additions. For the integration, which is effectively a summation, we need $2 \times N_{pilot} \times L$ real additions. Finally, for the scaling operation per Rake, we need $2 \times L$ real multiplications.

As for the load/store operations, the despreading requires the loading of the received pilots (in chip rate) for all paths, which leads to $N_{pilot} \times SF_{pilot} \times L$ load operations. We also have to load all the codes, requiring $N_{pilot} \times SF_{pilot}$ load operations. Furthermore, the store operation of the despread data (in data symbol rate) requires $N_{pilot} \times L$ store operations. Once despread, the CPICH data have to be reloaded by means of $N_{pilot} \times L$ load operations. We also need to load the pilot data symbols, leading to N_{pilot} load operations. Finally, the resulting channel gain estimates have to be stored by means of L store operations.

As for the memory requirements, the channel gain coefficients estimation is done in sequential manner, that is coefficient by coefficient. Hence, the memory required to store the despread data does not depend on the number of coefficients to be estimated, and this only needs $4 \times N_{pilot}$ bytes. The memory needed to store the resulting coefficients estimates is hence $4 \times L$ bytes.

In summary, the number of total arithmetic operations $\mathcal{O}$ assuming equal complexity of multiplication and addition, as well as the load/store operations $\mathcal{L}$, and memory requirements $\mathcal{M}$ in bytes per slot for the channel estimation are given as:

$$\mathcal{O} = 2 \times N_{pilot} \times L \times (SF + 2) + 2 \times L, \tag{5.11}$$

$$\mathcal{L} = N_{pilot} \times (SF_{pilot} \times (1 + L) + L) + N_{pilot}(1 + L) + L, \tag{5.12}$$

$$\mathcal{M} = 4 \times (N_{pilot} + L). \tag{5.13}$$

Note that for the MPIC detector, these operations are common for all stages of the MPIC, even for the final one, but are performed only once every stage, that is it does not depend on the number of iterations of the iterative channel estimation procedure.

5.5.2.4 MRC and MPIC Symbol Detectors

The detector is central to the performance of a receiver and hence of the relay in general. The role of the detector is to obtain the symbol stream to be decoded from the chip stream with the aid of the side information provided by the searcher. The theory behind detection and estimation processes is vast and well explored, with a range of available detectors trading complexity against performance [521].

Clearly, the optimum choice of any detection is the Maximum Likelihood (ML) detector, which searches through all possible transmitted sequences, compares them to some processed received data and decides thereupon on the most likely transmitted sequence. The metric which is to be maximized is $\eta = ||\mathbf{x} - \mathbf{Fs}||^2_{\mathbf{R}^{-1}}$, where $\mathbf{x}$ is the incoming data stream, $\mathbf{F}$ is a detection filter including the spreading and scrambling codes, $\mathbf{s}$ is the candidate signal stream with the most likely to be chosen, and $\mathbf{R}$ is the temporal–spatial correlation matrix. Needless to say, with an exponential dependency of complexity from sequence length and number of bits per symbol, such detector is prohibitively complex. Less complex detectors with the caveat of inferior performance are hence used, such as those explored below which are based on a simple maximum ratio combining (MRC) or multipath interference cancelation (MPIC).

The computational complexity of these two detectors depends upon the following parameters: SF_{data} is the spreading factor of the code used to spread the data burst (for example, $SF_{data} = 16$ for HSDPA); N_{data} is the number of data symbols per slot (for example, $N_{data} = 160$ for HSDPA); $SF_{pilot} = 256$ is the spreading factor for the pilot symbols on the CPICH; $N_{pilot} = 10$ is the number of CPICH pilot symbols per slot; MC is the number of simultaneous channelization codes per user; L is the number of identified MPCs; W is the maximum channel delay time (in chips); OSF is the over sampling factor; L_{RRC} is the length of the root-raised cosine pulse shape (in chips); R is the number of iterations of the channel estimation process at each MPIC stage; and $P \geq 2$ is the number of MPIC stages.

As for the *MRC detector*, it is clearly one of the simplest realizations of a detector. It uses a single receive antenna with a Rake receiver and is based on the MRC principle. Such a receiver is classified as a linear receiver, which ignores multiple access interference (MAI) in the detection process. A classical Rake architecture is depicted in Figure 5.6, where the receiver RF chain ends with an analog-to-digital converter that provides the signal stream at sampling rate. The stream is then receiver matched filtered by means of a root-raised cosine filter. The filtered stream is then fed into the acquisition unit, which estimates the relative delays between the MPCs at sample resolution. The acquisition unit then informs each Rake finger about the delays, which can then delay the incoming stream appropriately and sample it at chip rate. Thereafter, the despreading and descrambling operations are performed at chip level for the data stream and the pilot stream, both of which are then averaged over all chips per symbol. The CPICH symbols are utilized to estimate the complex channel coefficients, which are used to calculate the combined weight coefficients w according to the MRC algorithm, that is $w = h$, where h is the estimated channel coefficient on the channel path of interest. The complex conjugate of the combing weight w^* is then multiplied with the despread data stream at symbol rate. Finally, the outputs of the Rake fingers are added and fed into the symbol estimators and the outer modem.

The complexity of the MRC symbol detector can be estimated as follows: The Rake receiver has first to despread the incoming signal stream and then to perform the MRC combining per Rake finger. The complexities can be calculated similarly to the case of channel estimation, which shows that the correlation requires $2 \times N_{data} \times SF_{data} \times L$ real additions, the integration $2 \times N_{data} \times SF_{data} \times L \times MC$ real additions, the scale factor multiplication $2 \times N_{data} \times L \times MC$ real multiplications, and the channel compensation $2 \times N_{data} \times L \times MC$ real additions and $4 \times N_{data} \times L \times MC$ real multiplications.

As for the load/store operations, to perform the despreading operation, we have to load the received data (in chip rate) for all paths, which leads to $N_{data} \times SF_{data} \times L$ load operations. We also have to load all the codes, requiring $N_{data} \times SF_{data}$ load operations (in the case of multicodes, the corresponding channelization codes can be assumed to be loaded only once in the transmission, that is they do not have to be reloaded at each slot). We also need to load the channel gain coefficients, which requires L load operations. Finally, the store operation of the despread data (in data symbol rate) requires $N_{data} \times L \times MC$ store operations.

As for the memory requirements, we have to store the data resulting symbols after the despreading operation, that is $N_{data} \times L \times MC$. In addition, each time we have to store the $N_{data} \times MC$ resulting data symbol estimates.

In summary, the number of total arithmetic operations $\mathcal{O}$ assuming equal complexity of multiplication and addition, as well as the load/store operations $\mathcal{L}$, and memory requirements $\mathcal{M}$ in bytes per slot for the MRC detector are given as:

$$\mathcal{O} = 2 \times N_{data} \times L \times (2 \times SF_{data} + MC) + 6 \times N_{data} \times L \times MC, \qquad (5.14)$$

$$\mathcal{L} = N_{data} \times (SF_{data} \times (L+1) + L \times MC) + L, \qquad (5.15)$$

$$\mathcal{M} = 4 \times N_{data} \times (1+L) \times MC. \qquad (5.16)$$

As for the *MPIC detector*, it is a fairly powerful and also complex detector that has was introduced by Higuchi *et al.* [518]. It basically comprises several channel estimation and interference replica generation units (CEIGU), the number of which corresponds to the number of stages. In and after the second stage, the multipath interference (MPI) replica estimated in the previous stage is removed from the received signal for the input signal to the next stage. Let $\hat{I}_l^{(p)}$ be the estimated received signal of the *l*th path (called MPI replica for which $1 \leq l \leq L$) at the *p*th stage ($1 \leq p \leq P$). The received signal sequence is directly embedded in the first stage, and for the incoming signal sequence at the *p*th stage in and after the second stage, the MPI replica of all code channels except for its own path generated in the previous stage, $\hat{I}_l^{(p-1)}$, are removed from the received signal sequence. The structure of the CEIGU has also been shown and discussed [518]. In each CEIGU, the input sample sequence of each antenna is despread by means of a matched filter (MF) into the resolved multipath components.

It is important to note that the operation of despreading includes the descrambling operation also. The channel is estimated by using the common pilot symbols and decision feedback data symbols belonging to the same packet. Then, the phase variation of each path is compensated and coherently Rake combined. The MPI replica is generated using the decision data sequence and channel estimates. Since the channel estimation and data decision are updated at each stage, the accuracy of the MPI replica is improved from the resulting enhancement of channel estimation and decreasing data decision error [518]. The channel decoding and re-encoding is a very complex operation since it is performed at every iteration within each stage. Although the omission of a channel decoder/encoder would still yield acceptable results, a turbo decoder/encoder is assumed in this study which, strictly speaking, requires the inclusion of the entire outer modem into the decoding and encoding process. Note that for the complexity estimation of the MPIC, we have neglected the complexity of the encoding stage.

The complexity calculation is significantly more involved but has been demonstrated [518]. In summary, the number of total arithmetic operations $\mathcal{O}$ assuming equal complexity of multiplication and addition, as well as the load/store operations $\mathcal{L}$, and memory requirements $\mathcal{M}$ in bytes per slot for the MPIC detector are given as [518]:

$$
\begin{aligned}
\mathcal{O}_{\text{QPSK}} = {}& 2 \times OSF \times (OSF \times L_{RRC} + 1) \\
& \times (2 \times (N_{data} \times SF_{data} + W - 1) + L_{RRC}) \\
& + P \times N_{data} \times MC \times L \times (2 + 4 \times SF_{data}) \\
& + P \times 8 \times N_{data} \times MC \times L \times R + P \times N_{pilot} \times L \times (2 + 4 \times SF_{pilot}) \\
& + P \times L \times (2 + 4 \times N_{pilot}) + P \times 2 \times L \\
& + (P - 1) \times 4 \times L \times R \times (1 + N_{data} \times MC) \\
& + (P - 1) \times 2 \times N_{data} \times L \times (MC \times (1 + 2 \times SF_{data}) + 2 \times SF_{data}) \\
& + (P - 1) \times 2 \times L \times (1 + N_{pilot}) \\
& + (P - 1) \times 4 \times L \times N_{data} \times SF_{data} \times (2 \times OSF \times L_{RRC} + 1) \\
& + (P - 1) \times 2 \times L^2 \times OSF \times (N_{data} \times SF_{data} + L_{RRC} + W - 1) \\
& + 4 \times L \times (R - 1) \times (1 + N_{data} \times MC),
\end{aligned}
\tag{5.17}
$$

$$
\begin{aligned}
\mathcal{O}_{\text{16QAM}} = {}& 2 \times OSF \times (OSF \times L_{RRC} + 1) \\
& \times (2 \times (N_{data} \times SF_{data} + W - 1) + L_{RRC}) \\
& + P \times N_{data} \times MC \times L \times (2 + 4 \times SF_{data}) \\
& + P \times 8 \times N_{data} \times MC \times L \times R + P \times N_{pilot} \times L \times (2 + 4 \times SF_{pilot}) \\
& + P \times L \times (2 + 4 \times N_{pilot}) \\
& + (P - 1) \times 4 \times N_{data} \times MC \times R \times (1 + 2 \times L) \\
& + (P - 1) \times 2 \times L \times (1 + 2 \times R) \\
& + (P - 1) \times 2 \times N_{data} \times L \times (MC \times (3 + 2 \times SF_{data}) + 2 \times SF_{data}) \\
& + (P - 1) \times 2 \times L \times (1 + N_{pilot}) \\
& + (P - 1) \times 4 \times L \times N_{data} \times SF_{data} \times (2 \times OSF \times L_{RRC} + 1) \\
& + (P - 1) \times 2 \times L^2 \times OSF \times (N_{data} \times SF_{data} + L_{RRC} + W - 1) \\
& + 4 \times N_{data} \times MC \times (R - 1) \times (1 + 2 \times L) + 2 \times L \times (2 \times R - 1),
\end{aligned}
\tag{5.18}
$$

$$\mathcal{L} = 8 \times OSF \times (N_{data} \times SF_{data} + L_{RRC} + W - 1) + 4 \times OSF \times L_{RRC} + 4$$

$$+ 4 \times P \times N_{data} \times MC \times (SF_{data} \times (1 + L) + L)$$

$$+ 4 \times P \times R \times (N_{data} \times MC \times (1 + L) + L)$$

$$+ 4 \times P \times N_{pilot} \times (SF_{pilot} \times (1 + L) + L)$$

$$+ 4 \times P \times N_{pilot} \times (1 + L)$$

$$+ 4 \times P \times L + 4 \times (P - 1) \times R \times (2 \times L + N_{data} \times MC \times (1 + L))$$

$$+ 4 \times (P - 1) \times L \times (1 + N_{pilot} + N_{data} \times (MC \times (1 + SF_{data}) + 2 \times SF_{data}))$$

$$+ 4 \times (P - 1) \times L \times (N_{data} \times SF_{data} \times (1 + OSF) + OSF \times (2 \times L_{RRC} - 1) + 1)$$

$$+ 4 \times (P - 1) \times (2 \times OSF \times L_{RRC} + 1)$$

$$+ 4 \times (P - 1) \times (L + 1) \times OSF \times (N_{data} \times SF_{data} + L_{RRC} + W - 1)$$

$$+ 4 \times (P - 1) \times L \times (OSF \times (N_{data} \times SF_{data} + 2 \times L_{RRC} - 1) + 1)$$

$$+ 4 \times (R - 1) \times (2 \times L + N_{data} \times MC \times (1 + L)), \tag{5.19}$$

$$\mathcal{M} = 8 \times OSF \times (N_{data} \times SF_{data} + L_{RRC} \times + W - 1) + 8 \times (OSF \times L_{RRC} + 1)$$

$$+ 4 \times N_{data} \times MC \times (1 + L) + 4 \times (N_{pilot} + L) + 4 \times (N_{data} + L)$$

$$+ 4 \times N_{data} \times SF_{data} \times (3 + 2 \times L) + 4 \times L \times (OSF \times (2 \times L_{RRC} - 1) + 1)$$

$$+ 8 \times (2 \times OSF \times L_{RRC} + 1)$$

$$+ 4 \times (2 + L) \times OSF \times (N_{data} \times SF_{data} + L_{RRC} + W - 1). \tag{5.20}$$

Note that the arithmetic complexities of QPSK and 16QAM are different.

5.5.2.5 Outer Modem

The outer relay modem has to perform the following operations: soft demodulation, de-mapping of the physical channels, second de-interleaving, demultiplexing of traffic channels, desegmentation, first de-interleaving, rate matching, channel decoding, bit-to-byte conversion, and CRC computation.

For the complexity evaluation, note that the outer modem of HSDPA differs slightly from the above outlined proceedings; however, since the baseband complexity is dominated by the channel decoder we assume for simplicity that the complexity of both systems coincides. Furthermore, the cost of the index loop (hardware loop in the DSP) is not considered. Also, during the interleaving process, we do not generate the interleaver on the fly; we consider that it is already in the memory. Conceptually, bits are written into a two-dimensional matrix row wise and read from it column wise. The size of the dedicated control channel (DCCH), which is needed for the operation of relayed dedicated channels, is 100 bits and is encoded by a convolutional code of 1/3 rate: the size of the DCCH is hence 360 bits with CRC and redundancy included. Note that we only focus on the decoding of the dedicated traffic channel (DTCH), the decoding of the DCCH being neglected. Also, the assumed interleaver depth is 20 ms; note that the shorter interleaver depth of HSDPA insignificantly diminishes the outer modem complexity.

The evaluation of the complexity of the outer modem depends upon the following parameters: N_{rx} is the number of bits per received transport block before the physical channel demapping; N_{TTI} is the number of bits of a transport block; $N_{slots,TTI}$ is the number of slots per transport block TTI duration (for example, $N_{slots,TTI} = 30$ for a 20 ms TTI or $N_{slots,TTI} = 3$ for HSDPA); R_c is the channel code rate; $R_{p,dtch}$ is the puncturing level of the DTCH; $R_{p,dcch}$ is the puncturing level of the DCCH; and

N_{iter} is the number of iterations for the turbo-decoder. Note that above parameters are not independent; for instance, if a 1/3 channel code is used, then $N_{TTI} = N_{rx}/3$.

To estimate the complexity, we merely need to follow the 3G outer modem receiver stack and apply a similar reasoning to that already done for the blocks of the inner modem. This leads to a number of total arithmetic operations $\mathcal{O}$, assuming equal complexity of multiplication and addition, as well as the load/store operations $\mathcal{L}$, and memory requirements $\mathcal{M}$ in bytes per transport block assuming *convolutional decoding*:

$$
\begin{aligned}
\mathcal{O} = {} & q \times N_{rx}/2 \times 20 + N_{rx} \times 3 \\
& + (N_{rx} - 360 \times (1 - R_{p,dcch})/2) \times (2 + R_{p,dtch}/(1 - R_{p,dtch})) \\
& + (N_{TTI} + 16)/16 \times 256 \times 2 + (N_{TTI} + 16) \times 2 + N_{TTI}/8 \times 2,
\end{aligned}
\tag{5.21}
$$

$$
\begin{aligned}
\mathcal{L} = {} & q \times N_{rx} + N_{rx} \times 4 + (N_{rx} - 360 \times (1 - R_{p,dcch})/2) \times 2 \\
& + (N_{rx} - 360 \times (1 - R_{p,dcch})/2) \times (1 + R_{p,dtch}/(1 - R_{p,dtch})/2) \\
& + (N_{TTI} + 16)/16 \times 256 \times 2 + (N_{TTI} + 16) \times (1/2 + 1/8) \\
& + N_{TTI}/8 \times 2,
\end{aligned}
\tag{5.22}
$$

$$
\begin{aligned}
\mathcal{M} = {} & (N_{rx} + (N_{rx} - 360 \times (1 - R_{p,dcch})/2))/8 \\
& + \max(N_{rx} - 360 \times (1 - R_{p,dcch})/2, \\
& (N_{rx} - 360 \times (1 - R_{p,dcch})/2)/(1 - R_{dtch}))/8 \\
& + ((N_{rx} - 360 \times (1 - R_{p,dcch})/2)/R_{p,dtch} - 12) \times R_c/8 + N_{TTI}/4,
\end{aligned}
\tag{5.23}
$$

where $q = 1$ for QPSK and $q = 4$ for 16QAM. Note that to obtain the approximate complexity per slot, we have to divide the above numbers by the number of slots the transport block occupies, that is $N_{slots,TTI}$.

Finally, the number of total arithmetic operations $\mathcal{O}$ assuming equal complexity of multiplication and addition, as well as the load/store operations $\mathcal{L}$, and memory requirements $\mathcal{M}$ in bytes per transport block assuming *turbo decoding* is:

$$
\begin{aligned}
\mathcal{O} = {} & q \times N_{rx}/2 \times 20 + N_{rx} \times 3 \\
& + (N_{rx} - 360 \times (1 - R_{p,dcch})/2) \times (2 + R_{p,dtch}/(1 - R_{p,dtch})) \\
& + 2 \times N_{iter} \times (N_{TTI} + 16) \times (3/2 + 37 + 16/24) + (N_{TTI} + 16) \times 2 \\
& + N_{TTI}/8 \times 2,
\end{aligned}
\tag{5.24}
$$

$$
\begin{aligned}
\mathcal{L} = {} & q \times N_{rx} + N_{rx} \times 4 + (N_{rx} - 360 \times (1 - R_{p,dcch})/2) \times 2 \\
& + (N_{rx} - 360 \times (1 - R_{p,dcch})/2) \times (1 + R_{p,dtch}/(1 - R_{p,dtch})/2) \\
& + 2 \times N_{iter} \times (N_{TTI} + 16) \times (23 + 16/24) + (N_{TTI} + 16) \times (1/2 + 1/8) \\
& + N_{TTI}/8 \times 2,
\end{aligned}
\tag{5.25}
$$

$$
\begin{aligned}
\mathcal{M} = {} & (N_{rx} + (N_{rx} - 360 \times (1 - R_{p,dcch})/2))/8 \\
& + \max(N_{rx} - 360 \times (1 - R_{p,dcch})/2, \\
& (N_{rx} - 360 \times (1 - R_{p,dcch})/2)/(1 - R_{dtch}))/8 \\
& + ((N_{rx} - 360 \times (1 - R_{p,dcch})/2)/R_{p,dtch} - 12) \times R_c/8 + N_{TTI}/4.
\end{aligned}
\tag{5.26}
$$

5.5.3   Power Consumption

The power in a relay terminal is consumed by the transceiver RF front-end, by the baseband processing unit (including memory maintenance) and, if any, by higher layer processing and screen/keyboard/etc. The consumption of these elements is estimated in this section.

5.5.3.1   RF Front-End Consumption

Despite the load on the signal processing part, as identified in Section 5.5.2, the front end of the relay is actually also fairly constrained. This is mainly due to the following reasons:

- **PAPR.** WCDMA signals have a high peak-to-average power ratio (PAPR), which is due to the QPSK modulation exhibiting a nonconstant envelope after filtering. The amount of PAPR depends on the number of codes transmitted simultaneously on one carrier frequency. The higher the PAPR, the more stringent the requirement on the linearity of the power amplifier to prevent compression or clipping of the signal.
- **Signal Dynamics.** A CDMA system requires a high output power dynamic, mainly due to the need for power control to ease the near–far effect.
- **FDD Operation.** Assuming the simultaneity of the transmitter and the receiver path in FDD mode, we have to consider higher filter losses for a full duplex system compared to TDD, and separate Tx and Rx chip sets to decrease interference problems. Also, the isolation in the duplexer between Tx and Rx ought to be at least 50 dB.
- **Relay Bands.** A supportive relay may decide to relay more than one FDD connection or a cooperative relay may decide to handle its own data via one FDD connection and the relayed one via another one. In either of these cases, the relay would need to support more than one RF transmit and more than one receive RF chain.

A typical RF UMTS FDD transceiver is based around a conventional heterodyne architecture using a single IF at about 380 MHz (on the Tx path) and 190 MHz (on the Rx path). With this receiver topology, the adjacent channel selectivity (ACS) constraint of 33 dB at 5 MHz is commonly achieved with a combination of an IF SAW filter and analog baseband filtering, and the dynamics of the output power can be distributed over the global transmitter path. The choice of this intermediary frequency is related to the duplex of 190 MHz (in band 1), and the transceiver requires a single local oscillator for the up/down RF conversion. We also need another oscillator, called an IF oscillator, for the I/Q modulator/demodulator. However, this architecture is complex and needs a large number of external components and stages (image rejection, up/down mixer, IF stage, I/Q stage, etc.), which leads to high power consumption and consequently shorter battery life; manufacturers use this architecture for quick time to market (TTM).

For better integration, it could also be possible to use direct conversion for the receiver path and a heterodyne (IF = 190 MHz) approach for the transmitter path. In fact, for the receiver path, the tendency is toward direct conversion, in spite of the important constraints of this type of topology (DC offset, large bandwidth required to the ADC, most of the gain is contributed by the baseband amplifier, etc.), but it is favorable for high integration, cost and power consumption. Some technical solutions are possible; for example, the use of high-pass filtering to resolve the problem of the DC offset, with an acceptable degradation of the system performance by the wideband nature of the signal. A lot of published work on receiver W-CDMA is based on direct-conversion [522–527]. In this case, it is possible to use an IF frequency equal to 190 MHz in the transmitter, favorable for a single RF local oscillator for up/down conversion.

To obtain an estimate on the power consumption, we assume the following:

- insertion losses of the global RF filtering (duplexer filter and, if applicable, diplexer) in the transmit path of about 3.5 dB;

- maximum $P_{out} = 21$ dBm and about 90% transmission time;
- average receiver chain power consumption $P_{Rx} = 85$ mW [523];
- average transmitter chain power consumption $P_{Tx} = 215$ mW [523];
- power consumption is 60 mW for the ADC and 30 mW for the DAC.

As thoroughly quantified by Raju [528], the power consumption of the RF front end heavily depends on the transmit power distribution. This distribution, in turn, depends on the UMTS service, data rate, communication environment, etc. To simplify this study, we will assume a Gaussian distribution as shown in Figure 5.8 [529]. The limit at around 21 dBm transmission power is due to the saturation of power control. Note that this study can easily be repeated using the more sophisticated models available [528].

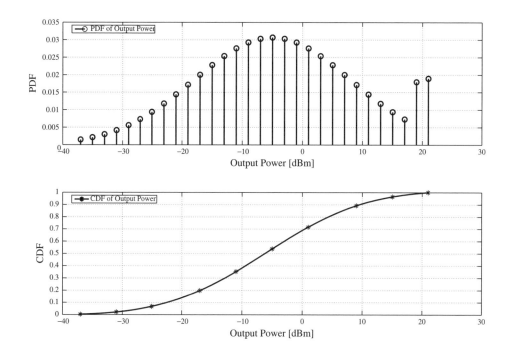

Figure 5.8 Typical probability and cumulative distribution of RF output power [529]

Given the PDF and resulting CDF in Figure 5.8, we immediately obtain a required average RF output power of 10 dBm assuming 90% communication time. Then, given the insertion losses of 3.5 dB, the amplifier has to provide 13.5 dBm, which is equivalent to 22.4 mW. For such an output power, the power amplifier efficiency is fairly low and will be around 6% [530]. Note that the efficiency of the power amplifier is significantly higher at maximum transmission powers of 21 dBm, where it can reach 40–50%. However, at 6% average efficiency, the average power required by the amplifier is 22.4 mW/0.06 = 373 mW.

Note that, as said before, the PAPR of the modulated signal depends on the number of codes transmitted simultaneously on one carrier; to evaluate the necessary output saturation power of the power amplifier, we need to consider the worst case, that is for a user bit rate of 384 kbps where 3 codes can be used in parallel and transmitted at maximum power. For this, we estimated the PAPR of about 4.5 dB [525]. The saturation power of the power amplifier hence needs to be at least 21 dBm + 4.5 dB = 25.5 dBm.

In some cases, for example, in the case of a cooperative relay that has its own data to transmit, the relay is powered by a battery. A major challenge is to power the power amplifier by a constant battery voltage, which is very difficult to achieve as voltage decreases nonlinearly over time. A voltage stabilizer is hence vital, which usually operates at an efficiency of 85%. The total power consumption of the power amplifying unit hence increases to 373 mW / 0.85 = 440 mW.

In addition to the power amplifier consumption, we have the average power consumed in the RF transceiver chains. We consume 215 mW in the transmitter chain plus 30 mW for the DAC and 85 mW in the receiver chain plus 60 mW for the ADC. The total average power consumed, excluding the power amplifier, is hence 390 mW for a single transmitter and single receiver architecture.

The power consumption of the RF transceiver chain are summarized in Table 5.7, where these numbers are independent of data rate and approximately independent of the year of manufacturing (with our current horizon). Note that in the case of more than one FDD link, these numbers simply need to be multiplied by the number of maintained FDD links. A similar procedure can also be applied if more transmit or receive antennas are added to the relay.

Table 5.7 Typical RF power consumption (mW) of a nonbattery and battery powered UMTS relay

	Non battery powered	Battery powered
Power amplifier Tx	373 mW	440 mW
Tx/Rx RF chains	390 mW	390 mW
Total power	763 mW	830 mW
Increase in %	–	8 %

5.5.3.2 Digital Baseband Consumption

The baseband consumes power for maintaining arithmetic operations and associated memory. The power consumption depends on the technological advances, an approximate projection to which is listed in Table 5.8 [adapted from 520]. This allows us to estimate the approximate power consumption at baseband, which mainly comprises the following elements:

- **Maintaining Memory.** The power required to maintain the memory is negligible compared to the RF chain(s) as confirmed by current data sheets. We will therefore henceforth neglect the power consumption due to memory maintenance.
- **Arithmetic Processor Operations.** As for the power consumption of the digital processor, various models exist that match the number of operations to the power consumption where traditionally a

Table 5.8 Technology roadmap for various important hardware parameters

	2001	2006	2011
DRAM [Gb]	0.256	16	512
Memory input impedance [pF]	–	8	3
DSP consumption [mW/Mips]	0.5	0.03	0.01
4 ALU DSP frequency [MHz]	800	1500	1900

static and dynamic power contribution is accounted for. The static contribution is usually around 10% of the dynamic one, and is hence neglected in our analysis. The dynamic power consumption can be obtained by means of the required mW/Mips, given in Table 5.8. With the projected power consumption of 0.01 mW/Mips in 2011 and some specific numbers for the number of arithmetic operations derived in Section 5.5.2, we can obtain the power consumptions for the respective baseband algorithms for which we divide given arithmetic complexity numbers by 1e6 to obtain the required million instructions per second (Mips) and multiply by 0.01 mW/Mips; note that the power consumption is given as per instruction and not per cycle. A quick check with the numbers given in Section 5.5.4 shows that the baseband consumption in the processor is negligible in comparison with the RF chain power consumption. We will henceforth neglect the power consumption due to processor activities.

- **Load/Store Operations.** The power consumed during the load/store processes, however, is not negligible. This is mainly due to the fact that the data bus system with attached memory and processor exhibits a capacitive behavior, with capacitances given in Table 5.8. To obtain the consumed power, a common approach is to model each storage bit as a capacitance of $C_{in} = 3$ pF (in 2011) to which a voltage of $V = 3$ V is applied. The average power per slot is then given by $1/2 \cdot 8 \cdot C_{in} \cdot V^2 \cdot f$, where the '8' results from 8 bits/byte and f is the frequency of the load/store operation, that is the number in bytes given through $\mathcal{L}$ in Section 5.5.2 divided by the slot duration. We will quantify the power consumption in a later section providing the case studies.

It is important to realize that the power consumption of the above will heavily depend on the hardware architecture of choice. A reliable estimation of the power consumption for coming years is hence very difficult. Nonetheless, exposed qualitative and quantitative tendencies give a rough indication of the power consumption of the digital baseband modem.

5.5.3.3 Higher Layers and Peripheral Hardware

Measurements on a UMTS terminal performed in the laboratory in France Telecom R&D, Grenoble, France, have revealed that multimedia screen, keypad, LEDs, speech amplifier and higher layers consume approximately 600 mW of the total power of a single receive antenna terminal. While we expect this consumption to decrease until 2011, we also expect new services and peripherals to be introduced that will potentially keep this number approximately constant. Therefore, in the case of the cooperative relay, which requires higher layer algorithms (MAC, RRM, application, etc.) and peripheral hardware (multimedia screen, keypad, etc.), we will consider this extra power consumption to be approximately 600 mW.

5.5.4 Case Studies

We will assume that a relay needs to support both down- and uplink. Although being differently structured and encoded, the up- and downlink UMTS frames incur the same order of complexity. For simplicity, we will assume that they require approximately the same memory and the same number of arithmetic and load/store operations. We will now quantify the approximate complexity and power consumptions of the four UMTS relay cases identified in the previous section.

5.5.4.1 Supportive AF with Analog Hardware

The supportive AF relay with analog hardware needs to maintain two FDD links on four frequency bands, which, as has been explained in Section 5.2.1, is due the fact that as of today FDR needs to be used in a small and simple relay instead of DFR. The complexities and power consumptions hence become:

- **Arithmetic Complexity.** The supportive AF relay with analog hardware has clearly no arithmetic complexity or any load/store operations, that is $\mathcal{O} = 0$, $\mathcal{L} = 0$ and $\mathcal{M} = 0$.
- **Power Consumption.** In the absence of arithmetic complexity, the power consumption associated with it is hence zero. As for the RF front end, neglecting the power consumption of the frequency translation, the total power consumption is twice that provided in Table 5.7 due to two FDD links being maintained. Since some constant power supply would typically be available to such an AF relay, the total power consumption mounts to approximately $\mathcal{P} = 2 \times 763\,\text{mW} = 1.53\text{W}$.
- **Important Trends.** The complexity and power consumption is independent of the number and type of relayed traffic. The power consumption doubles for every doubling of FDD connections used. Adding more antennas would yield noise gains at the receiver and beamforming gains at the transmitter, none of which is very useful because the former yields little gain and the latter is difficult to achieve; however, if more antennas are added then a doubling of antennas approximately doubles the power consumption.

This has been summarized and compared to other cases in Table 5.9.

Table 5.9 Summary of important relay requirement for the considered UMTS case studies

Case study	Memory (kb)	Frequency (MHz)	Power (mW)
s-AF$_a$	0	0	1,530
s-AF$_d$	53	–	765
Speech, MRC, s-DF$_d$	105	930	859
HSDPA, MRC, s-DF$_d$	110	580	818
HSDPA, MPIC, s-DF$_d$	340	3400	1450
Speech, MRC, c-DF$_d$	89	320	1442
HSDPA, MRC, c-DF$_d$	96	420	1460

5.5.4.2 Supportive AF with Digital Hardware

The supportive AF relay with digital hardware is able to use TDR and, without quantifying the spectral efficiency, we will henceforth assume that only one FDD link for up- and downlink needs to be supported. Also, the supportive AF with digital hardware does not require any arithmetic complexity in terms of additions or multiplications, but needs to load/store the oversampled data stream and also keep it in memory. Typically, the analog data stream needs to be sampled at at least twice its bandwidth, which requires a sampling rate SR of at least $SR = 10\,\text{MHz}$. This is approximately equivalent to an oversampling factor of $OSF = 2$ in the signal processing domain, although higher values are more common. The complexities and power consumptions hence become:

- **Arithmetic Complexity.** There is no arithmetic complexity, that is $\mathcal{O} = 0$. The number of store operations per slot at the relay receiver is $10 \cdot 10^6$ samples/s $\times 10 \cdot 10^{-3}$ s/frame $\div$ 15 slots/frame $\approx 6.7 \cdot 10^3$ samples/slot; the number of load operations at the relay transmitter is the same, yielding a total of $\mathcal{L} = 13.3 \cdot 10^3$ load/store operations per slot. The total memory requirements needed to support up- and downlinks amount to $\mathcal{M} = 4$ bytes/sample $\times 13.3 \cdot 10^3$ samples/slot $= 53.2\,\text{kb}$ per slot.
- **Power Consumption.** Using the approach taken in Section 5.5.3, the total power consumption due to the load/store operations can be calculated as 2.2 mW. The RF frontend power consumption is taken from Table 5.7 and amounts to roughly 763 mW assuming a nonbattery-powered node, which

is clearly significantly higher than the power consumed by the load/store operations. The total power consumption is thus $\mathcal{P} = 765\,\text{mW}$.

- **Important Trends.** The complexity and power consumption is again independent of the number and type of relayed traffic. The number of load/store operations as well as memory, however, doubles for every doubling of the sampling rate. The power consumption also almost doubles for every doubling of FDD connections used. Adding more antennas yields the same insights and trends as for the analog hardware case.

This has been summarized and compared to other cases in Table 5.9.

5.5.4.3 Supportive DF with Digital Hardware

The supportive DF relay with digital hardware is able to use any form of access, multiplexing and relaying protocol. We will henceforth assume that TDR is used with only one FDD link for up- and downlink. The supportive DF with digital hardware does incur arithmetic complexity as detailed in the previous section. This complexity heavily depends on the choice of general system as well as UMTS service parameters, where we will study speech and HSDPA data services. Concerning the general system parameters, we have assumed the following: channel delay spread $DS = 12$; $N_{chips} = 2560$; matched filter depth $L_{RCC} = 8$; number of pilot bits $N_{bit,pilot} = 20$; number of pilot symbols $N_{pilot} = 256$; pilot spreading factor $SF_{pilot} = 256$; number of pilot chips $N_{chip,pilot} = 2560$; number of Rake fingers $L = 4$; and the ratio between full and partial channel acquisition $R_{acq} = 0.9$. As for the oversampling factor OSF, a typical manufacturer's choice is to sample the received signal at twice the chip rate, that is $SF = 2$, and then perform interpolation up to 8 samples/chip in order to track the channel delays.

As for the *speech service* operating at 12.2 kbps, we have assumed the following: QPSK modulation; overall number of data symbols per slot $N_{data} = 20$; spreading factor $SF_{data} = 128$; number of multicodes $MC = 1$; convolutional channel code at rate $R_c = 1/3$; puncturing level for DTCH $R_{p,dtch} = 0.2$; puncturing level for DCCH $R_{p,dcch} = 0.2$; number of bits before demapping $N_{rx} \approx 840$; and the size of the transport block $N_{data} = 840/3$. We also assume that the relay needs to support $U = 40$ speech users in a coverage-limited hot-spot in parallel. Furthermore, only the MRC detector will be considered in the study. We are now in a position to quantify the complexities and power consumptions of the relay supporting UMTS speech services:

- **Arithmetic Complexity.** With reference to Figure 5.6, the complexity in the receiver is obtained as the complexity of the RRC filter + the complexity of the channel acquisition + $U \times$ the complexity of the remaining building blocks in Section 5.5.2. The complexity of the relay transmitter is $U \times$ the one of the outer modem (replacing the complexity of the convolutional decoder by the one of the encoder) + $U \times$ the complexity of the spreading operations + the complexity of the RRC filter. Plugging the above numbers into the complexity expressions of Section 5.5.2, yields for the total number of arithmetic operations $\mathcal{O} \approx 2500000$ and the number of load/store operations $\mathcal{L} \approx 594000$. The required memory is $\mathcal{M} \approx 105000$ bytes per slot, where we have simply added the memory requirements of each identified building block; however, more sophisticated implementations could facilitate only the maximum to be taken. To obtain the processing frequency $\mathcal{R}$ of the relay, the number of arithmetic operations $\mathcal{O}$ per slot needs to be divided by 4 (number of operations per cycle) and the duration of a slot (10 ms/15). We obtain $\mathcal{R} \approx 930\,\text{MHz}$ which – as per Table 5.8 – allows us to use current and future 4 ALU DSPs.
- **Power Consumption.** Using the approach described in Section 5.5.3 with the above numbers, the total power consumption due to the load/store operations can be calculated as 96 mW. The RF front end power consumption is taken from Table 5.7 and amounts to approximately 763 mW assuming a nonbattery-powered node. The total power consumption is thus $\mathcal{P} = 859\,\text{mW}$.

- **Important Trends.** Doubling the number of speech users requires the processing frequency to be increased by a factor of approximately 1.5 and increases the total power consumption by approximately 5%. Doubling the RRC filter length requires the processing frequency to be increased by approximately 25% but has no impact onto the power consumption. Again, the power consumption almost doubles for every doubling of FDD connections used. Finally, although not explicitly considered in the above complexity study, doubling the antennas requires approximately double the processing frequency and 10% more power.

As for the *HSDPA data service* operating at approximately 1.5 Mbps, we have assumed the following: 16QAM modulation; overall number of data symbols per slot $N_{data} = 160$; spreading factor $SF_{data} = 16$; number of multicodes $MC = 5$; turbo channel code at rate $R_c = 1/3$; number of turbo decoding iterations $N_{iter} = 4$; iterations per MPIC stage $R = 1$; number of MPIC stages $P = 4$; puncturing level for DTCH $R_{p,dtch} = 0$; puncturing level for DCCH $R_{p,dcch} = 1$; number of bits before demapping $N_{rx} \approx 2 \times 960 \times MC$; and the size of the transport block $N_{data} = 2 \times 960 \times MC/3$. We also assume that the relay needs to support only $U = 1$ HSDPA user, which amounts approximately to the same code use as for the speech case. Furthermore, both MRC and MPIC detectors will be considered in the study. We are now in a position to quantify the complexities and power consumption of a MRC-based relay supporting one UMTS HSDPA service:

- **Arithmetic Complexity.** Following the same recipe as for the speech service, yields for the total number of arithmetic operations $\mathcal{O} \approx 1600000$, the number of load/store operations $\mathcal{L} \approx 340000$, the required memory $\mathcal{M} \approx 110000$ in bytes per slot, and the processing frequency $\mathcal{R} \approx 580$ MHz.
- **Power Consumption.** Using the approach already taken in Section 5.5.3 with the above numbers, the total power consumption due to the load/store operations can be calculated as 55 mW. The RF front end power consumption is taken from Table 5.7 and amounts to approximately 763 mW assuming a nonbattery-powered node. The total power consumption is $\mathcal{P} = 818$ mW. Clearly, the power consumption due to load/store has increased w.r.t. the speech service and accounts for about 7% of the total power budget.
- **Important Trends.** Doubling the number of HSDPA users or the number of multicodes requires the processing frequency to be increased by a factor of approximately 1.5 and increases the total power consumption by approximately 5%. Going from 16QAM to QPSK lowers the requirement on the processing frequency by approximately 20% and on power by 3%. Doubling the number of turbo decoding iterations or the length of the RRC filter has almost no impact on the complexity. Again, the power consumption almost doubles for every doubling of FDD connections used. Finally, doubling the antennas requires approximately double the processing frequency but has almost no impact on the power.

As for the complexities and power consumptions of a MPIC-based relay supporting one UMTS HSDPA service, we obtain:

- **Arithmetic Complexity.** Following the same recipe as for the speech service, yields are for the total number of arithmetic operations $\mathcal{O} \approx 9000000$, the number of load/store operations $\mathcal{L} \approx 4200000$, the required memory $\mathcal{M} \approx 340000$ in bytes per slot, and the processing frequency $\mathcal{R} \approx 3400$ MHz. MPIC detection is clearly a very complex process, which either requires the use of a very powerful digital architecture or MRC-based detectors.
- **Power Consumption.** The total power consumption due to the load/store operations can be calculated as 686 mW. The RF front end power consumption is taken from Table 5.7 and amounts to approximately 763 mW assuming a nonbattery-powered node. The total power consumption is thus $\mathcal{P} = 1450$ mW. The digital processing hence consumes approximately the same as the RF front end.
- **Important Trends.** Doubling the number of HSDPA users or the number of multicodes requires the processing frequency and total power to be increased by a factor of approximately 1.3. Going from 16QAM to QPSK lowers the requirement on the processing frequency by approximately 12%

and on power by 5%. Doubling the number of turbo decoding iterations has almost no impact on the complexity. Doubling the length of the RRC filter requires the processing frequency to be increased by a factor of 1.5 but does not impact on power. Doubling the number of MPIC stages doubles both complexity as well as total power consumption. Doubling the number of iterations per MPIC stage increases the frequency by approximately 33% and the power consumption by 20%. Again, the power consumption almost doubles for every doubling of FDD connections used. Finally, doubling the antennas requires approximately double the processing frequency and 15% more power.

This has been summarized and compared to other cases in Table 5.9.

5.5.4.4 Cooperative DF with Digital Hardware

Whilst supportive DF relays rather act as (planned) base stations, cooperative DF relays typically are mobile stations with own traffic to transmit. The complexity due to relaying hence needs to be kept at bay, which is why we only consider the cooperative relay of one speech and one HSDPA data connection in addition to the relay's own speech and HSDPA connection. In the previous section we demonstrated that the use of a MPIC detector incurs high complexities, which is why we concentrate on the MRC-based detector only. As for HSDPA, we also have assumed 16QAM but no multicode transmission, yielding an effective data rate of approximately 320 kbps per user. Otherwise, we have assumed the same parameters as in the previous section for the supportive DF relay with digital hardware.

The complexities and power consumptions of the cooperative relay supporting its own and a relayed UMTS *speech service* are as follows:

- **Arithmetic Complexity.** Following the same recipe as before, yields for the total number of arithmetic operations $\mathcal{O} \approx 850000$, the number of load/store operations $\mathcal{L} \approx 72000$, the required memory $\mathcal{M} \approx 89000$ in bytes per slot, and the processing frequency $\mathcal{R} \approx 320\,\text{MHz}$. The cooperative relay can hence easily support the extra speech connection.
- **Power Consumption.** The total power consumption due to the load/store operations can be calculated to be 12 mW. The RF frontend power consumption is taken from Table 5.7 and amounts to approximately 830 mW assuming a battery-powered node. Since the relay is a mobile station itself, the higher layers as well as peripheral hardware consume another 600 mW. The total power consumption is thus $\mathcal{P} = 1442\,\text{mW}$.
- **Important Trends.** Every addition of a relayed speech user increases the required processing frequency by approximately 5% but has a negligible impact on the power consumption. Doubling the RRC filter length requires the processing frequency to be increased by approximately 25% but has again no impact on the power consumption. The power consumption increases by approximately 40–50% for every addition of FDD connection. Finally, doubling the antennas requires approximately double processing frequency but has almost no impact on the power.

The complexities and power consumptions of the cooperative relay supporting its own and a relayed UMTS *HSDPA service* are as follows:

- **Arithmetic Complexity.** Following the same recipe as before, yields for the total number of arithmetic operations $\mathcal{O} \approx 1100000$, the number of load/store operations $\mathcal{L} \approx 180000$, the required memory $\mathcal{M} \approx 96000$ in bytes per slot, and the processing frequency $\mathcal{R} \approx 420\,\text{MHz}$. The cooperative relay can hence support this extra HSDPA connection.
- **Power Consumption.** The total power consumption due to the load/store operations can be calculated as 30 mW. The RF front end power consumption amounts to approximately 830 mW and the higher layers as well as peripheral hardware consume another 600 mW. The total power consumption is $\mathcal{P} = 1460\,\text{mW}$.

- **Important Trends.** The same trends as for the HSDPA MRC supportive relay hold with the only exception being that the impact on the power consumption is significantly lower due to the additional 600 mW power consumption.

This has been summarized and compared to other cases in Table 5.9. In summary, it can also be observed that memory is not a constraining factor in any of these case studies. This is because not much memory is needed nor is it a very expensive building block. It can further be seen that the processing frequency can generally be realized by 4 ALU DSP architectures (or equivalents). The only exception with a large difference is the case where a MPIC detector is used, which is therefore not recommended from a performance point of view. Finally, the power consumption of the analog relay is very high, even exceeding the power needs of complex digital architecture. This is because the RF chains with the amplifiers have been, until now, the most power-hungry building blocks. The power consumption of the cooperative relays is also high but a large part of this is consumed by higher layers and external peripheries. Finally, a general trend is that doubling the number of antennas at the relay, approximately doubles the requirement on the processing frequency but has only a minor impact on the total power consumption.

5.6 Complexity of LTE/WiMAX Relay

The focus of this section is to evaluate the complexity of a relay based on OFDM technology, which is known to offer numerous advantages w.r.t. CDMA technologies but also many challenges. The two major telecommunications systems using OFDM are the 3GPP-driven LTE and IEEE-driven WiMAX initiatives. Both are regarded as beyond 3G (B3G) systems but are clearly not 4G since they do not fulfill the requirements set out by the ITU for 4G next generation mobile networks (NGMN). The latter requires downlink rates of 100 Mbps for mobile and 1 Gbps for fixed-nomadic users at bandwidths of around 100 MHz, which are, however, being targeted by LTE Advanced and WiMAX II. The key to such high spectral efficiencies in these systems is the combined use of OFDM, adaptive coding and modulation (ACM) and MIMO at PHY layer and many more sophisticated techniques at access and system level. The key components influencing the complexity of a potential relay are very similar in both LTE and WiMAX, which is why they will be subject to the same study.

To evaluate the complexity of a relay based on LTE or WiMAX specifications, we will follow the same steps as for the UMTS-based relay. We will hence focus on the number of arithmetic operations, number of load/store operations, required memory, processing frequency and total power consumption. Since the derivations of these complexities are very similar to the ones already exposed in the context of the UMTS-based relay, this section will be significantly shorter with some emphasis on the major differences between UMTS and LTE/WiMAX.

5.6.1 LTE versus WiMAX

To understand the rational behind the complexity analysis as well as the choice of some key parameters, we need to understand the major commonalities and differences between both systems.

5.6.1.1 Background

LTE is backed by 3GPP and builds on UMTS/HSxPA and GSM/EDGE which together account for more than 80% of global mobile subscriber numbers. Work on LTE, also referred to as enhanced UTRA (E-UTRA), commenced in 1998 and has reached maturity through a set of standard documents forming today's Release 8 and 9 [see for example, 531–534]. LTE enjoys strong support from numerous cellular manufacturers, operators and service providers. The experience of these companies has triggered a few important key decisions on the choice of technologies and parameters, which may give it a competitive advantage over WiMAX. First LTE equipment emerged in 2009 but significant rollouts are not expected before 2011.

Although not even yet deployed, proposals for improvements of LTE have commenced recently, which aim at responding to the requirements set out by the ITU for 4G NGMNs. The decision was taken at a 3GPP workshop in April 2008 and a first set of 3GPP requirements on LTE Advanced was approved in June 2008 [535, 536]. It requires peak data rates of 1 Gbps in the downlink using approximately 70 MHz, and 500 Mbps in the uplink using approximately 40 MHz; it hence requires a very high spectral efficiency along with spectrum flexibility.

WiMAX is standardized within the IEEE under its Part 16 working group on Air Interface for Fixed and Mobile Broadband Wireless Access Systems. It is supported by the WiMAX Forum, which was formed in June 2001, to promote conformity and interoperability of the standard. Initially thought to be just a last-mile and backbone solution, it has lately been enhanced with the support for mobility and other key elements. The former has been known as IEEE 802.16-2004 [537] (also IEEE 802.16d or fixed WiMAX), and the latter as IEEE 802.16e-2005 [538] (also IEEE 802.16e or mobile WiMAX). Mobile WiMAX has recently been adopted as IP-OFDMA for inclusion as the sixth wireless link system under IMT-2000, which is a clear success for the WiMAX camp.

Only a few selected countries enjoy WiMAX roll-outs, most notably Pakistan and the USA. It lacks behind 2G/3G, which dominates cellular rollouts but has advanced LTE which yet does not have running networks. The response to ITU for 4G NGMNs will be WiMAX II, also known as IEEE 802.16 m.

5.6.1.2 Commonalities

Both standardized LTE and WiMAX families enjoy some key similarities at PHY, which are summarized below:

- **Modulation.** Both use the principle of OFDM to transmit information. This implies that both transceivers need to support FFT and IFFT.
- **Downlink Multiple Access.** Both facilitate OFDMA in the downlink where different subcarriers within an OFDM symbol are allocated to different users.
- **Link Adaptation.** Both allow link adaptation depending on channel conditions through ACM, that is adaptive modulation is used per subcarrier and also different channel codes are available.
- **MIMO.** Both require MIMO-enabled transceivers that are capable of switching between diversity and multiplexing modes.
- **Duplexing Method.** Originally, WiMAX only supported TDD whilst LTE supported both TDD and FDD. However, the latest developments within WiMAX also include the option of FDD [538, 539].

This list of similarities is likely to increase over time as WiMAX will profit from LTE's expertise on mobility support and LTE on WiMAX's expertise on broadband delivery.

5.6.1.3 Differences

There are however also some key differences between both standardized LTE and WiMAX families at PHY, which mainly relate to how up- and downlinks are handled. They are summarized below:

- **Uplink Multiple Access.** Whilst WiMAX uses also OFDMA in the uplink, LTE has opted for SC-FDMA. The latter, being a single-carrier approach, eases the requirements on the amplifier in the mobile terminal and is hence more power efficient. This is because the amplifier has to be designed to handle peak power; the WiMAX uplink OFDMA has a peak-to-average ratio of about 10 dB, whilst the LTE SC-FDMA peak-to-average ratio is only about 5 dB.

- **FFT Size.** WiMAX schedules users on their best available subcarriers, which requires each user to perform a FFT over the entire OFDM symbol, often involving 1000-point FFT operations. LTE, on the other hand, gives each user a variable chunk of subcarriers in the form of physical resource blocks (PRBs), which together form one OFDM symbol. This allows each user to extract just the relevant data part. Whilst less efficient than the WiMAX solution, it reduces the complexity at the terminal side since the handset can typically get away with a 16-point FFT.
- **Relays.** WiMAX has been working hard lately on a relay-enabled system, also referred to as IEEE 802.16j [540], which was ratified in 2009. LTE, on the other hand, does not support a relay-enabled mode in its Releases 8 and 9. However, LTE Advanced through Release 10 onwards has outlined that relays will be part of the system design.

There are numerous other differences, such as those relating to the signalling overhead, efficiency to handle handovers, etc., but this is only loosely related to the complexity analysis.

5.6.2 System Assumptions

As per above, the most complex elements in either systems are more or less the same. We will therefore now concentrate on WiMAX because it has a relaying framework as already specified [540]. We then briefly justify and present the choice of scenarios, access method and physical layer.

5.6.2.1 Choice of Scenarios

In this study, we will not deal with purely analog hardware because it is not able to support TDR needed for a proper WiMAX functioning. We will hence concentrate on the following cases:

- **Supportive AF with Digital Hardware (s-AF$_d$).** Even though IEEE 802.16j does not support AF protocols, this case study is useful if some simple repeaters are to be used in the WiMAX cell to boost coverage. Note that this is not related to the notion of a transparent relay station in WiMAX, where the relay does not transmit the preamble and other information at the beginning of the frame. Similarly to the UMTS complexity study, we will quantify the memory requirements and power consumption.
- **Supportive DF with Digital Hardware (s-DF$_d$).** The core of the study will relate to the DF protocol obeying the IEEE 802.16j specifications. It will quantify complexity, memory and power consumption.
- **Cooperative DF with Digital Hardware (c-DF$_d$).** Finally, we will also investigate the complexity of a relay that is a terminal itself and hence has its own traffic to transmit. Note that this is different to the notion of cooperative relaying as used in the specifications of IEEE 802.16j.

We will assume that the relay consists of the hardware elements discussed in Section 5.2.1 and Section 5.3.1.

5.6.2.2 Choice of Access Method

WiMAX supports TDD, FDD and half-duplex FDD. The option of FDD has been included lately to match regulatory requirements. Another advantage of FDD over TDD is that the guard times between down- and uplink phases can be neglected, which is a detrimental factor in TDD designs for large coverage areas. We will henceforth concentrate on the TDD case. A system-wide WiMAX synchronization is generally needed which, among others, allows minimizing of interference between WiMAX base stations. Synchronization is typically achieved using the global positioning system (GPS) reference at the BS. TDD, as has been discussed before, enables asymmetric down- and uplink traffic flows and thus exhibits a better spectral efficiency than does FDD. Due to the single TDD channel for down- and uplink, hardware transceiver design is greatly simplified. It offers yet another advantage, in that the down and up link channels are reciprocal hence greatly benefiting MIMO techniques. The reciprocity would, for instance, facilitate MIMO precoding techniques, or an optimum choice of transmit antennas or the allocation of optimum power at the transmitting side.

The current draft of IEEE 802.16j [540] supports in-band TDR, although other options are also being considered [82, 83]. An example of the thus resulting frame structure used by the WiMAX relay is shown in Figure 5.9. Here, the access is realized over frequency (given number of subcarriers per OFDM symbol) and time (given number of OFDM symbols). According to channel conditions, users are allocated a specified set of subcarriers within each OFDM symbol over a frame duration. These OFDM symbols are communicated within frames, where the relay first receives information from the base station in the relay downlink receive frame and then retransmits it to the destination mobile station in the relay downlink transmit frame. Thereupon the mobile station communicates with the relay in the relay uplink receive frame and finally the data is conveyed to the base station in the relay uplink transmit frame. All these frames are separated by temporal gaps to facilitate signal propagation, frame processing and switching between the transmit and receive (or vice versa) phases.

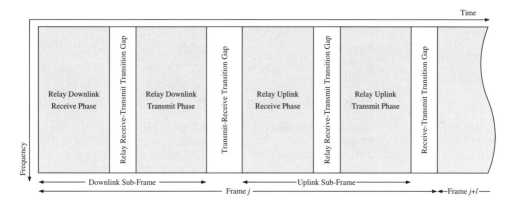

Figure 5.9   Relay frame structure as per IEEE 802.16j [540]

5.6.2.3   Choice of Link Layer

WiMAX facilitates channel bandwidths ranging from 1.25 to 20 MHz, which is realized in practice by suitably adopting the number of subcarriers and their spacing. The useful duration of an OFDM symbol is equal to the inverse of the subcarrier spacing. The full symbol duration, however, contains the actual useful duration and the cyclic prefix (CP). The latter is simply a copy from the end of the useful symbol inserted before the start of the symbol and, if chosen to be longer than the channel delay spread, allows intersymbol interference (ISI) introduced by the multipath components to be eliminated.

Since it is not focus of this book, we will not deal with all possible PHY configurations of the WiMAX standard but rather focus on a few selected ones. Most notably, we will use a bandwidth of 10 MHz with a 1024-point FFT sampled at a sampling frequency of 11.2 MHz, which yields a subcarrier spacing of $11.2\,\text{MHz}/1024 = 10.9375\,\text{kHz}$. This, in turn, allows us to use approximately

10 MHz/10.9375 kHz $= 914$ subcarriers and yields a useful symbol duration of $1/10.9375$ kHz $=$ 91.43 μs. As for the length of the CP, the standard specifies a set of CP values but the initial profile specifies a CP value of 1/8 of the useful symbol duration, that is the duration of the CP is 91.4 μs/8 $=$ 11.43 μs. This gives a total OFDM symbol duration of approximately 102.86 μs. The standard stipulates a frame duration of 5 ms, which would allow us to transmit 5 ms/102.86 μs $= 48.61$ symbols. However, to account for the total of all transition times, we will only fill the frame with 44 symbols; this gives a total transition time of $(48.61-44)\times 102.86$ μs $= 474.18$ μs. This seems to be very large but may just be needed to account for hardware switching times, frame processing times, relay synchronization, etc.

We assume a fully loaded system and almost symmetrical down- and uplink traffic, where we assign to the down- and uplink transmit and receive phases 11 OFDM symbols, respectively. We will neglect the signal and control traffic, except for the pilot symbols needed to estimate the channel. Also, we will not differentiate between partially loaded subcarriers (PUSC) and fully loaded subcarriers (FUSC), nor do we apply the concept of ACM. To convey user data, the standard stipulates the use of data bursts, which are composed of a given number of subcarriers and symbols in the frequency–time grid. Each of these bursts can be assigned to a given user. We will assume up to five bursts, but we are not explicitly interested in how user data is mapped into these bursts. We will assume a fixed modulation of 16QAM and 1/2-rate turbo coding only. We further assume an approximate pilot density of 15%, although this is very flexible and varies between FUSC/PUSC/ACM as well as up- and downlinks.

Table 5.10   Choice of WiMAX PHY parameters

Parameter	Value
Bandwidth	10 MHz
Sampling frequency	11.2 MHz
FFT size	1024
Subcarrier spacing	10.9375 kHz
Number of useful subcarriers	914
Pilot density	15 %
Useful symbol duration	92.43 μs
CP duration	11.43 μs
Total symbol duration	102.86 μs
Frame duration	5 ms
Total transition duration	474.18 μs
Total useful data duration	4.5258 ms
Total number of symbols	44
Downlink relay receive symbols	11
Downlink relay transmit symbols	11
Uplink relay receive symbols	11
Uplink relay transmit symbols	11
Number of downlink bursts	Five
Number of uplink bursts	Five
Modulation	16QAM
Channel code	1/2 turbo code
Effective downlink rate	6.84 Mbps
Effective uplink rate	6.84 Mbps
Number of receive antennas	Two
Number of retransmit antennas	Two
Spatial receive method	Diversity MRC
Spatial transmit method	Alamouti STBC

This yields an approximate useful up- and downlink throughput of $\log_2(16)$ bits/subcarrier $\times$ 914 subcarriers/symbol $\times$ 85% useful subcarriers $\times$ 11 symbols $\div$ 2 encoded/useful bits $\div$ (5 ms total frame duration / 2 subframes / 2 phases) $\approx$ 13.67 Mbps per phase. Since a half-duplex relay is used and the relayed frame does not contain any new information, the effective end-to-end up and downlink throughput is 13.67 Mbps $\div 2 = 6.84$ Mbps.

Finally, we will assume that the relay possesses two antennas, which it uses for MRC diversity reception and Alamouti STBC at retransmission. All these assumptions have been summarized in Table 5.10.

5.6.2.4 Digital Modem Design

Based on above link layer specs, a typical realization of a digital WiMAX relay is shown in Figure 5.10. The analog signal is received in each of the RF front ends and then converted into the digital domain by means of sampling. Thereupon, the cyclic prefix is removed and the FFT applied, which yields the 1024 complex samples in parallel, of which only the useful 914 subcarriers will be used further. In practical applications, it has been reported that a matched filtering to the OFDM symbol is not needed since the spectral mask of the 1024-FFT with only 914 loaded subcarriers resembles an RRC filter.

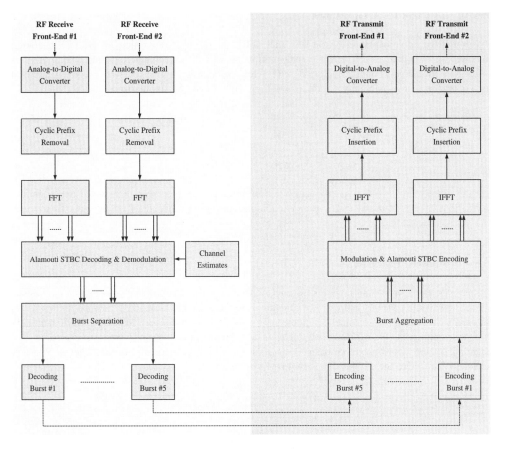

Figure 5.10 Most essential building blocks of a WiMAX digital relay with MRC detector and STBC Alamouti transmission

Per subcarrier, the channel is estimated and compensated for by using Alamouti space–time block decoding (if applicable). Each subcarrier is then demodulated. The bursts are finally separated and decoded separately. The relay transmitter performs exactly the same steps but in inverse order.

As for the digital hardware architecture, we follow the same assumptions as for the UMTS-based relay. We thus assume that the arithmetic logical unit of the signal processing board can do four instructions per cycle, the data is represented by 16 bits per complex dimension, and additions and multiplications have more or less the same associated complexity. Moreover, we assume again a single digital architecture for the sake of analytical tractability, thus excluding the use of the specialized circuitry that can be found in mobile terminals.

5.6.3 Algorithmic Complexity

For subsequent complexity analysis, we will concentrate on the building blocks which consume most complexity and power. We will hence only deal with the receive and transmit fast Fourier transforms, channel estimation, Alamouti space-time decoding, combining, demodulation and turbo decoding. We thus omitted the complexity of the cyclic removal, CRC check, etc., on the receiving side and all building blocks except the IFFT operation on the transmitting side.

5.6.3.1   FFT and IFFT Operations

Assuming the implementation of the split-radix FFT algorithm of dimension N, the FFT operation requires $4 \times N \times \log_2 N - 6 \times N + 8$ real multiplications and additions per symbol. The number of load/store operations is approximately $N \times (1 + \log_2 N)$ and the number of bytes needed to perform this operation is approximately $4 \times N \times (1 + \log_2 N)$. This can be summarized as:

$$\mathcal{O} = N_{rx} \times N_{tx} \times (4 \times N \times \log_2 N - 6 \times N + 8), \tag{5.27}$$

$$\mathcal{L} = N_{rx} \times N_{tx} \times 2 \times N \times \log_2 N, \tag{5.28}$$

$$\mathcal{M} = N_{rx} \times N_{tx} \times 4 \times N \times \log_2 N, \tag{5.29}$$

for the entire relay with N_{rx} receive and N_{tx} transmit antennas. Note that more sophisticated implementations of the Alamouti space–time encoding requiring only one IFFT at the transmission side are not considered here.

5.6.3.2   Channel Estimation

To estimate the narrowband complex channel realization in each time–frequency bin of the OFDM signal, one would typically estimate the channel using the available pilot and then use optimum filters to estimate the remaining channel coefficients. The known time–frequency Wiener and Kalman filters are very complex and we will hence only consider some linear interpolation between the estimated pilot symbols on a symbol-by-symbol basis. Neglecting the exact pilot structure of the different WiMAX channels, we derive the complexity per OFDM symbol to be:

$$\mathcal{O} = N_{rx} \times N_{tx} \times N_{sub} \times (7 - D_p), \tag{5.30}$$

$$\mathcal{L} = N_{rx} \times N_{tx} \times N_{sub} \times (3 \times D_p + 1), \tag{5.31}$$

$$\mathcal{M} = 4 \times N_{rx} \times N_{tx} \times N_{sub} \times (D_p + 1), \tag{5.32}$$

where D_p is the pilot density.

5.6.3.3 Alamouti Space–Time Block Decoding

We consider here the complexity of the channel equalization, the Alamouti space–time block decoding and the 16QAM soft demodulation, which is easily derived to be per symbol:

$$\mathcal{O} = N_{sub} \times (8 \times N_{rx} + 2 \times (N_{rx} - 1) + 40 \times \log_2 16), \tag{5.33}$$

$$\mathcal{L} = N_{sub} \times (8 \times N_{rx} + 4 \times \log_2 16), \tag{5.34}$$

$$\mathcal{M} = 4 \times 4 \times N_{sub} \times N_{rx}, \tag{5.35}$$

where N_{sub} is the number of useful data subcarriers.

5.6.3.4 Turbo Decoding

Finally, we assume that the data is separated into coded bursts of equal length and decoded using iterative turbo decoding. Using the same complexity analysis as for a UMTS-based relay, the complexity and load/store per symbol as well as the needed memory can be estimated as:

$$\mathcal{O} = N_{bursts} \times 2 \times N_{iter} \times (R_c \times N_{sub} \times \log_2 16 \times N_{subframe}/N_{bursts} + 16)$$
$$\times (3/2 + 37 + 16/24)/N_{subframe}, \tag{5.36}$$

$$\mathcal{L} = N_{bursts} \times 2 \times N_{iter} \times (R_c \times N_{sub} \times \log_2 16 \times N_{subframe}/N_{bursts} + 16)$$
$$\times (23 + 16/24)/N_{subframe}, \tag{5.37}$$

$$\mathcal{M} = N_{bursts} \times (N_{sub} \times \log_2 16/N_{bursts} - 12) \times R_c/8, \tag{5.38}$$

where N_{burst} is the number of encoded data bursts, N_{iter} the number of turbo decoding iterations, R_c the coding rate and $N_{subframe}$ the number of symbols that convey the given number of bursts.

5.6.4 Power Consumption

Again, the power in a WiMAX-type relay terminal is consumed by the transceiver RF front end, by the baseband processing unit (including memory maintenance) and, if any, by higher layer processing and screen/keyboard/etc. The consumption of these elements is estimated in this section using the approach already taken for the UMTS relay.

5.6.4.1 RF Front-End Consumption

Despite the load on the signal processing part, as identified in Section 5.6.3, the front end of the WiMAX relay is also constrained which is mainly due to the following reasons:

- **PAPR.** OFDM signals suffer from high PAPRs, even more than WCDMA systems. An OFDM signal theoretically exhibits a PAPR of around $10 \times \log_{10} N$ dB, where N is the number of FFT points. Recent approaches, however, manage to reduce this to, for example, 10-12 dB for a 1024-point FFT [541, 542].
- **Power Amplifier.** The large PAPR requires the system to work and the power amplifier therefore to support a large backoff to prevent compression or clipping of the signal. Given the large backoff, the majority of the signal will be amplified at medium power where power amplifiers are typically less efficient. Therefore, the amplifier for OFDM systems needs to be more linear and more efficient when compared with CDMA or narrowband solutions.

- **Local Oscillators.** The local oscillators need to be of superior quality, exhibiting very low phase noises to guarantee a proper reception and retransmission of the OFDM wideband signal.

The hardware architecture of the WiMAX relay is very similar to that discussed in the context of a UMTS extended to multiple receive and transmit antennas. To obtain an estimate of the power consumption, we assume the same as for the UMTS relay with the following modifications:

- negligible insertion losses since the TDD operation does not require a duplexer;
- average transmission power of $P_{out} = 23$ dBm [543].

The above average output power of 23 dBm is equivalent to 200 mW, where the power amplifier efficiency is around 30%. This means that the amplifier needs to be powered by 200 mW/0.3 = 665 mW. If the WiMAX relay is battery operated, then the voltage stabilizer requires a total of 665 mW / 0.85 = 782 mW.

In addition to the power amplifier consumption, we have the average power consumed in the RF transceiver chains. We consume 215 mW in the transmitter chain + 30 mW for the DAC and 85 mW in the receiver chain + 60 mW for the ADC. Assuming equal time sharing between receiver and transmitter, the average power per antenna element excluding the power amplifier is 195 mW. The power consumption of the RF transceiver chain assuming two receive and two transmit antennas, is summarized in Table 5.11, where these numbers are independent of data rate and approximately independent of the year of manufacture (with the current production horizon).

Table 5.11 Typical RF power consumption (in mW) of a nonbattery and battery powered WiMAX relay with two antennas

	Nonbattery powered	Battery powered
Power amplifier Tx	2×665 mW	2×782 mW
Tx/Rx RF chains	2×195 mW	2×195 mW
Total power	1720 mW	1954 mW
Increase in %	–	14 %

5.6.4.2 Digital Baseband Consumption

Similar to the UMTS relay, the following can be said concerning the power consumption in the digital baseband:

- **Maintaining Memory.** The power required to maintain the memory is again neglected.
- **Arithmetic Processor Operations.** As for the power consumption of the digital processor, it will also be neglected in subsequent studies.
- **Load/Store Operations.** The power consumed during the load/store processes, however, is not neglected. Using the capacitances given in Table 5.8 and the same approach as used before, we will quantify the power consumption induced by the load and store operations.

Again, since the power consumption of the above will heavily depend on the hardware architecture of choice, the exposed qualitative and quantitative tendencies give a rough indication of the power consumption of the digital WiMAX baseband modem.

5.6.4.3 Higher Layers and Peripheral Hardware

If WiMAX user equipment is used to act as a relay, then the power consumption of higher layers and peripheral hardware needs to be taken into account. Typical WiMAX user equipment is a notebook, the power consumption of which can vary significantly. As of 2008, the idle state of a notebook consumes around 8 W and the operational state up to 30 W. We will henceforth assume an average consumption of 15 W.

5.6.5 Case Studies

We will assume that a relay needs to support both down- and uplink in TDD. Although being differently structured and encoded, the up- and downlink WiMAX frames incur the same order of complexity. For simplicity, we will assume that they require approximately the same memory and the same number of arithmetic and load/store operations. We will now quantify the approximate complexity and power consumption of the three WiMAX relay cases identified in the previous section.

5.6.5.1 Supportive AF with Digital Hardware

As in the case of the UMTS relay study, the supportive AF with digital hardware does not require any arithmetic complexity in terms of addition or multiplication, but needs to load/store the oversampled data stream and also keep it in memory. We assume a sampling rate of $SR = 2 \times 11.2$ MHz, which corresponds to an oversampling factor of $OSF = 2$ in the signal processing domain, although higher values are more common. The complexities and power consumptions hence become:

- **Arithmetic Complexity.** There is no arithmetic complexity, that is $\mathcal{O} = 0$. The approximate number of store operations per phase at the relay receiver is $22.4 \cdot 10^6$ samples/s $\times$ 11 symbols $\times$ $102.86\,\mu$s/symbol, yielding a total of $\mathcal{L} = 2 \times 25.3 \cdot 10^3$ load and store operations per phase. The total memory requirements needed to support each phase mount to $\mathcal{M} = 4$ bytes/sample $\times 50.6 \cdot 10^3$ samples/phase $= 202$ kb per phase, which is not negligible.
- **Power Consumption.** Using the approach taken in Section 5.5.3, the total power consumption due to the load/store operations can be calculated as 87 mW. The RF front end power consumption is taken from Table 5.11 and amounts to approximately 1720 mW assuming a nonbattery-powered node which, when taking load/store operations into account, amounts to approximately 95% of the entire power budget. The total power consumption is $\mathcal{P} = 1807$ mW.
- **Important Trends.** The complexity and power consumption is independent of the number and type of relayed traffic since the data stream is only sampled. The number of load/store operations as well as memory and power, however, doubles for every doubling of the sampling rate.

This has been summarized and compared to other cases in Table 5.12.

Table 5.12 Summary of important relay requirement for the considered WiMAX case studies

Case study	Memory (kb)	Frequency (MHz)	Power (mW)
s-AF$_d$	202	–	1807
s-DF$_d$	210	2192	2763
c-DF$_d$	210	2191	18 000

5.6.5.2   Supportive DF with Digital Hardware

Given the PHY specification summarized in Table 5.10 and following the same procedure as before, we are in a position to quantify complexity, memory and power consumption of a supportive WiMAX relay using digital hardware:

- **Arithmetic Complexity.** The average number of arithmetic operations per OFDM symbol is $\mathcal{O} \approx 902\ 000$, the number of load/store operations is $\mathcal{L} \approx 596\ 000$ and the required memory is $\mathcal{M} \approx 210\ 000$ in bytes. The processing frequency is hence $\mathcal{R} \approx 2192\,\text{MHz}$, which – as per Table 5.8 – requires the use of more than one 4 ALU DSP.
- **Power Consumption.** The total power consumption due to the load/store operations can be calculated as 1043 mW. The RF front-end power consumption is taken from Table 5.11 and amounts to approximately 1720 mW assuming a nonbattery-powered node. The total power consumption is thus $\mathcal{P} = 2763\,\text{mW}$, where the load/store operations consume about 30% of the entire power budget.
- **Important Trends.** Doubling the number of supported bursts and/or pilot density has virtually no influence on the complexity. Doubling the number of receive or transmit antennas increases the complexity by about 20%. Doubling the number of subcarriers almost doubles the memory requirements and increases the remaining complexity figures by about 20%. Doubling the modulation order has almost no impact on the memory requirements but increases arithmetic complexity and load/store operations by about 60% and total power consumption by about 20%.

This has been summarized and compared to other cases in Table 5.12.

5.6.5.3   Cooperative DF with Digital Hardware

The most important difference for this case is that the relay is assumed to be battery driven and that higher-layer power consumption has to be taken into account since the cooperative terminal is likely to belong to an end user. This also implies that the relay inserts its own traffic, where we will assume that one of the five bursts belongs to the cooperative relay. This implies that the received frame is only filled $4/5 = 80\%$, whilst the transmitted frame is full. The complexities and power consumptions can be calculated as follows:

- **Arithmetic Complexity.** Following the same recipe as before, yields for the total number of arithmetic operations $\mathcal{O} \approx 902\ 000$, the number of load/store operations $\mathcal{L} \approx 596\ 000$, the required memory $\mathcal{M} \approx 210\ 000$ in bytes per slot, and the processing frequency $\mathcal{R} \approx 2191\,\text{MHz}$. The impact w.r.t. the supportive case is hence negligible under the considered system assumptions.
- **Power Consumption.** The total power consumption due to the load/store operations can be calculated to be 1043 mW. The RF front-end power consumption is taken from Table 5.11 and amounts to approximately 1954 mW assuming a battery-powered node. Since the relay is likely to be a laptop or similar, the higher layers as well as peripheral hardware consume another 15 W. The total power consumption is thus $\mathcal{P} \approx 18\,\text{W}$, where only about 15% of this power is due to WiMAX.
- **Important Trends.** The same trends as for the supportive WiMAX relay hold with the only difference being that the impact of power consumption is significantly lower since most of the power is consumed by the cooperative's relay higher layers and peripheral hardware.

This has been summarized and compared to other cases in Table 5.12.

5.7   Hardware Demonstrators

As of 2009, a large amount of theory has been produced in the context of cooperative communication systems; however, the amount of real-world hardware testbeds and demonstrators is scarce. In this section, we will briefly discuss the realization of a few (nonexhaustive list of) demonstrators.

5.7.1   MIT's Commodity Hardware Demonstrator

Probably one of the first to report a functioning hardware platform realizing a cooperative sensing system was at MIT's Media Lab [544–546], dating back to 2004. The prime motivation for the demonstration was to come as close as possible to the theoretical DMT performance of distributed space–time coding in Nick Laneman's thesis [102]. It turned out that it was enough to use simple, low-cost, embedded radios built from commodity hardware without distributed space–time coded or beamforming protocols but only with a smart relay selection procedure [546]. That was deemed important given the simplicity of the algorithm where just one node transmits and so hardware can in principle be simple. Subsequent analysis [547] has proved that under some sum power constraints, relay selection and signal regeneration can outperform a class of space–time coding algorithms, showing that smart sensing can be (in special cases) superior to space–time coding.

The purpose of the demo was hence to show that intelligent sensing of the wireless environment can provide improved performance, even with ultra-low cost hardware. To achieve such performance, distributed and adaptive routing algorithms close to the physical layer were exploited, and algorithms at networking level provided gains even with individually-dumb terminals. In more detail, the aim of the demonstrator was threefold:

- **Improved Reliability.** As outlined in Chapter 1, relaying yields power gains and the choice of the optimum relay from a set of available relays yields additional diversity gains. This translates directly into improved reliability of the communication link, that is fewer errors and fewer link outages. The first aim was, therefore, to demonstrate that these gains were viable in a nonergodic propagation environment.
- **Link Adaptability.** The fact that the system chooses the best relay from a set of available relays allows the system to adapt to any channel dynamics, such as the temporal disappearance of a particular link, by switching between available links. The second aim was hence to show that the system is able to adapt to such channel variability.
- **Feasibility of Implementation.** The system components needed to facilitate such a simple cooperative system are actually all available. The third aim was thus to demonstrate that available commodity hardware can easily be modified to realize such a cooperative system.

Since complete wireless solutions, such as those built on IEEE 802.15.4, have predefined and non-modifiable physical (PHY) and medium access control (MAC) layers, the hardware demonstrator was built from scratch using a microcontroller unit (MCU) connected to a RF module. As per Figure 5.11, the MCU and RF have the following properties:

- **MCU Module.** The architecture of the MCU is the 8-bit 8051 MCU at a clock speed of 22.1184 MHz. The ADC used a 12-bit representation and the DAC a 10-bit representation. The reason why more bits are needed to capture the analog input is because it has a much larger variability than the digital output due to wireless channel variations. The unit was powered by 2 AA batteries giving 3V. All communication functionalities, such as frame transmission, frame

 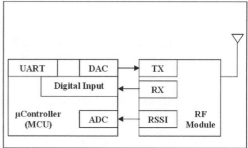

Figure 5.11   Hardware realization of cooperative nodes (left) and schema of used hardware components (right) [544, 545]

synchronization, frame reception, data detection, cyclic redundancy check (CRC), and other upper-level link access and routing protocols, were facilitated by the MCU.

- **RF Module.** The RF module is composed of a transmitter, receiver and a received signal strength indicator (RSSI). The aim of the latter is to estimate the channel quality between transmitter and receiver. The module is built on RF monolithics, operates at a frequency of 916.5 MHz and realizes an effective baud rate of 115 kBd using simple on–off keying.

The built hardware cost around 20 Euros per node as of 2006. This is a very low price and shows that low-cost implementation of cooperative systems is feasible.

The protocol used to facilitate cooperation is based on prior known concepts, such as preamble sampling, antenna selection, metric-dependent response delay, etc. The fully distributed cooperative relay protocol is based on the following phases:

- **Channel Power Estimation.** The channel powers of all wireless links are determined first using a technique similar to preamble sampling [548]. Prior to the transmission of information, the source node broadcasts a preamble in the form of a ready-to-send (RTS) message that is answered by the destination with a clear-to-send (CTS) message. Both RTS and CTS are sampled by the relay nodes which – using the RSSI module – allows each of the relays to estimate the channel power $\gamma_{s,j}$ between source and relay j as well as the power $\gamma_{j,d}$ between relay j and destination (due to channel reciprocity for the latter).
- **Channel Quality Metric.** The knowledge on the quality of the wireless channels of the entire network allows us to implement a proactive selection of the best relay. The metric for the best choice of relay is not unique; for instance, for transparent architectures, a good solution is to choose the relay which maximizes the metric $\gamma_j = \gamma_{s,j}\gamma_{j,d}/(\gamma_{s,j} + \gamma_{j,d})$; for regenerative relaying, a good choice is the metric $\gamma_j = \min(\gamma_{s,j}, \gamma_{j,d})$ [437]. Since the demonstrator uses a regenerative approach, the latter metric has been used.
- **Proactive Relay Selection.** The local knowledge of the channel quality metric needs to be intelligently distributed to the network so that the best relay can be used for cooperation and the remaining ones put to sleep. This is implemented in an entirely distributed fashion where, after the transmission of CTS, the relay signals its availability by transmitting a ready-to-relay (RTR) message, which is delayed for a time that is inversely proportional to the local channel quality metric γ_j. Therefore, a node with excellent conditions is responding before a node with bad conditions and hence is elected as relaying node. The probability that RTR messages collide is not zero and suitable collision resolution mechanisms thus need to be implemented, which has not been done in this particular demonstrator.

Note that choosing the best relay out of a set of available relays is equivalent to the problem of choosing the best antenna out of a set of available antennas and is known to yield the same diversity gains as if all antennas had been used [549, 550].

Furthermore, as already alluded to in Chapter 1, proactive relay selection contrasts a reactive approach. In the former, only the relay participates that is deemed to have best channel conditions and hence be of highest use to the cooperative network. In the latter, all relays receive the data broadcast by the source, decode it and then only that relay or those relays are chosen for cooperation that managed to decode the information correctly. Proactive relay selection clearly requires more transmission energy due to the preambles but consumes less reception energy due to the use of a single or subset of relays only. An optimal solution would clearly depend on the number of available relays and the energy spent in transmission and reception.

- **Data Communication.** Once the relay is determined, the source transmits its data which is then received by the relay and the destination. Each transmitted frame includes header (source ID, destination ID, CRC, etc.) and the actual data. The destination decodes the information and keeps it, given that the CRC flags no decoding error. The relay also decodes the data and, given that the CRC indicates that there is no error, re-encodes it and sends it to the destination, which then decodes this relayed data. In this particular demonstration, no maximum ratio combining prior to decoding has been implemented; instead, the destination keeps only the one or one of the two copies that yielded error-free decoding.

The above has realized a completely distributed operation of a cooperative relaying protocol at very low complexity but with some overhead needed for relay selection. The protocol effectively guarantees very reliable communication even with variations and dynamics in the wireless channel. As exemplified in Figures 5.12 and 5.13, the best relay is being chosen whilst the other relay paths are being blocked. Once the chosen relay path is blocked, the protocol automatically chooses another, better, relay path. That particular realization of a supportive relaying protocol hence significantly lowers the end-to-end link outage probability.

The implementation of this simple relaying protocol has exposed a few problems and design constraints that ought to be taken into account when building cooperative relaying systems:

- **Refreshing Rate.** The choice of the optimum relay needs to be constantly verified and updated if needed. The refreshing interval of the relay selection hence needs to be lower than the channel's

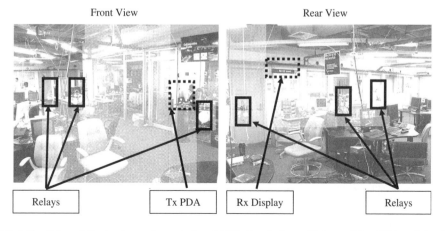

Front View   Rear View

| Relays | Tx PDA | Rx Display | Relays |

Figure 5.12 Setup of the hardware demonstrator at MIT. Front view with transmitting PDA and the three relays (left). Rear view with receiving display and the same three relays (right) [544, 545]

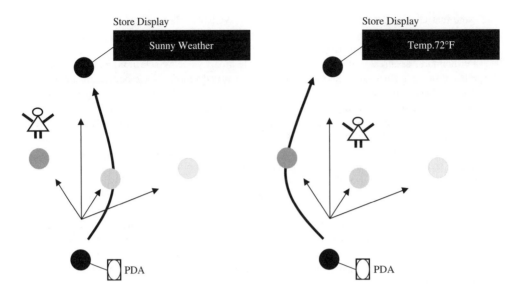

Figure 5.13 Schema of cooperative communication with direct link and relay selection. Shadowing of one relay node and hence relaying via another relaying node (left) and vice versa (right), where also the context can be changed depending on which node has been chosen as relay [544, 545]

coherence time. In this particular indoor realization, the channel is changing very slowly and hence has a large coherence time. At a walking speed of about 1 m/s, the channel coherence time is in the order of 300 ms, which means that the relay selection algorithm needs to be run about three times per second. Environments with greater mobility will require higher refreshing rates, which in turn lowers the efficiency of the system. Note however that this overhead is something that cannot be avoided in any communication scheme since coherent reception requires channel estimation and noncoherent reception requires the channel to remain constant, which implies that the channel must be sensed periodically to verify that this holds true.

- **Limited ADC Resolution.** Whilst the DAC resolution of 10 bits was more than sufficient due to the low dynamic range of the output signal, the 12-bit ADC resolution was just about enough to capture the dynamics of the wireless channel. Outdoors deployments will certainly experience stronger channel fluctuations, which in turn requires a higher ADC resolution.
- **8-Bit Architecture.** The 8 bits available to perform calculations, channel estimation, etc., turned out to be a limiting factor. Floating point operations are hence not feasible and quantization noise also becomes a detrimental factor, which has already been discussed in Section 5.3.2.

Despite these problems, MIT's commodity hardware decode-and-forward supportive relaying demonstrator is pioneering in that it is one of the first to demonstrate and quantify performance gains due to relaying in realistic environments.

5.7.2 ETH's RACooN Demonstrator

ETH has revealed its Radio Access with Cooperating Nodes (RACooN) custom-built radio testbed back in 2004 [551] with novel results constantly emerging [552–556]. As of 2009, it comprises up to ten single antenna nodes able to relay in half-duplex manner. Each node is mounted on a cart that can be easily moved around. As per Figure 5.14, the used configuration comprised two sources, two destinations and three relay nodes. Based on prior and in-house developed algorithms, coherent

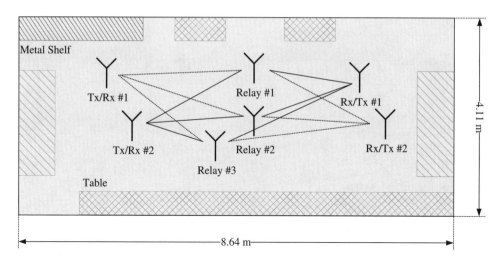

Figure 5.14 Topological set-up with two source nodes located on opposite ends of the room, two destination nodes and three relay nodes

and noncoherent forwarding in this multiuser environment have been implemented. The aim of the demonstration was the following:

- **Feasibility of Implementation.** The first aim was to demonstrate that the algorithms developed by Wittneben [554] can be successfully implemented in hardware and yield viable performance gains.
- **Improved Power Gains.** The derived algorithms facilitate the minimization of interference from other users and relaying nodes. This has theoretically been shown to yield significant power gains in terms of an improved effective signal-to-interference ratio (SIR). The second aim was hence to quantify these gains in a real-world deployment.
- **No Global Phase Reference.** The third aim was to demonstrate that it is practically possible to implement a coherent forwarding protocol onto a distributed cooperative relaying system without the necessity for a global phase reference. The latter was deemed to be needed to synchronize the local oscillators of the relaying node.

Facing a similar problem to the developer of the MIT demonstrator, the hardware has been built from scratch. It consists of the following building blocks:

- **Baseband Module.** The used architecture does not include digital processing but rather the possibility to sample the data, store it and feed it via an Ethernet connection to a processing unit where the actual processing will be performed. The baseband clock rate is 80 MHz leading to a symbol duration of 12.5 ns. The ADC/DAC have a resolution of 16 bits which – given the above discussion of the MIT demonstrator – is deemed to be sufficient to handle a wide range of wireless channel variations. The timeslot synchronization is facilitated by (USB) cables and synchronization is performed prior to communication. The baseband module is powered by an AC or DC supply.
- **RF Module.** The RF module operates at a carrier frequency ranging from 5.1 to 5.9 GHz at tunable steps of 1 MHz. Maximum transmission power is in the range of 25 dBm but tunable to facilitate transmit power control. A highly sensitive LNA is used. The clock is provided by an ultrastable 10 MHz rubidium clock which has a $5 \cdot 10^{-4}$ Hz frequency drift due to aging per month or 0.01 Hz in 10 years. The half-duplex operation requires switching between transmission and reception mode, which takes about 5–10 clock cycles or 62–125 ns of forced idle time. The used antennas are dipoles mounted at 1.5 m height.

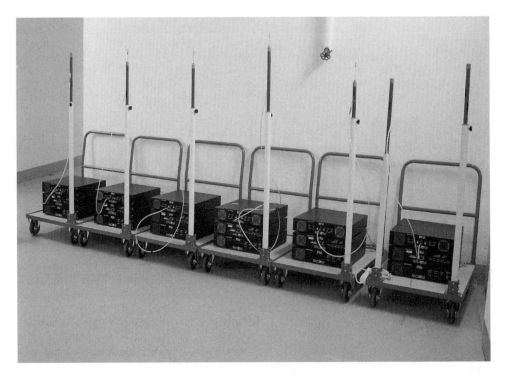

Figure 5.15 ETH's RACooN platform consisting of ten nodes

The RACooN hardware demonstrator is shown in Figure 5.15. It is in fact a highly flexible and state-of-the art platform that allows for flexible software setup and dynamic configuration of a variety of parameters, such as transmit data, LNA gain, local oscillator frequency, transmit power level, phase shift, choice of preamble, etc. Each node is given a set of up to 512 instructions, where one instruction is executed per slot and all slots between nodes are synchronized. Each instruction can consist of three possible commands: transmit (transmission of test signal or stored data); receive (measurement of power and phase of test signal or storage of complex baseband data); idle (pausing for a configurable time during which data can be stored, loaded, etc).

The RACooN platform allows therefore the implementation of fairly sophisticated relaying protocols. An example realization, which is composed of a training and an evaluation phase, is realized as follows:

- **Training Phase.** The aim of the training phase is to estimate the complex gains of all wireless fading channels. The training phase is initiated in that firstly all source nodes and then all relaying nodes transmit mutually orthogonal training sequences in the form of cyclically shifted m-sequences with a length of 511 samples. This allows all relays and destinations to estimate the channel across the 80-MHz bandwidth in bin sizes of $80\,\text{MHz}/511 = 0.16\,\text{MHz}$. Since the algorithms were developed assuming a frequency-flat channel, the channel of a single frequency bin is used. Given that the calculated coherence bandwidth is in the order of 1.6 MHz, the channel over the bin of 0.16 MHz can be assumed to be flat. (Note that an algorithmic extension to the frequency-selective case is possible and it [556] has already been indicated that FIR filters be used to do so.) A central processing unit calculates the optimum linear-process (LP) relaying gain matrix/function according [554] which suppress interuser interference. Finally, these complex coefficients are communicated to each relay node.

- **Evaluation Phase.** During this phase, the sources transmit their data, which is then received by the relays. The relays multiply the received complex baseband data by the prior determined complex gain coefficient, which minimizes interuser interference. Since the hardware platform has a finite switching time between reception and retransmission, the previously calculated gain matrix will be outdated and phase noise will prevent all interference being removed. Finally, for the given frequency bin, the destination calculates the SIR, which is used below to quantify the performance gains.

Clearly, both phases use a regenerative relaying approach that is influenced by the nature of the algorithms as well as the fact that the platform operates in half-duplex in the time domain. Note however that the relays are not able to decode the data. Since they employ single antennas, they also cannot separate the sources in space. The relays merely use an FFT to shift their received signals to the frequency domain, after which they apply their respective gain factors to the specified subchannel. After shifting the resulting signal back to the time domain again, it is retransmitted. The lessons learned so far from this hardware deployment are:

- **Refreshing Rate.** Similarly to MIT's hardware platform, the refreshing rate of the training phase at which the channel gain matrix is calculated has a profound impact on the performance of the distributed relaying protocol. As such, large refresh intervals yield large phase drifts, hence outdates of the gain matrix and thus deterioration of the protocol. On the other hand, having small refresh intervals yields low efficiencies, as during the training phase no useful data is transmitted. It is important to note however that varying relay phases does not require the gain matrix to be updated. Only the change of the propagation environment makes it essentially necessary to update the gain factors. As per Remark 3 of Berger and Wittneben [555] and Equation (11) [556], the relay phases do not have an impact on the received signal at the destinations as long as they stay constant for a single transmission cycle. A phase drift over multiple transmission cycles does not, therefore, affect the system performance.
- **Narrowband Architecture.** To extract all the gains that such multiuser relaying architecture can provide, the approach needs to be extended from narrowband to wideband algorithms. As alluded to in the previous chapter, the diversity gains provided by the relays in addition to the ones already provided due to the wideband channel are small. However, relays may prove useful in mitigating the created interuser interference by effectively employing distributed wideband beamforming and thus achieving a distributed spatial multiplexing gain.

In summary, ETH's hardware demonstration has pioneered the proof-of-concept of multiuser relaying algorithms. It has successfully demonstrated that performance gains with relays are viable and of interest to real-world network deployments.

5.7.3 Easy-C Project

One of the largest ongoing projects aimed at demonstrating the benefits as well as shortcomings of cooperative communications is the Easy-C project [557–559]. It is a research and development project funded by the German Federal Ministry for Education and Research (BMBF) and arguably the largest running cellular testbed project worldwide.

The aim of the project stretches beyond cooperative systems. As such, the project serves to test, benchmark and develop key technologies for the next generation of cellular networks. Particular attention will be paid to high spectral efficiency; low latency; fairness; and low capital and operational cost per bit [557, 558]. All of these will be key in enabling incumbent and emerging services. One of the main limiting factors to achieving the above, however, is interference. Easy-C thus takes a different cellular design approach in that interference is intelligently shaped, which is achieved by stipulating intercell cooperative techniques. The following techniques are thus investigated and quantified through the project's testbeds: advanced multi-antenna techniques; multicell joint transmission for interference mitigation; multicell joint detection for interference cancelation; multicell interference coordination; and cooperative relaying [558]. Since these developments are well aligned with 3GPP and ITU standardization activities, there is likely to be a major impact on the cellular community.

All of the above-mentioned techniques are somehow pertinent to the context of this book, since Easy-C proposes cooperation at various levels:

- **Cooperative MTs.** The obvious choice is to let MTs relay and cooperate in order to boost system capacity. Other ideas include using relaying techniques to localize interference and hence appropriately deal with it. This concept, however, will not be the focus of the Easy-C project but might be considered in future project realizations.
- **Cooperative BSs.** A more advanced approach, and very much the focus of Easy-C, is to let BSs cooperate, which requires advanced cooperative radio resource management (RRM) algorithms but facilitates interference management through cell cooperation. The thus-established multicell cooperation is also referred to as Network MIMO. Requiring a meshed backbone, adjacent BSs cooperate to establish a distributed MIMO system, to which signal processing methods such as macro diversity, joint detection and joint transmission can be applied.

These approaches are jointly being investigated by two major testbeds, the Easy-C testbed in Dresden and the LTE-A testbed in Berlin. The Easy-C testbed in Dresden is illustrated in Figure 5.16 and has the following key features [558]:

- **Testbed Specifications.** The focus of the testbed is on the PHY layer based on 3GPP LTE parameterizations. A proprietary MAC is used. Transmission takes place in UMTS band 7. The testbed is fully remotely controlled. There are currently 10 sites with 28 sectors that are connected by microwave backhaul links. A quintessential feature is that the BSs are synchronized w.r.t. frame timing and oscillator frequency. Each of these BSs comprises the following elements: TES 100-W power amplifier and duplexer; meinberg GPS device for synchronization (10 MHz reference clock + 1s pulse); control computer; network switch; Signalion Sorbas SDR platform (8 × Virtex IV FPGA); and a Kathrein antenna, which is cross polarized (2×2) at 60° beamwidth. The MT is mounted on a rickshaw and comprises the following elements: control computer; Signalion Sorbas SDR Radio platform (8 × Virtex IV FPGA); battery pack of 120 Ah for approximately 8 h of operation; a remote connection with the testbed's infrastructure via UMTS; and omnidirectional antennas (2×2).

Figure 5.16 Dresden's Easy-C platform consists of cooperating BSs (left) and nomadic MSs on the ground (right) [558]

- **Downlink.** For the downlink network MIMO approach closed loop techniques with real time CSI feedback loop are being used. The current focus however is on cooperative linear predistortion techniques with the use of adaptive coding and modulation. The configuration to-date includes up to three BSs serving up to three MSs jointly on the same time frequency resource with up to two spatial streams per MS. The data is then processed offline at the MS.
- **Uplink.** The uplink currently does not include any realtime feedback loop. Observations are drawn after some offline signal processing. Also, the cooperative detection algorithms based on the received data from the testbed are processed offline. Furthermore, modulation and coding is preselected. In summary, the uplink is fairly constrained for real-time networking.

The LTE-A testbed in Berlin is illustrated in Figure 5.17 and has the following key features [560]:

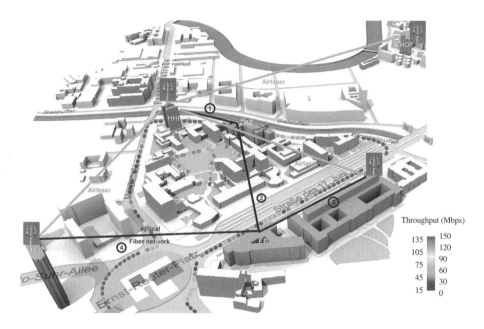

Figure 5.17   Berlin's LTE-A platform consists of cooperating BSs [560]

- **Testbed Specifications.** The key features of the Berlin field trial include intercell interference coordination; synchronized base stations; multicell channel estimation; interference-aware scheduling; interference cancelation algorithms; high-speed, low-latency, backbone network; distributed cooperation between BSs, and the exchange of data and channel knowledge. The tested sector of interest is surrounded by a ring of six interfering sectors contributing mostly to the expected interference. While there has been an experimental LTE site on top of FhG HHI since 2007, three new sites were set up in 2008. Sites are linked over 500 or 700 m by 1 Gbit/s free-space optical links. In addition, two fiber links are available from FhG HHI to T-Labs and TU Berlin for experiments with high backbone demands. At each site, there is a GPS-locked rubidium reference clock. All base stations can be operated coherently as a single-frequency network (SFN). The propagation environment has been thoroughly characterized by channel measurements. The 5 µs cyclic prefix of LTE is not violated both in isolated and SFN modes. The statistical distribution of signal and interference channels is similar to multicell simulations if the down-tilt of BS antennas is properly taken into account.

- **Cooperative Base Stations.** The potential of cooperative base-station concepts to reduce the interference between the cells in a full-coverage mobile network will be given close attention. Cooperation between adjacent BSs promises a significant increase in performance of cellular networks, and this has also been corroborated by recent measurements conducted by these project partners. The potential gain benefits, however, require a distributed cooperative signal processing of adjacent BSs, which are thus synchronized and require a multicell channel estimation. Moreover, the highspeed backbone architecture with low latency enables cooperative interference cancelation over time-varying signal and interference channels. Base stations and terminals in multiple cells are considered as cooperative inputs and outputs respectively of an enlarged multiple-input–multiple-output (MIMO) system.

Measured performance results were not available at the Time of writing. However, viable results are likely to appear in the second half of 2009.

5.8   Concluding Remarks

This chapter has introduced hardware issues related to the realization of cooperative relaying systems. We have touched upon hardware building blocks, their complexity, power consumption and cost. In more detail, we dealt with the following issues:

- **Baseline Analog and Digital Relaying Architectures.** We introduced purely analog and digital hardware architectures, discussing in great detail their building blocks and design peculiarities. A comparison between both has revealed that the purely analog hardware architecture is significantly cheaper than the digital counterpart. However, as advocated in this book, digital processing is, in general, superior to purely analog hardware architecture with the latter only facilitating transparent relaying. In addition, even transparent relays need to be provided with control information related to the choice of frequency bands, transmission power, etc. Due to today's communication systems being digital, this requires at least some simple digital processing capabilities within the relay that one could equally use to deploy (some simple) regenerative protocols. Therefore, except for very simple or unplanned and uncontrolled relaying rollouts, it is deemed unlikely that transparent architectures with purely analog hardware will be used in real-world rollouts at all.
- **Complexity and Power Consumption of UMTS and WiMAX/LTE Relay.** Another important cornerstone of this chapter was the complexity evaluation of the relay assuming two technologies, namely 3G UMTS based on W-CDMA and WiMAX based on OFDM. Given some prior system details, we have derived the number of arithmetic operations, required memory and consumed power in minute detail for relays based on either of the two technologies. We have successfully quantified the required complexity and, with the given technological roadmap, effectively confirmed that simple as well as cooperative relaying is generally technologically viable. The only exception is the use of very sophisticated PHY techniques, such as interference cancelers, etc.
- **Available Relay Hardware Demonstrators.** The chapter closed with a description of a few available hardware platforms dedicated to demonstrating the benefits of cooperative communication systems. Our focus was on the simple but pioneering MIT laboratory testbed, ETH's very sophisticated RACooN lab testbed, and Germany's large outdoor cellular Easy-C testbed. Other sophisticated testbeds demonstrating the benefits and shortcomings of cooperative communication paradigms are either already available or will appear soon, such as Eurecom's OpenAir Interface platform [561], CTTC's GEDOMIS platform [562], University of Paderborn's RailCab SOR-BAS programmable IEEE 802.11g WLAN platform [563], etc. Finally, for the interested reader, component-wise insights or theoretical hardware design challenges for cooperative systems have been exposed [353, 564–573].

From the above it becomes clear that hardware implementations of simple as well as more sophisticated cooperative communication systems remain largely unexplored. This comes along with numerous unsolved and open problems. An example is the issue of distributed synchronization where cooperative nodes are sufficiently synchronous to allow for space–time transmissions, beamforming, etc. We will again highlight some more open issues in Chapter 6.

6

Conclusions and Outlook

We conclude this book by summarizing and highlighting its contributions to the area of hardware, channel and PHY layer algorithms of distributed cooperative communication systems. We shall also discuss how certain simplifying assumptions throughout this book could be relaxed to reflect better the real-world deployments of such systems. We will then discuss in some detail open research topics that we identified whilst writing this book. Finally, even though not really related to the technical content of this book, we shall also briefly discuss some open business challenges when deploying cooperative systems.

6.1 Contributions

Throughout this book, we have tried to maintain a common thread. Most notably, we have divided hardware, channel and protocol designs into transparent versus regenerative approaches. Furthermore, for the channel and protocol design, we have rigorously distinguished simple relaying versus space–time processing approaches. Throughout, we have tried to give basics, taxonomies and foundations, together with first key milestones and latest developments in this dispersive field. We now review the book's contributions on a per-chapter basis.

6.1.1 Chapter 1

The aim of this chapter was to give a general introduction and taxonomy of the diverse field of cooperative communication systems. To this end, we have dealt with the following:

- **Quintessentials.** First and foremost, we have seen that cooperation facilitates pathloss, diversity and multiplexing gains where, surprisingly, pathloss gains are typically by an order of magnitude larger than the other gains. These gains essentially stimulated the design of suitable PHY layer techniques as well as the study of how the wireless channel impacts their performance and how to incorporate them into some tailored hardware – essentially inspiring the writing of this book.

  Without aspiring for a generic taxonomy, we then introduced some canonical architectures, which formed the foundation for the entire book. A superposition of these topologies allows us to construct arbitrary cooperative systems. We observed that the prime protocol design driver in said systems is due to the choice between transparent and regenerative relaying approaches; the former simply amplifies the received signal, whereas the latter performs some form of digital processing. The importance of this choice also inspired the structure of this book where all technical chapters deal with either approach separately. The second most important factor is the choice between traditional

Cooperative Communications   Mischa Dohler and Yonghui Li
© 2010 John Wiley & Sons, Ltd

relaying and space–time processing relaying architectures; the former is realized by means of an arbitrary number of serial and/or parallel relays, whereas the latter is realized by means of an arbitrary number of nodes performing one of the many possible forms of distributed space–time processing. Some other parameters were also found to be of importance, such as the availability of a direct link, the degree of cooperation, etc.

We then gave some example application scenarios, including cellular, WLAN and WSN systems. It could generally be observed that cellular and WLAN systems can benefit from cooperation to boost their coverage or capacity or both. WSNs and embedded systems, on the other hand, would typically only make use of the coverage gains.

- **Pros and Cons of Cooperation.** We further discussed some high level advantages and disadvantages of using cooperation in a wireless system. Generally, one would gain in performance, get a better and more balanced QoS throughout the system, be able to make an unplanned rollout, and generally also enjoy significant cost savings. There are numerous detrimental issues, however, such as more complex schedulers, a generally increased overhead, the difficulty of finding suitable relay partners on the fly, the accommodation of the extra relay traffic, an increase in end-to-end delay, and the need to provide sufficient synchronization. These factors essentially lead to tradeoffs which we also discuss, such as coverage versus capacity, algorithmic versus hardware complexity, interference, ease-of-deployment and cost versus performance.

  We then quantified the performance gains of cooperation in a Shannon sense, where we looked at capacity over ergodic channels and outage over nonergodic channels. As such, cooperation improves both the rate regions as well as outage behavior and thus introduces information-theoretical inspiration into realistic PHY layer protocol designs. We have further discussed the DMT for cooperative systems, which essentially states that diversity gains achieved due to cooperation can be utilized to boost the rate, and vice versa.

- **Thorough Taxonomy.** An important part of this chapter is the thorough taxonomy related to cooperative systems, which pertained to the relaying node itself, a neighborhood's first-hop access resolution approach and then networking aspects at large. Concerning the node taxonomy, we focused on the node behavior (egoistic, supportive, cooperative), transparent relaying mechanisms (AF, LF, etc.) and regenerative relaying mechanisms (EF, CF, DF, etc.); for each of these, we also highlighted the most important design parameters. Concerning the access resolution taxonomy, we dealt with duplexing methods for a point-to-point link (FDD, TDD and the novel division free duplexing, DFD), duplexing methods for the incoming and outgoing relaying links (FDR, TDR and DFR), access protocols (TDMA, FDMA, CDMA, CSMA, etc.) and resource allocation strategies (over time, frequency, code, etc.). Concerning cooperative networking aspects, we dealt with canonical information flows (direct, serial, space–time links and hybrids thereof) and also highlighted important design parameters pertaining to networking aspects. Finally, we also dealt with system analysis and synthesis, where we showed which parameters are generally of importance, how they can be analyzed and synthesized. The latter is of particular importance for system designers.

- **Early Key Milestones.** The remainder of the chapter is dedicated to first key milestones in the respective areas of supportive, cooperative, distributed space–time relaying. Even though relaying protocols have been known in the engineering community for almost a century, they have been revived with the seminal information and communication theoretical works of Cover, Gamal and Laneman, among others.

This chapter thus naturally paves the way for the subsequent technical chapters on channel modeling, PHY layers and hardware implementations.

6.1.2   Chapter 2

The wireless channel is key to understanding the performance behavior of cooperative systems properly, which is why an entire chapter is dedicated to this subject. We commenced with an introduction to channel modeling and the behavior of the most important parameters. We then deal with important channel modeling issues pertaining to regenerative, transparent and space–time relaying channels:

- **Fundamental Channel Characteristics.** In the section summarizing general channel characteristics, we commence with a summary on propagation principles, including EM wave properties, propagation mechanisms (free-space, reflection, refraction, scattering, diffraction) and signal distortions (Doppler, multipath). We then moved on to describing typical approaches to propagation modeling, which are based on the observation that these effects result in three multiplicative effects, that is pathloss, shadowing and fading. We then discuss typical pathloss modeling approaches (free-space, single-slope, dual-slope, ray-tracing, empirical, etc), properties of shadowing (log-normal distribution, ACF, CCF) and finally properties of fading. The latter is then expanded through a discussion on important channel modeling approaches pertaining to its most important underlying parameters (envelope, phase, delay profile, etc.) and selectivity issues (temporal, spectral and spatial selective versus nonselective channels). We then also give a taster on the trends of regenerative and transparent relaying channels, which are greatly expanded in the subsequent sections of this chapter.

- **Regenerative Relaying Channel.** We then embark upon the description and quantification of the behavior of regenerative wireless relay channels. The quintessential property of said channels is that the relay essentially decouples the incoming and outgoing wireless channel through its regenerative operation. Having said this, some important differences to traditional channel models remain, such as the mobility of both communication ends, etc. We thus commence our discussion on propagation modeling issues of regenerative channels, that is pathloss, shadowing and fading. As for pathloss, we generally observe a significant change of breakpoint behavior and aggregated power gain. As for shadowing, we generally observe a significant change in shadowing standard deviation and its aggregated behavior as seen by the destination, as well as a change in the correlation behavior. As for fading, we see that the amplitude and resulting power statistics generally do not change, whereas there are significant changes in the temporal, spatial and spectral correlation functions.

  These are hence dealt with separately throughout the remainder of this section. We commence with the temporal domain only, where we depart from a double-bounced two-ring geometrical model for SISO regenerative relaying channels to derive some temporal key quantities, such as the autocorrelation function, Doppler spectrum, level crossing rate, average fade duration, etc. The initially simple scenario is then also extended to more complicated propagation environments, such as non-isotropic scattering, Ricean fading conditions, etc. We then extend the temporal insights to the spatial domain by considering a single- and double-bounced two-ring model for MIMO regenerative relaying channels; again, we calculate key second-order channel characteristics. Finally, we extend these insights to the spectral domain by considering a three-dimensional single and double-bounced two-cylinder model for MIMO wideband regenerative relaying channels.

  We then also dwel on how properly to simulate a regenerative relaying channel, where we first highlight typical modeling approaches (Akki and Habers simulation model, discrete line spectrum methods, deterministic and random sum-of-sinusoids methods, modified method of equal areas, L_p-norm method, etc.), and then look in more detail on some specific simulation scenarios (MIMO narrowband and wideband regenerative relay channels). We also give a state-of-the-art summary on issues pertaining to channel estimation in regenerative networks. Finally, we also summarize conducted measurement campaigns and the resulting empirical models, such as M-VCE and WINNER outdoors and indoors relaying models.

- **Transparent Relaying Channel.** Exactly the same exercise is then repeated for the transparent relaying channel. The key difference for this channel w.r.t. the regenerative one is the change of fading amplitudes and thus resulting powers. This is due to the relay, and hence the entire network topology, becoming truly part of the wireless channel. In addition to above second-order moments, we have hence also look in great detail on the amplitude statistics resulting from the cascaded fading channels.

- **Distributed Space–Time Relaying Channel.** Finally, the distributed space–time channel is been very briefly dealt with. We show that one can essentially reduce the derivation of its properties by superimposing the analysis conducted in the preceding sections on regenerative and transparent relaying channels. Numerous problems, however, remain open for such channel configuration.

This chapter thus naturally paves the way for the subsequent technical chapters on transparent and regenerative PHY layer protocols.

6.1.3   Chapter 3

This chapter is exclusively dedicated to PHY layer protocols tailored to transparent relay deployment, where a relay receives a signal, transparently processes it and then retransmits it. Transparent processing essentially implies that the relay is not accessing the information content but only performs analog changes to the waveform by means of, for instance, analog amplification, analog phase shifting, etc. This chapter is roughly split into three parts, that is transparent relaying protocols, transparent distributed space–time processing protocols and their distributed optimization to yield the best possible system-wide gains:

- **Transparent Relaying Protocols.** As for the relaying protocols, we analyze four canonical topologies that differ in the number of parallel relaying branches and serial relaying segments, that is single-branch dual-hop, single-branch multihop, multi-branch single-hop and multibranch multihop. These can be used to construct any type of transparent relaying topology. The exposed mathematical tools can also be used to analyze and synthesize more complex system settings. For example, one would typically use the CDF analysis to obtain the CDF and thus PDF and MGF of the end-to-end SNR of serially concatenated transparent multihop relaying channels. Parallel transparent multi-branch relaying systems, on the other hand, are directly dealt with by using the MGF approach. A generic framework for any type of relaying topology and arbitrary fading channels (obeying some loose requirements) is also discussed as it only involves the knowledge of the channel PDF at zero SNR. From a performance point of view, it could generally be observed that the multihop fading channel performed worse than the direct channel unless the fading (and shadowing) gains are taken into account. The multibranch channel, on the other hand, already enjoys a diversity gain at fading level and hence outperforms systems without relays.
- **Transparent Distributed Space–Time Processing Protocols.** As for transparent distributed space–time processing protocols, we have shown how the traditional space–time processing techniques originally developed for spatially collocated antenna elements can be adapted to topologies with spatially distributed antenna elements. We have thus dealt with canonical algorithms pertaining to distributed space–time block coding, space–time trellis coding, spatial multiplexing and beamforming.

  The distributed transparent STBC part has been dedicated to linear dispersion distributed space–time codes (DLD-STC). In the DLD-STC protocol, upon receiving signals from the source, each relay performs a linear transformation of the previously received T signals. All the relays then forward the linearly transformed signals to the destination at the same time. The signals received at the destination then form a DLD-STC. There are two major differences between the LD codes in MIMO systems with collocated antennas and DLD-STC for relay networks. In MIMO systems, since all antennas are collocated, the LD codes can be designed in a centralized manner. In wireless relay networks, however, relay nodes are spatially distributed so that the DLD-STC has to be implemented in a distributed manner. Secondly, in MIMO systems, all transmit antennas have perfect knowledge of the information symbols. However, in relay networks, each relay only has access to a noisy copy of the original information symbols. These differences essentially result in the different designs between DLD-STC and traditional LD codes, which has been the inspiration for this section.

The distributed transparent STTC part is dedicated to the establishment of general and topology-tailored STTC design criteria. As for the general criteria, it could be proved that the sufficient upper bound on the pairwise error probability, governed by the determinant of the codeword difference matrix, does not depend on the actual channel fading statistics. Therefore, a STTC optimized for Rayleigh channel is sufficient for Ricean, double Rayleigh, log-normal and any other channel PDF. It is, however, also pointed out that this bound is not tight; better codes can be derived using other criteria and approaches. An example of such better codes results from designs tailored to specific protocols and topologies, where we assume that the source is equipped with one transmit antenna, each of the relays is equipped with one antenna and the destination is equipped with multiple receive antennas. We then establish the STTC design criteria together with the resulting generator matrixes for the three different relaying protocols typically encountered in the open literature. It is noted that significant gains can be achieved by such tailored design over the general design known from collocated antenna elements.

The distributed transparent spatial multiplexing (SM) part is dedicated to a network with multiple source, relay and destination nodes, where several source nodes wish to send multiple data streams to several of the destination nodes without causing interference between data streams. Therefore, a major challenge is to design the system such that an interference free signal is obtained at the destination nodes. This has been achieved by means of interference cancelation techniques, which have been applied at the source, relay and destination nodes or any combination thereof. The level of processing at the source, relay and destination nodes has been shown to be dependent on several factors, such as whether there is cooperation between nodes, or if nodes have stringent energy requirements and thus cannot perform much processing. We thus analyze two-hop AF relay networks for four different system setups based on zero forcing precoders and/or receivers for processing at the (i) source and relay, (ii) relay, (iii) relay and destination and (iv) destination nodes. We consider the zero forcing approach because it is a practical and simple scheme often considered for interference cancelation. For these systems, we designed the precoder matrices such that they minimized outage and then demonstrate their advantageous performance. Some surprising insights could be drawn in that adding relays does not always improve performance, which essentially is due to a fundamental tradeoff between increase in reliability and increase in mutual interference.

The distributed transparent beamforming (BF) part has been dedicated to systems that allow to controlling of the phase and relative amplitude of the signal transmitted through each spatially distributed antenna so as to form the strongest possible beam in the direction of the receiver. The thus realized transmit beamforming provides significant array gains and is proven to be the optimal transmission scheme under given conditions. It thus improves the performance of cooperative transmission when the full CSI at the relay is exploited, leading to the novel concepts of distributed beamforming and distributed MMSE. We have therefore discussed the beamforming designs under two power constraints: (i) a global sum power constraint, for which the total power consumption of all relays is not greater than a given power level, and (ii) an individual relay power constraint, for which each relay has its own transmission power limit. The latter is often observed in practical systems, where each relay is equipped with its own battery and thus rather suffers from individual power limits. When changing from global to individual power constraint, the power control algorithm at each relay is modified accordingly to ensure that its own transmission power limits are met. We established that under the individual relay power constraint, some relays do not use their maximum power in order for the destination to achieve the optimal SNR.

- **Transparent Distributed System Optimization.** A cornerstone in system design is the optimization of all or at least the most influential parameters. From prior conducted analysis in this chapter, we see that a key parameter is the allocated transmission power.

One of the challenges we addressed, therefore, was how to allocate distributively the total transmission power among source and all relays or distributively control the transmission power of each relay to satisfy the individual relay power constraint. This is of paramount importance in, for instance, wireless sensor networks, so as to prolong the life time of sensor nodes. We thus found the power allocation solutions that maximize the overall received SNR. Since the bit error rate can be described as a monotonic decreasing function of the destination SNR for a diversity system, the maximization of the overall received SNR is also equivalent to the minimization of bit error rate. We found that the exact solutions are not tractable, thus forcing us to consider a received SNR upper bound, from which we calculated the power allocation coefficients. Simulation results demonstrate, however, that the proposed power allocation scheme based on this bound can yield considerable SNR gains compared with the equal power allocation. This gain monotonically increases as the number of relays increases. This demonstrates that the distributed adaptive power allocation is very important in optimizing system performance and reducing the overall consuming power of a wireless relay network. In a practical system, the power allocation factors can be cal-culated at the destination, after which the reverse link channels can be used to feedback/broadcast these factors to the source and relays. The source and relays then adjust their transmit power based on these feedback values.

Another very important issue we touch upon related to relay selection in distributed transparent communication systems. Among the numerous ways of carrying out relay selection to the benefit of the overall system performance, we have introduced a simple relay selection algorithm based on an AF relay protocol. We then performed a lengthy and in-depth analysis using the harmonic mean as well as the novel modified harmonic mean theories to characterize, dimension and optimize said systems. By means of simulations, this protocol is then shown to facilitate system design for nonorthogonal two-hop multiple relay networks as it considerably improves the system performance and capacity compared to the all-participation orthogonal AF relay schemes.

A general assumption throughout this chapter is that synchronization is perfect and other system parameters are generally at least as good as for collocated MIMO antenna arrays.

6.1.4 Chapter 4

This chapter is exclusively dedicated to PHY layer protocols tailored to regenerative relay deployment, where a relay receives a signal, processes it and then retransmits it. Regenerative processing essentially implies that the relay changes the waveform, information representation and/or information contents in digital domain. This chapter is roughly split into three parts, namely regenerative relaying protocols, regenerative distributed space–time coding protocols and some advanced topics in distributed network coding:

- **Regenerative Relaying Protocols.** We primarily deal with decode-and-forward (DF) protocol, compress-and-forward (CF) protocol, soft information relaying (SIR) and adaptive relaying protocol (ARP) as well as the selective decode-and-forward (S-DF) protocol. While we also revise some early state-of-the-art contributions here, most of the treated material has emerged only recently.

  As for the DF protocol, we commenced by introducing an equivalent one-hop communication model that simplifies subsequent calculations. In particular, it allows the use of a near-optimal combining technique based on MRC, which turns optimal if there are no decoding errors at the regenerative relay. This greatly simplifies detection at the destination compared with ML decoding and derivatives.

  As for the CF protocol, its use is motivated by the fact that the relay and the destination receive different noisy versions of essentially the same source signal. CF would use this correlation to compress the received signals. We consider the CF protocol based on the Wyner–Ziv Coding scheme, which consists of a quantizer followed by an index encoder. The quantizer converts the input analog signals into the digital signals, which are then processed by the succeeding index encoder to get further compression. We thus concentrate on the quantizer design for said Wyner–Ziv CF protocol. We then move on to CF protocol designs based on Slepian–Wolf coding, which essentially deals with the compression of two or more correlated data streams. This technique has been used independently by the relay directly to compress the binary sequence obtained by making a hard decision on the relay received signals.

As for the SIR protocol, this is based on the well known insight that the main performance deterioration in DF protocols comes from the error propagation during the decoding and re-encoding process when decoding errors occur at the relay. One way of avoiding such an error propagation is to calculate and forward the corresponding soft information instead of making a decision on the transmitted information symbols at the relay. Forwarding soft information at the relays provides additional information to the destination decoder to make decisions, instead of making premature decisions at the relay decoder. We introduce the basic principles of SIR and also discuss some mathematical models in great detail, such as SIR based on soft symbol estimation, SIR based on log-likelihood ratios, derivation of the mean square errors of signal estimation at the relay, etc.

As for ARP, it trades the advantages and disadvantages of DF and AF protocols for different channel conditions between source, relay(s) and destination. Here, each relay adaptively selects the AF or DF protocol based on whether its decoding result is correct or not. All the relays that fail to decode correctly use the AF protocol to amplify the received signals and forward them to the destination. On the other hand, all the relays, which can successfully decode the received signals, use the DF protocol. The signals received at the destination, forwarded from all relays, by using either the AF or DF protocol, are combined into one signal to recover the source information. In a practical system, in order to determine whether a relay can decode correctly or not, it has been suggested that some CRC bits appended to each information block be used. Whilst the decoding procedure needed to determine if the frame is in error is complex, one could make use of the threshold Bhattacharyya code parameter to simplify said procedure. In a practical setting, the ARP can automatically adapt to the channel quality by simply switching between the AF and the DF protocols without requiring the CSI to be fed back from the destination to the relays or the source. This feature is very important in practical relay networks, especially in a large multihop network in which the feedback of CSI for adaptation is very expensive. Another important feature is that the processing at relays and destination for the ARP scheme is the same as for the AF and DF and it does not add much extra complexity to the system. We thus introduce and analyze the ARP scheme as well as evaluating its performance and comparing it to other relaying protocols.

As for the S-DF protocol, those relays that can decode the received signals correctly use DF protocols to forward the signals to the destination and the remaining relays will stay in an idle state. The only difference from ARP is therefore that S-DF has idle relays which in ARP would use the AF protocol. In that section, we briefly quantify the performance of the S-DF protocol.

- **Regenerative Distributed Space–Time Coding Protocols.** We deal with distributed space–time block coding, distributed space–time trellis coding and distributed turbo coding protocols in great detail with many novel design paradigms and performance insights.

    As for distributed space–time block coding, we first introduce the generic VAA relaying architecture where relaying nodes in each relaying stage are generally allowed to cooperate. After describing the distributed encoding and decoding procedure, we derive in closed form the symbol error rates for orthogonal STBCs operating over general Nakagami-m fading channels. The analysis caters for the generic case where each channel spanned between any distributed transmit and any distributed receive antenna can obey an arbitrary fading factor m. We then move on to derive resource allocation procedures, which yield near-optimal performance at minimal computational complexity. That is, depending on average fading conditions in the network, each relaying node decides how much power, bandwidth/frame length and modulation index to utilize. The cases of full and partial cooperation at each VAA relaying stage are discussed and the superiority of derived algorithms w.r.t. non-optimized solutions demonstrated.

    As for distributed space–time trellis coding, we assume that each codeword is split into two parts, that is one transmitted by the source and the other by the relay nodes. Generally, we observe that distributed space–time coding is based on incremental redundancy and thus allows a much more flexible distribution of channel symbols between source and relay nodes compared with repetition algorithms based on, for example, AF protocols. We analyze two different distributed STTC schemes, namely one based on the DF protocol assuming error free decoding at the relay and the other based on the EF protocol taking into account the erroneous detections at the relay. We discuss in great detail the design approach for both protocols based on the pairwise error probability analysis. To this end, we provide the encoder structure, error analysis over slow and quasi-slow fading channels, resulting design criteria and resulting codes with their performance.

As for distributed turbo coding, these protocols borrow the concept from the conventional LDPC and turbo codes and apply them in distributed networks. One prime challenge which we address in this section is in designing a practical capacity approaching distributed coding schemes with decoding errors and avoiding error propagation. We concentrate on three relaying protocols: (i) the DF protocol; (ii) the SIR protocol, and (iii) the generalized distributed turbo coding protocol. For all these protocols, we introduce the distributed turbo encoder and decoder structure, and also conduct a detailed performance analysis, which has been corroborated by numerous performance curves.

- **Regenerative Distributed Network Coding Protocols.** We finally dealt with distributed network coding, an advanced regenerative distributed space–time processing approach. We first deal with distributed network coding for a single source-destination pair and then with network coding division multiplexing for multiple source–destination pairs.

  As for distributed network–channel coding, its design has been influenced by the observation that network coding is an effective technique for increasing the network's spectral efficiency, among other things, by using simple coding and routing processes through the network. A properly designed network coding protocol can also provide strong error correcting capabilities for packets flowing through lossy networks. We thus introduce the distributed network and channel coding scheme designed for general wireless relay networks where the coding schemes are based on the general graph codes, such as low density generator matrix (LDGM) codes, low density parity check (LDPC) codes, etc. To aid a better understanding, we give a brief review of graph codes and their decoding. The principle of graph code based distributed network and channel coding schemes is then been introduced and dealt with in great detail. The performance of these schemes and their superiority has been evaluated through computer simulations.

  As for network coding based multiple access division multiplexing protocols, these explore both network and channel coding gains as well as enabling multiple source groups to communicate with multiple destination nodes independently. In the said design paradigm, each relay performs a linear network coding and a graph code is formed at each destination. A code-nulling algorithm is then proposed in order to eliminate the intergroup interference at each destination. The outlined approach thus enables multiple groups of source nodes to communicate with multiple destination nodes simultaneously without any interference between the groups.

Again, a general assumption throughout this chapter is that synchronization is perfect and other system parameters are generally at least as good as for collocated MIMO antenna arrays.

6.1.5 Chapter 5

In this chapter we introduce hardware issues related to the realization of cooperative relaying systems. We deal with canonical elements of purely analog and digital hardware architectures, carry out a very thorough complexity and power consumption analysis for relays based on UMTS and LTE/WiMAX specifications, and also give a summary of available hardware platforms realizing cooperative relaying systems:

- **Analog and Digital Hardware Architectures.** We discuss the important building blocks needed to realize a purely analog as well as digital relaying hardware architecture. For both architectures, we highlight design peculiarities and potential pitfalls. We also conduct a cost comparison between both that reveals that the purely analog hardware architecture is significantly cheaper than the digital counterpart. However, we also clearly highlight that not only do digital architectures usually outperform their analog counter part designs, but also that analog architectures need some form of control information which, in the end, still requires some digital modems to be present in the analog relay.

- **UMTS/LTE/WiMAX Complexity and Power Consumption Analysis.** We also conduct a thorough complexity evaluation of various relaying hardware architectures based on two standards, that is 3 G UMTS based on W-CDMA and WiMAX/LTE based on OFDM. Given some prior detailed system details, we derive the number of arithmetic operations, required memory and consumed power in minute detail for relays based on either of the two technologies. We successfully quantify the required complexity and, with the given technological roadmap, effectively confirm that simple as well as cooperative relaying is generally technologically viable. The only exception, as has been shown, is the use of very sophisticated PHY techniques in the form of, for example, interference cancelers.

- **Cooperative Relaying Hardware Testbeds.** We then describe some first milestone hardware platforms, which were built to demonstrate the benefits of cooperative relaying systems. Most notably, we describe in some depth the pioneering MIT laboratory testbed, ETH's very sophisticated RACooN laboratory testbed, and Germany's large outdoor cellular Easy-C testbed. We discuss their aims, approaches and various tradeoffs.

Whilst this chapter often maintains a high-level perspective on hardware issues, its prime aim is to familiarize researchers designing PHY layer algorithms with inherent constraints and opportunities related to the implementation of these algorithms into real-world hardware.

6.2   Real-World Impairments

The techniques and analysis described in this book are based on a varying set of assumptions, some of which are clearly very different from real-world operating conditions. Whilst such an approach is justified in that fundamentals are conveyed in a much clearer fashion, extending the said analysis and protocols to more realistic operating conditions is vital. We will now discuss how prior approaches can be applied to real-world impairments. Note that much of this is still entirely open and hence naturally forms a significant part of open problems and research challenges the discussed in Section 6.3.

6.2.1   Going Wideband

With the exception of the channel modeling chapter, we have exclusively dealt with narrowband communication systems. With shortening symbol/chip/sample durations, however, existing and emerging communication systems tend to become more and more wideband. This has a profound impact on the system analysis and design. System designers have a wide range of tools at their disposal for dealing with frequency-selective systems, such as (i) equalizer (typically used in 2 G systems); (ii) Rake receiver (typically used in 3 G systems); and (iii) OFDM transceivers (to be used in 4 G systems).

Using an equalizer significantly complicates the analysis and, since of limited interest as of today, is not further considered here. Using a Rake-type receiver allows one to extend the techniques and protocols discussed earlier fairly easily. As such, one could approximate the performance of the wideband system by taking a narrowband system with diversity reception where the number of diversity branches equates the number of Rake fingers. Using OFDM-type systems essentially reduces the wideband system to a set of parallel narrowband systems, to each of which prior analysis is applicable. Having said this, caution has to be used with the channel statistics, as the statistics before and after the (I)FFT might differ. Furthermore, whilst ergodic approaches evaluating, for instance, the Shannon capacity or average error rates do not significantly change, the nonergodic approaches evaluating, say, the outage probability largely remain an open topic.

6.2.2   Impact of Shadowing

Whilst pathloss is a fairly deterministic effect and fading is usually picked up by good channel codes, shadowing remains essentially the most detrimental performance factor in real-world systems. Its nonergodic characteristics, namely its slow changes over a given communication window, lead to large outages that an operator usually needs to cater for by means of large power margins. Except for the chapter on channel modeling, we have not incorporated shadowing into our analysis. One of the main reasons why shadowing is not encountered in even contemporary literature is that its underlying statistics are difficult to deal with. Indeed, its log-normal behavior as per Equation (2.3), generally does not lead to average closed form expressions.

Different approaches can and have already been taken in the past to incorporate shadowing. One approach is to approximate the log-normal distribution by another, easier to deal with, distribution. Examples of such distribution are the gamma distribution or the log-normal distribution's expansion through Hermite polynomials [91, 574]. Another approach is to express the PDF jointly with the fading process by means of the Meijer-G function for which closed form expression for various performance metrics has recently been derived [348, 349, 360].

Assuming an ergodic fading channel for which average error rates can be obtained in closed form using prior analysis, one can obtain the symbol or frame outage probabilities over nonergodic shadowing channels [92, 93, 575–581]. Indeed, one would typically find an invertible Laplace approximation of the said error rates. This has the advantage of allowing the outage probabilities to be obtained in closed form and thus analytical optimization and synthesis of important system parameters becomes feasible. This approach is essentially of utmost importance to operators but still in its complete infancy.

6.2.3   Impact of Interference

With the exception of some sections in the chapter on regenerative relaying protocols, we have generally not considered interference. Depending on the deployment mode of a particular communication system with respect to itself and other systems, the following interference scenarios may arise: (i) cochannel interference (CCI); (ii) adjacent channel interference (ACI); (iii) partial channel interference (PCI). CCI typically occurs in any wireless system reusing frequency bands, that is two spatially separated cells (in cellular network) or links (for example, in cooperative networks) use exactly the same band–but not necessarily the same technology–to communicate; interference hence occurs, which decreases as the spatial separation increases. ACI occurs in wireless systems operating in the same place and time but being allocated adjacent frequency bands; the amount of ACI usually depends on the spectral masks and guard bands imposed by the regulators. For instance, the UMTS-TDD and DECT systems suffer from some ACI in the region of 2 GHz. PCI occurs when two systems have partially overlapping frequency bands, that is the occurring interference is not as strong is in the case of CCI and not as mild as in the case of ACI. PCI is not typical in current wireless communication systems, but will gain in importance over the coming years as the frequency bands will become more and more congested.

Numerous techniques have been researched for a plethora of access schemes that facilitate dealing with interference. Broadly, these techniques can be categorized into: (i) interference avoidance (IA); (ii) interference mitigation (IM), and (iii) interference cancelation (IC). The prime aim of IA is to avoid interference between competing systems, where such mechanisms can usually be implemented at MAC layer. For instance, the frequency-hopping pattern of current Bluetooth systems avoids certain frequency bands if some interfering source, such as a microwave oven, is detected. Similarly, base stations in 4 G systems may coordinate their transmissions in a cooperative fashion so as to minimize interference or avoid it altogether. The aim of IM is to mitigate occurring interference in a best possible manner, which can be achieved both at PHY and MAC layers. For instance, beamforming is known to mitigate interference successfully–albeit not entirely eliminating or avoiding it. The aim of IC, also know as interference suppression, is to cancel all occurring interferences. It is known to be capacity optimum, albeit being very complex if applied optimally. Suboptimum and hence less complex solutions, however, exist with some of them having been discussed in Chapter 5.

With the stringent capacity requirements of next-generation broadband systems, of prime importance will be the design of an air interface to which low-complexity high-performance IC techniques are applicable in conjunction with other capacity boosting techniques, such as cooperative relaying. A large body of research in this direction has already been conducted but not yet introduced in prior analysis, where the following nonexclusive set of approaches can be differentiated: filter based approaches; transform methods; joint detection/multiuser detection; cyclostationary approaches; neural networks; higher order statistics and source separation; spatial processing; and analog techniques. An often used subset of these is: maximum likelihood (ML) detector; successive interference canceler (SIC); parallel interference canceler (PIC); iterative interference canceler (IIC); and interference subspace rejection (ISR), to mention a few. The inclusion of these techniques into interference-limited cooperative relaying systems is however largely unexplored and certainly deserves highest attention.

6.2.4   Inclusion of Channel Coder

With the exception of some sections in the chapter on regenerative relaying protocols, we have generally not considered capacity-approaching outer channel coder. However, modern communication systems all make use of channel coders and different forms of interleavers. To include these into the analysis and design in a rigorous fashion is fairly intricate. However, various heuristics have been put forward that mainly work in asymptotic regime. For instance, a good approximation is to assume that the SNR at the output of the decoder is equal to the SNR at the input multiplied by the code rate and minimum free distance of the used code. Another viable approach has been outlined [580] for various block and trellis codes, which we shall briefly discuss. For the sake of illustration, we consider only a single-input single-output (SISO) but coded system. The channel coder is a linear convolutional encoder or a binary block encoder. We also consider an interleaving scheme at bit level, which facilitates the avoidance of burst errors and hence leads to a constant average bit error probability (BEP).

First, we consider Hamming and Golay Codes [278]. The word error probability (WEP) with a hard decision decoder is a function of the uncoded BEP at the input of the decoder and is given by [278]:

$$P_w\left(E\right) = \sum_{j=t+1}^{J} \binom{J}{j} P_b^j \left(1 - P_b\right)^{J-j}, \tag{6.1}$$

where J is the word length in bits, t the number of errors that can be corrected by the code, $d_{\min}$ the minimal Hamming distance of the code, P_b the uncoded bit error probability of the channel, which we have essentially dealt with in Chapters 3 and 4, and E the error event. The parameters J, t and $d_{\min}$ are summarized in Table 6.1 for the considered block codes. To incorporate coding into cooperative systems, we are finally interested in the packet error probability (PEP). A packet is considered as erroneous when at least one bit is not recovered. Under the assumption that a perfect interleaving is performed, the errors are uniformly spread over the entire packet and are independent; hence, the

Table 6.1   Parameters for used codes

Code	J	k	$d_{\min}$	Code rate	t
Hamming	7	4	3	0.57	1
Golay	23	12	7	0.52	3

WEP is the same for all coded words over the packet duration. The PEP can therefore be expressed as a function of the WEP as:

$$P_p(E) = 1 - (1 - P_w(E))^{N/k}, \tag{6.2}$$

with N being the packet length in bits and k the number of information bits in one code word. Interestingly, a useful approximation has also been described [580]:

$$P_w(E) \approx k_{code} \frac{P_b^{t+1}}{(1 - P_b \tilde{y})^{1+t-J}} ; \forall P_b, \tag{6.3}$$

where $\tilde{y} = (t + 1) / (t + 2)$, and k_{code} is a code dependent constant given as:

$$k_{code} = \frac{1}{(t + 1) B (t + 1, J - t)}. \tag{6.4}$$

It can hence be observed that the symbol error rates derived throughout this book can easily be converted to bit error rates, which in turn allow the packet error rates for a block coded system to be derived.

Secondly, we consider convolutional codes. For convolutional codes, we are interested in the first event error probability (EEP), that is $P(E_1)$, which is defined as the probability that the decoded path deviates from the correct path for the first time in the jth arbitrary trellis section. This probability is supposed to be the same for all trellis sections. The EEP can be upper bounded as [582]:

$$P(E_1) \leq \frac{1}{T} \sum_{k=d_{\min}}^{\infty} a_k P_k, \tag{6.5}$$

where T is the puncturing period, P_k is the probability that an erroneous sequence with a Hamming distance k was chosen instead of the correct sequence, and a_k is the weight for the Hamming distance k. Considering a hard decision decoder, this quantity can be expressed for $k \geq d_{\min}$ as [583]:

$$P_k = \begin{cases} \displaystyle\sum_{e=\frac{k+1}{2}}^{k} \binom{k}{e} P_b^e (1 - P_b)^{k-e} & \text{for odd } k \\[4mm] \displaystyle\frac{1}{2} \binom{k}{\frac{k}{2}} P_b^{\frac{k}{2}} (1 - P_b)^{\frac{k}{2}} + \sum_{e=\frac{k}{2}+1}^{k} \binom{k}{e} P_b^e (1 - P_b)^{k-e} & \text{for even } k \end{cases} \tag{6.6}$$

The decoder is implemented with a Viterbi algorithm over a block with length N. The block error probability is defined as the probability that the likelihood sequence given by the Viterbi algorithm is not the coded sequence for the considered block. The probability that the likelihood sequence deviates from the correct path at the jth section of the trellis is upper bounded by Equation (6.5). A block error can occur anywhere in the trellis with the same probability. Hence, considering a block equal to a packet, the PEP is upper bounded by:

$$P_p(E) \leq \frac{N}{T} \sum_{k=d_{\min}}^{\infty} a_k P_k. \tag{6.7}$$

Again, a useful approximation has been derived [580], where the approximation of the first event error probability obtained [584] has been used:

$$\lim_{P_b \to 0} P_p(E) \approx N \lim_{P_b \to 0} P(E_1), \tag{6.8}$$

with the asymptotic first event error probability given by:

$$\lim_{P_b \to 0} P\left(E_1\right) \approx \begin{cases} \dfrac{1}{2T} a_{d_{\min}} \left(\dfrac{d_{\min}}{\frac{d_{\min}}{2}}\right) P_b^{\frac{d_{\min}}{2}} & \text{for } d_{\min} \text{ even} \\[2ex] \dfrac{1}{T} \left(a_{d_{\min}} + a_{d_{\min}+1}\right) \left(\dfrac{d_{\min}}{\frac{d_{\min}+1}{2}}\right) P_b^{\frac{d_{\min}+1}{2}} & \text{for } d_{\min} \text{ odd,} \end{cases} \tag{6.9}$$

where $a_{d_{\min}}$ is the number of paths with the minimal distance $d_{\min}$. Hence, in asymptotic regime $\overline{\gamma}_s \to \infty$, that is $P_b \to 0$, the PEP is expressed as a single power of the uncoded BEP at the input of the decoder. Table 6.2 summarizes the asymptotic expression of the PEP for two convolutional codes without puncturing, that is $T = 1$. It can hence again be observed that the error rates derived throughout this book can easily be inserted into above coded error rate expressions.

Table 6.2 Asymptotic expression of PEP with two convolutional encoder $K = 4$ and $K = 7$ with generator polynomials $15)_8$, $17)_8$ and $133)_8$, $171)_8$ in octal representation, respectively

K	R_c	T	$d_{\min}$	$a_{d_{\min}}$	$a_{d_{\min}+1}$	Asymptotic PEP
4	1/2	1	6	1	–	$10N P_b^3$
7	1/2	1	10	11	–	$1386N P_b^5$

6.2.5 Systems in Outage

With the exception of some sections in the chapter on transparent relaying protocols, we have generally not considered nonergodic systems. These systems can not generally be characterized by averages and the notion of outage hence plays a central role. Nonergodicity, for example, arises in the context of shadowing or slow-fading channels, or composites thereof. The calculation of outages is often much more involved than the calculation of averages, which explains why the amount of literature is scarce compared to the average analysis. Having said this, numerous tools have been published in the literature that can be directly used or adapted to treat outages in an analytical manner. Most notably, the already above-cited references [92, 93, 575–581] can be of use when introducing outage behavior into cooperative relaying systems.

We shall briefly illustrate a recently derived approach for quantifying the system's outage behavior. The PEO, $P_p(O)$, is defined as the probability of a PER exceeding a given threshold, P_p^* [92]:

$$P_p(O) = Pr(P_p(E|\overline{\gamma}_s) \geq P_p^*). \tag{6.10}$$

The PER is a nonincreasing function in $\overline{\gamma}_s$, hence Equation (6.10) is easily shown to be equivalent to:

$$P_p(O) = \int_0^{\overline{\gamma}_s\left(P_p^*\right)} p_{\overline{\gamma}_s}(\xi) d\xi, \tag{6.11}$$

where $\overline{\gamma}_s\left(P_p^*\right)$ is the required SNR to reach the target PEP P_p^* and $p_{\overline{\gamma}_s}(\overline{\gamma}_s)$ is the PDF of $\overline{\gamma}_s$. As said before, the average SNR $\overline{\gamma}_s$ is generally considered to be lognormally distributed with mean μ_{dB} and standard deviation σ_{dB}. This allows the PEO to be derived in closed form as [577]:

$$P_p(O) = Q\left(\frac{\mu_{dB} - 10\log_{10} \overline{\gamma}_p(P_p^*)}{\sigma_{dB}}\right), \tag{6.12}$$

where $Q(x)$ is the Gaussian Q-Function. From Equation (6.12) it is clear that the average SNR versus PEP is needed in order to obtain the PEO analytically. This SNR expression depending on the PEP and eventually on the BEP can be obtained by: (i) numerically inverting the PEP expression thanks to a set of look-up tables that need to cater for all possible shadowing and channel code parameterizations, which is very tedious; (ii) deriving an invertible approximation of the PEP in order to obtain a general closed form of the SNR as a function of the target PEP. Following the latter approach, we express first the SNR as a function of the BEP and then the BEP as a function of the PEP, assuming coded systems with the channel codes discussed above.

To execute the first step, we to obtain [580]:

$$\overline{\gamma}_s(P_b(E)) = \frac{m}{g_{mod}} \left[\left[\frac{P_b(E)}{k_{mod}} \sqrt{1 - \tilde{t} \sqrt[m]{\frac{P_b(E)}{k_{mod}}}} \right]^{-\frac{1}{m}} - 1 \right], \tag{6.13}$$

where $\tilde{t} = m/(m+1)$ and m is the Nakagami-m fading factor. Furthermore, k_{mod} is equal to k_{psk} or k_{qam} and g_{mod} is equal to g_{psk} or g_{qam} according to M-PSK or M-QAM signaling:

$$k_{psk} = \frac{1}{\sqrt{\pi} \max\left(\log_2 M, 2\right)} \frac{\Gamma\left(m + \frac{1}{2}\right)}{\Gamma\left(m + 1\right)}, \tag{6.14a}$$

$$k_{qam} = \frac{2g}{\sqrt{\pi} \log_2 M} \frac{\Gamma\left(m + \frac{1}{2}\right)}{\Gamma\left(m + 1\right)}, \tag{6.14b}$$

$$g_{psk} = \sin^2\left(\pi/M\right), \tag{6.14c}$$

$$g_{qam} = 3/(2(M-1)), \tag{6.14d}$$

and $g = 1 - 1/\sqrt{M}$ with M being the modulation order.

To execute the second step, we obtain for block codes [580]:

$$P_b = \left(\frac{P_w^*(E)}{k_{code}}\right)^{\frac{1}{t+1}} \left(1 - \tilde{y}\left(\frac{P_w^*(E)}{k_{code}}\right)^{\frac{1}{t+1}}\right)^{\frac{t+1-J}{t+1}}, \tag{6.15}$$

with $P_w^*(E)$ being the target word error probability. For convolution codes, the inversion on the approximate expression Equation (6.8) is straightforward and not shown here. It has been shown [580] that the thus obtained bounds, inversions and outage expressions hold sufficiently tightly in the operational region of interest. This approach can be extended to other systems and configurations, and allows as to take into account interference, etc.

6.2.6 Asymptotics

Finally, almost all chapters in this book have used asymptotic expressions as well as upper and lower bounds to characterize the performance of various relaying protocols. Such expressions are very useful as they are often the only ones available in closed form. In addition, they often yield clear insights into performance trends, such as the diversity order, etc. The use of these bounds and asymptotics is thus justified, as long as the rate of convergence is sufficiently high so as to exhibit small deviations in the region of interest. For instance, a typical region of interest for error rates before the channel decoder as well as outage probabilities is around 10%.

6.3 Open Research Problems

As with any thriving research area, numerous research problems remain open. We have alluded to some of them throughout the book as well as in the previous section. We will now dwell on some of such open challenges in more details.

6.3.1 Taxonomy

The arguably largest problem in research on cooperative relaying systems is the evident lack of a well-accepted taxonomy. Some first steps have been done through past journal papers published in the area, and Boyer *et al.* [78] as well as our first chapter. However, whilst we have tried to build a coherent and viable taxonomy, having read several hundred papers in this area, we feel that there is still a lot of improvement possible:

- **Definition of Protocols.** Whilst concepts, such as DF, EF, etc., are well established in the community, there is no common understanding on the precise functionality of each protocol. For instance, some works have advocated that DF implies decoding and retransmission without re-encoding, whereas other works assume re-encoding. It would therefore be desirable to have a unique definition of the functionality of each relaying protocol, together with a unique (short) acronym.
- **Inclusion of Real-World Impairments.** Considering real-world impairments, such as the above discussed shadowing, coding, interference, etc., profoundly impacts protocol design and therefore inherently any underlying taxonomy. It would therefore be desirable to have a unique and unambiguous taxonomy to hand that accommodates these factors.
- **Inclusion of Emerging Paradigms.** Considering emerging design paradigms, such as network coding, etc., also profoundly impacts protocol design. It would therefore also be desirable to have a unique and unambiguous taxonomy to hand that accommodates these emerging techniques.
- **Inclusion of Higher Layers.** In Chapter 1, we have briefly alluded to the role and impact of MAC and routing protocols onto the cooperative system. We had established that not all MAC protocols fit with all PHY techniques, which clearly implies that any taxonomy is impacted when considering higher OSI layers. It would therefore be desirable to have a taxonomy at hand that *a priori* incorporates the impact of said layers.

Ideally, any such taxonomy should be generic and flexible enough to be able to accommodate any approaches not foreseen at this point.

6.3.2 Wireless Channel

Whilst the community channel modeling is large, comparably little effort has been dedicated to the wireless channel for cooperative relaying systems. So far, only singular research groups have contributed to the topic, the majority of which forms the content of Chapter 2. Numerous research problems hence remain open, some of which are summarized below:

- **Closed Form Expressions.** Reading Chapter 2 reveals immediately that numerous expressions for first and second order statistics remain in integral form. It would therefore be desirable to obtain closed form expressions, either in exact or approximate form, for these open problems.
- **Distributed MIMO Channel.** We have only briefly touched upon the truly distributed MIMO channel in Section 2.5. Numerous analytical problems remain there, such as the second order

correlation functions given that all antenna elements move with different speeds, are embedded in different clutter environments, etc. Here, the incorporation of shadowing on a per-link basis is quite importants since each distributed MIMO subchannel is expected to suffer from different shadowing statistics.

- **Wideband Models.** Whilst we had touched upon some wideband modeling approaches, very few contributions are available today that cater for the wideband nature of the cooperative relaying channel. Most systems, however, are wideband in nature today. It would hence be desirable to have a complete mathematical body to hand that allows the modeling of the frequency dependency of the relaying channel for various propagation conditions and scenarios.

- **Polarization.** Another important issue we have not touched upon at all relates to polarization. As such, it would be desirable to have polarization models to hand that cater for the transparent and regenerative relaying channel.

- **Channel Prediction.** Whilst channel modeling, simulation and estimation are explored in quite some detail, very few results are available on the prediction of the cooperative relaying channel. Given that the temporal, spatial and spectral correlation functions of a cooperative system are very different from a noncooperative system, and we would expect a change in the design of said estimation algorithms.

- **Comprehensive Measurement Campaigns.** Whilst quite a few measurement campaigns have already been conducted to corroborate the gains of the cooperative relaying channel, very few measurement campaigns conduct a rigorous statistical hypothesis testing on obtained results. Furthermore, the truly distributed channel, polarization, etc., have not yet been characterized by means of real-world measurements.

An exact modeling of the wireless relaying channel is really of utmost importance as it is known to have a bigger impact on real-world system deployments than the actual choice of PHY algorithm(s) [66].

6.3.3 Transparent PHY Techniques

Given that a regenerative PHY layer is more likely to be used by network operators than transparent ones, not many issues of viable importance remain. Some of them are summarized below:

- **Closed Form Expressions.** Similarly to the wireless channel, in the context of transparent PHY layer techniques, numerous expressions are either not available at all or remain in integral form. It would hence be desirable to obtain closed form expressions, either in exact or approximate form, for these open problems.

- **Soft Information Relaying.** SIR is essentially a transparent relaying approach, which we have analyzed in the context of regenerative PHY techniques since it was used as a hybrid with regenerative protocols. Some open research issues pertaining to SIR are summarized below.

- **Asynchronous Designs.** We have assumed throughout this book that all nodes are synchronized, that is signals arrive at the destination synchronously. Given typically occurring clock-drifts and other real-world factors, however, causes possibly severe asynchronisms in the system. It would therefore be desirable to have relaying and space–time techniques to hand that cater for such asynchronisms and thereby facilitate robust communication irrespective of such impairments.

Whilst this list is fairly short, most of the open problems to be discussed in the context regenerative PHY techniques are also applicable in direct or modified form to the transparent one.

6.3.4 Regenerative PHY Techniques

The regenerative PHY layer is arguably the more viable approach and hence deserves more detailed attention. We summarize below a nonexhaustive list of issues that still deserve attention as they remain largely unsolved:

- **Exact Distribution of Equivalent Noise in SIR.** When we discussed soft information relay (SIR) protocols, we could see that the SIR based on the soft symbol estimate (SIR-SSE) achieved the unconstraint minimum mean square errors (MSE) among all existing relaying protocols and had a much smaller MSE compared with the SIR based on the log likelihood ratio (LLR). However, the performance of these two SIR schemes was almost the same. The reason for this is because we used a Gaussian distribution to approximate the equivalent noise in the SSE in SIR-SSE scheme, which is not a very accurate approximation. Although we have rectified the variance of this equivalent noise in the SSE to approximate the Gaussian distribution, such an approximation is still far from the accurate distribution. Recently, a very interesting error model has been proposed [585, 586] to model the decoding errors in the DF and the soft information in the MIMO relaying system. Considerable performance improvements have been observed compared with the Gaussian approximation models. The performance of SIR-SSE can possibly be further improved if a more accurate approximation of the equivalent noise in the SIR-SSE scheme can be found. How to represent and formulate mathematically such an accurate distribution is very important, but also very challenging. Unfortunately this has not been addressed so far.
- **Exact Distribution of Equivalent Noise in Distributed Turbo Coding.** The same issues exist for distributed turbo coding with soft information relaying (DTC-SIR). In the DTC-SIR, the relay needs to calculate the soft outputs for parity symbols, corresponding to the interleaved information sequence, by using the probability inference method. Although we can plot its distribution through simulations, this requires quite a considerable amount of work and simulation time. It is even harder to formulate its distribution compared as to the SIR-SSE. In the existing works, a Gaussian distribution has been used to approximate the SSE in the DTC-SIR. Its performance can be further improved by accurately modeling the distribution and designing the decoders accordingly. It is still an ongoing issue to find such an optimal decoder.
- **Optimal Retransmission of Soft Information.** Soft information is strictly speaking an analog signal. In practical systems, to transmit such analog signals, compression and/or quantization or other modulation techniques should be performed at the relays. The compression and forward protocol based on Wyner–Ziv coding (CF-WZC) can be used to quantize analog soft information into digital signals and compress the quantized signals before transmitting. Therefore, soft information relaying can be combined with compression and forward into a single relay protocol to be used in a practical system. Furthermore, the soft information can also be directly transmitted by using an analog phase modulation scheme [587], called continuous modulation. Besides these approaches, there will be other ways to transmit the soft information. Unfortunately, the optimal way to transmit the soft information for achieving the optimal performance is still an open problem.
- **Inclusion of Decoding Errors.** Most existing distributed coding schemes are developed on the basis on the conventional channel coding schemes, such as the space–time coding, turbo coding and LDPC coding. Furthermore, most distributed coding schemes rely on some ideal assumptions, such as error free decoding at relays, etc. Some initial work on modeling detection errors in the demodulation and forward have been done [442]. This model has been applied to construct distributed space–time trellis codes [482]. It has been shown that detection errors do have some effect in constructing practical distributed coding. However, there is still a lack of an accurate mathematical representation for modeling the decoding errors. As a result, no paper has actually

formulated accurate design criteria for distributed coding with DF protocols when decoding errors occur at relays.

- **Adaptive Coding and Modulation.** Additionally, most existing distributed coding schemes mainly concentrate on the fixed code rates and power allocations. Design of a distributed coding scheme, which can adaptively allocate the rate and power and distribute the code bits between the source and relays, as well as adaptively to select the relay protocols, has a potential to increase the overall network throughput, reduce the network power consumption and improve its reliability.
- **MIMO Relaying Networks.** Furthermore, not much has been done on the design of distributed coding schemes for MIMO relay networks. MIMO relay networks are much more complex than a single MIMO system or a SISO relay network. Multiple antennas at each node provide an additional dimension for designing distributed coding. Many open issues reside in this area, such as joint precoding design at source and relay for spatial multiplexing transmission, interference cancelation at relays and destination, design of optimum vector relay protocols, coding structures and code design criteria for MIMO relay networks, etc.
- **Prevention of Short-Cycles for Distributed LDGM/LDPC.** In generating a distributed LDGM or LDPC code, we should note that there is no coordination between relay nodes. Therefore, each relay randomly selects source symbols to generate the parity check. It may possibly happen that some source symbols are not included in any parity checks. Then the columns corresponding to these source symbols in the parity check matrix $\mathbf{H}$ will be all-zero vectors. That is, those source symbols that are not involved in parity checks are not protected by any parity symbols; therefore, the whole system performance will be seriously affected by these symbols. Furthermore, these distributed graph codes are constructed in a decentralized way; as a result, a significant number of short cycles may exist. This will also affect the system performance. To solve these problems, certain rules have to be applied to each relay to make sure that each source node has participated in at least one check equation and short cycles can thus be eliminated. How to design such a centralized or distributed mechanism is significant in practical systems, but is still an ongoing problem.
- **Consideration of Real-World Impairments.** There are also other practical issues that should be taken into account in implementing cooperative communications in wireless networks, such as synchronization between relays, signaling design, channel estimations, user coordination, resource management, and interference. Some initial work has been done to solve some of above-mentioned individual problems. For example, some asynchronous relaying schemes have been proposed to avoid synchronization problems; differential modulation schemes were developed to overcome channel estimation issues; partner selection schemes and adaptive resource (power, frequency, time slots) allocations have been proposed for effective user coordination and efficient resource management, etc. However, different schemes rely on different system assumptions and were developed under different system models. It is important to develop a unified system model and approach, which can solve all these issues in practice. In practical cooperative systems, there are also some security issues that should be considered, like attacks from malicious relays, denial of service from selfish relays for the sake of saving their energy, etc. How to ensure a secure transmission is also an important issue in cooperative wireless networks.

Above list of existing problems is fairly long but far from complete. Many more problems remain unsolved and certainly many more will appear once prior stated PHY problems are solved.

6.3.5 Hardware Considerations

Hardware designed for cooperative relaying systems has not received too much attention. Many important issues hence remain largely unsolved, some of which are discussed below:

- **Division Free (Full Duplex) Operation.** One of the most detrimental performance factors is due to the half duplex operation at the relay since it cannot transmit and receive at the same time and on the same band. Recent developments suggest that a division free operation is feasible using some sophisticated signal processing and/or hardware design. The availability of such a full duplex system would certainly be a quantum leap forward but remains so far an unsolved problem.
- **Synchronization and Clock Drifts.** While not of paramount importance for point-to-point systems, cooperative and particular distributed space–time processing systems require a high degree of cooperation between nodes. This applies with precise clocks within each relaying node and mechanisms that ensure that these remain mutually synchronized.
- **Purely Transparent Operation.** Transparent relaying protocols would certainly gain in ground w.r.t. regenerative ones if the relaying radios were capable of processing control information without digital radios. It would therefore be desirable to have a hardware architecture in place that allows analog control information to be received and executed without the need to have a digital transceiver. The cost (and sometimes performance) benefits would then make purely analog architectures a true competitor to digital approaches.

This completes the section on open research challenges and we will now very briefly allude to some business issues.

6.4   Business Challenges

Although not the focus of this book and hence not considered here at all, we would like to draw the attention of the interested reader to Timus [66], who gives an excellent treatment of business challenges in the context of relaying systems. In particular, he focuses on the large scale use of low cost relays with a focus on signal processing and radio resource allocation, rather than on hardware or network planning. A key question addressed in this dissertation is which relay cost is low enough for a relaying architecture to be viable from an economic point of view. The author develops a rigorous analytical framework for evaluating the viability of relaying solutions. This framework is based on a comparison between the relaying architectures and traditional single-hop cellular architectures. This comparative analysis is done from an operator perspective, and is formulated as a network-dimensioning problem. The associated investment decisions are based on financial measures (cost or profit) and taken under technical constraints (throughput, coverage, etc.).

The author first considers a large number of traditional dimensioning scenarios, in which the radio network is designed for a predefined traffic demand and target quality of service level. It has essentially been shown that the use of low cost relays can indeed be viable, but that the cost savings vary greatly from case to case and often are only modest. Due to the half-duplex nature of the low cost relays, they are best suited for providing coverage to guaranteed data rates, at low end-to-end spectral efficiency, and in environments with strong shadow fading. Interestingly, the type of environment and the placement of relays are more important than the specific protocols and algorithms used in the network; this corroborates the above discussed importance of shadowing, etc. According Timus [66], network planning remains an essential and challenging task, which is unlikely to be replaced by large-scale (unplanned) use of relays.

The author then also suggests a new direction of research in which the viability of relays is judged considering the entire life cycle of a radio network. Several commercially viable examples are given in which the temporary use of relays is economically viable, especially if the service uptake is slow or the uncertainty about the future demand is high. This is particularly relevant if the last-mile cost of a network is dominated by the backhaul transmission cost, and if relaying is implemented as a feature of an access point, rather than as a new device type.

References

[1] J. Gray, *Straw Dogs: Thoughts on Humans and Other Animals*. London, UK: Granta, 2002.

[2] M. Dohler, D. E. Meddour, S. M. Senouci, and A. Saadani, "Cooperation in 4g–Hype or Ripe?," *IEEE Technology and Society Magazine*, vol. 27, no. 1, pp. 13–17, 2008.

[3] F. Fitzek and M. Katz, *Cooperation In Wireless Networks: Principles And Applications – Real Egoistic Behavior Is To Cooperate!* Springer, 2006.

[4] Y. Zhang, H.-H. Chen, and M. Guizani, *Cooperative Wireless Communications*. CRC Press, 2009.

[5] M. Uysal, *Cooperative Communications for Improved Wireless Network Transmission: Frameworks for Virtual Antenna Array Applications*. IGI Global, 2009.

[6] F. Fitzek and M. Katz, *Cooperative Communications and Networking*. Cambridge University Press, 2009.

[7] V. Chandrasekhar, J. Andrews, and A. Gatherer, "Femtocell networks: a survey," in *Communications Magazine, IEEE*, vol. 46, pp. 59–67, Sept. 2008.

[8] D. Tse and P. Viswanath, *Fundamentals of Wireless Communication*, pp. 383–424. Cambridge University Press, 2005.

[9] R. Pabst, B. H. Walke, D. C. Schultz, P. Herhold, H. Yanikomeroglu, S. Mukherjee, H. Viswanathan, M. Lott, W. Zirwas, M. Dohler, H. Aghvami, D. D. Falconer, and G. P. Fettweis, "Relay-based deployment concepts for wireless and mobile broadband radio," *IEEE Communications Magazine*, vol. 42, pp. 80–89, Sept. 2004.

[10] 3rd Generation Partnership Project, "Technical specification group radio access network; opportunity driven multiple access," *3G TR 25.924 V1.0.0*, 1999.

[11] T. Rouse, S. McLaughlin, and H. Haas, "Coverage-capacity analysis of opportunity driven multiple access (ODMA) in UTRA TDD," in *3G Mobile Communication Technologies, 2001. Second International Conference (Conf. Publ. No. 477)*, pp. 252–256, Mar. 26–28, 2001.

[12] T. Rouse, I. Band, and S. McLaughlin, "Capacity and power investigation of opportunity driven multiple access (ODMA) networks in Tdd-cdma based systems," in *Communications, 2002. ICC 2002. IEEE International Conference*, vol. 5, pp. 3202–3206, Apr. 28–May 2, 2002.

[13] M. Ouertani, H. Besbes, and A. Bouallegue, "A New Design Approach for ODMA Systems," in *Control, Communications and Signal Processing, 2004. First International Symposium*, pp. 343–346, 2004.

[14] R.-G. Cheng, S.-M. Cheng, and P. Lin, "Power-efficient routing (PER) mechanism for ODMA systems," in *Wireless Networks, Communications and Mobile Computing, 2005 International Conference*, vol. 1, pp. 129–134, June 13–16, 2005.

[15] R. G. Cheng, S.-M. Cheng, and P. Lin, "Power-efficient routing mechanism for ODMA systems," *IEEE Transactions on Vehicular Technology*, vol. 55, pp. 1311–1319, July 2006.

[16] F. Peyrard, T. Val, and J. J. Mercier, "Simulations of ad-hoc WLAN with or without relay," in *Universal Personal Communications, 1998. ICUPC '98. IEEE 1998 International Conference*, vol. 1, pp. 711–715, Oct. 1998.

[17] N. Esseling, E. Weiss, A. Kramling, and W. Zirwas, "A multi hop concept for HiperLAN/2: Capacity and interference," in *European Wireless 2002*, vol. 1, pp. 1–7, February 2002.

[18] C. Casetti, C. F. Chiasserini, and L. Previtera, "Fair relaying and cooperation in multi-rate 802.11 networks," in *Vehicular Technology Conference, 2005. VTC 2005-Spring. 2005 IEEE 61st*, vol. 3, pp. 2033–2036, May/June 2005.

[19] A. So and B. Liang, "Exploiting spatial diversity in rate adaptive WLANs with relay infrastructure," in *Global Telecommunications Conference, 2005. GLOBECOM '05. IEEE*, vol. 5, Nov./Dec. 2005.

[20] A. So and B. Liang, "Effect of relaying on capacity improvement in wireless local area networks," in *Wireless Communications and Networking Conference, 2005 IEEE*, vol. 3, pp. 1539–1544, Mar. 13–17, 2005.

[21] K. N. Ting, Y. F. Ko, and M. L. Sim, "Voice performance study on single radio multihop IEEE 802.11b Systems With Chain Topology," in *Networks, 2005. Jointly held with the 2005 IEEE 7th Malaysia International Conference on Communication, 2005 13th IEEE International Conference*, vol. 1, Nov. 16–18, 2005.

[22] M. Kuhn, A. Ettefagh, I. Hammerstrom, and A. Wittneben, "Two-way communication for IEEE 802.11n WLANs Using Decode and Forward Relays," in *Signals, Systems and Computers, 2006. ACSSC '06. Fortieth Asilomar Conference*, (Pacific Grove, CA, USA), pp. 681–685, Oct./Nov. 2006.

[23] A. Ettefagh, M. Kuhn, I. Hammerstrom, and A. Wittneben, "On the range performance of decode-and-forward relays in IEEE 802.11 WLANs," in *Personal, Indoor and Mobile Radio Communications, 2006 IEEE 17th International Symposium*, pp. 1–5, Sept. 2006.

[24] L. Guo, X. Ding, H. Wang, Q. Li, S. Chen, and X. Zhang, "Cooperative relay service in a wireless LAN," *IEEE Journal on Selected Areas in Communications*, vol. 25, pp. 355–368, Feb. 2007.

[25] M.-H. Lu, P. Steenkiste, and T. Chen, "Time-aware opportunistic relay for video streaming over WLANs," in *Multimedia and Expo, 2007 IEEE International Conference*, pp. 1782–1785, July 2007.

[26] A. So and B. Liang, "Enhancing WLAN capacity by strategic placement of tetherless relay points," *IEEE Transactions on Mobile Computing*, vol. 6, pp. 474–487, May 2007.

[27] D. Niyato and E. Hossain, "Integration of IEEE 802.11 WLANs with IEEE 802.16-based multihop infrastructure mesh/relay networks: A game-theoretic approach to radio resource management," *IEEE Network*, vol. 21, pp. 6–14, May/June 2007.

[28] A. Gkelias, M. Dohler, and H. Aghvami, "HIPERLAN/2 for vehicle-to-vehicle communication," in *Personal, Indoor and Mobile Radio Communications, 2002. The 13th IEEE International Symposium*, vol. 3, pp. 1058–1062, Sept. 15–18, 2002.

[29] X. Yang, L. Liu, N. H. Vaidya, and F. Zhao, "A vehicle-to-vehicle communication protocol for cooperative collision warning," in *Mobile and Ubiquitous Systems: Networking and Services, 2004. MOBIQUITOUS 2004. The First Annual International Conference*, pp. 114–123, Aug. 22–26, 2004.

[30] M. Guo, M. H. Ammar, and E. W. Zegura, "V3: a vehicle-to-vehicle live video streaming architecture," in *Pervasive Computing and Communications, 2005. PerCom 2005. Third IEEE International Conference*, pp. 171–180, Mar. 8–12, 2005.

[31] L. Weixin, W. Ning, Z. Zhongpei, L. Shaoqian, and J. Na, "The differential detection OFDM cooperative diversity system in vehicle-to-vehicle communications," in *ITS Telecommunications Proceedings, 2006 6th International Conference*, pp. 1118–1121, June 2006.

[32] L. Lin, T. Osafune, and M. Lenardi, "Floating car data system enforcement through vehicle to vehicle communications," in *ITS Telecommunications Proceedings, 2006 6th International Conference*, pp. 122–126, June 2006.

[33] T. Yashiro, "A new paradigm of V2v communication services using Nomadic Agent," in *Mobile and Ubiquitous Systems–Workshops, 2006. 3rd Annual International Conference*, pp. 1–6, July 17–21, 2006.

[34] B. Wang, I. Sen, and D. W. Matolak, "Performance evaluation of 802.16e in vehicle to vehicle channels," in *Vehicular Technology Conference, 2007. VTC-2007 Fall. 2007 IEEE 66th*, pp. 1406–1410, Sept. 30–Oct. 2007, 2007.

[35] M. Jerbi, P. Marlier, and S. M. Senouci, "Experimental assessment of V2v and I2v communications," in *Mobile Adhoc and Sensor Systems, 2007. MASS 2007. IEEE Internatonal Conference*, pp. 1–6, Oct. 8–11, 2007.

[36] B. Mourllion and S. Glaser, "V2v Communication analysis by a probabilistic approach," in *Vehicular Technology Conference, 2007. VTC2007-Spring. IEEE 65th*, pp. 2575–2579, Apr. 22–25, 2007.

[37] P. Patwa and R. Dutta, "Joint modeling of mobility and communication in a V2v network for congestion amelioration," in *Computer Communications and Networks, 2007. ICCCN 2007. Proceedings of 16th International Conference*, pp. 575–582, Aug. 13–16, 2007.

[38] J. M. Lee, M. J. Yu, Y. H. Yoo, and S. G. Choi, "A new scheme of global mobility management for inter-vanets handover of vehicles in V2v/v2i network environments," in *Networked Computing and Advanced Information Management, 2008. NCM '08. Fourth International Conference*, vol. 2, pp. 114–119, Sept. 2–4, 2008.

[39] F. Ye, M. Adams, and S. Roy, "V2v Wireless communication protocol for rear-end collision avoidance on highways," in *Communications Workshops, 2008. ICC Workshops '08. IEEE International Conference*, (Beijing), pp. 375–379, May 19–23, 2008.

[40] J. Min, J. Ha, S. Yun, I. Kang, and H. Kim, "Secure vehicular communication for safety applications–a measurement study," in *Vehicular Technology Conference, 2008. VTC Spring 2008. IEEE*, pp. 3016–3020, May 11–14, 2008.

[41] A. Iyer, A. Kherani, A. Rao, and A. Karnik, "Secure V2v communications: Performance impact of computational overheads," in *Computer Communications Workshops, 2008. INFOCOM. IEEE Conference*, pp. 1–6, Apr. 13–18, 2008.

[42] H. Ilhan, I. Altunbas, and M. Uysal, "Performance analysis and optimization of relay-assisted vehicle-to-vehicle (v2v) cooperative communication," in *Signal Processing, Communication and Applications Conference, 2008. SIU 2008. IEEE 16th*, pp. 1–4, Apr. 20–22, 2008.

[43] S. Oyama, "Vehicle safety communications: progresses in Japan," in *Vehicular Electronics and Safety, 2008. ICVES 2008. IEEE International Conference*, pp. 241–241, Sept. 22–24, 2008.

[44] J. Santa, A. Moragon, and A. F. Gomez-Skarmeta, "Experimental evaluation of a novel vehicular communication paradigm based on cellular networks," in *Intelligent Vehicles Symposium, 2008 IEEE*, pp. 198–203, June 4–6, 2008.

[45] M. L. Sichitiu and M. Kihl, "Inter-vehicle communication systems: a survey," *IEEE Communications Surveys & Tutorials*, vol. 10, pp. 88–105, second quarter 2008.

[46] B. Atwood, B. Warneke, and K. S. J. Pister, "Preliminary circuits for smart dust," in *Mixed-Signal Design, 2000. SSMSD. 2000 Southwest Symposium*, pp. 87–92, Feb. 27–29, 2000.

[47] D. Estrin, D. Culler, K. Pister, and G. Sukhatme, "Connecting the physical world with pervasive networks," *IEEE Pervasive Computing*, vol. 1, pp. 59–69, Jan./Mar. 2002.

[48] I. F. Akyildiz, W. Su, Y. Sankarasubramaniam, and E. Cayirci, "A survey on sensor networks," *IEEE Communications Magazine*, vol. 40, pp. 102–114, Aug. 2002.

[49] K. S. J. Pister, "Smart dust-hardware limits to wireless sensor networks," in *Distributed Computing Systems, 2003. Proceedings. 23rd International Conference*, May 19–22, 2003.

[50] K. Xu, H. Hassanein, G. Takahara, and Q. Wang, "Relay node deployment strategies in heterogeneous wireless sensor networks: single-hop communication case," in *Global Telecommunications Conference, 2005. GLOBECOM '05. IEEE*, vol. 1, Nov. 28–Dec. 2, 2005.

[51] M. Dohler, A. H. Aghvami, Z. Zhou, Y. Li, and B. Vucetic, "Near-optimum transmit power allocation for space–time block encoded wireless communication systems," in *Communications, IEE Proceedings*, vol. 153, pp. 459–463, June 2, 2006.

[52] M. Dohler, A. Gkelias, and A. H. Aghvami, "Capacity of distributed PHY-layer sensor networks," *IEEE Transactions on Vehicular Technology*, vol. 55, pp. 622–639, Mar. 2006.

[53] M. Dohler, Y. Li, B. Vucetic, A. H. Aghvami, M. Arndt, and D. Barthel, "Performance analysis of distributed space–time block-encoded sensor networks," *IEEE Transactions on Vehicular Technology*, vol. 55, pp. 1776–1789, Nov. 2006.

[54] I. F. Akyildiz, T. Melodia, and K. R. Chowdury, "Wireless multimedia sensor networks: A survey," *IEEE [see also IEEE Personal Communications] Wireless Communications*, vol. 14, pp. 32–39, Dec. 2007.

[55] E. Bjornemo, M. Johansson, and A. Ahlen, "Two hops is one too many in an energy-limited wireless sensor network," in *Acoustics, Speech and Signal Processing, 2007. ICASSP 2007. IEEE International Conference*, vol. 3, (Honolulu, HI), pp. 181–184, Apr. 15–20, 2007.

[56] M. Zhao, M. Ma, and Y. Yang, "Mobile data gathering with multiuser MIMO technique in wireless sensor networks," in *Global Telecommunications Conference, 2007. GLOBECOM '07. IEEE*, pp. 838–842, Nov. 26–30, 2007.

[57] A. del Coso, U. Spagnolini, and C. Ibars, "Cooperative distributed MIMO channels in wireless sensor networks," *IEEE Journal on Selected Areas in Communications*, vol. 25, pp. 402–414, Feb. 2007.

[58] J.-L. Lu, F. Valois, D. Barthel, and M. Dohler, "Low-energy address allocation scheme for wireless sensor networks," in *Personal, Indoor and Mobile Radio Communications, 2007. PIMRC 2007. IEEE 18th International Symposium*, pp. 1–5, Sept. 3–7, 2007.

[59] T. Watteyne, I. Auge-Blum, M. Dohler, and D. Barthel, "Geographic forwarding in wireless sensor networks with loose position-awareness," in *Personal, Indoor and Mobile Radio Communications, 2007. PIMRC 2007. IEEE 18th International Symposium*, (Athens), pp. 1–5, Sept. 3–7, 2007.

[60] M. Dohler, D. Barthel, R. Maraninchi, L. Mounier, S. Aubert, C. Dugas, A. Buhrig, R. Paugnat, M. Renaudin, A. Duda, M. Heusse, and R. Valois, "The ARESA project: facilitating research, development and commercialization of WSNs," in *Sensor, Mesh and Ad Hoc Communications and Networks, 2007. SECON '07. 4th Annual IEEE Communications Society Conference*, pp. 590–599, June 18–21, 2007.

[61] C. Lima and G. T. F. de Abreu, "Game-theoretical relay selection strategy for geographic routing in multi-hop WSNs," in *Positioning, Navigation and Communication, 2008. WPNC 2008. 5th Workshop*, (Hannover), pp. 277–283, Mar. 27–27, 2008.

[62] J. Kim and W. Lee, "Cooperative relaying strategies for multi-hop wireless sensor networks," in *Communication Systems Software and Middleware and Workshops, 2008. COMSWARE 2008. 3rd International Conference*, pp. 103–106, Jan. 6–10, 2008.

[63] R. C. Palat, A. Annamalau, and J. R. Reed, "Cooperative relaying for ad-hoc ground networks using swarm UAVs," in *Military Communications Conference, 2005. MILCOM 2005. IEEE*, pp. 1588–1594, Oct. 17–20, 2005.

[64] P. J. Vincent, M. Tummnala, and J. Mceachen, "A new method for distributing power usage across a sensor network," in *Sensor and Ad Hoc Communications and Networks, 2006. SECON '06. 2006 3rd Annual IEEE Communications Society*, vol. 2, pp. 518–526, Sept. 28–28, 2006.

[65] P. J. Vincent, M. Tummala, and J. McEachen, "An energy-efficient approach for information transfer from distributed wireless sensor systems," in *System of Systems Engineering, 2006 IEEE/SMC International Conference*, Apr. 24–26, 2006.

[66] B. Timus, *Studies on the Viability of Cellular Multihop Networks with Fixed Relays*. PhD Dissertation, KTH, Stockholm, Sweden, 2009.

[67] G. Kramer, I. Maric, and R. D. Yates, *Cooperative Communications*. Foundations and Trends in Networking, Hanover, MA: NOW Publishers Inc., vol. 1, no. 3-4, 2006.

[68] C. Shannon, "A mathematical theory of communication," *The Bell System Technical Journal*, vol. 27, pp. 379–423/623–656, July/Oct. 1948.

[69] W. C. Y. Lee, "Estimate of channel capacity in Rayleigh fading environment," *IEEE Transactions on Vehicular Technology*, vol. 39, pp. 187–189, Aug. 1990.

[70] L. H. Ozarow, S. Shamai, and A. D. Wyner, "Information theoretic considerations for cellular mobile radio," *IEEE Transactions on Vehicular Technology*, vol. 43, pp. 359–378, May 1994.

[71] I. Gradshteyn and I. Ryshik, *Table of Integrals, Series, and Products*. Academia Press, sixth edition, 2000.

[72] A. Sendonaris, E. Erkip, and B. Aazhang, "User cooperation diversity. Part I. System description," *IEEE Transactions on Communications*, vol. 51, pp. 1927–1938, Nov. 2003.

[73] A. Sendonaris, E. Erkip, and B. Aazhang, "Increasing uplink capacity via user cooperation diversity," in *Information Theory, 1998. Proceedings. 1998 IEEE International Symposium*, Aug. 16–21, 1998.

[74] A. Sendonaris, E. Erkip, and B. Aazhang, "User cooperation diversity. Part II. Implementation aspects and performance analysis," *IEEE Transactions on Communications*, vol. 51, pp. 1939–1948, Nov. 2003.

[75] T. E. Hunter and A. Nosratinia, "Performance analysis of coded cooperation diversity," in *Communications, 2003. ICC '03. IEEE International Conference*, vol. 4, pp. 2688–2692, May 11–15, 2003.

[76] T. E. Hunter, S. Sanayei, and A. Nosratinia, "The outage behavior of coded cooperation," in *Information Theory, 2004. ISIT 2004. Proceedings. International Symposium*, June 27–July 2, 2004.

[77] T. E. Hunter, S. Sanayei, and A. Nosratinia, "Outage analysis of coded cooperation," *IEEE Transactions on Information Theory*, vol. 52, pp. 375–391, Feb. 2006.

[78] J. Boyer, D. D. Falconer, and H. Yanikomeroglu, "Cooperative connectivity models for wireless relay networks," *IEEE Transactions on Wireless Communications*, vol. 6, pp. 1992–2000, June 2007.

[79] I. Abou-Faycal and M. Medard, "Optimal uncoded regeneration for binary antipodal signaling," in *Communications, 2004 IEEE International Conference*, vol. 2, pp. 742–746, June 20–24, 2004.

[80] K. S. Gomadam and S. A. Jafar, "Optimal relay functionality for SNR maximization in memoryless relay networks," *IEEE Journal on Selected Areas in Communications*, vol. 25, pp. 390–401, Feb. 2007.

[81] S. Chen, M. A. Beach, and J. P. McGeehan, "Division-free duplex for wireless applications," *Electronics Letters*, vol. 34, pp. 147–148, Jan. 22, 1998.

[82] T. Yu, S. Han, S. Jung, J. Son, Y. Chang, H. Kang, and R. Taori, *Proposal for Full Duplex Relay*. IEEE 802.16 Broadband Wireless Access Working Group, May 2008.

[83] M. Hart, Y. Zhou, P. Zhu, and P. Santos, *Out-of-band Relay Clarification*. IEEE 802.16 Broadband Wireless Access Working Group, Mar. 2008.

[84] AxellWireless, *Axell Wireless Launches a New MultiBand Repeater Platform at CommunicAsia 2008*. Press Room, Axell Wireless, June 2008.

[85] 3GPP, *Universal Terrestrial Radio Access (UTRA) Repeater Planning Guidelines and System Analysis*. TR 25.956, Mar. 2008.

[86] 3GPP, *UTRA Repeater Radio Transmission and Reception*. TR 25.106, Mar. 2008.

[87] Juni, *Juni introduces new range of ICS Repeaters*. Press Room, Juni Australia Pty Ltd, July 2006.

[88] S. W. Kim, Y. T. Lee, S. I. Park, H. M. Eum, J. H. Seo, and H. M. Kim, "Equalization digital on-channel repeater in the single frequency networks," *IEEE Transactions on Broadcasting*, vol. 52, pp. 137–146, June 2006.

[89] A. Gkelias, M. Dohler, V. Friderikos, and A. H. Aghvami, "Wireless multi-hop CSMA/CA with cross-optimised PHY/MAC," in *Global Telecommunications Conference Workshops, 2004. GlobeCom Workshops 2004. IEEE*, pp. 39–43, Nov. 29–Dec. 3, 2004.

[90] A. Ozgur, O. Leveque, and D. Tse, "Hierarchical cooperation achieves linear capacity scaling in ad hoc networks," in *INFOCOM 2007. 26th IEEE International Conference on Computer Communications. IEEE*, pp. 382–390, May 6–12, 2007.

[91] M. Simon and M.-S. Alouini, *Digital Communication over Fading Channels*. John Wiley & Sons, Inc., second edition, 2005.

[92] A. Conti, M. Z. Win, and M. Chiani, "On the inverse symbol-error probability for diversity reception," *IEEE Transactions on Communications*, vol. 51, pp. 753–756, May 2003.

[93] P. Mary, M. Dohler, J. M. Gorce, G. Villemaud, and M. Arndt, "BPSK bit error outage over Nakagami-m fading channels in lognormal shadowing environments," *IEEE Communications Letters*, vol. 11, pp. 565–567, July 2007.

[94] E. van der Meulen, *Transmission of information in a t-terminal discrete memoryless channel*. Department of Statistics, University of California, Berkeley, CA, Technical Report, 1968.

[95] E. van der Meulen, "Three-terminal communication channels," *Adv. Appl. Prob.*, vol. 3, pp. 120–154, 1971.

[96] T. Cover and A. E. Gamal, "Capacity theorems for the relay channel," *IEEE Transactions on Information Theory*, vol. 25, pp. 572–584, Sept. 1979.

[97] T. J. Harrold and A. R. Nix, "Intelligent relaying for future personal communication systems," in *Capacity and Range Enhancement Techniques for the Third Generation Mobile Communications and Beyond (Ref. No. 2000/003), IEE Colloquium*, pp. 1–9, Feb. 11, 2000.

[98] T. J. Harrold and A. R. Nix, "Capacity enhancement using intelligent relaying for future personal communication systems," in *Vehicular Technology Conference, 2000. IEEE VTS-Fall VTC 2000. 52nd*, vol. 5, pp. 2115–2120, Sept. 24–28, 2000.

[99] R. Wang and D. C. Cox, "A step toward ad hoc networks: can relays really improve the performance of cellular networks?" in *Signals, Systems and Computers, 2003. Conference Record of the Thirty-Seventh Asilomar Conference*, vol. 2, pp. 1743–1747, Nov. 9–12, 2003.

[100] J. N. Laneman and G. W. Wornell, "Energy-efficient antenna sharing and relaying for wireless networks," in *Wireless Communications and Networking Conference, 2000. WCNC. 2000 IEEE*, vol. 1, pp. 7–12, Sept. 23–28, 2000.

[101] J. N. Laneman, G. W. Wornell, and D. N. C. Tse, "An efficient protocol for realizing cooperative diversity in wireless networks," in *Information Theory, 2001. Proceedings. 2001 IEEE International Symposium*, June 24–29, 2001.

[102] J. Laneman, *Cooperative Diversity in Wireless Networks: Algorithms and Architectures*. PhD Disseration, MIT, USA, September 2002.

[103] T. E. Hunter and A. Nosratinia, "Coded cooperation under slow fading, fast fading, and power control," in *Signals, Systems and Computers, 2002. Conference Record of the Thirty-Sixth Asilomar Conference*, vol. 1, pp. 118–122, Nov. 3–6, 2002.

[104] T. E. Hunter and A. Nosratinia, "Cooperation diversity through coding," in *Information Theory, 2002. Proceedings. 2002 IEEE International Symposium*, 2002.

[105] A. Stefanov and E. Erkip, "Cooperative coding for wireless networks," in *Mobile and Wireless Communications Network, 2002. 4th International Workshop*, pp. 273–277, Sept. 9–11, 2002.

[106] M. Dohler, *A Novel Statistical Indoor Model*. Diploma in Electrical Engineering, TU-Dresden, Dresden, Germany, 2000.

[107] M. Dohler, A. Aghvami, F. Said, and S. Ghorashi, *Improvements in or Relating to Electronic Data Communication Systems*. Patent Publication No. WO 03/003672, priority date 28 June 2001.

[108] J. N. Laneman and G. W. Wornell, "Distributed space–time coded protocols for exploiting cooperative diversity in wireless networks," in *Global Telecommunications Conference, 2002. GLOBECOM '02. IEEE*, vol. 1, pp. 77–81, Nov. 17–21, 2002.

[109] A. Stefanov and E. Erkip, "On the performance analysis of cooperative space–time coded systems," in *Wireless Communications and Networking, 2003. WCNC 2003. 2003 IEEE*, vol. 2, pp. 729–734, Mar. 20–20, 2003.

[110] A. Stefanov and E. Erkip, "Cooperative space–time coding for wireless networks," in *Information Theory Workshop, 2003. Proceedings. 2003 IEEE*, pp. 50–53, Mar. 31–Apr. 4, 2003.

[111] H. Sato, *Information Transmission through a Channel with Relay*. The Aloha System, University of Hawaii, Honolulu, Tech. Rep. B76-7, March 1976.

[112] T. Cover and J. Thomas, *Elements of Information Theory*. John Wiley & Sons, Inc., 1991.

[113] P. Gupta and P. R. Kumar, "The capacity of wireless networks," *IEEE Transactions on Information Theory*, vol. 46, pp. 388–404, Mar. 2000.

[114] G. Kramer, M. Gastpar, and P. Gupta, "Cooperative strategies and capacity theorems for relay networks," *IEEE Transactions on Information Theory*, vol. 51, pp. 3037–3063, Sept. 2005.

[115] M. Grossglauser and D. N. C. Tse, "Mobility increases the capacity of ad hoc wireless networks," *IEEE/ACM Transactions on Networking*, vol. 10, pp. 477–486, Aug. 2002.

[116] I. Telatar, *Capacity of Multiantenna Gaussian Channels*. AT&T Bell Laboratories Internal Technical Memo, 1995.

[117] I. Telatar, "Capacity of multi-antenna Gaussian channels," *European Transactions on Telecommunication*, vol. 10, pp. 585–595, Nov. 1999.

[118] G. Foschini and M. Gans, "On limits of wireless communications in a fading environment when using multiple antennas," *Wireless Personal Communications (Springer)*, vol. 6, pp. 311–335, Mar. 1998.

[119] G. Foschini, "Layered space–time architecture for wireless communications in a fading environment when using multi-element antennas," *Bell Labs Technical Journal*, vol. 1, pp. 41–59, autumn 1996.

[120] S. M. Alamouti, "A simple transmit diversity technique for wireless communications," *IEEE Journal on Selected Areas in Communications*, vol. 16, pp. 1451–1458, Oct. 1998.

[121] V. Tarokh, H. Jafarkhani, and A. R. Calderbank, "Space–time block codes from orthogonal designs," *IEEE Transactions on Information Theory*, vol. 45, pp. 1456–1467, July 1999.

[122] V. Tarokh, N. Seshadri, and A. R. Calderbank, "Space–time codes for high data rate wireless communication:performance criterion and code construction," *IEEE Transactions on Information Theory*, vol. 44, pp. 744–765, Mar. 1998.

[123] B. Vucetic and J. Yuan, *Space Time Coding*. John Wiley & Sons, Inc., 2003.

[124] P. Larsson, "Large-scale cooperative relaying network with optimal combining under aggregate relay power constraint," in *Proc. Future Tele. Conf.*, 2003.

[125] A. Ozgur, O. Leveque, and D. N. C. Tse, "Hierarchical cooperation achieves optimal capacity scaling in ad hoc networks," *IEEE Transactions on Information Theory*, vol. 53, pp. 3549–3572, Oct. 2007.

[126] Kateyeva, "Report on television satellite relay," in *Radio Technika (in Russian)*, vol. 14, p. 67, Jan. 1959.

[127] M. Handelsman, "Performance equations for a "stationary" passive satellite relay (22,000-mile altitude) for communication," *IRE Transactions on Communications Systems*, vol. 7, pp. 31–37, May 1959.

[128] S. P. Brown and G. F. Senn, "Project SCORE," in *Proceedings of the IRE*, vol. 48, pp. 624–630, Apr. 1960.

[129] W. Hagan, "Communication by polar-orbit satellite relay," *IRE Transactions on Communications Systems*, vol. 8, pp. 250–254, Dec. 1960.

[130] F. E. Bond, C. R. Cahn, and H. F. Meyer, "Interference and channel allocation problems associated with orbiting satellite communication relays," in *Proceedings of the IRE*, vol. 48, pp. 608–612, Apr. 1960.

[131] H. H. Beverage, "The New York–Philadelphia ultra-high-frequency facsimile relay system," in *RCA Rev.*, vol. 1, pp. 15–31, July 1936.

[132] C. W. Hansell, "Radio-relay-systems development by the Radio Corporation of America," in *Proceedings of the IRE*, vol. 33, pp. 156–168, Mar. 1945.

[133] S. R. Saunders, *Antennas and Propagation for Wireless Communication Systems*. New York: John Wiley & Sons, Inc., 1999.

[134] M. Patzold, *Mobile Fading Channels*. John Wiley & Sons Inc., 2002.

[135] R. Vaughan and J. Andersen, *Channels, Propagation and Antennas for Mobile Communications*. The IEE, UK, 2003.

[136] A. Matveev, *Electrodynamics and Relativity Theory*. Moscow Press (in Russian) 1964.

[137] R. C. French, "The effect of fading and shadowing on channel reuse in mobile radio," *IEEE Transactions on Vehicular Technology*, vol. 28, pp. 171–181, Aug. 1979.

[138] C. Namislo, "Analysis of mobile radio slotted ALOHA networks," *IEEE Journal on Selected Areas in Communications*, vol. 2, pp. 583–588, July 1984.

[139] D. J. Goodman and A. A. M. Saleh, "The near/far effect in local ALOHA radio communications," *IEEE Transactions on Vehicular Technology*, vol. 36, pp. 19–27, Feb. 1987.

[140] C. S. Patel, G. L. Stuber, and T. G. Pratt, "Statistical properties of amplify and forward relay fading channels," *IEEE Transactions on Vehicular Technology*, vol. 55, pp. 1–9, Jan. 2006.

[141] H. L. Bertoni, *Radio Propagation for Modern Wireless Systems*. Prentice Hall Professional Technical Reference, 1999.

[142] J. Parsons, *The Mobile Radio Propagation Channel*. second edition, New York: John Wiley & Sons, 2000.

[143] T. Rappaport, *Wireless Communications: Principles and Practice*. Upper Saddle River, NJ, USA: Prentice Hall PTR, 2001.

[144] *Propagation data and prediction methods for the planning of short-range outdoor radiocommunication systems and radio local area networks in the frequency range 300 MHz and 100 GHz*. Rec. ITU-R P. 1411-3.

[145] IEEE 802.15 Working Group for WPAN, http://www.ieee802.org/15.

[146] H. L. Bertoni, W. Honcharenko, L. R. Macel, and H. H. Xia, "UHF propagation prediction for wireless personal communications," *Proceedings of the IEEE*, vol. 82, pp. 1333–1359, Sept. 1994.

[147] N. Patwari, G. D. Durgin, T. S. Rappaport, and R. J. Boyle, "Peer-to-peer low antenna outdoor radio wave propagation at 1.8 GHz," in *Vehicular Technology Conference, 1999 IEEE 49th*, vol. 1, pp. 371–375, May 16–20, 1999.

[148] T. J. Harrold, A. R. Nix, and M. A. Beach, "Propagation studies for mobile-to-mobile communications," in *Vehicular Technology Conference, 2001. VTC 2001 Fall. IEEE VTS 54th*, vol. 3, pp. 1251–1255, Oct. 7–11, 2001.

[149] J. Foerster, *Channel Modeling Sub-committee Report Final (doc.: IEEE 802-15-02/490r1-SG3a)*. submitted to IEEE P802.15 Working Group for Wireless Personal Area Networks (WPANs), Feb. 2002. Available: http://grouper.ieee.org/groups/802/15/pub/2002/Nov02, 2002.

[150] S. S. Ghassemzadeh, R. Jana, C. W. Rice, W. Turin, and V. Tarokh, "A statistical path loss model for in-home UWB channels," in *Ultra Wideband Systems and Technologies, 2002. Digest of Papers. 2002 IEEE Conference*, pp. 59–64, May 21–23, 2002.

[151] I. Z. Kovacs, P. C. F. Eggers, K. Olesen, and L. G. Petersen, "Investigations of outdoor-to-indoor mobile-to-mobile radio communication channels," in *Vehicular Technology Conference, 2002. Proceedings. VTC 2002-Fall. 2002 IEEE 56th*, vol. 1, pp. 430–434, Sept. 24–28, 2002.

[152] J. Maurer, T. Fugen, and W. Wiesbeck, "Narrow-band measurement and analysis of the inter-vehicle transmission channel at 5.2 GHz," in *Vehicular Technology Conference, 2002. VTC Spring 2002. IEEE 55th*, vol. 3, pp. 1274–1278, May 6–9, 2002.

[153] D. B. Green and A. S. Obaidat, "An accurate line of sight propagation performance model for ad-hoc 802.11 wireless LAN (WLAN) devices," in *Communications, 2002. ICC 2002. IEEE International Conference*, vol. 5, pp. 3424–3428, Apr. 28–May 2, 2002.

[154] A. F. Molisch, J. R. Foerster, and M. Pendergrass, "Channel models for ultrawideband personal area networks," *IEEE [see also IEEE Personal Communications] Wireless Communications*, vol. 10, pp. 14–21, Dec. 2003.

[155] S. S. Ghassemzadeh, R. Jana, C. W. Rice, W. Turin, and V. Tarokh, "Measurement and modeling of an ultra-wide bandwidth indoor channel," *IEEE Transactions on Communications*, vol. 52, pp. 1786–1796, Oct. 2004.

[156] G. Acosta, K. Tokuda, and M. A. Ingram, "Measured joint Doppler-delay power profiles for vehicle-to-vehicle communications at 2.4 GHz," in *Global Telecommunications Conference, 2004. GLOBECOM '04. IEEE*, vol. 6, pp. 3813–3817, Nov. 29–Dec. 3, 2004.

[157] Z. Wang, E. K. Tameh, and A. R. Nix, "Statistical peer-to-peer channel models for outdoor urban environments at 2 GHz and 5 GHz," in *Vehicular Technology Conference, 2004. VTC2004-Fall. 2004 IEEE 60th*, vol. 7, pp. 5101–5105, Sept. 26–29, 2004.

[158] A. Fort, C. Desset, J. Ryckaert, P. De Doncker, L. van Biesen, and S. Donnay, "Ultra wide-band body area channel model," in *Communications, 2005. ICC 2005. 2005 IEEE International Conference*, vol. 4, pp. 2840–2844, May 16–20, 2005.

[159] A. F. Molisch, K. Balakrishnan, D. Cassioli, C.-C. Chong, S. Emami, A. Fort, J. Karedal, J. Kunisch, H. Schantz, and K. Siwiak, "A comprehensive model for ultrawideband propagation channels," in *Global Telecommunications Conference, 2005. GLOBECOM '05. IEEE*, vol. 6, Dec. 2–2, 2005.

[160] J. Ryckaert, C. Desset, A. Fort, M. Badaroglu, V. De Heyn, P. Wambacq, G. van der Plas, S. Donnay, B. van Poucke, and B. Gyselinckx, "Ultra-wide-band transmitter for low-power wireless body area networks: design and evaluation," *IEEE Transactions on Circuits and Systems I: Regular Papers*, vol. 52, pp. 2515–2525, Dec. 2005.

[161] A. Molisch, K. Balakrishnan, D. Cassioli, C.-C. Chong, S. Emami, A. Fort, J. Karedal, J. Kunisch, H. Schantz, U. Schuster, and K. Siwiak, *IEEE 802.15.4a channel model – final report*. IEEE802.15.4a working group, 2005, 2005.

[162] K. Konstantinou, S. Kang, and C. Tzaras, "A measurement-based model for mobile-to-mobile UMTS links," in *Vehicular Technology Conference, 2007. VTC2007-Spring. IEEE 65th*, pp. 529–533, Apr. 22–25, 2007.

[163] C. R. Anderson, H. I. Volos, W. C. Headley, F. C. B. Muller, and R. M. Buehrer, "Low antenna ultra wideband propagation measurements and modeling in a forest environment," in *Wireless Communications and Networking Conference, 2008. WCNC 2008. IEEE*, pp. 1229–1234, Mar. 31–Apr. 2008, 2008.

[164] J. N. Laneman, D. N. C. Tse, and G. W. Wornell, "Cooperative diversity in wireless networks: Efficient protocols and outage behavior," *IEEE Transactions on Information Theory*, vol. 50, pp. 3062–3080, Dec. 2004.

[165] M. O. Hasna and M. S. Alouini, "A performance study of dual-hop transmissions with fixed gain relays," in *Acoustics, Speech, and Signal Processing, 2003. Proceedings. (ICASSP '03). 2003 IEEE International Conference*, vol. 4, pp. 189–92, Apr. 6–10, 2003.

[166] R. U. Nabar, H. Bolcskei, and F. W. Kneubuhler, "Fading relay channels: performance limits and space–time signal design," *IEEE Journal on Selected Areas in Communications*, vol. 22, pp. 1099–1109, Aug. 2004.

[167] M. O. Hasna and M. S. Alouini, "A performance study of dual-hop transmissions with fixed gain relays," *IEEE Transactions on Wireless Communications*, vol. 3, pp. 1963–1968, Nov. 2004.

[168] G. K. Karagiannidis, "Performance bounds of multihop wireless communications with blind relays over generalized fading channels," *IEEE Transactions on Wireless Communications*, vol. 5, pp. 498–503, Mar. 2006.

[169] Z. Wang, E. K. Tameh, and A. R. Nix, "Joint shadowing process in urban peer-to-peer radio channels," *IEEE Transactions on Vehicular Technology*, vol. 57, pp. 52–64, Jan. 2008.

[170] G. J. Byers and F. Takawira, "Spatially and temporally correlated MIMO channels: modeling and capacity analysis," *IEEE Transactions on Vehicular Technology*, vol. 53, pp. 634–643, May 2004.

[171] A. S. Akki and F. Haber, "A statistical model of mobile-to-mobile land communication channel," *IEEE Transactions on Vehicular Technology*, vol. 35, pp. 2–7, Feb. 1986.

[172] M. Patzold, B. O. Hogstad, and N. Youssef, "Modeling, analysis, and simulation of MIMO mobile-to-mobile fading channels," *IEEE Transactions on Wireless Communications*, vol. 7, pp. 510–520, Feb. 2008.

[173] A. G. Zajic and G. L. Stubber, "Space–time correlated mobile-to-mobile channels: modelling and simulation," *IEEE Transactions on Vehicular Technology*, vol. 57, pp. 715–726, Mar. 2008.

[174] A. S. Akki, "Statistical properties of mobile-to-mobile land communication channels," *IEEE Transactions on Vehicular Technology*, vol. 43, pp. 826–831, Nov. 1994.

[175] A. S. Akki, "The influence of mobile-to-mobile land communication channel spectrum on the error rate of binary DPSK and NFSK matched filter receivers," *IEEE Transactions on Vehicular Technology*, vol. 43, pp. 832–836, Nov. 1994.

[176] J. Salz and J. H. Winters, "Effect of fading correlation on adaptive arrays in digital mobile radio," *IEEE Transactions on Vehicular Technology*, vol. 43, pp. 1049–1057, Nov. 1994.

[177] M. D. Austin and G. L. Stuber, "Velocity adaptive handoff algorithms for microcellular systems," *IEEE Transactions on Vehicular Technology*, vol. 43, pp. 549–561, Aug. 1994.

[178] A. Abdi, J. A. Barger, and M. Kaveh, "A parametric model for the distribution of the angle of arrival and the associated correlation function and power spectrum at the mobile station," *IEEE Transactions on Vehicular Technology*, vol. 51, pp. 425–434, May 2002.

[179] P. Petrus, J. H. Reed, and T. S. Rappaport, "Effects of directional antennas at the base station on the Doppler spectrum," *IEEE Communications Letters*, vol. 1, pp. 40–42, Mar. 1997.

[180] C. Q. Xu, C. L. Law, and S. Yoshida, "On the Doppler power spectrum at the mobile unit employing a directional antenna," *IEEE Communications Letters*, vol. 5, pp. 13–15, Jan. 2001.

[181] Y. R. Zheng, "A non-isotropic model for mobile-to-mobile fading channel simulations," in *Military Communications Conference, 2006. MILCOM 2006*, pp. 1–7, Oct. 23–25, 2006.

[182] F. Vatalaro and A. Forcella, "Doppler spectrum in mobile-to-mobile communications in the presence of three-dimensional multipath scattering," *IEEE Transactions on Vehicular Technology*, vol. 46, pp. 213–219, Feb. 1997.

[183] A. G. Zajic and G. L. Stuber, "A three dimensional parametric model for wideband MIMO mobile-to-mobile channels," in *Global Telecommunications Conference, 2007. GLOBECOM '07. IEEE*, pp. 3760–3764, Nov. 26–30, 2007.

[184] L.-C. Wang and Y.-H. Cheng, "A statistical mobile-to-mobile Rician fading channel model," in *Vehicular Technology Conference, 2005. VTC 2005-Spring. 2005 IEEE 61st*, vol. 1, pp. 63–67, May 30–June 1, 2005.

[185] L.-C. Wang, W.-C. Liu, and Y.-H. Cheng, "Statistical analysis of a mobile-to-mobile Rician fading channel model," *IEEE Transactions on Vehicular Technology*, vol. 58, pp. 32–38, Jan. 2009.

[186] M. Patzold, U. Killat, and F. Laue, "An extended Suzuki model for land mobile satellite channels and its statistical properties," *IEEE Transactions on Vehicular Technology*, vol. 47, pp. 617–630, May 1998.

[187] A. G. Zajic and G. L. Stuber, "Three-dimensional modeling and simulation of wideband MIMO mobile-to-mobile channels," *IEEE Transactions on Wireless Communications*, vol. 8, pp. 1260–1275, Mar. 2009.

[188] H. Kang, G. L. Stuber, T. G. Pratt, and M. A. Ingram, "Studies on the capacity of MIMO systems in mobile-to-mobile environment," in *Wireless Communications and Networking Conference, 2004. WCNC. 2004 IEEE*, vol. 1, pp. 363–368, Mar. 21–25, 2004.

[189] M. Patzold, B. O. Hogstad, N. Youssef, and D. Kim, "A MIMO mobile-to-mobile channel model: Part I–The reference model," in *Personal, Indoor and Mobile Radio Communications, 2005. PIMRC 2005. IEEE 16th International Symposium*, vol. 1, pp. 573–578, Sept. 11–14, 2005.

[190] B. O. Hogstad, M. Patzold, N. Youssef, and D. Kim, "A MIMO mobile-to-mobile channel model: Part II–The simulation model," in *Personal, Indoor and Mobile Radio Communications, 2005. PIMRC 2005. IEEE 16th International Symposium*, vol. 1, pp. 562–567, Sept. 11–14, 2005.

[191] A. G. Zajic and G. L. Stuber, "Space–time correlated MIMO mobile-to-mobile channels," in *Personal, Indoor and Mobile Radio Communications, 2006 IEEE 17th International Symposium*, pp. 1–5, Sept. 11–14, 2006.

[192] L.-C. Wang and Y.-H. Cheng, "Modelling and capacity analysis of MIMO Rician fading channels for mobile-to-mobile," in *Vehicular Technology Conference, 2005. VTC-2005-Fall. 2005 IEEE 62nd*, vol. 2, pp. 1279–1283, Sept. 25–28, 2005.

[193] X. Cheng, C.-X. Wang, D. I. Laurenson, H.-H. Chen, and A. V. Vasilakos, "A generic geometrical-based MIMO mobile-to-mobile channel model," in *Wireless Communications and Mobile Computing Conference, 2008. IWCMC '08. International*, pp. 1000–1005, Aug. 6–8, 2008.

[194] X. Cheng, C.-X. Wang, D. I. Laurenson, and A. V. Vasilakos, "Second order statistics of non-isotropic mobile-to-mobile Ricean fading channels," in *Communications, 2009. ICC '09. IEEE International Conference*, June 2009.

[195] X. Cheng, C.-X. Wang, D. I. Laurenson, S. Salous, and A. V. Vasilakos, "An adaptive geometry-based stochastic model for MIMO mobile-tomobile channels," *IEEE Transactions on Wireless Communications*, under revision.

[196] A. Chelli and M. Patzold, "A MIMO mobile-to-mobile channel model derived from a geometric street scattering model," in *Wireless Communication Systems, 2007. ISWCS 2007. 4th International Symposium*, pp. 792–797, Oct. 17–19, 2007.

[197] A. G. Zajic and G. L. Stuber, "A three-dimensional MIMO mobile-to-mobile channel model," in *Wireless Communications and Networking Conference, 2007.WCNC 2007. IEEE*, pp. 1883–1887, Mar. 11–15, 2007.

[198] A. G. Zajic and G. L. Stuber, "Three-dimensional modeling, simulation, and capacity analysis of space–time correlated mobile-to-mobile channels," *IEEE Transactions on Vehicular Technology*, vol. 57, pp. 2042–2054, July 2008.

[199] J. Maurer, T. Fugen, T. Schafer, and W. Wiesbeck, "A new inter-vehicle communications (IVC) channel model," in *Vehicular Technology Conference, 2004. VTC2004-Fall. 2004 IEEE 60th*, vol. 1, pp. 9–13, Sept. 26–29, 2004.

[200] G. Acosta and M. A. Ingram, "Model development for the wideband expressway vehicle-to-vehicle 2.4 GHz channel," in *Wireless Communications and Networking Conference, 2006. WCNC 2006. IEEE*, vol. 3, pp. 1283–1288, Apr. 3–6, 2006.

[201] A. G. Zajic and G. L. Stuber, "Statistical properties of wideband MIMO mobile-to-mobile channels," in *Wireless Communications and Networking Conference, 2008. WCNC 2008. IEEE*, pp. 763–768, Mar. 31–Apr. 2008, 2008.

[202] A. G. Zajic, G. L. Stuber, T. G. Pratt, and S. T. Nguyen, "Wideband MIMO mobile-to-mobile channels: Geometry-based statistical modeling with experimental verification," *IEEE Transactions on Vehicular Technology*, vol. 58, pp. 517–534, Feb. 2009.

[203] A. Chelli and M. Patzold, "A wideband multiple-cluster MIMO mobile-to-mobile channel model based on the geometrical street model," in *Personal, Indoor and Mobile Radio Communications, 2008. PIMRC 2008. IEEE 19th International Symposium*, pp. 1–6, Sept. 15–18, 2008.

[204] J. D. Parsons and A. M. D. Turkmani, "Characterisation of mobile radio signals: model description," in *Communications, Speech and Vision, IEE Proceedings I*, vol. 138, pp. 549–556, Dec. 1991.

[205] A. Kuchar, J. P. Rossi, and E. Bonek, "Directional macro-cell channel characterization from urban measurements," *IEEE Transactions on Antennas and Propagation*, vol. 48, pp. 137–146, Feb. 2000.

[206] R. Wang and D. Cox, "Channel modeling for ad hoc mobile wireless networks," in *Vehicular Technology Conference, 2002. VTC Spring 2002. IEEE 55th*, vol. 1, pp. 21–25, May 6–9, 2002.

[207] C. S. Patel, S. L. Stuber, and T. G. Pratt, "Simulation of Rayleigh faded mobile-to-mobile communication channels," in *Vehicular Technology Conference, 2003. VTC 2003-Fall. 2003 IEEE 58th*, vol. 1, pp. 163–167, Oct. 6–9, 2003.

[208] C. S. Patel, G. L. Stuber, and T. G. Pratt, "Simulation of Rayleigh-faded mobile-to-mobile communication channels," *IEEE Transactions on Communications*, vol. 53, pp. 1876–1884, Nov. 2005.

[209] A. G. Zajic and G. L. Stuber, "Efficient simulation of Rayleigh fading with enhanced de-correlation properties," *IEEE Transactions on Wireless Communications*, vol. 5, pp. 1866–1875, July 2006.

[210] A. G. Zajic and G. L. Stuber, "Simulation models for MIMO Mobile-to-mobile channels," in *Military Communications Conference, 2006. MILCOM 2006*, pp. 1–7, Oct. 23–25, 2006.

[211] A. G. Zajic and G. L. Stuber, "A new simulation model for mobile-to-mobile Rayleigh fading channels," in *Wireless Communications and Networking Conference, 2006. WCNC 2006. IEEE*, vol. 3, pp. 1266–1270, Apr. 3–6, 2006.

[212] A. G. Zajic and G. L. Stuber, "3-D simulation models for wideband MIMO mobile-to-mobile channels," in *Military Communications Conference, 2007. MILCOM 2007. IEEE*, pp. 1–5, Oct. 29–31, 2007.

[213] A. G. Zajic and G. L. Stuber, "3-D MIMO mobile-to-mobile channel simulation," in *Mobile and Wireless Communications Summit, 2007. 16th IST*, pp. 1–5, July 1–5, 2007.

[214] Y. Liu, J. Zhang, and Y. R. Zheng, "Simulation of doubly-selective compound K fading channels for mobile-to-mobile communications," in *Wireless Communications and Networking Conference, 2008. WCNC 2008. IEEE*, pp. 1020–1025, Mar. 31–Apr. 2008, 2008.

[215] A. G. Zajic, G. L. Stuber, T. G. Pratt, and S. Nguyen, "Statistical modeling and experimental verification of wideband MIMO mobile-to-mobile channels in highway environments," in *Personal, Indoor and Mobile Radio Communications, 2008. PIMRC 2008. IEEE 19th International Symposium*, pp. 1–5, Sept. 15–18, 2008.

[216] M. Patzold, U. Killat, F. Laue, and Y. Li, "On the statistical properties of deterministic simulation models for mobile fading channels," *IEEE Transactions on Vehicular Technology*, vol. 47, pp. 254–269, Feb. 1998.

[217] C. A. Gutierrez-Diaz-de Leon and M. Patzold, "Sum-of-sinusoids-based simulation of flat fading wireless propagation channels under non-isotropic scattering conditions," in *Global Telecommunications Conference, 2007. GLOBECOM '07. IEEE*, pp. 3842–3846, Nov. 26–30, 2007.

[218] K. V. Mardia and P. E. Jupp, *Directional Statistics*. John Wiley & Sons, Inc., 1999.

[219] P. Harley, "Short distance attenuation measurements at 900 MHz and 1.8 GHz using low antenna heights for microcells," *IEEE Journal on Selected Areas in Communications*, vol. 7, pp. 5–11, Jan. 1989.

[220] I. Davis, J. S. and J. P. M. G. Linnartz, "Vehicle to vehicle RF propagation measurements," in *Signals, Systems and Computers, 1994. 1994 Conference Record of the Twenty-Eighth Asilomar Conference*, vol. 1, pp. 470–474, Oct. 31–Nov. 2, 1994.

[221] G. D. Durgin, V. Kukshya, and T. S. Rappaport, "Joint angle and delay spread statistics for 1920 MHz peer-to-peer wireless channels," in *Antennas and Propagation Society International Symposium, 2001. IEEE*, vol. 2, pp. 182–185, July 8–13, 2001.

[222] G. D. Durgin, V. Kukshya, and T. S. Rappaport, "Wideband measurements of angle and delay dispersion for outdoor and indoor peer-to-peer radio channels at 1920 MHz," *IEEE Transactions on Antennas and Propagation*, vol. 51, pp. 936–944, May 2003.

[223] Mobile Virtual Centre of Excellence, UK. [Online]. Available: http://www.mobilevce.com.

[224] P. Hafezi, D. Wedge, M. A. Beach, and M. Lawton, "Propagation measurements at 5.2 GHz in commercial and domestic environments," in *Personal, Indoor and Mobile Radio Communications, 1997. "Waves of the Year 2000". PIMRC '97., The 8th IEEE International Symposium*, vol. 2, pp. 509–513, Sept. 1–4, 1997.

[225] C. A. Tan, A. R. Nix, and M. A. Beach, "Dynamic spatial-temporal propagation measurement and super-resolution channel characterisation at 5.2 GHz in a corridor environment," in *Vehicular Technology Conference, 2002. Proceedings. VTC 2002-Fall. 2002 IEEE 56th*, vol. 2, pp. 797–801, Sept. 24–28, 2002.

[226] C.-C. Chong, D. I. Laurenson, and S. McLaughlin, "Statistical characterization of the 5.2 GHz wideband directional indoor propagation channels with clustering and correlation properties," in *Vehicular Technology Conference, 2002. Proceedings. VTC 2002-Fall. 2002 IEEE 56th*, vol. 1, pp. 629–633, Sept. 24–28, 2002.

[227] C.-C. Chong, D. I. Laurenson, C. M. Tan, S. McLaughlin, M. A. Beach, and A. R. Nix, "Modelling the dynamic evolution of paths of the wideband indoor propagation channels using the M-step, 4-state Markov model," in *Personal Mobile Communications Conference, 2003. 5th European (Conf. Publ. No. 492)*, pp. 181–185, Apr. 22–25, 2003.

[228] C.-C. Chong, C.-M. Tan, D. I. Laurenson, S. McLaughlin, M. A. Beach, and A. R. Nix, "A new statistical wideband spatio-temporal channel model for 5-GHz band WLAN systems," *IEEE Journal on Selected Areas in Communications*, vol. 21, pp. 139–150, Feb. 2003.

[229] C.-C. Chong, D. I. Laurenson, and S. McLaughlin, "A wideband dynamic spatio-temporal Markov channel model for typical indoor propagation environments," in *Vehicular Technology Conference, 2003. VTC 2003-Spring. The 57th IEEE Semiannual*, vol. 1, pp. 6–10, Apr. 22–25, 2003.

[230] C.-C. Chong, C.-M. Tan, D. I. Laurenson, S. McLaughlin, M. A. Beach, and A. R. Nix, "A novel wideband dynamic directional indoor channel model based on a Markov process," *IEEE Transactions on Wireless Communications*, vol. 4, pp. 1539–1552, July 2005.

[231] J. Kivinen, X. Zhao, and P. Vainikainen, "Empirical characterization of wideband indoor radio channel at 5.3 GHz," *IEEE Transactions on Antennas and Propagation*, vol. 49, pp. 1192–1203, Aug. 2001.

[232] I. Kovacs, *Radio Channel Characterisation for Private Mobile Radio Systems: Mobile-to-Mobile Radio Link Investigations*. PhD Dissertation, Aalborg University, Denmark, 2002.

[233] Y. Wang, I. B. Bonev, J. O. Nielsen, I. Z. Kovacs, and G. F. Pedersen, "Characterization of the indoor multiantenna body-to-body radio channel," *IEEE Transactions on Antennas and Propagation*, vol. 57, pp. 972–979, Apr. 2009.

[234] J. Maurer, T. Fugen, and W. Wiesbeck, "Physical layer simulations of IEEE802.11a for vehicle-to-vehicle communications," in *Vehicular Technology Conference, 2005. VTC-2005-Fall. 2005 IEEE 62nd*, vol. 3, pp. 1849–1853, Sept. 25–28, 2005.

[235] T. M. Schafer, J. Maurer, J. von Hagen, and W. Wiesbeck, "Experimental characterization of radio wave propagation in hospitals," *IEEE Transactions on Electromagnetic Compatibility*, vol. 47, pp. 304–311, May 2005.

[236] G. Acosta-Marum and M. A. Ingram, "Six time- and frequency-selective empirical channel models for vehicular wireless LANs," *IEEE Vehicular Technology Magazine*, vol. 2, pp. 4–11, Dec. 2007.

[237] G. Acosta-Marum and M. A. Ingram, "Six time- and frequency-selective empirical channel models for vehicular wireless LANs," in *Vehicular Technology Conference, 2007. VTC-2007 Fall. 2007 IEEE 66th*, pp. 2134–2138, Sept. 30–Oct. 2007, 2007.

[238] G. Acosta-Marum, *Measurement, Modeling, and OFDM Synchronization for the Wideband Mobile-to-Mobile Channel*. PhD Dissertation, Georgia Institute of Technology, USA, 2007.

[239] P. Kyritsi, P. Eggers, R. Gall, and J. M. Lourenco, "Measurement based investigation of cooperative relaying," in *Vehicular Technology Conference, 2006. VTC-2006 Fall. 2006 IEEE 64th*, pp. 1–5, Sept. 25–28, 2006.

[240] P. Kyritsi, P. Popovski, P. Eggers, Y. Wang, D. A. Khan, A. E. Bouaziz, B. Pietrarca, and G. Sasso, "Cooperative transmission: a reality check using experimental data," in *Vehicular Technology Conference, 2007. VTC2007-Spring. IEEE 65th*, pp. 2281–2285, Apr. 22–25, 2007.

[241] I. Sen and D. W. Matolak, "Vehicle–vehicle channel models for the 5-GHz band," *IEEE Transactions on Intelligent Transportation Systems*, vol. 9, pp. 235–245, June 2008.

[242] D. W. Matolak, I. Sen, W. Xiong, and N. T. Yaskoff, "5 GHZ wireless channel characterization for vehicle to vehicle communications," in *Military Communications Conference, 2005. MILCOM 2005. IEEE*, pp. 3016–3022, Oct. 17–20, 2005.

[243] D. W. Matolak, "Channel modeling for vehicle-to-vehicle communications," in *Communications Magazine, IEEE*, vol. 46, pp. 76–83, May 2008.

[244] IST-4-027756 Wireless World Initiative Radio, WINNER. [Online]. Available: www.ist-winner.org.

[245] D1.1.2 WINNER II Channel Models, IST-4-027756 WINNER. [Online]. Available: http://www.ist-winner.org/WINNER2-Deliverables/D1.1.2.zip.

[246] European Cooperation in Science and Technology, COST. [Online]. Available: http://www.cost.esf.org/domains_actions/ict.

[247] Y. Chen and V. K. Dubey, "Dynamic simulation model of indoor wideband directional channels," *IEEE Transactions on Vehicular Technology*, vol. 55, pp. 417–430, Mar. 2006.

[248] A. Zaidi and L. Vandendorpe, "Sensitivity of achievable rates for the relay channel. application to relaying with channel estimation error," in *Acoustics, Speech and Signal Processing, 2007. ICASSP 2007. IEEE International Conference*, vol. 3, pp. 497–500, Apr. 15–20, 2007.

[249] J. Chen, X. Yu, and C. C. J. Kuo, "V-BLAST receiver for MIMO relay networks with imperfect CSI," in *Global Telecommunications Conference, 2007. GLOBECOM '07. IEEE*, pp. 571–575, Nov. 26–30, 2007.

[250] C. Tellambura and Z. Zhang, "OFDMA based cooperative relay networks," in *Signals, Systems and Computers, 2008 42nd Asilomar Conference*, pp. 2014–2018, Oct. 26–29, 2008.

[251] B. Niu, M. C. Beluri, Z. Lin, and P. Chitrapu, "Relay assisted cooperative OSTBC communication with SNR imbalance and channel estimation errors," in *Vehicular Technology Conference, 2009. VTC Spring 2009. IEEE 69th*, pp. 1–5, Apr. 26–29, 2009.

[252] A. El-Keyi and B. Champagne, "Collaborative uplink transmit beamforming with robustness against channel estimation errors," *IEEE Transactions on Vehicular Technology*, vol. 58, pp. 126–139, Jan. 2009.

[253] B. Yi, S. Wang, and S. Y. Kwon, "On MIMO relay with finite-rate feedback and imperfect channel estimation," in *Global Telecommunications Conference, 2007. GLOBECOM '07. IEEE*, pp. 3878–3882, Nov. 26–30, 2007.

[254] Z. Zhang, W. Zhang, and C. Tellambura, "Cooperative OFDM channel estimation with frequency offsets," in *Global Telecommunications Conference, 2008. IEEE GLOBECOM 2008. IEEE*, pp. 1–5, Nov. 30–Dec. 2008.

[255] Z. Zhang, W. Zhang, and C. Tellambura, "Cooperative OFDM channel estimation in the presence of frequency offsets," *IEEE Transactions Vehicular Technology*, in press.

[256] I. Kammoun and M. Siala, "MAP channel estimation for Alamouti-based cooperative networks," in *Personal, Indoor and Mobile Radio Communications, 2008. PIMRC 2008. IEEE 19th International Symposium*, pp. 1–5, Sept. 15–18, 2008.

[257] F. Gao, T. Cui, and A. Nallanathan, "Training signal design for channel estimation in decode and forward relay networks," in *Communications, 2008. ICC '08. IEEE International Conference*, (Beijing), pp. 4306–4310, May 19–23, 2008.

[258] F. Gao, T. Cui, and A. Nallanathan, "Maximum likelihood channel estimation in decode-and-forward relay networks," in *Information Theory, 2008. ISIT 2008. IEEE International Symposium*, pp. 1233–1237, July 6–11, 2008.

[259] F. Gao, T. Cui, and A. Nallanathan, "Optimal training design for channel estimation in decode-and-forward relay networks with individual and total power constraints," *IEEE Transactions on Signal Processing*, vol. 56, pp. 5937–5949, Dec. 2008.

[260] A. S. Lalos, A. A. Rontogiannis, and K. Berberidis, "Channel estimation techniques for half-duplex cooperative communication systems," in *Wireless Pervasive Computing, 2008. ISWPC 2008. 3rd International Symposium*, pp. 591–595, May 7–9, 2008.

[261] J. Zhao, M. Kuhn, A. Wittneben, and G. Bauch, "Self-interference aided channel estimation in two-way relaying systems," in *Global Telecommunications Conference, 2008. IEEE GLOBECOM 2008. IEEE*, pp. 1–6, Nov. 30–Dec. 2008, 2008.

[262] J. Zhao, M. Kuhn, A. Wittneben, and G. Bauch, "Achievable rates of MIMO bidirectional broadcast channels with self-interference aided channel estimation," in *Wireless Communications and Networking Conference, 2009. WCNC 2009. IEEE*, pp. 1–6, Apr. 5–8, 2009.

[263] J. Boyer, D. D. Falconer, and H. Yanikomeroglu, "On the aggregate SNR of amplified relaying channels," in *Global Telecommunications Conference, 2004. GLOBECOM '04. IEEE*, vol. 5, pp. 3394–3398, Nov. 29–Dec. 3, 2004.

[264] W. Honcharenko, H. L. Bertoni, and J. L. Dailing, "Bilateral averaging over receiving and transmitting areas for accurate measurements of sector average signal strength inside buildings," *IEEE Transactions on Antennas and Propagation*, vol. 43, pp. 508–512, May 1995.

[265] V. Erceg, S. J. Fortune, J. Ling, J. Rustako, A. J., and R. A. Valenzuela, "Comparisons of a computer-based propagation prediction tool with experimental data collected in urban microcellular environments," *IEEE Journal on Selected Areas in Communications*, vol. 15, pp. 677–684, May 1997.

[266] J. Salo, H. M. El-Sallabi, and P. Vainikainen, "The distribution of the product of independent Rayleigh random variables," *IEEE Transactions on Antennas and Propagation*, vol. 54, pp. 639–643, Feb. 2006.

[267] B. Talha and M. Patzold, "On the statistical properties of double rice channels," in *WPMC2007*, pp. 517–522, Dec. 3–6, 2007.

[268] I. Gradshteyn and M. Ryzhik, *Table of Integrals, Series, and Products*. Sixth edition, Academic Press, 2000.

[269] V. S. Adamchik and O. I. Marichev, "The algorithm for calculating integrals of hypergeometric type functions and its realization in REDUCE system," in *International Conference Symbolic and Algebraic Computing*, pp. 212–224, 1990.

[270] G. K. Karagiannidis, N. C. Sagias, and P. T. Mathiopoulos, "Nakagami: A novel stochastic model for cascaded fading channels," *IEEE Transactions on Communications*, vol. 55, pp. 1453–1458, Aug. 2007.

[271] M. O. Hasna and M. S. Alouini, "Performance analysis of two-hop relayed transmissions over Rayleigh fading channels," in *Vehicular Technology Conference, 2002. Proceedings. VTC 2002-Fall. 2002 IEEE 56th*, vol. 4, pp. 1992–1996, Sept. 24–28, 2002.

[272] P. A. Anghel and M. Kaveh, "Exact symbol error probability of a cooperative network in a Rayleigh-fading environment," *IEEE Transactions on Wireless Communications*, vol. 3, pp. 1416–1421, Sept. 2004.

[273] M. O. Hasna and M. S. Alouini, "Outage probability of multihop transmission over Nakagami fading channels," *IEEE Communications Letters*, vol. 7, pp. 216–218, May 2003.

[274] M. O. Hasna and M. S. Alouini, "Harmonic mean and end-to-end performance of transmission systems with relays," *IEEE Transactions on Communications*, vol. 52, pp. 130–135, Jan. 2004.

[275] A. Ribeiro, X. Cai, and G. B. Giannakis, "Symbol error probabilities for general cooperative links," *IEEE Transactions on Wireless Communications*, vol. 4, pp. 1264–1273, May 2005.

[276] G. K. Karagiannidis, T. A. Tsiftsis, and R. K. Mallik, "Bounds for multihop relayed communications in Nakagami-m fading," *IEEE Transactions on Communications*, vol. 54, pp. 18–22, Jan. 2006.

[277] G. K. Karagiannidis, "Moments-based approach to the performance analysis of equal gain diversity in Nakagami-m fading," *IEEE Transactions on Communications*, vol. 52, pp. 685–690, May 2004.

[278] J. Proakis, *Digital Communications*. Fourth edition, McGraw Hill, 2000.

[279] T. A. Tsiftsis, G. K. Karagiannidis, P. T. Mathiopoulos, and S. A. Kotsopoulos, "Nonregenerative dual-hop cooperative links with selection diversity," *EURASIP Journal on Wireless Communications and Networking*, vol. 2006, Article ID 17862, 2006. doi:10.1155/WCN/2006/17862.

[280] B. Talha and M. Patzold, "A geometrical channel model for MIMO mobile-to-mobile fading channels in cooperative networks," in *Vehicular Technology Conference, 2009. VTC Spring 2009. IEEE 69th*, pp. 1–7, Apr. 26–29, 2009.

[281] Z. Hadzi-Velkov, N. Zlatanov, and G. K. Karagiannidis, "On the second order statistics of the multihop rayleigh fading channel," *IEEE Transactions on Communications*, vol. 57, pp. 1815–1823, June 2009.

[282] B. Talha and M. Patzold, "On the statistical properties of mobile-to-mobile fading channels in cooperative networks under line-of-sight conditions," in *WPMC2007*, pp. 388–393, Dec. 3–6, 2007.

[283] B. Talha and M. Patzold, "A novel amplify-and-forward relay channel model for mobile-to-mobile fading channels under line-of-sight conditions," in *Personal, Indoor and Mobile Radio Communications, 2008. PIMRC 2008. IEEE 19th International Symposium*, pp. 1–6, Sept. 15–18, 2008.

[284] B. Talha and M. Patzold, "Level-crossing rate and average duration of fades of the envelope of mobile-to-mobile fading channels in cooperative networks under line-of-sight conditions," in *Global Telecommunications Conference, 2008. IEEE GLOBECOM 2008. IEEE.*, pp. 1–6, Nov. 30–Dec. 2008, 2008.

[285] N. Zlatanov, Z. Hadzi-Velkov, and G. Karagiannidis, "Level crossing rate and average fade duration of the double Nakagami-m random process and application in MIMO keyhole fading channels," *IEEE Communications Letters*, vol. 12, pp. 822–824, Nov. 2008.

[286] J. C. S. S. Filho and M. Yacoub, "On the simulation and correlation properties of phase-envelope Nakagami fading processes," *IEEE Transactions on Communications*, vol. 57, pp. 906–909, Apr. 2009.

[287] Y. Jia and A. Vosoughi, "Training design for information rate maximization over amplify and forward relay channels," in *Signals, Systems and Computers, 2008 42nd Asilomar Conference*, pp. 1448–1452, Oct. 26–29, 2008.

[288] C. Lee, J. Joung, and Y. H. Lee, "A pilot emitting amplify-and-forward relay and its application to hop-by-hop beamforming," in *Personal, Indoor and Mobile Radio Communications, 2008. PIMRC 2008. IEEE 19th International Symposium*, pp. 1–5, Sept. 15–18, 2008.

[289] P. Lioliou and M. Viberg, "Least-squares based channel estimation for MIMO relays," in *Smart Antennas, 2008. WSA 2008. International ITG Workshop*, pp. 90–95, Feb. 26–27, 2008.

[290] G. Wang and C. Tellambura, "Super-imposed pilot-aided channel estimation and power allocation for relay systems," in *Wireless Communications and Networking Conference, 2009. WCNC 2009. IEEE*, pp. 1–6, Apr. 5–8, 2009.

[291] D. Neves, C. Ribeiro, A. Silva, and A. Gameiro, "Channel estimation schemes for OFDM relay-assisted systems," in *Vehicular Technology Conference, 2009. VTC Spring 2009. IEEE 69th*, pp. 1–5, Apr. 26–29, 2009.

[292] F. S. Tabataba, P. Sadeghi, T. Lamahewa, and M. R. Pakravan, "Statistical properties of amplify and forward relay links with channel estimation errors," in *Communications Theory Workshop, 2009. AusCTW 2009. Australian*, pp. 44–49, Feb. 4–7, 2009.

[293] B. Gedik and M. Uysal, "Impact of imperfect channel estimation on the performance of amplify-and-forward relaying," *IEEE Transactions on Wireless Communications*, vol. 8, pp. 1468–1479, Mar. 2009.

[294] Y. Wu and M. Patzold, "Parameter optimization for amplify-and-forward relaying with imperfect channel estimation," in *Vehicular Technology Conference, 2009. VTC Spring 2009. IEEE 69th*, pp. 1–5, Apr. 26–29, 2009.

[295] H. Muhaidat, M. Uysal, and R. Adve, "Pilot-symbol-assisted detection scheme for distributed orthogonal space–time block coding," *IEEE Transactions on Wireless Communications*, vol. 8, pp. 1057–1061, Mar. 2009.

[296] H. Dogan, "Maximum *a posteriori* channel estimation for cooperative diversity orthogonal frequency division multiplexing systems in amplify-and-forward mode," *IET Communications*, vol. 3, pp. 501–511, Apr. 2009.

[297] C. S. Patel and G. L. Stuber, "Channel estimation for amplify and forward relay based cooperation diversity systems," *IEEE Transactions on Wireless Communications*, vol. 6, pp. 2348–2356, June 2007.

[298] K. Kim, H. Kim, and H. Park, "OFDM channel estimation for the amplify-and-forward cooperative channel," in *Vehicular Technology Conference, 2007. VTC2007-Spring. IEEE 65th*, pp. 1642–1646, Apr. 22–25, 2007.

[299] T. Cui, F. Gao, and A. Nallanathan, "Optimal training design for channel estimation in amplify and forward relay networks," in *Global Telecommunications Conference, 2007. GLOBECOM '07. IEEE*, pp. 4015–4019, Nov. 26–30, 2007.

[300] F. Gao, T. Cui, and A. Nallanathan, "On channel estimation and optimal training design for amplify and forward relay networks," *IEEE Transactions on Wireless Communications*, vol. 7, pp. 1907–1916, May 2008.

[301] M. Herdin and G. Auer, "Pilot design for OFDM amplify-and-forward with chunk reordering," in *Wireless Communications and Networking Conference, 2007.WCNC 2007. IEEE*, pp. 1400–1405, Mar. 11–15, 2007.

[302] K. S. Woo, H. Lee, H. K. Chung, and Y. S. Cho, "MMSE channel estimation for OFDM systems with transparent multi-hop relays," in *Personal, Indoor and Mobile Radio Communications, 2007. PIMRC 2007. IEEE 18th International Symposium*, pp. 1–4, Sept. 3–7, 2007.

[303] K. S. Woo, H. I. Yoo, Y. J. Kim, H. Lee, H. K. Chung, and Y. S. Cho, "Synchronization and channel estimation for OFDM systems with transparent multi-hop relays," in *Vehicular Technology Conference, 2007. VTC2007-Spring. IEEE 65th*, pp. 2414–2418, Apr. 22–25, 2007.

[304] K. S. Woo, H. I. Yoo, and Y. J. Kim, "Channel estimation for OFDM systems with transparent multi-hop relays," *IEICE Transactions on Communications*, vol. 90, no. 6, pp. 1555–1558, 2007.

[305] F. Gao, R. Zhang, and Y.-C. Liang, "On channel estimation for amplify-and-forward two-way relay networks," in *Global Telecommunications Conference, 2008. IEEE GLOBECOM 2008. IEEE*, pp. 1–5, Nov. 30–Dec. 2008.

[306] F. Gao, R. Zhang, and Y. C. Liang, "Channel estimation for OFDM modulated two-way relay networks," *IEEE Transactions on Signal Processing*: Accepted for publication.

[307] A. S. Lalos, A. A. Rontogiannis, and K. Berberidis, "Channel estimation techniques in amplify and forward relay networks," in *Signal Processing Advances in Wireless Communications, 2008. SPAWC 2008. IEEE 9th Workshop*, pp. 446–450, July 6–9, 2008.

[308] F. Liu, Z. Chen, X. Zhang, and D. Yang, "Channel estimation for amplify and forward relay in OFDM system," in *Wireless Communications, Networking and Mobile Computing, 2008. WiCOM '08. 4th International Conference*, pp. 1–4, Oct. 12–14, 2008.

[309] B. Gedik and M. Uysal, "Two channel estimation methods for amplify-and-forward relay networks," in *Electrical and Computer Engineering, 2008. CCECE 2008. Canadian Conference*, pp. 000615–000618, May 4–7, 2008.

[310] A. S. Behbahani and A. Eltawil, "On channel estimation and capacity for amplify and forward relay networks," in *Global Telecommunications Conference, 2008. IEEE GLOBECOM 2008. IEEE*, pp. 1–5, Nov. 30–Dec. 2008, 2008.

[311] T.-H. Pham, Y.-C. Liang, and A. Nallanathan, "Optimal training sequence design for bi-directional relay networks," in *Global Telecommunications Conference, 2008. IEEE GLOBECOM 2008. IEEE*, pp. 1–5, Nov. 30–Dec. 2008, 2008.

[312] T.-H. Pham, Y.-C. Liang, A. Nallanathan, and G. H. Krishna, "On the design of optimal training sequence for bi-directional relay networks," *IEEE Signal Processing Letters*, vol. 16, pp. 200–203, Mar. 2009.

[313] A. Firag and L. M. Garth, "Adaptive decoding and equalization for time reversal-space time block-coded cooperative diversity systems," in *Communications, 2008. ICC '08. IEEE International Conference*, pp. 531–537, May 19–23, 2008.

[314] K. Yan, S. Ding, Y. Qiu, Y. Wang, and H. Liu, "A low-complexity LMMSE channel estimation method for OFDM-based cooperative diversity systems with multiple amplify-and-forward relays," *EURASIP Journal on Wireless Communications and Networking*, vol. 2008, Article ID 149803, 2008.

[315] F. Hu, H. Zhang, X. You, H. Wang, and G. Wu, "Low complexity channel estimation for novel bi-directional relaying schemes," in *Vehicular Technology Conference, 2009. VTC Spring 2009. IEEE 69th*, pp. 1–5, Apr. 26–29, 2009.

[316] F. Roemer and M. Haardt, "Tensor-based channel estimation (TENCE) for two-way relaying with multiple antennas and spatial reuse," in *Acoustics, Speech and Signal Processing, 2009. ICASSP 2009. IEEE International Conference*, pp. 3641–3644, Apr. 19–24, 2009.

[317] L. B. Thiagarajan, S. Sun, and T. Q. S. Quek, "Joint carrier frequency offset and channel estimation in OFDM based non-regenerative wireless relay networks," in *Acoustics, Speech and Signal Processing, 2009. ICASSP 2009. IEEE International Conference*, pp. 2569–2572, Apr. 19–24, 2009.

[318] G. S. Rajan and B. S. Rajan, "Leveraging coherent distributed space–time codes for noncoherent communication in relay networks via training," *IEEE Transactions on Wireless Communications*, vol. 8, pp. 683–688, Feb. 2009.

[319] H. A. David and H. N. Nagaraja, *Order Statistics*. 3rd edition, Hoboken, New Jersey: John Wiley & Sons, 2003.

[320] P. M. Shankar, "Macrodiversity and microdiversity in correlated shadowed fading channels," *IEEE Transactions on Vehicular Technology*, vol. 58, pp. 727–732, Feb. 2009.

[321] M. Yuksel and E. Erkip, "Diversity in relaying protocols with amplify and forward," in *Global Telecommunications Conference, 2003. GLOBECOM '03. IEEE*, vol. 4, pp. 2025–2029, Dec. 1–5, 2003.

[322] O. Munoz, A. Agustin, and J. Vidal, "Cellular capacity gains of cooperative MIMO transmission in the downlink," in *Communications, 2004 International Zurich Seminar*, pp. 22–26, 2004.

[323] A. Ribeiro, X. Cai, and G. B. Giannakis, "Symbol error probabilities for general cooperative links," in *Communications, 2004 IEEE International Conference*, vol. 6, pp. 3369–3373, June 20–24, 2004.

[324] Q. Zhao and H. Li, "Performance of a differential modulation scheme with wireless relays in Rayleigh fading channels," in *Signals, Systems and Computers, 2004. Conference Record of the Thirty-Eighth Asilomar Conference*, vol. 1, pp. 1198–1202, Nov. 7–10, 2004.

[325] J. Boyer, D. D. Falconer, and H. Yanikomeroglu, "Multihop diversity in wireless relaying channels," *IEEE Transactions on Communications*, vol. 52, pp. 1820–1830, Oct. 2004.

[326] G. K. Karagiannidis, D. A. Zogas, N. C. Sagias, T. A. Tsiftsis, and P. T. Mathiopoulos, "Multihop communications with fixed-gain relays over generalized fading channels," in *Global Telecommunications Conference, 2004. GLOBECOM '04. IEEE*, vol. 1, pp. 36–40, Nov. 29–Dec. 3, 2004.

[327] G. K. Karagiannidis, T. A. Tsiftsis, R. K. Mallik, N. C. Sagias, and S. A. Kotsopoulos, "Closed-form bounds for multihop relayed communications in Nakagami-m fading," in *Communications, 2005. ICC 2005. 2005 IEEE International Conference*, vol. 4, pp. 2362–2366, May 16–20, 2005.

[328] Q. Zhao and H. Li, "Performance of differential modulation with wireless relays in Rayleigh fading channels," *IEEE Communications Letters*, vol. 9, pp. 343–345, Apr. 2005.

[329] T. Himsoon, W. Su, and K. J. R. Liu, "Differential transmission for amplify-and-forward cooperative communications," *IEEE Signal Processing Letters*, vol. 12, pp. 597–600, Sept. 2005.

[330] Q. Zhao and H. Li, "Performance analysis of an amplify-based differential modulation for wireless relay networks under Nakagami-m fading channels," in *Signal Processing Advances in Wireless Communications, 2005 IEEE 6th Workshop*, pp. 211–215, June 5–8, 2005.

[331] Y. Ding, J.-K. Zhang, and K. M. Wong, "The pairwise error probability and precoder design for amplify-and-forward half-duplex cooperative system," in *Computational Advances in Multi-Sensor Adaptive Processing, 2005 1st IEEE International Workshop*, pp. 149–152, Dec. 13–13, 2005.

[332] R. Annavajjala, P. C. Cosman, and L. B. Milstein, "On the performance of optimum noncoherent amplify-and-forward reception for cooperative diversity," in *Military Communications Conference, 2005. MILCOM 2005. IEEE*, pp. 3280–3288, Oct. 17–20, 2005.

[333] H. Mheidat and M. Uysal, "Impact of receive diversity on the performance of amplify-and-forward relaying under APS and IPS power constraints," *IEEE Communications Letters*, vol. 10, pp. 468–470, June 2006.

[334] T. Himsoon, W. Su, and K. J. R. Liu, "Differential modulation for multi-node amplify-and-forward wireless relay networks," in *Wireless Communications and Networking Conference, 2006. WCNC 2006. IEEE*, vol. 2, pp. 1195–1200, Apr. 3–6, 2006.

[335] Y. Zhao, R. Adve, and T. J. Lim, "Symbol error rate of selection amplify-and-forward relay systems," *IEEE Communications Letters*, vol. 10, pp. 757–759, Nov. 2006.

[336] H. Mheidat, M. Uysal, and N. Al-Dhahir, "Time-reversal space–time equalization for amplify-and-forward relaying," in *Communications, 2006. ICC '06. IEEE International Conference*, vol. 4, pp. 1705–1711, June 2006.

[337] Q. Zhao and H. Li, "Distributed modulation for cooperative wireless communications," in *Information Sciences and Systems, 2006 40th Annual Conference*, pp. 1068–1072, Mar. 22–24, 2006.

[338] S. Yang and J. C. Belfiore, "A novel two-relay three-slot amplify-and-forward cooperative scheme," in *Information Sciences and Systems, 2006 40th Annual Conference*, pp. 1329–1334, Mar. 22–24, 2006.

[339] C. Moolkum, C. Pirak, and S. Jitapunkul, "Tight approximate bounds on bit error rate for full-rate differentially encoded cooperative communications," in *Communications and Information Technologies, 2006. ISCIT '06. International Symposium*, pp. 384–388, Oct. 18–Sept. 2006, 2006.

[340] C. Moolkum, C. Pirak, and S. Jitapunkul, "On performance of full-rate differentially encoded cooperative communications:bit error rate bounds," in *TENCON 2006. 2006 IEEE Region 10 Conference*, pp. 1–4, Nov. 14–17, 2006.

[341] M. Elfituri and W. Hamouda, "Performance analysis of a new transmission scheme for multi-relay channels," in *Signal Processing Systems Design and Implementation, 2006. SIPS '06. IEEE Workshop*, pp. 34–38, Oct. 2–4, 2006.

[342] B. Maham and A. Hjorungnes, "Amplify-and-forward space–time coded cooperation via incremental relaying," in *Wireless Communication Systems, 2007. ISWCS 2007. 4th International Symposium*, pp. 407–411, Oct. 17–19, 2007.

[343] Y. Zhang, Y. Ma, and R. Tafazolli, "Tighter bounds of symbol error probability for amplify-and-forward cooperative protocol over Rayleigh fading channels," in *Personal, Indoor and Mobile Radio Communications, 2007. PIMRC 2007. IEEE 18th International Symposium*, pp. 1–5, Sept. 3–7, 2007.

[344] Y. Ding, J.-K. Zhang, and K. M. Wong, "The amplify-and-forward half-duplex cooperative system: pairwise error probability and precoder design," *IEEE Transactions on Signal Processing*, vol. 55, pp. 605–617, Feb. 2007.

[345] S. Ikki and M. H. Ahmed, "Performance analysis of cooperative diversity wireless networks over Nakagami-m fading channel," *IEEE Communications Letters*, vol. 11, pp. 334–336, Apr. 2007.

[346] H. A. Suraweera and G. K. Karagiannidis, "Closed-form error analysis of the non-identical Nakagami-m relay fading channel," *IEEE Communications Letters*, vol. 12, pp. 259–261, Apr. 2008.

[347] D. S. Michalopoulos and G. K. Karagiannidis, "PHY-layer fairness in amplify and forward cooperative diversity systems," *IEEE Transactions on Wireless Communications*, vol. 7, pp. 1073–1082, Mar. 2008.

[348] M. Di Renzo, F. Graziosi, and F. Santucci, "On the performance of CSI-assisted cooperative communications over generalized fading channels," in *Communications, 2008. ICC '08. IEEE International Conference*, pp. 1001–1007, May 19–23, 2008.

[349] M. Di Renzo, F. Graziosi, and F. Santucci, "Performance of cooperative multi-hop wireless systems over log-normal fading channels," in *Global Telecommunications Conference, 2008. IEEE GLOBECOM 2008. IEEE*, pp. 1–6, Nov. 30–Dec. 2008, 2008.

[350] G. Farhadi and N. C. Beaulieu, "On the performance of amplify-and-forward cooperative systems with fixed gain relays," *IEEE Transactions on Wireless Communications*, vol. 7, pp. 1851–1856,/May 2008.

[351] M. Safari and M. Uysal, "Cooperative diversity over log-normal fading channels: performance analysis and optimization," *IEEE Transactions on Wireless Communications*, vol. 7, pp. 1963–1972,/May 2008.

[352] B. Barua, H. Ngo, and H. Shin, "On the SEP of cooperative diversity with opportunistic relaying," *IEEE Communications Letters*, vol. 12, pp. 727–729, Oct. 2008.

[353] Q. Wang, D. Fan, Y. H. Lin, J. Chen, and Z. Zhu, "Design of BS transceiver for IEEE 802.16e OFDMA mode," in *Acoustics, Speech and Signal Processing, 2008. ICASSP 2008. IEEE International Conference*, pp. 1513–1516, Mar. 31–Apr. 2008, 2008.

[354] Y.-C. Liang and R. Zhang, "Optimal analogue relaying with multi-antennas for physical layer network coding," in *Communications, 2008. ICC '08. IEEE International Conference*, pp. 3893–3897, May 19–23, 2008.

[355] G. Farhadi and N. C. Beaulieu, "Amplify-and-forward cooperative systems with fixed gain relays," in *Communications, 2008. ICC '08. IEEE International Conference*, pp. 4300–4305, May 19–23, 2008.

[356] M. Ju and I.-M. Kim, "BER analysis for noncoherent FSK in two-hop relay networks," in *Electrical and Computer Engineering, 2008. CCECE 2008. Canadian Conference*, pp. 000223–000226, May 4–7, 2008.

[357] S. Gunawardena and N. Rajatheva, "SEP formula for single relay selection in a multiple relay environment over Rayleigh fading channels," in *Electrical and Computer Engineering, 2008. CCECE 2008. Canadian Conference*, pp. 001931–001936, May 4–7, 2008.

[358] J. P. K. Chu, A. W. Eckford, and R. S. Adve, "Characterization of relay channels using the Bhattacharyya parameter," in *Global Telecommunications Conference, 2008. IEEE GLOBECOM 2008. IEEE*, pp. 1–5, Nov. 30–Dec. 2008, 2008.

[359] T. Issariyakul and V. Krishnamurthy, "Amplify-and-forward cooperative diversity wireless networks: model, analysis, and monotonicity properties," *IEEE/ACM Transactions on Networking*, vol. 17, pp. 225–238, Feb. 2009.

[360] M. Di Renzo, F. Graziosi, and F. Santucci, "On the performance of cooperative systems with blind relays over Nakagami-m and Weibull fading," in *Wireless Communications and Networking Conference, 2009. WCNC 2009. IEEE*, pp. 1–5, Apr. 5–8, 2009.

[361] H. A. Suraweera, D. S. Michalopoulos, and G. K. Karagiannidis, "Performance of distributed diversity systems with a single amplify-and-forward relay," *IEEE Transactions on Vehicular Technology*, vol. 58, pp. 2603–2608, June 2009.

[362] H. A. Suraweera, D. S. Michalopoulos, and G. K. Karagiannidis, "Semi-blind amplify-and-forward with partial relay selection," *Electronics Letters*, vol. 45, pp. 317–319, Mar. 12, 2009.

[363] H. A. Suraweera, R. H. Y. Louie, Y. Li, G. K. Karagiannidis, and B. Vucetic, "Two hop amplify-and-forward transmission in mixed Rayleigh and Rician fading channels," *IEEE Communications Letters*, vol. 13, pp. 227–229, Apr. 2009.

[364] H. Suraweera, G. Karagiannidis, and P. Smith, "Performance analysis of the dual-hop asymmetric fading channel," *IEEE Transactions on Wireless Communications*, vol. 8, pp. 2783–2788, June 2009.

[365] S. S. Ikki and M. H. Ahmed, "Performance of cooperative diversity using equal gain combining (EGC) over Nakagami-m fading channels," *IEEE Transactions on Wireless Communications*, vol. 8, pp. 557–562, Feb. 2009.

[366] S. S. Ikki and M. H. Ahmed, "On the performance of amplify-and-forward cooperative diversity with the Nth best-relay selection scheme," in *Communications, 2009. ICC '09. IEEE International Conference*, pp. 1–6, June 14–18, 2009.

[367] A. Chowdhery and R. K. Mallik, "Linear detection for the nonorthogonal amplify and forward protocol," *IEEE Transactions on Wireless Communications*, vol. 8, pp. 826–835, Feb. 2009.

[368] B. Maham and A. Hjorungnes, "Performance analysis of amplify-and-forward opportunistic relaying in Rician fading," *IEEE Signal Processing Letters*, vol. 16, pp. 643–646, Aug. 2009.

[369] K.-S. Hwang, Y.-C. Ko, and M. S. Alouini, "Performance analysis of incremental opportunistic relaying over identically and non-identically distributed cooperative paths," *IEEE Transactions on Wireless Communications*, vol. 8, pp. 1953–1961, Apr. 2009.

[370] S. Chen, W. Wang, X. Zhang, and D. Zhao, "Performance of amplify-and-forward MIMO relay channels with transmit antenna selection and maximal-ratio combining," in *Wireless Communications and Networking Conference, 2009. WCNC 2009. IEEE*, pp. 1–6, Apr. 5–8, 2009.

[371] B. Maham and A. Hjorungnes, "Asymptotic performance analysis of amplify-and-forward cooperative networks in a Nakagami-m fading environment," *IEEE Communications Letters*, vol. 13, pp. 300–302, May 2009.

[372] Y. Ding and M. Uysal, "Multi-relay cooperative OFDM with amplify-and-forward relaying," in *Radio and Wireless Symposium, 2009. RWS '09. IEEE*, pp. 614–617, Jan. 18–22, 2009.

[373] X. Lei, N. C. Beaulieu, and P. Fan, "Precise MGF performance analysis of amplify-and-forward cooperative diversity in Nakagami-m fading," in *Information Theory, 2009. CWIT 2009. 11th Canadian Workshop*, pp. 13–16, May 13–15, 2009.

[374] M. S. Fazel, R. Hoshyar, and R. Tafazolli, "Flexible amplify and forward relaying protocol with optimized duplexing," in *Vehicular Technology Conference, 2009. VTC Spring 2009. IEEE 69th*, pp. 1–5, Apr. 26–29, 2009.

[375] V. Ganwani, B. K. Dey, G. V. V. Sharma, S. N. Merchant, and U. B. Desai, "Performance analysis of amplify and forward based cooperative diversity in MIMO relay channels," in *Vehicular Technology Conference, 2009. VTC Spring 2009. IEEE 69th*, pp. 1–5, Apr. 26–29, 2009.

[376] M. Torabi, W. Ajib, and D. Haccoun, "Performance analysis of amplify-and-forward cooperative networks with relay selection over Rayleigh fading channels," in *Vehicular Technology Conference, 2009. VTC Spring 2009. IEEE 69th*, pp. 1–5, Apr. 26–29, 2009.

[377] H. Ilhan, I. Altunbas, and M. Uysal, "Performance analysis of relay systems under cascaded Nakagami channels," in *Signal Processing and Communications Applications Conference, 2009. SIU 2009. IEEE 17th*, pp. 912–915, Apr. 9–11, 2009.

[378] S. Chen, W. Wang, and X. Zhang, "Performance analysis of multiuser diversity in cooperative multi-relay networks under Rayleigh-fading channels," *IEEE Transactions on Wireless Communications*, vol. 8, pp. 3415–3419, July 2009.

[379] E. Kocan, M. Pejanovic-Djurisic, D. S. Michalopoulos, and G. K. Karagiannidis, "BER performance of OFDM amplify-and-forward relaying system with subcarrier permutation," in *Wireless Communication, Vehicular Technology, Information Theory and Aerospace and Electronic Systems Technology, 2009. Wireless VITAE 2009. 1st International Conference*, pp. 252–256, May 17–20, 2009.

[380] D. Senaratne and C. Tellambura, "Unified exact performance analysis of two hop amplify and forward relaying," in *Communications, 2009. ICC '09. IEEE International Conference*, June 2009.

[381] D. Senaratne and C. Tellambura, "Unified exact performance analysis of two hop amplify and forward relaying in Nakagami fading," *IEEE Transactions on Vehicular Technology*, submitted for publication.

[382] K. Cho and D. Yoon, "On the general BER expression of one- and two-dimensional amplitude modulations," *IEEE Transactions on Communications*, vol. 50, pp. 1074–1080, July 2002.

[383] Z. Wang and G. B. Giannakis, "A simple and general parameterization quantifying performance in fading channels," *IEEE Transactions on Communications*, vol. 51, pp. 1389–1398, Aug. 2003.

[384] I. Hammerstroem, M. Kuhn, B. Rankov, and A. Wittneben, "Space–time processing for cooperative relay networks," in *Vehicular Technology Conference, 2003. VTC 2003-Fall. 2003 IEEE 58th*, vol. 1, pp. 404–408, Oct. 6–9, 2003.

[385] S. Yang and J.-C. Belfiore, "Optimal spacetime codes for the MIMO amplify-and-forward cooperative channel," *IEEE Transactions on Information Theory*, vol. 53, pp. 647–663, Feb. 2007.

[386] H. Mheidat, M. Uysal, and N. Al-Dhahir, "Equalization techniques for distributed space–time block codes with amplify-and-forward relaying," *IEEE Transactions on Signal Processing*, vol. 55, pp. 1839–1852,/May 2007.

[387] H. Mheidat and M. Uysal, "Space–time coded cooperative diversity with multiple-antenna nodes," in *Information Theory, 2007. CWIT '07. 10th Canadian Workshop*, pp. 17–20, June 6–8, 2007.

[388] K. Yan, S. Ding, Y. Qiu, Y. Wang, and H. Liu, "A simple Alamouti space–time transmission scheme for asynchronous cooperative communications over frequency-selective channels," in *Advanced Communication Technology, 2008. ICACT 2008. 10th International Conference*, vol. 3, pp. 1569–1572, Feb. 17–20, 2008.

[389] J. He and P. Y. Kam, "Adaptive cooperative space–time block coding with amplify-and-forward Strategy," in *Vehicular Technology Conference, 2008. VTC Spring 2008. IEEE*, pp. 1236–1240, May 11–14, 2008.

[390] Y. Han, S. H. Ting, C. K. Ho, and W. H. Chin, "High rate two-way amplify-and-forward half-duplex relaying with OSTBC," in *Vehicular Technology Conference, 2008. VTC Spring 2008. IEEE*, pp. 2426–2430, May 11–14, 2008.

[391] J. He and P. Y. Kam, "Exact bit error probability of cooperative space–time block coding with amplify-and-forward strategy," in *Communications, 2008. ICC '08. IEEE International Conference*, pp. 4591–4595, May 19–23, 2008.

[392] D. Sreedhar, A. Chockalingam, and B. S. Rajan, "High-rate, single-symbol decodable distributed STBCs for partially-coherent cooperative networks," in *Information Theory, 2008. ISIT 2008. IEEE International Symposium*, pp. 2538–2542, July 6–11, 2008.

[393] M. Badr and J. C. Belfiore, "Distributed space time codes for the amplify-and-forward multiple-access relay channel," in *Information Theory, 2008. ISIT 2008. IEEE International Symposium*, pp. 2543–2547, July 6–11, 2008.

[394] N. H. Vien, H. H. Nguyen, and T. Le-Ngoc, "Distributed space–time block coded OFDM with subcarrier grouping," in *Vehicular Technology Conference, 2008. VTC 2008-Fall. IEEE 68th*, pp. 1–5, Sept. 21–24, 2008.

[395] A. Firag and L. M. Garth, "Adaptive joint decoding and equalization for space–time block-coded amplify-and-forward relaying systems," *IEEE Transactions on Signal Processing*, vol. 57, pp. 1163–1176, Mar. 2009.

[396] X. Xu, B. Zheng, and J. Cui, "Performance analysis of distributed space–time block code in cooperative MIMO network," in *Intelligent Information Technology Application, 2008. IITA '08. Second International Symposium*, vol. 3, pp. 635–639, Dec. 20–22, 2008.

[397] N. Eltayeb, S. Kassim, and J. Chambers, "A two-hop amplify-and-forward scheme for extended orthogonal space time coding in cooperative networks," in *Information Theory and Its Applications, 2008. ISITA 2008. International Symposium*, pp. 1–5, Dec. 7–10, 2008.

[398] B. Maham, A. Hjorungnes, and G. Abreu, "Distributed GABBA space–time codes in amplify-and-forward relay networks," *IEEE Transactions on Wireless Communications*, vol. 8, pp. 2036–2045, Apr. 2009.

[399] D. Sreedhar, A. Chockalingam, and B. Rajan, "Single-symbol ML decodable distributed STBCs for partially coherent cooperative networks," *IEEE Transactions on Wireless Communications*, vol. 8, pp. 2672–2681, May 2009.

[400] S. Muhaidat, P. Ho, and M. Uysal, "Distributed differential space–time coding for broadband cooperative networks," in *Vehicular Technology Conference, 2009. VTC Spring 2009. IEEE 69th*, pp. 1–6, Apr. 26–29, 2009.

[401] Y. Song, H. Shin, and E.-K. Hong, "MIMO cooperative diversity with scalar-gain amplify-and-forward relaying," *IEEE Transactions on Communications*, vol. 57, pp. 1932–1938, July 2009.

[402] B. Hassibi and B. M. Hochwald, "High-rate codes that are linear in space and time," *IEEE Transactions on Information Theory*, vol. 48, pp. 1804–1824, July 2002.

[403] Y. Jing and B. Hassibi, "Distributed space–time coding in wireless relay networks," *IEEE Transactions on Wireless Communications*, vol. 5, pp. 3524–3536, Dec. 2006.

[404] M. Dohler, B. Rassool, and H. Aghvami, "Performance evaluation of STTCs for virtual antenna arrays," in *Vehicular Technology Conference, 2003. VTC 2003-Spring. The 57th IEEE Semiannual*, vol. 1, pp. 57–60, Apr. 22–25, 2003.

[405] O. Canpolat, M. Uysal, and M. M. Fareed, "Analysis and design of distributed space-time trellis codes with amplify-and-forward relaying," *IEEE Transactions on Vehicular Technology*, vol. 56, pp. 1649–1660, July 2007.

[406] S. K. Hong and J. M. Chung, "Improved design criterion for distributed space–time trellis codes with AF relaying," *Electronics Letters*, vol. 44, pp. 1212–1213, Sept. 25, 2008.

[407] Z. Zhong, S. Zhu, G. Lv, and J. Xu, "Analysis and design of distributed space–time trellis code with asynchronous amplify-and-forward relaying," in *Global Telecommunications Conference, 2008. IEEE GLOBECOM 2008. IEEE*, pp. 1–6, Nov. 30–Dec. 2008, 2008.

[408] J. Yuan, Z. Chen, B. Vucetic, and W. Firmanto, "Performance and design of space–time coding in fading channels," *IEEE Transactions on Communications*, vol. 51, pp. 1991–1996, Dec. 2003.

[409] Y. Zhang and M. G. Amin, "Distributed turbo-BLAST for cooperative wireless networks," in *Sensor Array and Multichannel Processing, 2006. Fourth IEEE Workshop*, pp. 452–455, July 12–14, 2006.

[410] A. Darmawan, S. W. Kim, and H. Morikawa, "LLR-based ordering in amplify-and-forward cooperative spatial multiplexing system," in *Wireless Communications and Networking Conference, 2007. WCNC 2007. IEEE*, pp. 819–824, Mar. 11–15, 2007.

[411] A. Darmawan, S. W. Kim, and H. Morikawa, "Amplify-and-forward scheme in cooperative spatial multiplexing," in *Mobile and Wireless Communications Summit, 2007. 16th IST*, pp. 1–5, July 1–5, 2007.

[412] P. Elia, K. Vinodh, M. Anand, and P. V. Kumar, "D-MG tradeoff and optimal codes for a class of AF and DF cooperative communication protocols," in *Information Theory, 2007. ISIT 2007. IEEE International Symposium*, pp. 681–685, June 24–29, 2007.

[413] P. Lusina, R. Schober, and L. Lampe, "Diversity-multiplexing trade-off of the hybrid non-orthogonal amplify-decode and forward protocol," in *Information Theory, 2008. ISIT 2008. IEEE International Symposium*, pp. 2375–2379, July 6–11, 2008.

[414] J. Li, W. Chen, C. Zhao, X. Yang, and X. Wang, "On the throughput-reliability tradeoff analysis in amplify-and-forward cooperative channels," in *Communications, 2008. ICC '08. IEEE International Conference*, pp. 1034–1038, May 19–23, 2008.

[415] Z. Zhang, W. Zhang, and C. Tellambura, "Improved OFDMA uplink frequency offset estimation via cooperative relaying: AF or Dcf?," in *Communications, 2008. ICC '08. IEEE International Conference*, pp. 3313–3317, May 19–23, 2008.

[416] R. H. Y. Louie, Y. Li, and B. Vucetic, "Zero forcing in general two hop relay networks," *IEEE Transactions on Vehicular Technology*, in press.

[417] J. A. Tague and C. I. Caldwell, "Expectations of useful complex wishart forms," *Multidimensional Syst. Signal Process.*, vol. 5, no. 3, pp. 263–279, 1994.

[418] T. K. Y. Lo, "Maximum ratio transmission," *IEEE Transactions on Communications*, vol. 47, pp. 1458–1461, Oct. 1999.

[419] A. Narula, M. J. Lopez, M. D. Trott, and G. W. Wornell, "Efficient use of side information in multiple-antenna data transmission over fading channels," *IEEE Journal on Selected Areas in Communications*, vol. 16, pp. 1423–1436, Oct. 1998.

[420] G. Barriac, R. Mudumbai, and U. Madhow, "Distributed beamforming for information transfer in sensor networks," in *Information Processing in Sensor Networks, 2004. IPSN 2004. Third International Symposium*, pp. 81–88, Apr. 26–27, 2004.

[421] R. Mudumbai, J. Hespanha, U. Madhow, and G. Barriac, "Scalable feedback control for distributed beamforming in sensor networks," in *Information Theory, 2005. ISIT 2005. Proceedings. International Symposium*, pp. 137–141, Sept. 4–9, 2005.

[422] R. Mudumbai, G. Barriac, and U. Madhow, "On the feasibility of distributed beamforming in wireless networks," *IEEE Transactions on Wireless Communications*, vol. 6, pp. 1754–1763, May 2007.

[423] Y. Jing and H. Jafarkhani, "Network beamforming using relays with perfect channel information," in *Acoustics, Speech and Signal Processing, 2007. ICASSP 2007. IEEE International Conference*, vol. 3, pp. 473–476, Apr. 15–20, 2007.

[424] V. Havary-Nassab, S. Shahbazpanahi, A. Grami, and Z.-Q. Luo, "Network beamforming based on second order statistics of the channel state information," in *Acoustics, Speech and Signal Processing, 2008. ICASSP 2008. IEEE International Conference*, pp. 2605–2608, Mar. 31–Apr. 2008, 2008.

[425] Z. Ding, W. H. Chin, and K. K. Leung, "Distributed beamforming and power allocation for cooperative networks," *IEEE Transactions on Wireless Communications*, vol. 7, pp. 1817–1822,/May 2008.

[426] Y. Jing and H. Jafarkhani, "Network beamforming with channel means and covariances at relays," in *Communications, 2008. ICC '08. IEEE International Conference*, pp. 3743–3747, May 19–23, 2008.

[427] Y. Jing and H. Jafarkhani, "Network beamforming using relays with perfect channel information," *IEEE Transactions on Information Theory*, vol. 55, pp. 2499–2517, June 2009.

[428] H. Chen, A. B. Gershman, and S. Shahbazpanahi, "Filter-and-forward distributed beamforming for relay networks in frequency selective fading channels," in *Acoustics, Speech and Signal Processing, 2009. ICASSP 2009. IEEE International Conference*, pp. 2269–2272, Apr. 19–24, 2009.

[429] V. Havary-Nassab, S. Shahbazpanahi, and A. Grami, "Optimal network beamforming for bi-directional relay networks," in *Acoustics, Speech and Signal Processing, 2009. ICASSP 2009. IEEE International Conference*, pp. 2277–2280, Apr. 19–24, 2009.

[430] N. Khajehnouri and A. H. Sayed, "A distributed MMSE relay strategy for wireless sensor networks," in *Signal Processing Advances in Wireless Communications, 2005 IEEE 6th Workshop*, pp. 796–800, June 5–8, 2005.

[431] N. Khajehnouri and A. H. Sayed, "Distributed MMSE relay strategies for wireless sensor networks," *IEEE Transactions on Signal Processing*, vol. 55, pp. 3336–3348,/July 2007.

[432] A. S. Behbahani, R. Merched, and A. Eltawil, "On signal processing methods for MIMO relay architectures," in *Global Telecommunications Conference, 2007. GLOBECOM '07. IEEE*, pp. 2967–2971, Nov. 26–30, 2007.

[433] R. Krishna, Z. Xiong, and S. Lambotharan, "A cooperative MMSE relay strategy for wireless sensor networks," *IEEE Signal Processing Letters*, vol. 15, pp. 549–552, 2008.

[434] A. S. Behbahani, R. Merched, and A. M. Eltawil, "Optimizations of a MIMO relay network," *IEEE Transactions on Signal Processing*, vol. 56, pp. 5062–5073,/Oct. 2008.

[435] V. Wong and C. Leung, "A transmit power control scheme for improving performance in a mobile packet radio system," *IEEE Transactions on Vehicular Technology*, vol. 43, pp. 174–180, Feb. 1994.

[436] J. Zander, "Radio resource management in future wireless networks: requirements and limitations," *IEEE Communications Magazine*, vol. 35, pp. 30–36, Aug. 1997.

[437] A. Bletsas, A. Khisti, D. P. Reed, and A. Lippman, "A simple cooperative diversity method based on network path selection," *IEEE Journal on Selected Areas in Communications*, vol. 24, pp. 659–672, Mar. 2006.

[438] R. H. Y. Louie, Y. Li, and B. Vucetic, "Performance analysis of beamforming in two hop amplify and forward relay networks," in *Communications, 2008. ICC '08. IEEE International Conference*, pp. 4311–4315, May 19–23, 2008.

[439] R. Louie, Y. Li, H. A. Suraweera, and B. Vucetic, "Beamforming with Antenna Correlation in Two Hop Amplify and Forward Relay Networks," in *Wireless Communications and Networking Conference, 2009. WCNC 2009. IEEE*, pp. 1–6, Apr. 5–8, 2009.

[440] R. H. Y. Louie, Y. Li, H. A. Suraweera, and B. Vucetic, "Performance analysis of beamforming in two hop amplify and forward relay networks with antenna correlation," *IEEE Transactions on Wireless Communications*, vol. 8, pp. 3132–3141, June 2009.

[441] M. O. Hasna and M. S. Alouini, "End-to-end performance of transmission systems with relays over Rayleigh-fading channels," *IEEE Transactions on Wireless Communications*, vol. 2, pp. 1126–1131, Nov. 2003.

[442] T. Wang, A. Cano, G. B. Giannakis, and J. N. Laneman, "High-performance cooperative demodulation with decode-and-forward relays," *IEEE Transactions on Communications*, vol. 55, pp. 1427–1438, July 2007.

[443] M. R. Souryal and H. You, "Quantize-and-forward relaying with M-ary phase shift keying," in *Wireless Communications and Networking Conference, 2008. WCNC 2008. IEEE*, pp. 42–47, Mar. 31–Apr. 2008, 2008.

[444] Y. Qi, R. Hoshyar, and R. Tafazolli, "A novel quantization scheme in compress-and-forward relay system," in *Vehicular Technology Conference, 2009. VTC Spring 2009. IEEE 69th*, pp. 1–5, Apr. 26–29, 2009.

[445] R. Hu and J. Li, "Practical compress-forward in user cooperation: Wyner-Ziv cooperation," in *Information Theory, 2006 IEEE International Symposium on*, (Seattle, WA), pp. 489–493, July 9–14, 2006.

[446] Z. Liu, V. Stankovic, and Z. Xiong, "Wyner-Ziv coding for the half-duplex relay channel," in *Acoustics, Speech, and Signal Processing, 2005. Proceedings. (ICASSP '05). IEEE International Conference*, vol. 5, Mar. 18–23, 2005.

[447] M. Uppal, Z. Liu, V. Stankovic, and Z. Xiong, "Compress-forward coding with BPSK modulation for the half-duplex gaussian relay channel," *IEEE Transactions on Signal Processing* in press.

[448] R. Hu and J. Li, "Exploiting Slepian-Wolf codes in wireless user cooperation," in *Signal Processing Advances in Wireless Communications, 2005 IEEE 6th Workshop*, pp. 275–279, June 5–8, 2005.

[449] M. Kuhn, J. Wagner, and A. Wittneben, "Cooperative processing for the WLAN uplink," in *Wireless Communications and Networking Conference, 2008. WCNC 2008. IEEE*, pp. 1294–1299, Mar. 31–Apr. 2008.

[450] G. Kramer, M. Gastpar, and P. Gupta, "Cooperative strategies and capacity theorems for relay networks," *IEEE Transactions on Information Theory*, vol. 51, pp. 3037–3063, Sept. 2005.

[451] A. Wyner and J. Ziv, "The rate-distortion function for source coding with side information at the decoder," *IEEE Transactions on Information Theory*, vol. 22, pp. 1–10, Jan. 1976.

[452] R. Liu, P. Spasojevic, and E. Soljanin, "Incremental redundancy cooperative coding for wireless networks: cooperative diversity, coding, and transmission energy gains," *IEEE Transactions on Information Theory*, vol. 54, pp. 1207–1224, Mar. 2008.

[453] D. Slepian and J. Wolf, "Noiseless coding of correlated information sources," *IEEE Transactions on Information Theory*, vol. 19, pp. 471–480, July 1973.

[454] Y. Li, B. Vucetic, T. F. Wong, and M. Dohler, "Distributed turbo coding with soft information relaying in multihop relay networks," *IEEE Journal on Selected Areas in Communications*, vol. 24, pp. 2040–2050, Nov. 2006.

[455] X. Bao and J. Li, "Efficient message relaying for wireless user cooperation: decode-amplify-forward (DAF) and hybrid DAF and coded-cooperation," *IEEE Transactions on Wireless Communications*, vol. 6, pp. 3975–3984, Nov. 2007.

[456] P. Weitkemper, D. Wuibben, V. Kuhn, and K.-D. Kammeyer, "Soft information relaying for wireless networks with error-prone source-relay link," in *Proc. Int. ITG Conf. on Source and Channel Coding*, 2008.

[457] H. H. Sneessens and L. Vandendorpe, "Soft decode and forward improves cooperative communications," in *3G and Beyond, 2005 6th IEE International Conference*, pp. 1–4, Nov. 7–9, 2005.

[458] S. Yang and R. Koetter, "Network coding over a noisy relay: a belief propagation approach," in *Information Theory, 2007. ISIT 2007. IEEE International Symposium*, (Nice), pp. 801–804, June 24–29, 2007.

[459] R. Thobaben and E. G. Larsson, "Sensor-network-aided cognitive radio: on the optimal receiver for estimate-and-forward protocols applied to the relay channel," in *Signals, Systems and Computers, 2007. ACSSC 2007. Conference Record of the Forty-First Asilomar Conference*, (Pacific Grove, CA), pp. 777–781, Nov. 4–7, 2007.

[460] K. Lee and L. Hanzo, "MIMO-assisted hard versus soft decoding-and-forwarding for network coding aided relaying systems," *IEEE Transactions on Wireless Communications*, vol. 8, pp. 376–385, Jan. 2009.

[461] K. S. Gomadam and S. A. Jafar, "Optimal relay functionality for SNR maximization in memoryless relay networks," *IEEE Journal on Selected Areas in Communications*, vol. 25, pp. 390–401, Feb. 2007.

[462] P. Weitkemper, D. Wubben, and K. D. Kammeyer, "Minimum MSE relaying in coded networks," in *Smart Antennas, 2008. WSA 2008. International ITG Workshop*, (Vienna), pp. 96–103, Feb. 26–27, 2008.

[463] M. Fu, "Stochastic analysis of turbo decoding," *IEEE Transactions on Information Theory*, vol. 51, pp. 81–100, Jan. 2005.

[464] W. Pu, C. Luo, S. Li, and C. W. Chen, "Continuous network coding in wireless relay networks," in *INFOCOM 2008. The 27th Conference on Computer Communications. IEEE*, pp. 1526–1534, Apr. 13–18, 2008.

[465] Y. Li and B. Vucetic, "On the performance of a simple adaptive relaying protocol for wireless relay networks," in *Vehicular Technology Conference, 2008. VTC Spring 2008. IEEE*, pp. 2400–2405, May 11–14, 2008.

[466] Y. Li, B. Vucetic, Z. Zhou, and M. Dohler, "Distributed adaptive power allocation for wireless relay networks," *IEEE Transactions on Wireless Communications*, vol. 6, pp. 948–958, Mar. 2007.

[467] M. Dohler, *Virtual Antenna Arrays*. PhD Dissertation, King's College, London, UK, 2003.

[468] W. M. Younis and A. H. Sayed, "Interference suppression of asynchronous multi-user space–time block coded transmissions," in *Acoustics, Speech, and Signal Processing, 2004. Proceedings. (ICASSP '04). IEEE International Conference*, vol. 4, pp. 789–792, May 17–21, 2004.

[469] F. Ng, J. Hwu, M. Chen, and X. Li, "Asynchronous space–time cooperative communications in sensor and robotic networks," in *Mechatronics and Automation, 2005 IEEE International Conference*, vol. 3, pp. 1624–1629, 2005.

[470] Y. Li, W. Zhang, and X.-G. Xia, "Distributive high-rate full-diversity space-frequency codes for asynchronous cooperative communications," in *Information Theory, 2006 IEEE International Symposium*, pp. 2612–2616, July 9–14, 2006.

[471] M. O. Damen and A. R. Hammons, "Delay-tolerant space–time block codes for asynchronous cooperative relaying," in *Wireless Pervasive Computing, 2007. ISWPC '07. 2nd International Symposium*, Feb. 5–7, 2007.

[472] H. Papadopoulos, "Asynchrony resilient space–time block codes," in *Information Theory Workshop, 2007. ITW '07. IEEE*, pp. 355–360, Sept. 2–6, 2007.

[473] D. Wang, H. Minn, and N. Al-Dhahir, "Robust asynchronous STBC transceiver for multiple-access frequency-selective channels," in *Sarnoff Symposium, 2007 IEEE*, pp. 1–5, Apr. 30–May 2007, 2007.

[474] A. R. Hammons and M. O. Damen, "On lattice decoding of space–time block codes for asynchronous cooperative relaying," in *Sarnoff Symposium, 2007 IEEE*, pp. 1–5, Apr. 30–May 2007, 2007.

[475] S. Annanab, T. Tobita, T. Taniguchi, and Y. Karasawa, "Multiuser asynchronous MIMO STBC adaptive array transmission scheme in fast fading channel," in *Vehicular Technology Conference, 2008. VTC 2008-Fall. IEEE 68th*, pp. 1–5, Sept. 21–24, 2008.

[476] K. Yan, S. Ding, Y. Qiu, Y. Wang, and H. Liu, "A simple single-carrier space–time transmission scheme for asynchronous cooperative communications over frequency-selective channels," in *Advanced Communication Technology, 2008. ICACT 2008. 10th International Conference*, vol. 2, pp. 875–878, Feb. 17–20, 2008.

[477] G. Susinder Rajan and B. Sundar Rajan, "OFDM based distributed space time coding for asynchronous relay networks," in *Communications, 2008. ICC '08. IEEE International Conference*, pp. 1118–1122, May 19–23, 2008.

[478] N. Bin Ramli, T. Taniguchi, and Y. Karasawa, "Spatio-temporal adaptive array for asynchronous multiuser MIMO-STBC transmission," in *Communications, Control and Signal Processing, 2008. ISCCSP 2008. 3rd International Symposium*, pp. 362–367, Mar. 12–14, 2008.

[479] Y. Li, W. Zhang, and X.-G. Xia, "Distributive high-rate space-frequency codes achieving full cooperative and multipath diversities for asynchronous cooperative communications," *IEEE Transactions on Vehicular Technology*, vol. 58, pp. 207–217, Jan. 2009.

[480] E. Larsson and P. Stoica, *Space–Time Block Coding for Wireless Communications*. Cambridge University Press, 2003.

[481] H. Shin and J. H. Lee, "Exact symbol error probability of orthogonal space–time block codes," in *Global Telecommunications Conference, 2002. IEEE GLOBECOM '02*, vol. 2, pp. 1197–1201, Nov. 17–21, 2002.

[482] J. Yuan, Z. Chen, Y. Li, and J. Chu, "Distributed space–time trellis codes for a cooperative system," *IEEE Transactions on Wireless Communications*, vol. 8, pp. 4897–4905, Oct. 2009.

[483] S. Yiu, R. Schober, and L. Lampe, "Decentralized distributed space–time trellis coding," *IEEE Transactions on Wireless Communications*, vol. 6, pp. 3985–3993, Nov. 2007.

[484] Y. Li and X.-G. Xia, "A family of distributed space–time trellis codes with asynchronous cooperative diversity," *IEEE Transactions on Communications*, vol. 55, pp. 790–800, Apr. 2007.

[485] A. Stefanov and E. Erkip, "Cooperative space–time coding for wireless networks," *IEEE Transactions on Communications*, vol. 53, pp. 1804–1809, Nov. 2005.

[486] W. Zhang, Y. Li, X.-G. Xia, P. C. Ching, and K. Ben Letaief, "Distributed space-frequency coding for cooperative diversity in broadband wireless ad hoc networks," *IEEE Transactions on Wireless Communications*, vol. 7, pp. 995–1003, Mar. 2008.

[487] K. Seddik and K. J. Liu, "Distributed space-frequency coding over broadband relay channels," *IEEE Transactions on Wireless Communications*, vol. 7, pp. 4748–4759, Nov. 2008.

[488] A. Chakrabarti, A. de Baynast, A. Sabharwal, and B. Aazhang, "Low density parity check codes for the relay channel," *IEEE Journal on Selected Areas in Communications*, vol. 25, pp. 280–291, Feb. 2007.

[489] J. Hu and T. M. Duman, "Low density parity check codes over wireless relay channels," *IEEE Transactions on Wireless Communications*, vol. 6, pp. 3384–3394, Sept. 2007.

[490] B. Zhao and M. C. Valenti, "Distributed turbo coded diversity for relay channel," *Electronics Letters*, vol. 39, pp. 786–787, May 15, 2003.

[491] M. Janani, A. Hedayat, T. E. Hunter, and A. Nosratinia, "Coded cooperation in wireless communications: space–time transmission and iterative decoding," *Signal Processing, IEEE Transactions [see also Acoustics, Speech, and Signal Processing, IEEE Transactions]*, vol. 52, pp. 362–371, Feb. 2004.

[492] Y. Li, B. Vucetic, and J. Yuan, "Distributed turbo coding with hybrid relaying protocols," in *Personal, Indoor and Mobile Radio Communications, 2008. PIMRC 2008. IEEE 19th International Symposium*, (Cannes), pp. 1–6, Sept. 15–18, 2008.

[493] Z. Zhang and T. M. Duman, "Capacity-approaching turbo coding and iterative decoding for relay channels," *IEEE Transactions on Communications*, vol. 53, pp. 1895–1905, Nov. 2005.

[494] S. X. Ng, Y. Li, and L. Hanzo, "Distributed turbo trellis coded modulation for cooperative communications," in *Communications, 2009. ICC '09. IEEE International Conference*, (Dresden, Germany), pp. 1–5, June 14–18, 2009.

[495] C. Berrou, A. Glavieux, and P. Thitimajshima, "Near Shannon limit error-correcting coding and decoding: Turbo-codes. 1," in *Communications, 1993. ICC 93. Geneva. Technical Program, Conference Record, IEEE International Conference*, vol. 2, pp. 1064–1070, May 23–26, 1993.

[496] L. Bahl, J. Cocke, F. Jelinek, and J. Raviv, "Optimal decoding of linear codes for minimizing symbol error rate (corresp.)," *IEEE Transactions on Information Theory*, vol. 20, pp. 284–287, Mar. 1974.

[497] S. Benedetto and G. Montorsi, "Unveiling turbo codes: some results on parallel concatenated coding schemes," *IEEE Transactions on Information Theory*, vol. 42, pp. 409–428, Mar. 1996.

[498] E. Malkamaki and H. Leib, "Evaluating the performance of convolutional codes over block fading channels," *IEEE Transactions on Information Theory*, vol. 45, pp. 1643–1646, July 1999.

[499] E. K. Hall and S. G. Wilson, "Design and analysis of turbo codes on Rayleigh fading channels," *IEEE Journal on Selected Areas in Communications*, vol. 16, pp. 160–174, Feb. 1998.

[500] R. Ahlswede, N. Cai, S. Y. R. Li, and R. W. Yeung, "Network information flow," *IEEE Transactions on Information Theory*, vol. 46, pp. 1204–1216, July 2000.

[501] D. Silva, F. R. Kschischang, and R. Koetter, "A Rank-metric approach to error control in random network coding," *IEEE Transactions on Information Theory*, vol. 54, pp. 3951–3967, Sept. 2008.

[502] S. W. Kim, "Concatenated network coding for large-scale multi-hop wireless networks," in *Wireless Communications and Networking Conference, 2007.WCNC 2007. IEEE*, pp. 985–989, Mar. 11–15, 2007.

[503] C. Hausl and P. Dupraz, "Joint Network-channel coding for the multiple-access relay channel," in *Sensor and Ad Hoc Communications and Networks, 2006. SECON '06. 2006 3rd Annual IEEE Communications Society*, vol. 3, (Reston, VA), pp. 817–822, Sept. 28–28, 2006.

[504] X. Bao and J. Li, "Adaptive network coded cooperation (ANCC) for wireless relay networks: matching code-on-graph with network-on-graph," *IEEE Transactions on Wireless Communications*, vol. 7, pp. 574–583, Feb. 2008.

[505] K. Misra, S. Karande, and H. Radha, "INPod: in-network processing over sensor networks based on code design," in *Sensor, Mesh and Ad Hoc Communications and Networks, 2007. SECON '07. 4th Annual IEEE Communications Society Conference*, pp. 324–333, June 18–21, 2007.

[506] R. G. Gallager, *Low-Density Parity-Check Codes*. PhD Dissertation, MIT, USA, 1963.

[507] D. J. C. MacKay and R. M. Neal, "Near Shannon limit performance of low density parity check codes," *Electronics Letters*, vol. 32, Aug. 29, 1996.

[508] D. J. C. MacKay, "Good error-correcting codes based on very sparse matrices," *IEEE Transactions on Information Theory*, vol. 45, pp. 399–431, Mar. 1999.

[509] Y. Li, Z. Lin, and B. Vucetic, "Joint network-channel coding for relay interference channels," *IET Elecctronic Letters*, submitted.

[510] A. Shokrollahi, "Ldpc codes: an introduction," *Technical Report, Digital Fountain, Inc.*, Apr. 2003.

[511] W. T. Slingsby and J. P. McGeehan, "Antenna isolation measurements for on-frequency radio repeaters," in *Antennas and Propagation, 1995, Ninth International Conference (Conf. Publ. No. 407)*, pp. 239–243, Apr. 4–7, 1995.

[512] S. Bunin, Y. Korzh, and A. Voiter, "Single frequency duplex repeater," in *Microwave and Telecommunication Technology, 2004. CriMico 2004. 2004 14th International Crimean Conference*, pp. 273–274, Sept. 13–17, 2004.

[513] R. N. Braithwaite and S. Carichner, "Adaptive echo cancellation for an on-frequency RF repeater using a weighted power spectrum," in *Wireless Technologies, 2007 European Conference*, (Munich), pp. 82–85, Oct. 8–10, 2007.

[514] P. Larsson and M. Prytz, "MIMO on-frequency repeater with self-interference cancellation and mitigation," in *8th Scandinavian Workshop on Wireless Ad Hoc Networks*, May 2008.

[515] R. van de Plassche, J. Huijsing, and W. Sansen, *Analog Circuit Design: RF Analog-to-Digital Converters; Sensor and Actuator Interfaces; Low-Noise Oscillators, PLLs and Synthesizers*. First edition, Springer, November 1997.

[516] J. Laiho, A. Wacker, and T. Novosad, *Radio Network Planning and Optimisation for UMTS*. John Wiley & Sons, Inc., 2002.

[517] M. Nawrocki, M. Dohler, and A. Aghvami, *Understanding UMTS Radio Network Modelling, Planning and Automated Optimisation – Theory and Practice*. John Wiley & Sons, Inc., 2006.

[518] K. Higuchi, A. Fujiwara, and M. Sawahashi, "Multipath interference canceller for high-speed packet transmission with adaptive modulation and coding scheme in W-CDMA forward link," *IEEE Journal on Selected Areas in Communications*, vol. 20, pp. 419–432, Feb. 2002.

[519] L. Fathi, *Complexity of MPIC receiver for HSDPA R5*. France Telecom R&D, internal report, October 2004.

[520] M. Dohler, C. Lerau, E. Hardouin, M. Arnd, L. Fathi, and P. Jayet, *UMTS FDD Multi-Antenna Receiver Complexity Estimation*. France Telecom R&D, internal report, June 2005.

[521] H. Poor, *Introduction to Signal Detection and Estimation*. Springer-Verlag, New York, LLC, 1994.

[522] A. Parssinen, J. Jussila, J. Ryynanen, L. Sumanen, and K. A. I. Halonen, "A 2-GHz wide-band direct conversion receiver for WCDMA applications," in *IEEE Journal of Solid-State Circuits*, vol. 34, (San Francisco, CA), pp. 1893–1903, Dec. 1999.

[523] A. Parssinen, J. Jussila, J. Ryynanen, L. Sumanen, K. Kivekas, and K. Halonen, "A wide-band direct conversion receiver with on-chip A/d converters," in *VLSI Circuits, 2000. Digest of Technical Papers. 2000 Symposium*, pp. 32–33, June 15–17, 2000.

[524] M. Goldfarb, W. Palmer, T. Murphy, R. Clarke, B. Gilbert, K. Itoh, T. Katsura, R. Hayashi, and H. Nagano, "Analog baseband IC for use in direct conversion W-CDMA receivers," in *Radio Frequency Integrated Circuits (RFIC) Symposium, 2000. Digest of Papers. 2000 IEEE*, pp. 79–82, June 11–13, 2000.

[525] J. Jussila, J. Ryynanen, K. Kivekas, L. Sumanen, A. Parssinen, and K. A. I. Halonen, "A 22-ma 3.0-db NF direct conversion receiver for 3 g WCDMA," in *IEEE Journal of Solid-State Circuits*, vol. 36, pp. 2025–2029, Dec. 2001.

[526] J. Ryynanen, K. Kivekasl, J. Jussila, A. Parssinen, and K. Halonen, "Direct conversion receiver for GSM900, DCS1800, PCS1900, and WCDMA," in *Electronics, Circuits and Systems, 2003. ICECS 2003. Proceedings of the 2003 10th IEEE International Conference*, vol. 2, pp. 942–945, Dec. 14–17, 2003.

[527] M. Goldfarb, E. Balboni, and J. Cavey, "Even harmonic double-balanced active mixer for use in direct conversion receivers," *IEEE Journal of Solid-State Circuits*, vol. 38, pp. 1762–1766, Oct. 2003.

[528] A. Raju, *Characterization of Uplink Transmit Power and Talk Time in WCDMA Networks*. Virginia Polytechnic Institute and State University, MSc thesis, July 2008.

[529] O. Jensen, T. Kolding, C. Iversen, S. Laursen, R. Reynisson, J. Mikkelsen, E. Pedersen, M. Jenner, and T. Larsen, "RF receiver requirements for 3G WCDMA mobile equipment," *Microwave Journal*, vol. 43, pp. 22–46, Feb. 2000.

[530] N. Ceylan, J. E. Mueller, T. Pittorino, and R. Weigel, "Mobile phone power amplifier linearity and efficiency enhancement using digital predistortion," in *33rd European Microwave Conference, 2003*. vol. 1, pp. 269–272, Oct. 7–9, 2003.

[531] 3GPP, *Feasibility study for EUTRA and EUTRAN*. TR 25.912, Aug. 2007.

[532] 3GPP, *Requirements for EUTRA and EUTRAN*. TR 25.913, Mar. 2006.

[533] 3GPP, *E-UTRA; Physical layer; General description*. TR 36.201, Dec. 2007.

[534] 3GPP, *E-UTRA; Physical layer procedures*. TR 36.213, Sept. 2008.

[535] 3GPP, *Requirements for further advancements for E-UTRA (LTE-Advanced)*. TR 36.913, June 2008.

[536] E. Seidel, *Progress on "LTE Advanced"–The new 4G Standard*. White Paper, Nomor Research GmbH, Munich, Germany, July 2008.

[537] IEEE Standard for Local and Metropolitan Area Networks, *Air Interface for Fixed Broadband Wireless Access Systems*. Std 802.16-2004, June 2004.

[538] IEEE Standard for Local and Metropolitan Area Networks, *Air Interface for Fixed and Mobile Broadband Wireless Access Systems–Amendment for Physical and Medium Access Control Layers for Combined Fixed and Mobile Operation in Licensed Bands*. Std 802.16e-2004, Feb. 2006.

[539] IEEE Standard for Local and Metropolitan Area Networks, *FDD and H-FDD frame structure for IEEE 802.16j Multihop Relay Networks*. Std 802.16e-2004, Sept. 2008.

[540] IEEE Standard for Local and Metropolitan Area Networks, *Air Interface for Fixed and Mobile Broadband Wireless Access Systems–Multihop Relay Specification*. Draft 802.16j, Aug. 2007.

[541] R. W. Bauml, R. F. H. Fischer, and J. B. Huber, "Reducing the peak-to-average power ratio of multicarrier modulation by selected mapping," *Electronics Letters*, vol. 32, pp. 2056–2057, Oct. 24, 1996.

[542] S. H. Han and J. H. Lee, "An overview of peak-to-average power ratio reduction techniques for multi-carrier transmission," *IEEE Wireless Communications*, vol. 12, pp. 56–65, Apr. 2005.

[543] WiMAX Forum, *WiMAX System Evaluation Methodology*. Version 2.1, July 2008.

[544] A. Bletsas, *Intelligent Antenna Sharing in Cooperative Diversity Wireless Networks*. PhD Dissertation Proposal, Media Arts & Sciences Program, MIT, USA, May 2004.

[545] A. Bletsas, *Intelligent Antenna Sharing in Cooperative Diversity Wireless Networks*. PhD Dissertation, Media Laboratory, MIT, USA, September 2005.

[546] A. Bletsas and A. Lippman, "Implementing cooperative diversity antenna arrays with commodity hardware," in *Communications Magazine, IEEE*, vol. 44, pp. 33–40, Dec. 2006.

[547] A. Bletsas, H. Shin, and M. Z. Win, "Cooperative communications with outage-optimal opportunistic relaying," *IEEE Transactions on Wireless Communications*, vol. 6, pp. 3450–3460, Sept. 2007.

[548] A. El-Hoiydi, "Aloha with preamble sampling for sporadic traffic in ad hoc wireless sensor networks," in *Communications, 2002. ICC 2002. IEEE International Conference*, vol. 5, pp. 3418–3423, 2002.

[549] R. S. Blum and J. H. Winters, "On optimum MIMO with antenna selection," *IEEE Communications Letters*, vol. 6, pp. 322–324, Aug. 2002.

[550] M. Dohler, A. H. Aghvami, Z. Zhou, Y. Li, and B. Vucetic, "Capacity and outage of TAS schemes over Nakagami channels," in *COST273, Duisburg, Germany*, 2004.

[551] A. Wittneben, "Cooperative relaying networks," in *34th European Microwave Conference 2004, Joint EuMC/ECWT Workshop, invited paper*, Oct. 2004.

[552] S. Berger and A. Wittneben, "Experimental performance evaluation of multiuser zero forcing relaying in indoor scenarios," in *Vehicular Technology Conference, 2005. VTC 2005-Spring. 2005 IEEE 61st*, vol. 2, pp. 1101–1105, May 30–June 1, 2005.

[553] I. Hammerstroem, J. Zhao, S. Berger, and A. Wittneben, "Experimental performance evaluation of joint cooperative diversity and scheduling," in *Vehicular Technology Conference, 2005. VTC-2005-Fall. 2005 IEEE 62nd*, vol. 4, pp. 2428–2432, Sept. 25–28, 2005.

[554] A. Wittneben, "Coherent multiuser relaying with partial relay cooperation," in *Wireless Communications and Networking Conference, 2006. WCNC 2006. IEEE*, vol. 2, pp. 1027–1033, 2006.

[555] S. Berger and A. Wittneben, "Impact of noisy carrier phase synchronization on linear amplify-and-forward relaying," in *Global Telecommunications Conference, 2007. GLOBECOM '07. IEEE*, pp. 795–800, Nov. 26–30, 2007.

[556] S. Berger and A. Wittneben, "A coherent amplify-and-forward relaying demonstrator without global phase reference," in *Personal, Indoor and Mobile Radio Communications, 2008. PIMRC 2008. IEEE 19th International Symposium*, pp. 1–5, Sept. 15–18, 2008.

[557] R. Irmer, H. P. Mayer, A. Weber, V. Braun, M. Schmidt, M. Ohm, N. Ahr, A. Zoch, C. Jandura, P. Marsch, and G. Fettweis, "Multisite field trial for LTE and advanced concepts," in *IEEE Communications Magazine*, vol. 47, pp. 92–98, 2009.

[558] The EASY-C Project. [Online]. Available: http://www.easy-c.com.

[559] C. Jandura, P. Marsch, A. Zoch, and G. Fettweis, "A testbed for cooperative multi cell algorithms," in *TRIDENTCOM'08*, March 2008.

[560] *Berlin Field Trial of LTE-A: Multi-cell testbed for distributed cooperation in mobile networks (flyer)*. Broadband Mobile Communications Department, Fraunhofer Institute for Telecommunications Heinrich-Hertz-Institut (FhG HHI), 2008.

[561] H. Anouar, C. Bonnet, D. Câmara, F. Filali, and R. Knopp, "An overview of open air interface wireless network emulation methodology," *SIGMETRICS Perform. Eval. Rev.*, vol. 36, no. 2, pp. 90–94, 2008.

[562] L. M. Ventura, X. Nieto, J.-P. Gregoire, W. de Win, and N. Lugil, "A Broadband wireless MIMO-OFDM demonstrator. design and measurement results," in *Personal, Indoor and Mobile Radio Communications, 2006 IEEE 17th International Symposium*, pp. 1–5, Sept. 2006.

[563] S. Valentin, H. S. Lichte, D. Warneke, T. Biermann, R. Funke, and H. Karl, "Mobile cooperative WLANs–MAC and transceiver design, prototyping, and field measurements," in *Vehicular Technology Conference, 2008. VTC 2008-Fall. IEEE 68th*, pp. 1–5, Sept. 21–24, 2008.

[564] B. Can, M. Portalski, H. S. D. Lebreton, S. Frattasi, and H. A. Suraweera, "Implementation issues for OFDM-based multihop cellular networks," in *Communications Magazine, IEEE*, vol. 45, pp. 74–81, Sept. 2007.

[565] B. Jia and M. R. Soleymani, "Virtual space–time coding for cooperative communication and down-link receiver hardware implementation," in *Electrical and Computer Engineering, 2007. CCECE 2007. Canadian Conference*, pp. 916–919, Apr. 22–26, 2007.

[566] A. Krohn, M. Beigl, C. Decker, and T. Riedel, "Syncob: collaborative time synchronization in wireless sensor networks," in *Networked Sensing Systems, 2007. INSS '07. Fourth International Conference*, pp. 283–290, June 6–8, 2007.

[567] R. A. Iltis and R. Cagley, "Nonlinear channel estimation issues in cooperative sensor MIMO," in *Information Theory and Applications Workshop, 2007*, pp. 207–213, Jan. 29–Feb. 2007, 2007.

[568] R. E. Cagley, B. T. Weals, S. A. McNally, R. A. Iltis, S. Mirzaei, and R. Kastner, "Implementation of the Alamouti OSTBC to a distributed set of single-antenna wireless nodes," in *Wireless Communications and Networking Conference, 2007.WCNC 2007. IEEE*, pp. 577–581, Mar. 11–15, 2007.

[569] S. Mirzaei, A. Irturk, R. Kastner, B. T. Weals, and R. E. Cagley, "Design space exploration of a cooperative MIMO receiver for reconfigurable architectures," in *Application-Specific Systems, Architectures and Processors, 2008. ASAP 2008. International Conference*, pp. 167–172, July 2–4, 2008.

[570] G. Qi, P. Song, and K. Li, "Wireless synchronous triggering technology special for wireless sensor networks," in *Mechtronic and Embedded Systems and Applications, 2008. MESA 2008. IEEE/ASME International Conference*, pp. 176–180, Oct. 12–15, 2008.

[571] G. Jakllari, S. V. Krishnamurthy, M. Faloutsos, P. V. Krishnamurthy, and O. Ercetin, "A cross-layer framework for exploiting virtual MISO links in mobile ad hoc networks," *IEEE Transactions on Mobile Computing*, vol. 6, pp. 579–594, June 2007.

[572] A. Sharma, V. Gelara, S. R. Singh, T. Korakis, P. Liu, and S. Panwar, "Implementation of a cooperative MAC protocol using a software defined radio platform," in *Local and Metropolitan Area Networks, 2008. LANMAN 2008. 16th IEEE Workshop*, pp. 96–101, Sept. 3–6, 2008.

[573] M. Kurth, A. Zubow, and J. P. Redlich, "Cooperative opportunistic routing using transmit diversity in wireless mesh networks," in *INFOCOM 2008. The 27th Conference on Computer Communications. IEEE*, pp. 1310–1318, Apr. 13–18, 2008.

[574] J. A. Gubner, "A new formula for lognormal characteristic functions," *IEEE Transactions on Vehicular Technology*, vol. 55, pp. 1668–1671, Sept. 2006.

[575] M. Chiani, A. Conti, and O. Andrisano, "Outage evaluation for slow frequency-hopping mobile radio systems," *IEEE Transactions on Communications*, vol. 47, pp. 1865–1874, Dec. 1999.

[576] A. Conti, M. Z. Win, and M. Chiani, "Qos-based outage probability for diversity reception," in *Communications, 2002. ICC 2002. IEEE International Conference*, vol. 3, pp. 1429–1433, 2002.

[577] A. Conti, M. Z. Win, M. Chiani, and J. H. Winters, "Bit error outage for diversity reception in shadowing environment," *IEEE Communications Letters*, vol. 7, pp. 15–17, Jan. 2003.

[578] A. Conti, M. Z. Win, and M. Chiani, "Tight bounds on outage and throughput for M-QAM in fading channels," in *Communications, 2004 IEEE International Conference*, vol. 6, pp. 3358–3363, June 20–24, 2004.

[579] P. Mary, M. Dohler, J. M. Gorce, G. Villemaud, and M. Arndt, "M-ary symbol error outage over Nakagami-m fading channels in shadowing environments," *IEEE Transactions on Communications*, in press.

[580] P. Mary, *Analytical Outage Expressions Considering Shadowing, MIMO, Coding and Interference*. PhD Dissertation, INSA, Lyon, France, 2008.

[581] P. Mary, M. Dohler, J. M. Gorce, and G. Villemaud, "Symbol error outage for spatial multiplexing systems in Rayleigh fading channel and lognormal shadowing," in *Signal Processing Advances in Wireless Communications, 2009. SPAWC '09. IEEE 10th Workshop*, pp. 349–353, June 21–24, 2009.

[582] J. Hagenauer, "Rate-compatible punctured convolutional codes (RCPC Codes) and their applications," *IEEE Transactions on Communications*, vol. 36, pp. 389–400, Apr. 1988.

[583] A. Viterbi, "Convolutional codes and their performance in communication systems," *IEEE Transactions on Communications*, vol. 19, pp. 751–772, Oct. 1971.

[584] S. Manji and N. Mandayam, "Block error probability using list Viterbi decoding with hard decisions." citeseer.ist.psu.edu/317496.html

[585] K. Lee and L. Hanzo, "Multiple antenna assisted hard versus soft decoding-and-forwarding for network coding aided relaying systems," in *Global Telecommunications Conference, 2008. IEEE GLOBECOM 2008. IEEE*, pp. 1–5, Nov. 30–Dec. 2008, 2008.

[586] K. Lee and L. Hanzo, "MIMO-assisted hard versus soft decoding-and-forwarding for network coding aided relaying systems," *IEEE Transactions on Wireless Communications*, vol. 8, pp. 376–385, Jan. 2009.

[587] W. Pu, C. Luo, S. Li, and C. W. Chen, "Continuous network coding in wireless relay networks," in *INFOCOM 2008. The 27th Conference on Computer Communications. IEEE*, pp. 1526–1534, Apr. 13–18, 2008.

Index

Cooperative Communications   Mischa Dohler and Yonghui Li
© 2010 John Wiley & Sons, Ltd